Springer Series in Statistics

Andersen/Borgan/Gill/Keiding: Statistical Models Based on Counting Processes.
Anderson: Continuous-Time Markov Chains: An Applications-Oriented Approach.
Andrews/Herzberg: Data: A Collection of Problems from Many Fields for the Student and Research Worker.
Anscombe: Computing in Statistical Science through APL.
Berger: Statistical Decision Theory and Bayesian Analysis, 2nd edition.
Bolfarine/Zacks: Prediction Theory for Finite Populations.
Brémaud: Point Processes and Queues: Martingale Dynamics.
Brockwell/Davis: Time Series: Theory and Methods, 2nd edition.
Choi: ARMA Model Identification.
Daley/Vere-Jones: An Introduction to the Theory of Point Processes.
Dzhaparidze: Parameter Estimation and Hypothesis Testing in Spectral Analysis of Stationary Time Series.
Fahrmeir/Tutz: Multivariate Statistical Modelling Based on Generalized Linear Models.
Farrell: Multivariate Calculation.
Federer: Statistical Design and Analysis for Intercropping Experiments.
Fienberg/Hoaglin/Kruskal/Tanur (Eds.): A Statistical Model: Frederick Mosteller's Contributions to Statistics, Science and Public Policy.
Good: Permutation Tests: A Practical Guide to Resampling Methods for Testing Hypotheses.
Goodman/Kruskal: Measures of Association for Cross Classifications.
Grandell: Aspects of Risk Theory.
Hall: The Bootstrap and Edgeworth Expansion.
Härdle: Smoothing Techniques: With Implementation in S.
Hartigan: Bayes Theory.
Heyer: Theory of Statistical Experiments.
Jolliffe: Principal Component Analysis.
Kotz/Johnson (Eds.): Breakthroughs in Statistics Volume I.
Kotz/Johnson (Eds.): Breakthroughs in Statistics Volume II.
Kres: Statistical Tables for Multivariate Analysis.
Leadbetter/Lindgren/Rootzén: Extremes and Related Properties of Random Sequences and Processes.
Le Cam: Asymptotic Methods in Statistical Decision Theory.
Le Cam/Yang: Asymptotics in Statistics: Some Basic Concepts.
Manoukian: Modern Concepts and Theorems of Mathematical Statistics.
Manton/Singer/Suzman (Eds.): Forecasting the Health of Elderly Populations.
Miller, Jr.: Simultaneous Statistical Inference, 2nd edition.
Mosteller/Wallace: Applied Bayesian and Classical Inference: The Case of *The Federalist Papers.*
Pollard: Convergence of Stochastic Processes.
Pratt/Gibbons: Concepts of Nonparametric Theory.
Read/Cressie: Goodness-of-Fit Statistics for Discrete Multivariate Data.
Rieder: Robust Asymptotic Statistics.
Reinsel: Elements of Multivariate Time Series Analysis.

(continued after index)

Springer Series in Statistics

Helmut Rieder

Robust Asymptotic Statistics

Springer-Verlag
New York Berlin Heidelberg London Paris
Tokyo Hong Kong Barcelona Budapest

Helmut Rieder
Lehrstühl VII für Mathematik
Universität Bayreuth
D-95440 Bayreuth
Germany

Library of Congress Cataloging-in-Publication Data
Rieder, Helmut.
Robust asymptotic statistics / Helmut Rieder.
p. cm. — (Springer series in statistics)
Includes bibliographical references (p. -) and index.
ISBN 0-387-94262-9 (New York). — ISBN 3-540-94262-9 (Berlin)
1. Robust statistics. 2. Mathematical statistics—Asymptotic theory. I. Title. II. Series.
QA276.R515 1994
519.5′4—dc20 94-1070

Printed on acid-free paper.

Production managed by Ellen Seham; manufacturing supervised by Gail Simon.
Photocomposed copy prepared from the author's LaTeX files.
Printed and bound by Edwards Brothers, Inc., Ann Arbor, MI.
Printed in the United States of America.

9 8 7 6 5 4 3 2 1

ISBN 0-387-94262-9 Springer-Verlag New York Berlin Heidelberg
ISBN 3-540-94262-9 Springer-Verlag Berlin Heidelberg New York

To

Sonja, Astrid, Georg,
and Anja

Preface

To the king, my lord, from your servant Balasî[1] *:*
... The king[2] *should have a look.*
Maybe the scribe who reads to the king did not understand.
... I shall personally show, with this tablet that I am sending
to the king, my lord, how the omen was written.
Really, he who has not followed the text with his finger[3]
cannot possibly understand it.

This book is about optimally robust functionals and their unbiased estimators and tests. Functionals extend the parameter of the assumed ideal center model to neighborhoods of this model that contain the actual distribution. The two principal questions are (F): Which functional to choose? and (P): Which statistical procedure to use for the selected functional?

Using a local asymptotic framework, we deal with both problems by linking up nonparametric statistical optimality with infinitesimal robustness criteria. Thus, seemingly separate developments in robust statistics are presented in a unifying way.

Question (F) has not received much systematic treatment in the literature. Referring to an inaccessible but, by all accounts, pioneering manuscript by Takeuchi (1967), Bickel and Lehmann (1975) formulate certain desirable properties (monotony, symmetry, equivariance) of location functionals which, however, exclude M functionals with bounded scores. Extending Hampel's (1971) qualitative notion of robustness, they let each (possibly nonrobust) functional define its own refinement of the weak topology in which it is continuous. For the three classes of M, L, and R functionals, they calculate the maximal/minimal asymptotic efficiencies of the corresponding estimators relative to the mean over large nonparametric classes of probabilities. These comparative notions of both robustness and efficiency finally lead them to trimmed means. Problem (P) does not arise for

[1] Assyrian scribe, whose datable letters range from 671 to 667 B.C.
[2] Assurbanipal, who succeeded Esarhaddon in 669 B.C.,
was priestly educated and fully literate.
[3] or: before whom the (teacher's?) finger has not gone.

these authors since 'without symmetry, there is typically only one natural estimator of the location parameter, namely, the functional evaluated at the empirical distribution' [Bickel and Lehmann (1975, p 1057)]. In their Note one page later, they mention the possibility of superefficiency phenomena but say that 'in view of the results of LeCam, Huber, and Hájek one might conjecture that such phenomena can be avoided by requiring continuity (in a suitable sense) of the asymptotic variance of the estimator'. Only adding a prayer as a measure of caution, Huber (1981, p 6) takes essentially the same point of view.

Problem (P) may be solved more mathematically by the nonparametric optimality theorems (convolution representation, asymptotic minimax bounds) for the estimation and testing of differentiable functionals. The nonparametric approach of estimating such functionals goes back to Koshevnik and Levit (1976) who, without assuming a parametric center model, have combined the classical parametric local asymptotic optimality results due to LeCam (1972, 1986) and Hájek (1970, 1972) with the determination of a least favorable tangent in the spirit of Stein (1956). In robust statistics, this approach has been strongly promoted by Beran (1981–1984) and Millar (1981–1984), in connection with (Hellinger, Cramér–von Mises) minimum distance functionals. Obvious advantages are:

(i) a clearly defined estimand outside the ideal model, even in the absence of symmetry, or if the ideal parameter value is no longer identifiable, which is the case for the usual neighborhoods of robust statistics[4];

(ii) consideration of arbitrary estimators in the asymptotic minimax bounds, in particular, no restriction to M, L, or R estimates;

(iii) a certain locally uniform convergence of asymptotic minimax procedures, which is also used in the convolution theorem as the only regularity requirement on estimators.

By equating 'robust' with 'locally asymptotic minimax' [Beran (1984; §2.3, §2.4) and Millar (1981, p 75; 1982, p 731; 1983, p 220)], these authors seemed to create a new powerful robustness notion. Question (F), however, and the subjective choice of particular minimum distance functionals, were declared to 'fall distinctly outside the purview of the decision-theoretic structure' [Millar (1981, p 83)], and instead to be a matter of principle, namely, that of 'best fit' [Beran (1984)].

The situation has been controversial. On the one hand, Huber (1972) as well as Bickel and Lehmann (1975), were accused of 'being driven' to the 'desperate measure' [Millar (1983, p 217)] and to the 'debilitating relativism' [Millar (1981, p 76)] of defining the true parameter of interest to be 'whatever the estimator in hand', or 'one's favorite estimator', estimates. On the other hand, the subjective choice has apparently not been accepted as a general solution to problem (F) either: Begun et al. (1983, p 434) go so far as to declare that 'the completely nonparametric approach to represen-

[4] In the semiparametric literature, identifiability is retained by assumption.

tation theorems and asymptotic minimax bounds for estimating functionals fails' because 'it is simply not clear which functional should be analyzed'.

Against principles like 'best fit' one may in fact argue that most such principles have become notorious because of possible pathologies in innocent looking situations; see Ferguson (1967, p 136) on the 'unbiasedness' principle, and LeCam (1986; §17.6, pp 621–625) on the 'asymptotic efficiency' usually associated with the 'maximum likelihood' principle. Although not for Hilbertian distances, Donoho and Liu (1988 b) have shown that minimum distance estimates also may fail drastically.

Instead of principles, we therefore prefer objective criteria that are applicable to a wide class of functionals. Moreover, we look more closely into the nonparametric optimality theorems, whose proofs completely rely on the parametric results applied to one least favorable local alternative. It is true that these abstract theorems are geared to, and are to some extent justified by, the utmost generality. As a side condition on mathematical abstraction, however, elementary procedures should still be dealt with according to original intentions. But for the class of asymptotically linear estimators, it turns out that the asymptotic statistical optimality results, which—in the parametric case—generalize the Cramér–Rao bound for asymptotic variance, are—in the nonparametric setup with full neighborhoods—dominated by bias alone:

Any asymptotically linear estimator is optimal for its limiting functional, simply because it is the only unbiased one and any other asymptotically linear estimator generates infinite bias relative to the given functional. Rather than settle the controversy mentioned above, however, the evaluation in Rieder (1985) somehow stirred up a hornets' nest; obviously, the bias degeneration has been overlooked in the decision-theoretic literature. Further limitations of the nonparametric optimality results are recorded in Subsection 4.3.3; for example, their independence of any specific amount of contamination. As the essence of nonparametric optimality then, apart from the coverage of estimators that are not asymptotically linear, a certain locally uniform convergence and the task of suitable estimator constructions remain [Chapter 6].

To be able to trace the proofs of the nonparametric optimality theorems, and because a sufficiently lucid and elementary reference on asymptotic statistics (in particular, on the asymptotic minimax theorem) has been lacking, it proved necessary to develop a self-contained treatment of the main statistical optimality results. Our presentation is fairly simple (assuming only weak convergence), so that students with a basic knowledge of probability and statistics can understand the proofs. As such, the corresponding Chapters 2, 3, 4, and 6 (supported by Appendix A) should be interesting also outside robust statistics. Chapter 1, a nonstatistical extra, merely supplies some asymptotically normal procedures via differentiation.

In the sequel, the nonparametric optimality results are used in just two ways: First, as unbiasedness statements, and secondly, to quantify the costs at which any functional can be estimated or tested. On this basis, we then determine that functional which, subject to certain robustness constraints, is easiest to estimate or test. As robustness side conditions, bounds are imposed on the infinitesimal oscillation (estimation), respectively, the inclusion of given capacity balls into nonparametric hypotheses is required (testing). It is at this instance where the classical robustness theories on bias and variance may be invoked to complement nonparametric unbiasedness by some further, nontrivial, optimality.

Optimality results for general parameter are contained in Chapter 5. Chapter 7 derives those for linear regression, which serves as a prototype example of a structured model, where additional variants of neighborhoods and optimality problems arise. The conditional centering constraint occurring in this context takes values in some infinite-dimensional function space, and makes it necessary to explicitly derive a suitable Lagrange multiplier theorem [Appendix B].

We must admit that the monolithic type of optimality known from parametric statistics cannot be achieved in robust statistics. The solutions depend on the choice of neighborhoods, and one hardly knows precisely what sort of neighborhood one would 'like to be robust over' or, in Tukey's phrase, what one 'should choose to fear' [Donoho and Liu (1988 a, p 554)]. But the variety of robust solutions at least constitute a complete class.

The book has no pretensions of treating all developments in robustness, nonparametrics, or asymptotic statistics. We only consider the local, infinitesimal, approach (that is, no qualitative robustness, no breakdown aspects), and only the case of independent observations. The restricted selection and comprehensive treatment of basic issues, however, defines the framework for further refinements, extensions, and specializations.

Early notes date back to the DMV seminar on 'Robust Statistics' held jointly with Dr. Rudolph J. Beran in Düsseldorf, in 1982, and from Rudy's work and lectures I learned many asymptotic ideas and techniques. Surprisingly, at that time, none of the seminar participants asked for the reason why, on the one hand, Beran's lectures recommended particular minimum distance estimates on optimality grounds while, on the other hand, some M estimates based on suitably clipped scores functions were proved optimum in my lectures. The investigation and relation of different optimality notions, however, is not only a commitment of theoretical statisticians but also a prerequisite for being able to decide, and finally settle, on some procedure to use.

The present book, which should ideally have been that DMV seminar volume, has actually become the revised version to Rieder (1985) [as a rule, my revisions grow longer!]. Obviously, it has taken some time to write things up. I am much indebted to Dr. Walter Alt for his help in getting started with LaTeX. Over the years, I have profitably used part of the material in some seminars and advanced statistics courses in Bayreuth, although the text does not contain exercises.

My visit to the Mathematics Department at M.I.T. during 1990–1991 provided another opportunity to concentrate on the project. I am foremost indebted to Dr. Peter J. Huber, himself the originator of most of the area of robust statistics, who always had the time for discussions and made many insightful comments. Also, the cuneiform quote [5] opening this preface I owe to Peter's suggestion. I thank him and his wife, Dr. Effi Huber–Buser, for their hospitality and their constant and friendly encouragement.

I am equally indebted to Dr. Richard M. Dudley for his kind interest in my lectures, and for his very valuable comments on weak convergence, differentiable functionals, and convex loss functions.

It was a pleasure to share an office with Dr. Walter Olbricht. Many thanks go to him and to Dr. Evangelos Tabakis, as well as to Harvard students Peter Hallinan and Abdul-Reza Mansouri, who attended my lectures.

My faithful students in Bayreuth, in particular, Tassilo Hummel and Martin Schlather, found a number of typing errors; I thank them sincerely. Thanks are also due to Mrs. Angelika Geisser for her secretarial assistance. I am particularly thankful to Tilmann Gneiting, Martin and Evangelos for their careful reading of the last 1993-output.

Finally, I would like to thank SPRINGER-Verlag, especially, Mrs. Ellen Seham, Mr. Steven Pisano, and Dr. Martin Gilchrist for their kind and helpful cooperation.

H. RIEDER

Heimbach, March 1994

[5] English translation taken from Parpola (1970)

Contents

Notation

The reader may first wish to skim over the following list of notation.

Abbreviations

a.e.	almost everywhere, almost surely
as.	asymptotic, asymptotically
eventually	for all sufficiently large sequence indices
i.i.d.	stochastically independent, identically distributed
l.s.c., u.s.c.	lower/upper semicontinuous, semicontinuity
s.t.	subject to
w.r.t.	with respect to, relative to
CvM	Cramér–von Mises
LSE	least squares estimate
MD	minimum distance
MLE	maximum likelihood estimate
M, L, R	maximum likelihood type, linear function of order statistics, and rank based, respectively
MSE	mean square error
ONB, ONS	orthonormal basis/system
RHS, LHS	right/left-hand side
////	QED

Sets and Functions

$\mathbb{N}$	the natural numbers $\{1, 2, \ldots\}$
$\mathbb{Z}$	the integers $\{\ldots, -1, 0, 1, \ldots\}$
$\mathbb{R}$	the real numbers $(-\infty, \infty)$
$\bar{\mathbb{R}}$	the extended real numbers $[-\infty, \infty]$, homeomorphic to $[-1, 1] \subset \mathbb{R}$ via the isometry $z \mapsto z/(1 + \lvert z\rvert)$
$\mathbb{C}$	the complex numbers
$\times$	Cartesian product of sets; $A^m = A \times \cdots \times A$ (m times)
$\mathbf{I}_A$, $\mathbf{I}(A)$	indicator function of a set or statement A; thus, for any set A, we may write $\mathbf{I}_A(x) = \mathbf{I}(x \in A)$

id_Ω	identity function on the set Ω
sign	$\mathrm{sign}(x) = -1, 0, 1$ for x negative/zero/positive
$f(x \mp 0)$	left/right-hand limit at x of a function f
π	projection map
Λ, $\mathcal{I}$	L_α derivative ($\alpha = 1, 2$) and Fisher information ($\alpha = 2$), respectively [pp 56, 58, 60]
Δ, $\mathcal{J}$	CvM derivative and CvM information [p 60]

σ Algebras

$\mathbb{B}$, $\bar{\mathbb{B}}$	Borel σ algebras on $\mathbb{R}$ and $\bar{\mathbb{R}}$, respectively
$\sigma(\mathcal{E})$	smallest σ algebra (on Ω) including a system $\mathcal{E} \subset 2^\Omega$
$\bigotimes$	product of σ algebras; $\mathcal{A}^m = \mathcal{A} \otimes \cdots \otimes \mathcal{A}$ (m times)

Measures

$\mathcal{M}(\mathcal{A})$	the (nonnegative) measures on a σ algebra $\mathcal{A}$
$\mathcal{M}_\sigma(\mathcal{A})$	the σ finite measures on $\mathcal{A}$
$\mathcal{M}_b(\mathcal{A})$	the finite (or *b*ounded) measures on $\mathcal{A}$
$\mathcal{M}_1(\mathcal{A})$	the probability measures (mass 1) on $\mathcal{A}$
support P	smallest closed subset A of Ω (separable, metric) such that $P(\Omega \setminus A) = 0$ [Parthasaraty (1967; II Def. 2.1)]
$\log dQ/dP$	log likelihood [p 39]
$\ll$	domination of measures, and contiguity of (bounded) sequences of measures [pp 40, 41]
$\bigotimes$	product of measures; $P^m = P \otimes \cdots \otimes P$ (m times)
$*$	convolution of measures
$\xrightarrow{\mathrm{w}}$	weak convergence of (bounded) measures [p 331]

Probability measures Q on $\mathbb{B}^m$ may be identified with their (right continuous) distribution functions: $Q(y) = Q(\{x \mid x \le y\})$. Thus, in case $m = 1$, we may write $Q(\{x\}) = Q(x) - Q(x - 0)$.

Random Variables and Expectation

Random variables are just measurable maps: $X\colon (\Omega, \mathcal{A}) \to (\Omega', \mathcal{A}')$, which indicates the measurability $X^{-1}(\mathcal{A}') \subset \mathcal{A}$ of the map $X\colon \Omega \to \Omega'$.

$\sim$	distributed according to
$\sigma(X)$	σ algebra generated by X; $\sigma(X) = X^{-1}(\mathcal{A}')$
$X(P)$, $\mathcal{L}(X)$	image measure of P under a random variable X, law of X under P and an unspecified Pr, respectively

Occasionally, random variables are identified with their laws; for example, in writing $\Pr(\chi^2 > c)$ for the χ^2 tail probability $\chi^2(c, \infty) = 1 - \chi^2(c)$.

$\mathrm{E}\,X$	expectation of X
$\mathrm{Var}\,X$	variance of X

$\mathrm{Cov}(X,Y)$	covariance of X and Y
$\mathrm{Cov}\,X$, $\mathcal{C}(X)$	covariance matrix of X
$\mathrm{E}(X\|\mathcal{A})$	conditional expectation given the σ algebra $\mathcal{A}$
$\mathrm{E}(X\|Y)$	conditional expectation given the random variable Y; that is, given $\sigma(Y)$
$\mathrm{E}(X\|Y=y)$	conditional expectation given the value y of Y; that is, $\mathrm{E}(X\|Y=.)\circ Y = \mathrm{E}(X\|Y)$
$\Pr(A\|\ldots)$	conditional probability; $\Pr(A\|\ldots) = \mathrm{E}(\mathbf{I}_A\,\|\ldots)$
$\mathrm{E}_\bullet$	conditional expectation given the regressor [p 266]

The underlying measure may be recorded as a subscript (as in E_P) or by taking over its own subscript (as in E_θ when $P = P_\theta$).

$\xrightarrow{P_n}$	stochastic convergence, convergence in P_n probability
o_{P_n}, O_{P_n}	stochastic Landau symbols; that is, $\mathrm{o}_{P_n}(r_n)/r_n \xrightarrow{P_n} 0$, respectively, the sequence $\big\|\mathrm{O}_{P_n}(r_n)/r_n\big\|(P_n)$ tight on $\mathbb{R}$
$\hat F_{u,n}$, $\hat F_{x,n}$	(rectangular) empirical distribution function [p 340]
$\bar F_{u,n}$, $\bar F_{x,n}$	smoothed (rectangular) empirical [pp 339, 341]
$x_{n:i}$	order statistics $x_{n:1} \le \cdots \le x_{n:n}$ of $x_1,\ldots,x_n \in \mathbb{R}$

Laws

$R(a,b)$	the rectangular distribution on (a,b); since 'uniform' has other meanings, we say 'rectangular' (by shape of density)
λ_0, F_0	synonyms of the rectangular $R(0,1)$ on the unit interval and of its distribution function
λ^m	Lebesgue measure on $(\mathbb{R}^m,\mathbb{B}^m)$; $\lambda = \lambda^1$
$\mathcal{N}(a,C)$	normal law on $(\mathbb{R}^m,\mathbb{B}^m)$ with mean $a\in\mathbb{R}^m$ and covariance matrix $C\in\mathbb{R}^{m\times m}$
$\varphi_C(x-a)$	Lebesgue density at x of $\mathcal{N}(a,C)$ for $\det C \ne 0$
φ, Φ	standard normal density and distribution function on $\mathbb{R}$
$\chi^2_n(\delta^2)$	chi square distribution with n degrees of freedom and noncentrality parameter δ^2
$\mathbf{I}_a$	(Dirac) one-point measure in a; $\mathbf{I}_a(A) = \mathbf{I}_A(a)$

Mathematical Symbols

$\#A$	cardinality of a set A
$\subset$, $\supset$	subset/supset, or equal
$\le$	less or equal, coordinatewise on $\mathbb{R}^m$
$\|.\|$	Euclidean norm on $\mathbb{R}^m$
$\lceil . \rceil$	largest integer less or equal to [p 16]
$\lfloor . \rfloor$	smallest integer larger or equal to [p 16]

x^+, x^-	positive, negative parts
$\wedge$, min	minimum
$\vee$, max	maximum
inf, sup	pointwise infimum/supremum
$\inf_P$, $\sup_P$	P essential infimum/supremum
$\inf_\bullet$, $\sup_\bullet$	conditional essential extrema given the regressor [p 266]
$\uparrow$, $\downarrow$	monotone convergence from below/above of numbers (w.r.t. $\leq$) and sets (w.r.t. inclusion; union/intersection)
$a \propto b$	a proportional to b
$a \rightsquigarrow b$	a replaced by b
o, O	the usual Landau symbols
A°, $\overline{A}$	open kernel and closed hull, respectively, of some subset A of a topological space
$\mathbb{I}_k$	the unit $k \times k$ matrix
$A \in \mathbb{R}^{p \times k}$	a real matrix, p rows and k columns
A'	transpose of a matrix A
rk A	rank of A
tr A	trace of A
minev, maxev	minimal/maximal eigenvalue
$A < B$	$B - A$ positive definite
$A \leq B$	$B - A$ positive semidefinite
$A^{1/2}$	unique symmetric, positive semidefinite root of any matrix $A = A' \geq 0$
$\lvert t \rvert_A$	norm $\left(t'A^{-1}t\right)^{1/2}$ standardized by some $A = A' > 0$
$d_\mathcal{R}$	$\mathcal{R}$ differential [p 2]; especially, d_F (Fréchet/bounded), d_H (Hadamard/compact), d_G (Gateaux–Lèvy/weak)
$\int \ldots \int$	multiple integrals; also written as E…E (in decreasing size), or simply as one single $\int$ or E

Function Spaces

X^*	the topological dual $\mathcal{L}(X, \mathbb{R})$ [p 355]
$\mathcal{L}(X, Y)$	the bounded linear operators: $X \to Y$ [pp 1, 2]
$\mathcal{R}(X, Y)$	the o remainder terms of an $\mathcal{R}$ differentiation [p 1]
$\mathcal{D}_\mathcal{R}(X, Y; x)$	the maps: $X \to Y$ that are $\mathcal{R}$ differentiable at x [p 2]

The latter two spaces may also be indexed by: F (Fréchet/bounded), H (Hadamard/compact), and G (Gateaux–Lévy/weak), respectively [p 2]

$\mathcal{C}^k(\Theta)$	the bounded continuous functions: $\Theta \to \mathbb{R}^k$, equipped with the sup norm; $\mathcal{C}(\Theta) = \mathcal{C}^1(\Theta)$
$\mathcal{C}_1^k(\Theta)$	the bounded continuous functions: $\Theta \to \mathbb{R}^k$ having a bounded continuous derivative on Θ°, equipped with sup norm plus sup operator norm [p 14]; $\mathcal{C}_1(\Theta) = \mathcal{C}_1^1(\Theta)$

$\mathcal{D}(\Theta)$	the bounded functions: $\Theta \subset \mathbb{R} \to \mathbb{R}$ which are right continuous and have left limits, with sup norm
$L^p_\alpha(P)$	space of (equivalence classes of) $\mathbb{R}^p$-valued functions f such that $\int |f|^\alpha \, dP < \infty$; $L_\alpha(P) = L^1_\alpha(P)$
$\mathcal{L}^k_2(\mathcal{A})$	the Hilbert space of (equivalence classes of) $\xi\sqrt{dP}$ with any $\xi \in L^k_2(P)$, $P \in \mathcal{M}_b(\mathcal{A})$ [p 48]; $\mathcal{L}_2(\mathcal{A}) = \mathcal{L}^1_2(\mathcal{A})$
$Z^p_\alpha(\theta)$	$L^p_\alpha(P_\theta) \cap \{\mathrm{E}_\theta = 0\}$, the tangents at P_θ [p 125]
$\Psi_\alpha(\theta)$, $\Psi^D_\alpha(\theta)$	set of square integrable $(\alpha = 2)$, and bounded $(\alpha = \infty)$, influence curves at P_θ; respectively, partial influence curves at P_θ, with some matrix $D \in \mathbb{R}^{p \times k}$ such that $\operatorname{rk} D = p \le k$ [p 130]
$\Phi_\mu(\theta)$	set of CvM influence curves at P_θ [p 132]
$Z^p_{\alpha\bullet}$	$L^p_\alpha(P) \cap \{\mathrm{E}_\bullet = 0\}$, conditionally centered tangents [266]
$\Psi_{\alpha\bullet}$, $\Psi^D_{\alpha\bullet}$	set of square integrable, and bounded, conditionally centered (partial) influence curves $(\alpha = 2, \infty)$, respectively [p 274]
$\mathbb{M}_\bullet$, $\mathbb{M}^+_\bullet$, $\mathbb{M}^-_\bullet$	the Borel measurable functions: $\mathbb{R}^k \to \bar{\mathbb{R}}$ (nonnegative, nonpositive, respectively) [p 262]
$\mathbb{L}$, $\mathbb{H}$, $\mathbb{H}_\bullet$	$\mathbb{L} = L^p_\infty(P)$, $\mathbb{H} = L^p_2(P)$ [p 196], $\mathbb{H}_\bullet = L^p_2(K)$ [p 274]
L	the loss functions [pp 78, 81] (symmetric, subconvex, upper semicontinuous at ∞)

Neighborhoods, Bias Terms, Optimality Problems

$* = c, v, h, \mu \dots$	type of balls and metric: contamination, total variation, Hellinger, CvM, ... [p 124]
$t = 0, \varepsilon, \alpha$	type of neighborhoods: (general parameter) unconditional, and conditional regression neighborhoods with fixed contamination curve $\varepsilon \in \mathbb{M}^+_\bullet$, respectively average conditional regression balls of exponent $\alpha \in [1, \infty]$ [pp 262, 263]
$\mathcal{U}_{*,t}(\theta)$	neighborhood system about P_θ [p 124]
$U_{*,t}(\theta, r)$	such a neighborhood about P_θ of radius $r \in (0, \infty)$; in the as. setup, usually $r = \mathrm{O}(1/\sqrt{n})$ [p 124]
$\mathcal{G}_*(\theta)$, $\mathcal{G}_{*,t}$	corresponding tangent classes [pp 171, 266, 267]
$\omega_{*,t;s}$	oscillation/bias terms [pp 171, 172, 268]
$s = 0, 2, \infty, e$	the oscillation/bias variant [pp 171, 172, 268, 272]
$O^{\mathrm{tr}}_{*,t;s}(b)$	minimum trace problem s.t. bias bound [pp 196, 274]
$O^{\mathrm{ms}}_{*,t;s}(\beta)$	mean square error (MSE) problem [pp 207, 284]
$\omega^M_{*,t;s}$	oscillation/bias terms, invariant due to standardization through M, using the norm $|.|_{M^{-1}}$ [pp 215, 217, 325]
$O^{M,\mathrm{tr}}_{*,t;s}(b)$	invariant, M standardized minimum trace problem, with equivariant solution [pp 215, 325]

$O^{M,\mathrm{ms}}_{*,t;s}(\beta)$ invariant, M standardized MSE problem, with equivariant solution [pp 215, 330]

$\omega^{\mathcal{C}}_{*}$ oscillation/bias terms, invariant by self-standardization (through own covariance) [p 217]

An additional superscript g (to bias terms, optimization problems, ...) indicates reparametrization by g of the ideal center model [pp 211, 319]. The subscripts $t = 0$ (unconditional balls) and $s = 0$ (exact bias variants) need not show up in the notation explicitly.

Numbered Displays

Outside the current Section *sec*, a display numbered (*eqno*) is referred to by number *sec* (*eqno*).

When referred to by number, the display may occasionally be supposed to include the logical quantors like 'there exists' ($\exists$) or 'for all' ($\forall$) immediately preceeding or succeeding the display. This may seem risky but saves a lot of space and will always be clear from the context.

A display may be numbered out of sequence, with indented equation number. Then the same display will occur again further below in the section, with identical (nonindented) equation number that finally fits.

Displays,

displayed formula (*eqno*)

will be written without the usual punctuation . at the right end—even so if (*eqno*) is missing.

Chapter 1

Von Mises Functionals

1.1 General Remarks

The asymptotic (as.) normality of several classical estimators can be derived by way of the differentiation of functionals. The approach goes back to von Mises (1947), and has been cast into the framework of compact differentiation by Reeds (1976).

Another general technique, not included here, is certainly Hájek's (1968) projection method, which has been developed for rank statistics, and applied to R estimates by Jurečková (1969), and by Stigler (1974) to L estimates. Sophisticatedly designed ad hoc arguments, like Huber's (1967) proof of the as. normality of M estimates, sometimes yield stronger results.

Nevertheless, since the literature on as. distribution theory is so vast, compact differentiation remains an appealing unifying concept, which also suffices for the beginning to provide us with (as. normal) estimators and test statistics.

1.2 Regular Differentiations

The following excerpt from Averbukh and Smolyanov (1967, 1968), who have systematized various methods of differentiation, motivates compact differentiation on mathematical grounds.

For every pair of normed real vector spaces (X, Y) let a subset $\mathcal{R}(X, Y)$ of Y^X, the functions from X to Y, be given. The following four conditions are imposed on this system $\mathcal{R}$, which will provide the (Landau) o remainder terms in the first-order Taylor approximation of an $\mathcal{R}$ differentiation:

$$\varrho(0) = 0, \qquad \varrho \in \mathcal{R}(X, Y) \tag{1}$$

$$\mathcal{R}(X, Y) \ \text{ a real vector subspace of } \ Y^X \tag{2}$$

$$\mathcal{R}(X, Y) \cap \mathcal{L}(X, Y) = \{0\} \tag{3}$$

where $\mathcal{L}(X,Y)$ is the space of continuous linear maps from X to Y, and 0 stands for the zero operator. Moreover, in case $X = \mathbb{R}$, it is required that

$$\mathcal{R}(\mathbb{R},Y) = \{\, \varrho\colon \mathbb{R} \longrightarrow Y \mid \lim_{t\to 0} \varrho(t)/t = 0 \,\} \tag{4}$$

Definition 1.2.1 *Let $\mathcal{R}$ fulfill (1)–(4). Then a map $T\colon X \to Y$ between two normed real vector spaces is called $\mathcal{R}$* differentiable *at $x \in X$ if there exist some $L \in \mathcal{L}(X,Y)$ and $\varrho \in \mathcal{R}(X,Y)$ such that for all $h \in X$,*

$$T(x+h) = T(x) + Lh + \varrho(h) \tag{5}$$

The continuous linear map $d_{\mathcal{R}}T(x) = L$ is called $\mathcal{R}$ derivative *of T at x. The set of all functions $T\colon X \to Y$ which are $\mathcal{R}$ differentiable at x is denoted by $\mathcal{D}_{\mathcal{R}}(X,Y;x)$.*

Remark 1.2.2

(a) By (2) and (3) the $\mathcal{R}$ derivative $d_{\mathcal{R}}T(x)$ is uniquely determined.

(b) Condition (4) ensures that $\mathcal{R}$ differentiability in case $X = \mathbb{R}$ coincides with the usual notion of differentiability.

(c) We can obviously write

$$\mathcal{R}(X,Y) = \{\, T \in \mathcal{D}_{\mathcal{R}}(X,Y;0) \mid T(0) = 0,\ d_{\mathcal{R}}T(0) = 0 \,\} \tag{6}$$

(d) $\mathcal{D}_{\mathcal{R}}(X,Y;x)$ is a vector subspace of Y^X, and $T \mapsto d_{\mathcal{R}}T(x)$ is a linear map from $\mathcal{D}_{\mathcal{R}}(X,Y;x)$ to $\mathcal{L}(X,Y)$. ////

$\mathcal{R}$ differentiations may be constructed in a special way by means of coverings $\mathcal{S}$, whose elements are naturally assumed to be bounded sets S (so that $th \to 0$ uniformly for $h \in S$ as $t \to 0$): For every normed real vector space X let a covering $\mathcal{S}_X$ of X be given which consists of bounded subsets of X. If Y is another normed real vector space, define

$$\mathcal{R}_{\mathcal{S}}(X,Y) = \Big\{\, \varrho\colon X \longrightarrow Y \;\Big|\; \lim_{t\to 0} \sup_{h\in S} \frac{\|\varrho(th)\|}{t} = \varrho(0) = 0 \ \ \forall\, S \in \mathcal{S}_X \Big\} \tag{7}$$

Then the class $\mathcal{R}_{\mathcal{S}}$ satisfies the conditions (1)–(4).

Definition 1.2.3 *With X ranging through all normed real vector spaces,*

(a) Gateaux–Lévy *or* weak *differentiation is defined by the choices*

$$\mathcal{S}_{\mathrm{G}X} = \{\, S \subset X \mid S \text{ finite} \,\} \tag{8}$$

(b) Hadamard *or* compact *differentiation is defined by the choices*

$$\mathcal{S}_{\mathrm{H}X} = \{\, S \subset X \mid S \text{ compact} \,\} \tag{9}$$

(c) Fréchet *or* bounded *differentiation is defined by the choices*

$$\mathcal{S}_{\mathrm{F}X} = \{\, S \subset X \mid S \text{ bounded} \,\} \tag{10}$$

The three differentiations will be indicated by the corresponding authors' initials. Thus, for example, $\mathcal{D}_{\mathrm{F}}(X,Y;x) \subset \mathcal{D}_{\mathrm{H}}(X,Y;x) \subset \mathcal{D}_{\mathrm{G}}(X,Y;x)$.

The chain rule, which is well-known for bounded differentiation [Dieudonné (1960; Theorem 8.2.1)], would generally be very useful.

Definition 1.2.4 *An $\mathcal{R}$ differentiation satisfying* (1)–(4) *is called* regular *if the* chain rule *holds: For all maps $T \in \mathcal{D}_{\mathcal{R}}(X,Y;x)$ and $U \in \mathcal{D}_{\mathcal{R}}(Y,Z;y)$ between three normed real vector spaces that are $\mathcal{R}$ differentiable at $x \in X$ and $y = T(x) \in Y$, respectively, the composition $U \circ T$ is $\mathcal{R}$ differentiable at x, and*

$$d_{\mathcal{R}}(U \circ T)(x) = d_{\mathcal{R}}U(y) \circ d_{\mathcal{R}}T(x) \tag{11}$$

Example 1.2.5 [Fréchet (1937)] Let $T\colon \mathbb{R} \to \mathbb{R}^2$ and $U\colon \mathbb{R}^2 \to \mathbb{R}$ be defined by

$$T(t) = (t, t^2)\,, \qquad U(x,y) = x\,\mathbf{I}(y = x^2) \tag{12}$$

Then we have $T \in \mathcal{D}_{\mathrm{F}}(\mathbb{R},\mathbb{R}^2;0)$ and $U \in \mathcal{D}_{\mathrm{G}}(\mathbb{R}^2,\mathbb{R};0)$, whereas

$$d_{\mathrm{G}}U(0) = 0, \qquad U \circ T = \mathrm{id}_{\mathbb{R}} \tag{13}$$

Hence (11) is violated. Thus, weak differentiation is not regular. ////

Proposition 1.2.6 *Compact differentiation is regular.*

PROOF Actually, we use *sequentially* compact sets in this proof, which in the framework of normed spaces of course does not make any difference. Thus, for $h_n \to h$ in X and $t_n \to 0$ in $\mathbb{R}$, we must show that

$$\frac{V(x + t_n h_n) - V(x)}{t_n} \longrightarrow d_{\mathrm{H}}V(x)h \tag{14}$$

with $V = U \circ T$ and $d_{\mathrm{H}}V(x) = d_{\mathrm{H}}U(y) \circ d_{\mathrm{H}}T(x)$. But

$$T(x + t_n h_n) = T(x) + t_n k_n\,, \qquad k_n \longrightarrow d_{\mathrm{H}}T(x)h = k$$

since $T \in \mathcal{D}_{\mathrm{H}}(X,Y;x)$. Likewise, since $U \in \mathcal{D}_{\mathrm{H}}(Y,Z;y)$, $y = T(x)$,

$$\begin{aligned} V(x + t_n h_n) &= U(y + t_n k_n) = U(y) + t_n d_{\mathrm{H}}U(y)k + \mathrm{o}(t_n) \\ &= V(x) + t_n d_{\mathrm{H}}U(y) \circ d_{\mathrm{H}}T(x)h + \mathrm{o}(t_n) \end{aligned}$$

and this proves (14). ////

Remark 1.2.7

(a) In case $h_1 = h_2 = \cdots = h$ the proof also shows this: If T is weakly differentiable at x, and U is compactly differentiable at $y = T(x)$, then the composition $U \circ T$ is weakly differentiable at x with derivative

$$d_{\mathrm{G}}(U \circ T)(x) = d_{\mathrm{H}}U(y) \circ d_{\mathrm{G}}T(x) \tag{15}$$

In particular, the function U of Example 1.2.5 cannot be compactly differentiable at 0.

(b) If the map $T: X \to Y$ between normed vector spaces is compactly differentiable at a point $x \in X$, then T is continuous at x. To see this, let $h_n \to 0$ in X and choose $s_n \in (0, \infty)$ such that $s_n \to \infty$ but $s_n h_n \to 0$. Writing

$$T(x + h_n) = T(x + t_n s_n h_n) = T(x) + d_{\mathrm{H}}T(x)h_n + \mathrm{o}(t_n)$$

with $t_n = 1/s_n$ shows that $T(x + h_n) \to T(x)$. ////

Example 1.2.8 The map $T(x, y) = \mathbf{I}(0 < |y| < x^2)$ from $\mathbb{R}^2$ to $\mathbb{R}$ is weakly differentiable with zero derivative at the origin, but is not continuous and hence cannot have a compact derivative there. ////

Compact differentiation is actually the weakest regular $\mathcal{R}$ differentiation.

Theorem 1.2.9 *Given a regular $\mathcal{R}$ differentiation, and a map $T: X \to Y$ between normed real vector spaces that is $\mathcal{R}$ differentiable at some $x \in X$. Then T is compactly differentiable at x with derivative*

$$d_{\mathrm{H}}T(x) = d_{\mathcal{R}}T(x) \tag{16}$$

PROOF Suppose the map T satisfying (5) does not have the compact derivative $d_{\mathcal{R}}T(x)$ at x. Then there exists a compact $S \subset X$ such that, as $t \to 0$, $\|\varrho(th)\|/t$ fails to converge to 0 uniformly in $h \in S$. Therefore, there exists a sequence $t_n \to 0$ in $\mathbb{R}$ satisfying $|t_n| > |t_{n+1}|$ for all $n \geq 1$ as well as a convergent sequence $h_n \to h$ in X such that

$$\frac{T(x + t_n h_n) - T(x)}{t_n} \not\longrightarrow d_{\mathcal{R}}T(x)h \tag{17}$$

Now consider the following curve $\varphi: \mathbb{R} \to X$,

$$\varphi(t) = \begin{cases} x + t_n h_n & \text{if } t = t_n \text{ for some } n \geq 1 \\ x + th & \text{otherwise} \end{cases} \tag{18}$$

By property (4) of $\mathcal{R}$ differentiations, φ has $\mathcal{R}$ derivative h at $t = 0$. Therefore, according to the chain rule, $T \circ \varphi$ would at $t = 0$ have $\mathcal{R}$ derivative $d_{\mathcal{R}}T(x)h$. But this would enforce convergence in (17). ////

Remark 1.2.10

(a) According to this proof, $T: X \to Y$ has compact derivative $d_{\mathrm{H}}T(x)$ at $x \in X$ iff for all curves $\varphi: \mathbb{R} \to X$, $\varphi(0) = x$, differentiable at 0, the composition $T \circ \varphi$ has derivative $d_{\mathrm{H}}T(x) \circ d\varphi(0)$ at 0. [Differentiability of φ and $T \circ \varphi$ at 0 is in the usual sense, due to (4).] This actually is Hadamard's original definition, and agrees with the notion of *quasi-differentiability* introduced by Dieudonné (1960; Problem 4, pp 151, 152).

(b) The previous proof in particular shows that $d_{\mathcal{R}}T(x) = d_{\mathrm{H}}T(x)$. In fact, due to (4), the curve $\varphi(t) = x + th$ from $\mathbb{R}$ to X has $\mathcal{R}$ derivative $d_{\mathcal{R}}\varphi(0) = h$ at $t = 0$. Then, provided the $\mathcal{R}$ differentiation is regular, the chain rule implies that for all $T \in \mathcal{D}_{\mathcal{R}}(X, Y; x)$,

$$d_{\mathcal{R}}T(x) = d_{\mathrm{G}}T(x) \tag{19}$$

By the same argument, every remainder term $\varrho \in \mathcal{R}(X, Y)$ of a regular differentiation converges pointwise: $\lim_{t\to 0} \varrho(th)/t = 0$ for each $h \in X$. ////

Property (4) lets all $\mathcal{R}$ differentiations coincide in case $X = \mathbb{R}$. If $X = \mathbb{R}^k$ for some finite dimension $k > 1$, bounded and compact differentiation are still the same, but stronger than weak differentiation. In infinite dimensions, bounded and compact differentiation become different, which is demonstrated right away by an example from the real analysis literature, but also follows from the statistical functionals we shall study later on.

Example 1.2.11 Let $\lambda_0 = R(0, 1)$ be the rectangular measure on $(0, 1)$.
(a) [Sova (1966)] The map

$$T: L_1(\lambda_0) \to \mathbb{R}, \qquad T(u) = \int \sin u(x)\, \lambda_0(dx) \tag{20}$$

is boundedly differentiable nowhere, but compactly differentiable everywhere, with derivative

$$d_{\mathrm{H}}T(u)v = \int v(x) \cos u(x)\, \lambda_0(dx) \tag{21}$$

(b) [Reeds (1976), Vainberg (1964)] For $r > s \geq 1$ the map

$$T: L_r(\lambda_0) \to L_s(\lambda_0), \qquad T(u)(x) = \sin u(x) \tag{22}$$

is a so-called Nemytsky operator, which is boundedly differentiable everywhere. For $r = s = 2$, T has a weak derivative everywhere,

$$d_{\mathrm{G}}T(u)v = v \cos u \tag{23}$$

but T is boundedly differentiable nowhere. ////

From the finite-dimensional case we know that *continuous* weak differentiability implies (continuous) bounded differentiability. To extend this result to infinite-dimensional normed real vector spaces X and Y, let $\mathcal{S}$ be a covering of X consisting of bounded subsets of X, and equip the continuous linear maps $\mathcal{L}(X, Y) = \mathcal{L}_{\mathcal{S}}(X, Y)$ from X to Y with the uniformity of $\mathcal{S}$ convergence.

Theorem 1.2.12 *Suppose that the map $T: X \to Y$ is weakly differentiable on an open subset $U \subset X$, and that $d_{\mathrm{G}}T: U \to \mathcal{L}_{\mathcal{S}}(X, Y)$ is continuous.*
Then T is $\mathcal{R}_{\mathcal{S}}$ differentiable on U.

PROOF The space Y being locally convex, the mean value theorem (as a consequence of the separating hyperplane theorem [Proposition B.1.3]) is in force: Given $x, h \in X$, let y^* be any continuous linear functional on Y. Then the composition $y^*T(x+th)$ is a real-valued differentiable function of $t \in \mathbb{R}$, to which the elementary mean value theorem applies:

$$y^* \frac{T(x+th) - T(x)}{t} = \frac{y^*T(x+th) - y^*T(x)}{t} = y^* \, d_{\mathrm{G}}T(x+\tau th)h$$

for some $\tau \in (0,1)$. By the separating hyperplane theorem, therefore, the difference

$$\frac{T(x+th) - T(x)}{t} - d_{\mathrm{G}}T(x)h \tag{24}$$

cannot fall outside the closed convex hull of the set

$$\big\{ \big(d_{\mathrm{G}}T(x+\tau h) - d_{\mathrm{G}}T(x)\big)h \bigm| -|t| \le \tau \le |t| \big\} \tag{25}$$

Since the elements $S \in \mathcal{S}$ are bounded sets, and $d_{\mathrm{G}}T$ is continuous relative to the uniformity of $\mathcal{S}$ convergence, the difference (24) tends to 0 as $t \to 0$, uniformly in $h \in S$. ////

Remark 1.2.13 If X is a Banach space, the uniform boundedness principle applies [Rudin (1974; Theorem 5.8)]. By this principle, continuity of the weak derivative $d_{\mathrm{G}}T: U \to \mathcal{L}_{\mathrm{s}}(X,Y)$ relative to pointwise convergence entails continuity of $d_{\mathrm{G}}T: U \to \mathcal{L}_{\mathrm{c}}(X,Y)$ relative to uniform convergence on compacts. Therefore, if T has a weak derivative which is continuous relative to pointwise convergence, then T is already compactly differentiable with derivative $d_{\mathrm{H}}T = d_{\mathrm{G}}T$ (continuous in the topology of uniform convergence on compacts). This in particular applies to Example 1.2.11 a. ////

In view of this converse, the fact that various notions of differentiability turn out useful can only mean that statistical functionals are not differentiable on full neighborhoods, or that their derivatives are discontinuous.

1.3 The Delta Method

Compact differentiability can be linked with tightness of random increments to obtain a suitable first-order approximation in probability; see (4) below.

Two modifications are useful for the applications. The first distinguishes the space in which the increments take their values, called *tangent space*, from the domain of a functional [Definition 1.3.1]. The second modification eases the tightness condition somewhat [Proposition 1.3.2].

Definition 1.3.1 *Let* $\mathcal{R}$ *be a class with properties* 1.2(1)–1.2(4), *and consider a real vector space* X_0, *a normed vector subspace* $X \subset X_0$, *and a* *normed real vector space* Y. *Then a map* $T: X_0 \to Y$ *is called* $\mathcal{R}$ differentiable along X *at some point* $x \in X_0$, *with* $\mathcal{R}$ *derivative* $d_{\mathcal{R}}T(x)$, *if the map* $h \mapsto T(x+h)$ *from* X *to* Y *is* $\mathcal{R}$ *differentiable at* $h = 0$ *with* $\mathcal{R}$ *derivative* $d_{\mathcal{R}}T(x)$.

Proposition 1.3.2 *Let* X_0 *be a real vector space,* X *a normed vector subspace of* X_0, *and* Y *a normed real vector space. Suppose* $T: X_0 \to Y$ *is compactly differentiable along* X *at some* $x \in X_0$. *Then*

$$\lim_{t,\delta\to 0} \sup\left\{ \left\| \frac{T(x+th)-T(x)}{t} - d_{\mathrm{H}}T(x)h \right\| \,\middle|\, h \in K^{\delta} \right\} = 0 \qquad (1)$$

for all compact sets $K \subset X$, *where* $K^{\delta} = \{ y \in X \mid \inf_{z\in K} \|y-z\| \le \delta \}$.

PROOF Assume sequences $t_n \to 0$, $\delta_n \downarrow 0$ in $\mathbb{R}$, and $h_n \in K^{\delta_n}$. Choose any $k_n \in K$ such that $\|h_n - k_n\| \le \delta_n$. As K is compact, every subsequence has a further subsequence (m) such that k_m tends to some limit $h \in K$. But then also $h_m \to h$ and

$$\frac{T(x+t_m h_m)-T(x)}{t_m} \longrightarrow d_{\mathrm{H}}T(x)h$$

since T is compactly differentiable at x along X. ////

This formulation of compact differentiability is adapted to the following tightness formulation, in which possibly nonmeasurable functions Z_n occur. Accordingly, inner probabilities P_{n*} are employed.

Theorem 1.3.3 *Let* X_0 *be a real vector space,* X *a normed vector subspace of* X_0, *and* Y *a normed real vector space. Suppose* $T: X_0 \to Y$ *is compactly differentiable along* X *at some* $x \in X_0$. *Consider a sequence of functions* Z_n *from probability spaces* $(\Omega_n, \mathcal{A}_n, P_n)$ *to* X. *Assume that for every* $\varepsilon \in (0,\infty)$ *there exist a compact* $K \subset X$ *and a sequence* $\delta_n \to 0$ *such that*

$$\liminf_{n\to\infty} P_{n*}\big(\sqrt{n}\, Z_n \in K^{\delta_n}\big) \ge 1-\varepsilon \qquad (2)$$

Then, for every neighborhood U *of* x,

$$\lim_{n\to\infty} P_{n*}(x+Z_n \in U) = 1 \qquad (3)$$

$$\sqrt{n}\,\big(T(x+Z_n)-T(x)\big) = d_{\mathrm{H}}T(x)\sqrt{n}\, Z_n + \mathrm{o}_{P_{n*}}(n^0) \qquad (4)$$

PROOF Given $\varepsilon \in (0,1)$, choose K, (δ_n) by (2), and let $m_K \in (0,\infty)$ be a norm bound for K. Then $\sqrt{n}\, Z_n \in K^{\delta_n}$ implies that $\|Z_n\|$ is bounded by $(m_K+\delta_n)/\sqrt{n}$, hence $x + Z_n \in U$ eventually. Therefore,

$$\liminf_{n\to\infty} P_{n*}(x+Z_n \in U) \ge 1-\varepsilon$$

which proves (3). Moreover, Proposition 1.3.2 applies with the identifications: $t^{-1} = \sqrt{n}$, $\delta = \delta_n$, $h = \sqrt{n}\, Z_n$. Then (1) says that the following implication becomes true eventually,

$$\sqrt{n}\, Z_n \in K^{\delta_n} \implies \left\|\sqrt{n}\,\bigl(T(x+Z_n) - T(x)\bigr) - d_{\mathrm{H}}T(x)\sqrt{n}\, Z_n\right\| \le \varepsilon$$

Hence

$$\begin{aligned} &\liminf_{n\to\infty} P_{n*}\Bigl(\left\|\sqrt{n}\,\bigl(T(x+Z_n) - T(x)\bigr) - d_{\mathrm{H}}T(x)\sqrt{n}\, Z_n\right\| \le \varepsilon\Bigr) \\ &\qquad \ge \liminf_{n\to\infty} P_{n*}\bigl(\sqrt{n}\, Z_n \in K^{\delta_n}\bigr) \ge 1 - \varepsilon \end{aligned}$$

which proves (4). ////

In the following statistical examples, Z_n will be the empirical minus the theoretical measure at sample size n, or some function of the empirical process; in any case, Z_n takes values in some infinite-dimensional function space. The choice of this vector space X and its norm may be adapted to the functional T of interest. The norm should not be too strong (many open sets, few compacts) to destroy tightness, and not too weak (few open sets, many compacts) to destroy compact differentiability of the functional. This certainly sounds plausible, but one should also think of the open mapping theorem [Rudin (1973; Theorem 2.11)] which, if the normed space should be complete, permits just one topology.

The linear approximation $d_{\mathrm{H}}T(x)\sqrt{n}\, Z_n$ plays a dominant role in the local theories of asymptotic statistics and robustness. In our applications it will turn out to be a sum of i.i.d. variables. The summands are called an *influence curve* informally, until the notion is made precise (and tied with location, scale, or more general parametric models) [Chapter 4].

1.4 M Estimates

Let $x_i \sim F$ be finite-dimensional $\mathbb{R}^m$-valued, i.i.d. observations that follow some law $F \in \mathcal{M}_1(\mathbb{B}^m)$. As parameter space, we assume a nonempty, compact subset Θ of some finite-dimensional $\mathbb{R}^k$ that is the (topological) closure of its interior,

$$\Theta = \overline{\Theta^\circ} \tag{1}$$

By $\mathcal{C}^k(\Theta)$ we denote the space of all bounded continuous $\mathbb{R}^k$-valued functions on Θ, equipped with the sup norm, and by $L_2^k(F)$ the space of (equivalence classes of) all $\mathbb{R}^k$-valued functions on $\mathbb{R}^m$ whose k coordinates are square integrable under F. Let the function $\varphi\colon \mathbb{R}^m \times \Theta \to \mathbb{R}^k$ have sections

$$\varphi(x, .\,) \in \mathcal{C}^k(\Theta) \qquad \text{a.e. } F(dx) \tag{2}$$

$$\varphi_\theta = \varphi(.\,,\theta) \in L_2^k(F), \qquad \theta \in \Theta \tag{3}$$

Consider the following function of the empirical measure $\hat{F}_n = 1/n \sum_i \mathbf{I}_{x_i}$,

$$Z_n(x_1,\ldots,x_n;\theta) = \int \varphi(x,\theta)\,\hat{F}_n(x_1,\ldots,x_n;\,dx) = \frac{1}{n}\sum_{i=1}^n \varphi_\theta(x_i) \tag{4}$$

Then, by (2), we have

$$Z_n(x_1,\ldots,x_n;.) \in \mathcal{C}^k(\Theta) \qquad \text{a.e. } F^n(dx_1,\ldots,dx_n) \tag{5}$$

Fix any $\theta_\infty \in \Theta$. By definition, the M *estimate* with score function φ at sample size n may be any Borel measurable zero $S_n\colon \mathbb{R}^{mn} \to \Theta$ of Z_n,

$$\begin{aligned} Z_n(x_1,\ldots,x_n;S_n) &= 0, && \text{if a zero exists} \\ S_n = S_n(x_1,\ldots,x_n) &= \theta_\infty, && \text{otherwise} \end{aligned} \tag{6}$$

Introduce a functional $T\colon \mathcal{C}^k(\Theta) \to \Theta$, called an M *functional*, to select a zero,

$$\begin{aligned} f(T(f)) &= 0, && \text{if } f \text{ has a zero} \\ T(f) &= \theta_\infty, && \text{otherwise} \end{aligned} \tag{7}$$

Then, up to measurability, $S_n = T \circ Z_n$ is an M estimate at sample size n.

Remark 1.4.1 This remark concerns measurability of the process

$$(x_1,\ldots,x_n) \longmapsto Z_n(x_1,\ldots,x_n;.) \tag{8}$$

from $\mathbb{R}^{mn}$ to the $\mathbb{R}^k$-valued functions on Θ.

(a) If A denotes the exceptional null set allowed in condition (2), such that $\varphi(x,.)$ is continuous for all $x \in \mathbb{R}^m \setminus A$, then the map $x \mapsto \varphi(x,.)$ from $\mathbb{R}^m \setminus A$ to $\mathcal{C}^k(\Theta)$ is Borel measurable. To see this, pick any countable dense subset $\Theta_0 \subset \Theta$. Then for all $f \in \mathcal{C}^k(\Theta)$ and $\varepsilon \in (0,\infty)$,

$$\bigl\{\, x \in \mathbb{R}^m \setminus A \bigm| \|\varphi(x,.) - f\| \le \varepsilon \,\bigr\} = \bigcap_{\theta\in\Theta_0} \bigl\{\, x \notin A \bigm| |\varphi(x,\theta) - f(\theta)| \le \varepsilon \,\bigr\}$$

where each set under the countable intersection, hence the intersection itself, is in $\mathbb{B}^m$, due to measurability of the θ sections of φ, which is part of assumption (3). Consequentially, there exists an F^n null set $A_n \in \mathbb{B}^{mn}$ such that $(x_1,\ldots,x_n) \mapsto Z_n(x_1,\ldots,x_n;.)$ from $\mathbb{R}^{mn} \setminus A_n$ to $\mathcal{C}^k(\Theta)$ is Borel measurable.

(b) By compactness of Θ we have

$$\bigl\{\, f \in \mathcal{C}^k(\Theta) \bigm| f \text{ has no zero} \,\bigr\} = \bigcup_{\nu=1}^{\infty} \bigcap_{\theta\in\Theta_0} \bigl\{\, f \in \mathcal{C}^k(\Theta) \bigm| |f(\theta)| \ge 1/\nu \,\bigr\}$$

where Θ_0, as before, is countable and dense in Θ. Since projections are continuous, the set under consideration is Borel measurable. In particular,

$$\bigl\{ (x_1,\ldots,x_n) \in \mathbb{R}^{mn} \setminus A_n \bigm| Z_n(x_1,\ldots,x_n;.) \text{ has no zero} \bigr\} \in \mathbb{B}^{mn}$$

for some $A_n \in \mathbb{B}^{mn}$ satisfying $F^n(A_n) = 0$. ////

We introduce the function

$$\eta : \Theta \to \mathbb{R}^k, \qquad \eta(\theta) = \int \varphi(x,\theta)\, F(dx) \tag{9}$$

Hadamard Differentiability

Existence and compact differentiability of M functionals are then settled by the following result.

Theorem 1.4.2 *Assume that $\eta(\theta_0) = 0$ for some $\theta_0 \in \Theta^\circ$, and suppose that $\eta \in \mathcal{C}^k(\Theta)$, that η is locally homeomorphic at θ_0 and that η has a bounded derivative $d\eta(\theta_0)$ at θ_0 of rank $\operatorname{rk} d\eta(\theta_0) = k$.*

(a) *Then there exist some neighborhood V of η in $\mathcal{C}^k(\Theta)$ and a functional $T: V \to \Theta$ such that*

$$f(T(f)) = 0, \qquad f \in V \tag{10}$$

(b) *Every such functional T is compactly differentiable at η with derivative*

$$d_{\mathrm{H}} T(\eta) = -d\eta(\theta_0)^{-1} \circ \pi_{\theta_0} \tag{11}$$

where π_{θ_0} denotes the evaluation at θ_0.

(c) *The functional T need not be boundedly differentiable at η.*

This result will be tied with the following central limit theorem in $\mathcal{C}^k(\Theta)$ due to Jain and Marcus (1975; Theorem 1), to prove as. normality.

Proposition 1.4.3 *Assume that φ fulfills (2) and (3). Suppose a pseudodistance d on Θ such that*

$$\zeta \to \theta \in \Theta \implies d(\zeta,\theta) \to 0 \tag{12}$$

and

$$\int_0^1 \sqrt{H(r)}\, dr < \infty \tag{13}$$

for the metric entropy H of the space (Θ, d). Moreover, assume that

$$|\varphi(x,\zeta) - \varphi(x,\theta)| \le d(\zeta,\theta)\, M(x) \qquad \text{a.e. } F(dx) \tag{14}$$

for all $\zeta, \theta \in \Theta$, and some function

$$M \in L_2(F) \tag{15}$$

Then there is a Gaussian process Z on $\mathcal{C}^k(\Theta)$ such that for all $\zeta, \theta \in \Theta$,

$$\mathrm{E}\, Z(\theta) = 0, \qquad \mathrm{E}\, Z(\zeta) Z(\theta) = \operatorname{Cov}_F(\varphi_\zeta, \varphi_\theta) \tag{16}$$

and, as $n \to \infty$,

$$\sqrt{n}\,(Z_n - \eta)(F^n) \xrightarrow{\mathrm{w}} \mathcal{L}(Z) \tag{17}$$

in $\mathcal{C}^k(\Theta)$, weakly. Hence the sequence $\sqrt{n}\,(Z_n - \eta)(F^n)$ is tight in $\mathcal{C}^k(\Theta)$.

Remark 1.4.4

(a) The entropy $H(r)$ of the pseudometric space (Θ, d) is defined as log the minimal number of d balls of radius r to cover Θ.

(b) Assumptions (12), (14), and (15) ensure the continuity a.e. $F(dx)$ of $\varphi(x,.)$, that is, condition (2), as well as the continuity of η needed for Theorem 1.4.2.

(c) If $\varphi_\theta \in L_2^k(F)$ for some $\theta \in \Theta$, (14) and (15) imply (3). ////

Corollary 1.4.5 *Under the assumptions of* Theorem 1.4.2 *and* Proposition 1.4.3, *the* M *estimator sequence* $S_n = T \circ Z_n$ *has the as. expansion*

$$\sqrt{n}\,(S_n - \theta_0) = \frac{1}{\sqrt{n}} \sum_{i=1}^{n} \psi(x_i) + \mathrm{o}_{(F^n)_*}(n^0) \tag{18}$$

with influence curve

$$\psi(x) = -d\eta(\theta_0)^{-1}\varphi(x, \theta_0) \tag{19}$$

In particular, if S_n *are measurable, then*

$$\sqrt{n}\,(S_n - \theta_0)(F^n) \xrightarrow{\;\mathrm{w}\;} \mathcal{N}(0, ACA') \tag{20}$$

with $A^{-1} = d\eta(\theta_0)$ *and* $C = \int \varphi_{\theta_0}\varphi_{\theta_0}{}'\,dF$.

PROOF Proposition 1.4.3 ensures that condition (2) of Theorem 1.3.3 can be fulfilled by setting $\delta_n = 0$ and $x = \eta$. Hence statement (3) of Theorem 1.3.3 holds true with $U = V$ taken from Theorem 1.4.2 a. In particular, S_n is defined with F^n probability tending to 1 as $n \to \infty$.

By Theorem 1.4.2 b, the M functional T is compactly differentiable at η, so statement (4) of Theorem 1.3.3 holds true with the derivative given by (11). As $\eta(\theta_0) = 0$, this yields (18). In view of (3) and $\eta(\theta_0) = 0$, the Lindeberg–Lévy theorem applies; thus (20) follows from (18). ////

Remark 1.4.6 The assumptions of this as. normality proof are comparable with Huber's (1967, 1981) more classical conditions.

(a) In view of Huber (1981; Lemma 2.1) and his consistency assumption, our compactness condition on Θ is not restrictive. Apart from the continuity requirement concerning $\varphi(x,.)$, the conditions almost literally coincide with Huber's (1981) assumptions (N–1) to (N–3) of his Theorem 3.1. Note in particular the similarity of condition (14) and Huber's (N–3).

(b) Reeds (1976) requires condition (14) to hold with

$$d(\zeta, \theta) = |\zeta - \theta|^\delta \tag{21}$$

for some $\delta \in (0, \infty)$. Then condition (12) is fulfilled as well as (13), because for this (pseudo)metric we have $H(r) = \mathrm{O}(|\log r|)$ as $r \to 0$.

A bit more generally, suppose $d(\zeta,\theta) = f(|\zeta - \theta|)$ for some function f which is increasing, continuous at 0, $f(0) = 0$, and whose right continuous pseudoinverse satisfies

$$\int_0^1 \left|\log f^{-1}(r)\right| dr < \infty \tag{22}$$

Then (12) and (13) are again fulfilled. This specification corresponds to the weakening of (N–3)(iii) mentioned by Huber (1981, p 133).

(c) Because S_n depends on $\hat{F}_n$ only through Z_n, we can drop $\hat{F}_n$ after having passed to Z_n. This simplification bypasses certain considerations of Reeds (1976, pp 100–105; e.g., his Lemma 5.2.4) completely. ////

PROOF [Theorem 1.4.2]

(a) Let $B(0,r)$ denote the closed ball in $\mathbb{R}^k$ of radius $r \in (0,\infty)$ and center 0. Set $W = \eta^{-1}(B(0,r))$ with $r > 0$ chosen so small that the restricted function $\eta: W \to B(0,r)$ is homeomorphic, that is, continuous in both directions. Thus, W is a compact neighborhood of θ_0.

Then, if $g \in \mathcal{C}^k(\Theta)$ is such that $\sup_{\theta\in W} |g(\theta)| \le r$, the composition $-g \circ \eta^{-1}$ maps $B(0,r)$ continuously into itself. By Brouwer's fixed point theorem, there exists some $|\zeta| \le r$ such that $-g(\eta^{-1}(\zeta)) = \zeta$. Hence the identity $\eta(\theta) + g(\theta) = 0$ holds for $\theta = \eta^{-1}(\zeta) \in W$.

Let $V = \{f \in \mathcal{C}^k(\Theta) \mid \|f - \eta\| \le r\}$ and put $g = f - \eta$ for $f \in V$. As we have just shown, there is some $T(f) \in W$ satisfying $f(T(f)) = 0$. By the axiom of choice, these values can be connected to a map $T: V \to W$.

(b) To prove compact differentiability of T at η, let $K \subset \mathcal{C}^k(\Theta)$ be any compact. Then the Arzéla–Ascoli theorem supplies an $m_K \in (0,\infty)$ and an increasing function M_K satisfying $M_K(0+) = 0$, such that for all $g \in K$ and $\zeta, \theta \in \Theta$ we have $\|g\| \le m_K$ and $|g(\zeta) - g(\theta)| \le M_K(|\zeta - \theta|)$.

For $t \in \mathbb{R}$, $|t| \le r/m_K$ and $g \in K$ define $\beta_t = \beta(tg)$ and T_t by

$$T_t = T(\eta + tg) = \theta_0 - tA\,g(\theta_0) + t\beta_t \tag{23}$$

At this point we note that $\eta + tg \in V$, hence $\eta(T_t) + t\,g(T_t) = 0$ and so $|\eta(T_t)| \le m_K|t|$ by part (a). We must show that, as $t \to 0$, β_t tends to 0 uniformly in $g \in K$. Contrary to Reeds (1976, p 107), we now invoke the inverse function theorem [Dieudonné (1960; Theorem 8.2.3)], whose assumptions are fulfilled. It gives us the following expansion of η^{-1} at 0:

$$\eta^{-1}(y) = \theta_0 + Ay + |y|\kappa(y)$$

where $\lim_{y\to 0} \kappa(y) = 0$. Plug in $y = \eta(T_t)$ to get

$$T_t - \theta_0 = A\eta(T_t) + |\eta(T_t)|\,\kappa(\eta(T_t)) \tag{24}$$

hence

$$|T_t - \theta_0| \le \|A\|\, m_K|t| + m_K|t|\,\bar{\kappa}(m_K|t|)$$

where $\bar{\kappa}(s) = \sup_{|y|\le s} |\kappa(y)|$ and $\lim_{s\to 0} \bar{\kappa}(s) = 0$. Therefore, there exists some $s_0 \in (0, r/m_K)$ such that for all $|t| \le s_0$ and $g \in K$,

$$|T_t - \theta_0| \le a_K |t| \tag{25}$$

with the finite constant $a_K = m_K(1 + \|A\|)$; in particular, $T_t \to \theta_0$ uniformly in $g \in K$, as $t \to 0$. Differentiability at θ_0 of η itself yields

$$- tg(T_t) = \eta(T_t) = A^{-1}(T_t - \theta_0) + |T_t - \theta_0| \sigma(T_t - \theta_0) \tag{26}$$

where $\lim_{x\to 0} \sigma(x) = 0$. Thus we obtain

$$\begin{aligned} -tg(\theta_0) + tA^{-1}\beta_t &= A^{-1}\bigl(-tA\, g(\theta_0) + t\beta_t\bigr) = A^{-1}(T_t - \theta_0) \\ &= \eta(T_t) - |T_t - \theta_0| \sigma(T_t - \theta_0) = -tg(T_t) - |T_t - \theta_0| \sigma(T_t - \theta_0) \end{aligned}$$

Hence for $t \neq 0$,

$$A^{-1}\beta_t = g(\theta_0) - g(T_t) - \sigma(T_t - \theta_0) \frac{|T_t - \theta_0|}{t} \tag{27}$$

and so

$$\begin{aligned} |A^{-1}\beta_t| &\le M_K(|T_t - \theta_0|) + |\sigma(T_t - \theta_0)| \Bigl| \frac{T_t - \theta_0}{t} \Bigr| \\ &\le M_K(a_K|t|) + a_K \bar{\sigma}(a_K|t|) \end{aligned} \tag{28}$$

as soon as $|t| \le s_0$, where $\bar{\sigma}(s) = \sup_{|x|\le s} |\sigma(x)|$ and $\lim_{s\to 0} \bar{\sigma}(s) = 0$. Applying A to $A^{-1}\beta_t$ yields

$$|\beta_t| \le \|A\|\, |A^{-1}\beta_t| \le \|A\| \bigl(M_K(a_K|t|) + a_K \bar{\sigma}(a_K|t|)\bigr) \tag{29}$$

and this upper bound indeed tends to 0 as $t \to 0$.

(c) By example, choose $\Theta = [-1, 1]$, $\eta = \mathrm{id}_{[-1,1]}$, $\theta_0 = 0$, and consider the following continuous, bounded maps $g_n\colon [-1,1] \to [-1,1]$:

$$g_n = -1 \text{ on } [-1, 0], \quad g_n = 1 \text{ on } [2/n, 1], \quad g_n \text{ linear in between} \tag{30}$$

Then $0 = T_n + (nT_n - 1)/n$ for $T_n = T(\eta + \frac{1}{n} g_n)$, hence $T_n = 1/(2n)$. Thus

$$n\bigl(T(\eta + \tfrac{1}{n} g_n) - T(\eta)\bigr) + d\eta(\theta_0)^{-1} g_n(\theta_0) = -\tfrac{1}{2} \tag{31}$$

which disproves bounded differentiability of T. ////

Fréchet Differentiability

Bounded differentiability of the functional T can however be achieved at the cost of *continuous* differentiability of the function η introduced in (9). Accordingly, the bounded continuous functions $\mathcal{C}^k(\Theta)$ will be replaced by the subspace $\mathcal{C}_1^k(\Theta)$ of all bounded continuous functions f from Θ to $\mathbb{R}^k$

that, on the interior Θ^o, have a continuous and bounded derivative. This space is equipped with the norm

$$\|f\|_1 = \|f\| + \|df\| \tag{32}$$

where

$$\|f\| = \sup\{|f(\theta)| \mid \theta \in \Theta\}$$
$$\|df\| = \sup\{|df(\theta)h| \mid |h| \le 1,\, \theta \in \Theta^o\}$$

denote the sup and sup operator norms, respectively.

Theorem 1.4.7 *Suppose that* $\eta \in \mathcal{C}_1^k(\Theta)$. *Assume that* $\eta(\theta_0) = 0$ *for some* $\theta_0 \in \Theta^o$, *and that the derivative has full rank:* $\operatorname{rk} d\eta(\theta_0) = k$.

(a) *Then there exists an open neighborhood* $V_0 \subset \mathcal{C}_1^k(\Theta)$ *of* η *such that for every open connected neighborhood* $V \subset V_0$ *of* η *there is a unique continuous map* $T\colon V \to \Theta$ *satisfying*

$$T(\eta) = \theta_0\,, \qquad f(T(f)) = 0, \quad f \in V \tag{33}$$

(b) T *is continuously boundedly differentiable on* V *with derivative*

$$d_{\mathrm{F}}T(f) = -df(T(f))^{-1} \circ \pi_{T(f)}\,, \qquad f \in V \tag{34}$$

where $\pi_{T(f)}$ *denotes the evaluation at* $T(f)$.

PROOF The evaluation map

$$\pi\colon \mathcal{C}_1^k(\Theta) \times \Theta \longrightarrow \mathbb{R}^k, \qquad \pi(f,\theta) = \pi_\theta(f) = f(\theta)$$

is boundedly differentiable on $\mathcal{C}_1^k(\Theta) \times \Theta^o$ with derivative

$$d_{\mathrm{F}}\pi(f,\theta)\,(g,t) = g(\theta) + df(\theta)\,t \tag{35}$$

In particular, the first and second partials read

$$d_{\mathrm{F}1}\,\pi(f,\theta)\,g = g(\theta)\,, \qquad d_{\mathrm{F}2}\,\pi(f,\theta)\,t = df(\theta)\,t \tag{36}$$

To verify continuity of these derivatives, write

$$\begin{aligned}
&\big\|d_{\mathrm{F}}\pi(h,\zeta) - d_{\mathrm{F}}\pi(f,\theta)\big\| \\
&\quad = \sup\big\{\big|d_{\mathrm{F}}\pi(h,\zeta)(g,t) - d_{\mathrm{F}}\pi(f,\theta)(g,t)\big| \bigm| \|g\|_1 \le 1, |t| \le 1\big\} \\
&\quad = \sup\big\{\big|g(\zeta) + dh(\zeta)t - g(\theta) - df(\theta)t\big| \bigm| \|g\|_1 \le 1, |t| \le 1\big\} \\
&\quad \le \sup\big\{|g(\zeta) - g(\theta)| \bigm| \|g\|_1 \le 1\big\} + \|df(\zeta) - df(\theta)\| + \|dh(\zeta) - df(\zeta)\|
\end{aligned}$$

By the mean value theorem [Dieudonné (1960; Theorem 8.5.4)], the first term in this upper bound is not larger than $|\zeta - \theta|$ since $\|dg\| \le \|g\|_1 \le 1$. As $\zeta \to \theta$, the second term also tends to zero because of the continuity in operator norm of the derivative of $f \in \mathcal{C}_1^k(\Theta)$. The third term is not larger than the sup operator norm $\|df - dh\|$, which is bounded by $\|df - dh\|_1$, and this norm tends to 0 iff $h \to f$ in $\mathcal{C}_1^k(\Theta)$.

Therefore, as $\text{п}(\eta,\theta_0) = 0$, and $d_{F2}\,\text{п}(\eta,\theta_0) = d\eta(\theta_0)$ also is a homeomorphic automorphism of $\mathbb{R}^k$, the classical implicit function theorem [Dieudonné (1960; Theorem 10.2.1)] applies and yields the result. ////

Remark 1.4.8 The implicit function theorem is in general not true for compact differentiation, even in the framework of normed spaces. Related to this is the failure of both versions of the inverse function theorem [assuming a pointwise, respectively continuous derivative; Theorems 8.2.3 and 10.2.5 of Dieudonné (1960)], for compact differentiation. For bounded differentiation, these theorems may break down in nonnormable (though still metrizable) topological vector spaces [Averbukh, Smolyanov (1967)]. The classical implicit function theorem has an extension to more general spaces [Keller (1974), Reeds (1976; Theorem A.4.3)]; namely, under the assumption of a continuous second partial bounded derivative, to a first factor space which may be arbitrarily locally convex, and to a second factor space which is Banach. ////

In connection with Theorem 1.4.7, the tightness argument must be adapted to the space $\mathcal{C}_1^k(\Theta)$. Because of Fréchet differentiability, boundedness of the processes $\sqrt{n}\,(Z_n - \eta)(F^n)$ in $\mathcal{C}_1^k(\Theta)$ would actually suffice. Then as. normality follows as in Corollary 1.4.5.

Proposition 1.4.9 *Let* $\Theta \subset \mathbb{R}^k$ *be a compact ball. Assume that the function* $\varphi\colon \mathbb{R}^m \times \Theta \to \mathbb{R}^k$ *satisfies* (3) *and*

$$\varphi(x,.) \in \mathcal{C}_1^k(\Theta) \qquad \text{a.e. } F(dx) \tag{37}$$

$$d_2\varphi(.,\theta) \in L_2^{k\times k}(F), \qquad \theta \in \Theta \tag{38}$$

Suppose there exists a pseudometric d *on* Θ *with the properties* (12) *and* (13), *and a function* (15): $M \in L_2(F)$, *such that for all* $\zeta,\theta \in \Theta$,

$$\|d_2\varphi(x,\zeta) - d_2\varphi(x,\theta)\| \le d(\zeta,\theta)\,M(x) \qquad \text{a.e. } F(dx) \tag{39}$$

Then assertions (16) *and* (17) *of* Proposition 1.4.3 *hold true with* $\mathcal{C}_1^k(\Theta)$ *in the place of* $\mathcal{C}^k(\Theta)$.

Remark 1.4.10

(a) By the argument of Remark 1.4.1 a, assumption (37) ensures that the process Z_n introduced in (4) is with F^n probability 1 a random element of the space $\mathcal{C}_1^k(\Theta)$.

(b) By (12), (15), and (39), $d_2\varphi(x,.)$ is continuous a.e. $F(dx)$.

(c) Assumption (37) implies that for all $\zeta,\theta \in \Theta^{\circ}$,

$$\begin{aligned}&\varphi(x,\zeta) - \varphi(x,\theta) - d_2\varphi(x,\theta)\,(\zeta-\theta)\\ &\quad= \int_0^1 \bigl(d_2\varphi(x,\theta + t(\zeta-\theta)) - d_2\varphi(x,\theta)\bigr)\,dt\,(\zeta-\theta) \qquad \text{a.e. } F(dx)\end{aligned}$$

Conditions (12), (15), and (39), and an integration w.r.t. F show that η is differentiable at $\theta \in \Theta^{\circ}$ with derivative $d\eta(\theta) = \int d_2\varphi(x,\theta)\, F(dx)$, which is continuous on Θ° [as required for Theorem 1.4.7].

(d) Under the assumptions of Theorem 1.4.7 and Proposition 1.4.9, $T: V \to \Theta$ is continuous, and Z_n, which satisfies $F^n(Z_n \in V) \to 1$, is Borel measurable according to Remark 1.4.1 a. Thus there exist a sequence of events $B_n \in \mathbb{B}^n$ such that $F^n(B_n) \to 1$ and $S_n = T \circ Z_n$ restricted to B_n are Borel measurable. ////

PROOF [Proposition 1.4.9] If Θ is a compact ball of center $a \in \mathbb{R}^k$, the map

$$\alpha: \mathcal{C}_1^k(\Theta) \longrightarrow \mathcal{C}^{k\times k}(\Theta^{\circ}) \times \mathbb{R}^k, \qquad \alpha(f) = (df, f(a)) \tag{40}$$

is a linear isomorphism to the subspace of exact differentials (satisfying the Schwartz condition) times $\mathbb{R}^k$. The inverse of α is obviously given by

$$\alpha^{-1}(g,y) = f\,, \qquad f(\theta) = y + \int_0^1 g\bigl(a + t(\theta - a)\bigr)\, dt\, (\theta - a) \tag{41}$$

Both α and α^{-1} are continuous. Now the assumptions have been formulated in such a way that Proposition 1.4.3 applies to $d_2\varphi$. Hence the first component of $\alpha\bigl(\sqrt{n}\,(Z_n - \eta)\bigr)$ converges weakly to a Gaussian measure on $\mathcal{C}^{k\times k}(\Theta^{\circ})$. To the second component, Lindeberg–Lévy applies. ////

Remark 1.4.11 The definition of α—hence, the result—may be extended via $\alpha(f) = (df; f(a_1), f(a_2), \dots)$ to the case that Θ is a finite, or possibly a countably infinite, disjoint union of connected regions. ////

Huber (1981; Chapter 3, pp 49–50, and Chapter 6, pp 127–133) has other proofs for the consistency and as. normality of M estimates. Locally uniform extensions of these results will be derived in Section 6.2.

1.5 Quantiles

Let $x_{n:1} \le \cdots \le x_{n:n}$ be the order statistics of real-valued i.i.d. random variables $x_1, \dots, x_n$ with some distribution function $F \in \mathcal{M}_1(\mathbb{B})$.

Location Quantiles

Fix $\alpha \in (0,1)$. The α *quantile statistic* at sample size n may be any Borel measurable function $S_n: \mathbb{R}^n \to \mathbb{R}$ that on $\mathbb{R}^n$ satisfies

$$x_{n:\lfloor n\alpha \rfloor} \le S_n(x_1, \dots, x_n) \le x_{n:\lceil n\alpha \rceil + 1} \tag{1}$$

where $\lfloor . \rfloor$ denotes the smallest integer larger or equal to, and $\lceil . \rceil$ the largest integer less or equal to. An α *quantile functional* T is any pseudoinverse distribution function evaluated at α,

$$T'(F) \le T(F) \le T''(F) \tag{2}$$

where, by definition,

$$T'(F) = \inf\{\, x \in \mathbb{R} \mid F(x) \ge \alpha \,\}, \quad T''(F) = \sup\{\, x \in \mathbb{R} \mid F(x) \le \alpha \,\}$$

Using the empirical distribution function $\hat{F}_n \colon \mathbb{R}^n \to \mathcal{D}(\mathbb{R})$,

$$\hat{F}_n(x_1, \ldots, x_n; y) = \hat{F}_{x,n}(y) = \frac{1}{n}\sum_{i=1}^{n} \mathbf{I}(x_i \le y) \tag{3}$$

we thus have

$$T'(\hat{F}_{x,n}) \le S_n \le T''(\hat{F}_{x,n}) \tag{4}$$

With the conventions that $\inf \emptyset = \infty$, $\sup \emptyset = -\infty$, definition (2) is good for arbitrary functions $F \colon \mathbb{R} \to \mathbb{R}$. We shall however stay in $\mathcal{D}(\mathbb{R})$, the space of all bounded right continuous real-valued functions on $\mathbb{R}$ with left limits, equipped with the sup norm. For any subset U of $\mathbb{R}$, let $\mathcal{C}(U)$ denote the space of all bounded continuous functions from U to $\mathbb{R}$, again equipped with the sup norm. Let us agree that

$$\mathcal{D}(\mathbb{R}) \cap \mathcal{C}(U) = \{\, G \in \mathcal{D}(\mathbb{R}) \mid G \text{ continuous on } U \,\}$$

and introduce the restriction operator

$$\Pi_U \colon \mathcal{D}(\mathbb{R}) \cap \mathcal{C}(U) \longrightarrow \mathcal{C}(U) \tag{5}$$

which is of norm $\|\Pi_U\| \le 1$, hence has bounded derivative $d_{\mathrm{F}}\Pi_U = \Pi_U$.

Theorem 1.5.1 *Suppose the distribution function F is continuous on some neighborhood U of $a = F^{-1}(\alpha)$ and has derivative $f(a) > 0$ at a.*

(a) *Then the α quantile functional T is compactly differentiable at F along $\mathcal{D}(\mathbb{R}) \cap \mathcal{C}(U)$ with derivative*

$$d_{\mathrm{H}} T(F)\Delta = -\frac{\Delta(a)}{f(a)} \tag{6}$$

(b) *T is not boundedly differentiable at F along $\mathcal{D}(\mathbb{R}) \cap \mathcal{C}(U)$, and T is not weakly differentiable at F along $\mathcal{D}(\mathbb{R})$.*

PROOF

(a) It is no restriction to choose $U = [a', a'']$, a compact neighborhood of a such that, on $U = [a', a'']$, F is continuous and $2|\kappa| < f(a)$, where

$$F(x) = \alpha + f(a)(x-a) + \kappa(x)(x-a) \tag{7}$$

Put $\delta = \min\{F(a'') - \alpha, \alpha - F(a')\}$. If $G \in \mathcal{D}(\mathbb{R}) \cap \mathcal{C}(U)$, $\|G\| < \delta$, then $(F+G)(x) < \alpha < (F+G)(y)$ for all $x < a'$, $y > a''$. Hence, by continuity,

$$G \in \mathcal{D}(\mathbb{R}) \cap \mathcal{C}(U),\ \|G\| < \delta \implies T(F+G) = T\big(\Pi_U(F+G)\big) \in U \tag{8}$$

As also $\Pi_U G(a) = G(a)$, the chain rule allows us to verify compact differentiability of T at $\Pi_U F$ along $\mathcal{C}(U)$.

Given a compact $K \subset \mathcal{C}(U)$, the Arzéla–Ascoli theorem supplies an increasing function $M_K\colon (0,\infty) \to (0,\infty)$ and some $m_K \in (0,\infty)$ such that $\lim_{u\to 0} M_K(u) = 0$ and, for all $G \in K$ and $u, v \in U$,

$$|G(u)| \le m_K\,, \qquad |G(v) - G(u)| \le M_K(|v-u|) \tag{9}$$

For $G \in K$ and $|t| < \delta/m_K$ put $F_t = \Pi_U F + tG$. Then $T(F_t) \in U$ and

$$\begin{aligned} \alpha &= F_t(T(F_t)) = F(T(F_t)) + tG(T(F_t)) \\ &\overset{(7)}{=} \alpha + f(a)(T(F_t) - a) + (T(F_t) - a)\,\kappa(T(F_t)) + tG(T(F_t)) \end{aligned} \tag{10}$$

As $2\big|\kappa(T(F_t))\big| \le f(a)$, (10) implies that

$$|T(F_t) - a| \le c_K |t|\,, \qquad c_K = \frac{2m_K}{f(a)} \tag{11}$$

Inserting this bound back into (10) again, we obtain

$$\begin{aligned} &\Big|f(a)\frac{T(F_t) - a}{t} + G(a)\Big| \\ &\quad\le \frac{|T(F_t) - a|}{|t|}\big|\kappa(T(F_t))\big| + \big|G(T(F_t)) - G(a)\big| \\ &\quad\le c_K \sup\big\{|\kappa(x)| \bigm| |x| \le c_K|t|\big\} + M_K(c_K|t|) \longrightarrow 0 \end{aligned} \tag{12}$$

uniformly in $G \in K$, as $t \to 0$.

(b) Define maps

$$G_n(x) = \begin{cases} 0 & \text{if } x \le a - \frac{1}{n} \\ 1 & \text{if } a \le x \\ \text{linear} & \text{in between.} \end{cases} \tag{13}$$

which make a bounded set in $\mathcal{C}(\mathbb{R})$ as $\|G_n\| = 1$. Fix $t \in (0,\infty)$. Then

$$(F + tG_n)\big(a - \tfrac{1}{n}\big) < \alpha \le (F + tG_n)(a)$$

hence

$$a - \tfrac{1}{n} < T(F + tG_n) \le a \tag{14}$$

so that $\sup_{n\ge 1} T(F + tG_n) = a$. Therefore,

$$\sup_{n\ge 1}\Big|\frac{T(F + tG_n) - a}{t} + G_n(a)\Big| \ge 1 \tag{15}$$

This disproves bounded differentiability of T at F along X, since the compact derivative is the only candidate for a bounded derivative.

To see that T is not even weakly differentiable at F along $\mathcal{D}(\mathbb{R})$, pick some increasing function $G \in \mathcal{D}(\mathbb{R})$ such that $1/2 \leq G \leq 1$ and G is continuous except at a where $G(a-0) \neq G(a+0)$. Then, for $t \in (0,\delta)$ and $F_t = F+tG$, we obtain that $F_t(a') < \alpha < F_t(a-0)$. By the continuity of F_t on $[a',a)$ it follows that $a' \leq T(F_t) < a$, and then (10) and (11) apply. Therefore, and by the corresponding argument for $t \in (-\delta,0)$,

$$f(a)\frac{T(F_t)-a}{t} = -G(T(F_t)) - \frac{T(F_t)-a}{t}\,\kappa(T(F_t)) \longrightarrow -G(a \mp 0) \quad (16)$$

as $t \to 0$ from above and below, respectively. This disproves Gateaux differentiability. ////

Remark 1.5.2 Reeds (1976) proves (a) for $F = F_0 = \mathrm{id}_{[0,1]}$, the rectangular on $[0,1]$. To derive the result for general F, he applies the finite-dimensional delta method to the inverse F^{-1}. It is not obvious, though, how to obtain compact differentiability of a fixed number of quantiles under general F this way [Example 1.5.6]. For the same reason, it is more flexible in the following proof not to insist on exact equality in (17), but to work with upper and lower bounds. ////

The product probability spaces $(\mathbb{R}^n, \mathbb{B}^n, F^n)$ are now the underlying sample spaces. As for randomization, consult Section A.3 [p 340].

Proposition 1.5.3 *Suppose F is continuous on some neighborhood U of $a = F^{-1}(\alpha)$ and has derivative $f(a) > 0$ at a. Let S_n and T', T'' satisfy (1) and (2), respectively. Then there exist randomized Borel measurable maps*

$$Z_n', Z_n'' \colon \mathbb{R}^n \longrightarrow \mathcal{D}(\mathbb{R}) \cap \mathcal{C}(U)$$

such that

$$T'(F+Z_n') \leq S_n \leq T''(F+Z_n'') \quad (17)$$

Both processes $Z_n = Z_n', Z_n''$ satisfy

$$\lim_{M\to\infty} \limsup_{n\to\infty} F^n\big(\sqrt{n}\,\|Z_n\| > M\big) = 0 \quad (18)$$

and, when restricted to U, they converge weakly in $\mathcal{C}(U)$

$$\Pi_U(\sqrt{n}\,Z_n)(F^n) \xrightarrow{\;w\;} \Pi_U(B \circ F) \quad (19)$$

to the restriction of Brownian Bridge B on $\mathcal{C}[0,1]$ composed with F. In particular, for all $\varepsilon, \delta \in (0,1)$ there exists a compact $K \subset \mathcal{C}(U)$ such that

$$\liminf_{n\to\infty} F^n\big(\|Z_n\| < \delta,\ \Pi_U(\sqrt{n}\,Z_n) \in K\big) \geq \varepsilon \quad (20)$$

PROOF We invoke some constructions from Section A.3: Let the sequence $u_i \sim F_0$ be i.i.d. rectangular on $[0,1]$, and denote the empirical distribution function based on $u_1, \ldots, u_n$ by $\hat{F}_{u,n}$. The smoothed rectangular empirical $\bar{F}_{u,n}$ interpolates the values 0 at 0, 1 at 1, and $i/(n+1)$ at $u_{n:i}$ linearly, where $u_{n:1} \le \cdots \le u_{n:n}$ denote the rectangular order statistics. By a well-known result [Donsker, Proposition A.3.1], the smoothed rectangular empirical process converges in $\mathcal{C}[0,1]$ to the Brownian Bridge B,

$$\sqrt{n}\,(\bar{F}_{u,n} - F_0)(F_0^n) \xrightarrow{\mathrm{w}} B \tag{21}$$

Two slightly different continuous modifications turn out useful in this proof:

$$n\,\bar{F}'_{u,n}(s) = \begin{cases} 0 & \text{if } s = 0 \\ i & \text{if } s = u_{n:i}\,, \quad i = 1, \ldots, n \\ 1 & \text{if } s = 1 \\ \text{linear} & \text{in between} \end{cases} \tag{22}$$

and

$$n\,\bar{F}''_{u,n}(s) = \begin{cases} 0 & \text{if } s = 0 \\ i-1 & \text{if } s = u_{n:i}\,, \quad i = 1, \ldots, n \\ 1 & \text{if } s = 1 \\ \text{linear} & \text{in between} \end{cases} \tag{23}$$

They enclose the original $\hat{F}_{u,n}$, and $\bar{F}_{u,n}$ as well, from above and below,

$$\bar{F}''_{u,n}(s) \le \hat{F}_{u,n}(s), \bar{F}_{u,n}(s) \le \bar{F}'_{u,n}(s) \le \bar{F}''_{u,n}(s) + \tfrac{1}{n} \tag{24}$$

So (21) implies that

$$\sqrt{n}\,(\bar{F}'_n - F_0)(F_0^n) \xrightarrow{\mathrm{w}} B\,, \qquad \sqrt{n}\,(\bar{F}''_n - F_0)(F_0^n) \xrightarrow{\mathrm{w}} B \tag{25}$$

For general F and i.i.d. variables $x_i \sim F$ we use construction A.3(8) of an i.i.d. rectangular sequence $u_i \sim F_0$ such that $x_i = F^{-1}(u_i)$ a.e. and the (smoothed) empiricals are related by A.3(7) and A.3(10). Now define

$$\bar{F}'_{x,n}(y) = \bar{F}'_{u,n}(F(y)), \qquad \bar{F}''_{x,n}(y) = \bar{F}''_{u,n}(F(y)) \tag{26}$$

Then (24) and A.3(7) and A.3(10) imply the bound

$$\bar{F}''_{x,n}(y) \le \hat{F}_{x,n}(y), \bar{F}_{x,n}(y) \le \bar{F}'_{x,n}(y) \le \bar{F}''_{x,n}(y) + \tfrac{1}{n} \tag{27}$$

This and (4) prove (17) for the randomized processes

$$\begin{aligned} Z'_n(x_1, \ldots, x_n; y) &= \bar{F}'_n(x_1, \ldots, x_n; y) - F(y) \\ Z''_n(x_1, \ldots, x_n; y) &= \bar{F}''_n(x_1, \ldots, x_n; y) - F(y) \end{aligned} \tag{28}$$

Moreover, we know from Proposition A.3.5 that

$$\lim_{M \to \infty} \limsup_{n \to \infty} \sup_{F \in \mathcal{F}} F^n\big(\sqrt{n}\,\|\hat{F}_n - F\| > M\big) = 0 \tag{29}$$

which, connected with bound (27), proves (18). To prove (19), we use the transform

$$g \longmapsto \Pi_U(g \circ F)$$

which is continuous (linear) from $\mathcal{C}[0,1]$ to $\mathcal{C}(U)$, and apply the continuous mapping theorem [Proposition A.1.1] to (25). This gives the result. The space $\mathcal{C}(U)$ being polish (separable, complete), tightness is implied by weak convergence [Proposition A.1.2]. ////

Corollary 1.5.4 *Assume a distribution function F on $\mathbb{R}$ that, on some neighborhood of $a = F^{-1}(\alpha) = T(F)$, is continuous, strictly increasing, and has derivative $f(a) > 0$ at a. Then the quantile sequence* (1) *satisfies*

$$\sqrt{n}\,(S_n - a) = \frac{1}{\sqrt{n}} \sum_{i=1}^{n} \psi(x_i) + \mathrm{o}_{F^n}(n^0) \tag{30}$$

with influence curve

$$\psi(x) = \frac{\alpha - \mathbf{I}(x \le a)}{f(a)} \tag{31}$$

Hence

$$\sqrt{n}\,(S_n - a)(F^n) \xrightarrow{\mathrm{w}} \mathcal{N}\Big(0, \frac{\alpha(1-\alpha)}{f^2(a)}\Big) \tag{32}$$

PROOF Combine compact differentiability [Theorem 1.5.1 a] of the particular α quantile functionals T' and T'' via (8) with boundedness and tightness (20) from Proposition 1.5.3. Then the general delta method [Theorem 1.3.3] and (17) yield

$$-\sqrt{n}\,Z'_n(a) \le \sqrt{n}\,f(a)(S_n - a) + \mathrm{o}_{F^n}(n^0) \le -\sqrt{n}\,Z''_n(a) \tag{33}$$

Since $\|Z_n - \hat{F}_n + F\| \le 1/n$, one can replace $Z_n = Z'_n, Z''_n$ in this approximation by $\hat{F}_n - F$, and thus obtain the result. ////

Remark 1.5.5 Under further assumptions on F (existence of a bounded second derivative in a neighborhood of a), Bahadur's (1966) representation of quantiles specifies the remainder term $\mathrm{o}_{F^n}(n^0)$ in (30) more precisely to be some $\mathrm{O}_{F^n}\big((\log n)/\sqrt[4]{n}\big)$. ////

Example 1.5.6 Given k probabilities α_j and weights $h_j \in \mathbb{R}$, consider the linear combination of α_j quantile functionals T_j,

$$T(F) = h_1 T_1(F) + \cdots + h_k T_k(F) \tag{34}$$

For example, T could be the *interquartile range*. Assume a distribution function F that has a derivative $f(a_j) > 0$ at the points $a_j = F^{-1}(\alpha_j)$

and is continuous in some neighborhood U of $\{a_1,\dots,a_k\}$. Then T is compactly differentiable at F along $\mathcal{D}(\mathbb{R})\cap\mathcal{C}(U)$ with derivative

$$d_{\mathrm{H}}T(F)\,\Delta = -h_1\,\frac{\Delta(a_1)}{f(a_1)} - \cdots - h_k\,\frac{\Delta(a_k)}{f(a_k)} \tag{35}$$

This follows from Theorem 1.5.1 a, using the linearity of the compact derivative and the inclusions $\mathcal{C}(U)\subset\mathcal{C}(U_j)$ for $U_j\supset U$. The corresponding linear combination of sample α_j quantiles has the as. expansion

$$\sqrt{n}\,(S_n-a) = \frac{1}{\sqrt{n}}\sum_{i=1}^n \psi(x_i) + \mathrm{o}_{F^n}(n^0) \tag{36}$$

with influence curve

$$\psi = \sum_{j=1}^k \psi_j\,,\qquad \psi_j(x) = \frac{h_j}{f(a_j)}\big(\alpha_j - \mathbf{I}(x\le a_j)\big) \tag{37}$$

Hence,

$$\sqrt{n}\,(S_n-a)(F^n) \xrightarrow{\mathrm{w}} \mathcal{N}(0,\sigma^2)\,,\qquad \sigma^2 = \sum_{j,l=1}^k h_j h_l\,\frac{\alpha_j\wedge\alpha_l - \alpha_j\alpha_l}{f(a_j)f(a_l)} \tag{38}$$

The influence curve ψ has jumps of height $h_j/f(a_j)$ at a_j. ////

Scale Quantiles

Employing two quantile location functionals T_α and T_β, satisfying (2) for some $\alpha,\beta\in(0,1)$, respectively, a *quantile scale functional* T is defined by

$$T(F) = T_\beta(|F_\alpha|)\,,\quad |F_\alpha|(y) = F\big(y+T_\alpha(F)\big) - F\big(-y+T_\alpha(F)-0\big) \tag{39}$$

For $\alpha=\beta=1/2$, the *median absolute deviation* (MAD) functional results.

Theorem 1.5.7 *Assume F is a distribution function that has a derivative f in $a=T_\alpha(F)$ and in $a+b$, $a-b$, where $b=T(F)$, such that*

$$f(a)>0,\qquad f(a+b)+f(a-b)>0 \tag{40}$$

and suppose F is continuous on some neighborhood U of $\{a,a+b,a-b\}$.

(a) *Then the functional T given by (39) is compactly differentiable at F along $\mathcal{D}(\mathbb{R})\cap\mathcal{C}(U)$ with derivative*

$$d_{\mathrm{H}}T(F)\,\Delta = -\frac{\Delta(a+b)-\Delta(a-b)}{f(a+b)+f(a-b)} + \frac{f(a+b)-f(a-b)}{f(a+b)+f(a-b)}\cdot\frac{\Delta(a)}{f(a)} \tag{41}$$

(b) *Assume the empirical process* $\hat{F}_n - F$ *is approximated by possibly randomized Borel measurable maps*

$$Z_n: \mathbb{R}^n \longrightarrow \mathcal{D}(\mathbb{R}) \cap \mathcal{C}(U)$$

such that

$$\sqrt{n}\,\big\|Z_n - \hat{F}_n + F\big\| \xrightarrow{F^n} 0 \tag{42}$$

Then the quantile scale estimator sequence

$$S = (S_1, S_2, \dots), \qquad S_n = T(F + Z_n) \tag{43}$$

has the as. expansion

$$\sqrt{n}\,\big(S_n - T(F)\big) = \frac{1}{\sqrt{n}} \sum_{i=1}^{n} \psi(x_i) + \mathrm{o}_{F^n}(n^0) \tag{44}$$

with influence curve

$$\psi(x) = \frac{\beta - \mathbf{I}(-b < x - a \le b)}{f(a+b) + f(a-b)} + \frac{f(a+b) - f(a-b)}{f(a+b) + f(a-b)} \cdot \frac{\mathbf{I}(x \le a) - \alpha}{f(a)} \tag{45}$$

In particular, S *is as. normal:* $\sqrt{n}\,\big(S_n - T(F)\big)(F^n) \xrightarrow{\ \mathrm{w}\ } \mathcal{N}\big(0, \int \psi^2\, dF\big)$.

PROOF The proof is related to the proofs to Theorem 1.5.1 a and Proposition 1.5.3, only it is a bit more complicated.

(a) Choose $b' < b < b''$ and then $a' < a < a''$ such that

$$\min\{b'' - b, b - b'\} > \max\{a'' - a, a - a'\}$$

and

$$\begin{aligned} 2|\kappa_1(x)| &< f(a), & x &\in [a', a''] \\ 4|\kappa_2(x)| &< f(a+b) + f(a-b), & x &\in [a' + b', a'' + b''] \\ 4|\kappa_3(x)| &< f(a+b) + f(a-b), & x &\in [a' - b'', a'' - b'] \end{aligned} \tag{46}$$

where κ_1, κ_2, and κ_3 denote the little o terms in the first-order Taylor expansion of F at a, $a{+}b$, and $a{-}b$, respectively. Then the four differences

$$\begin{gathered} \alpha - F(a'), \quad F(a'') - \alpha \\ \beta - F(a'' + b') + F(a' - b'), \quad F(a' + b'') - F(a'' - b'') - \beta \end{gathered}$$

are all strictly positive. Choose $\delta > 0$ smaller than the minimum of these four numbers. Without loss of generality, we may put

$$U = [a', a''] \cup [a' + b', a'' + b''] \cup [a' - b'', a'' - b']$$

assuming a', a'' and b', b'' are sufficiently close to a and b, respectively, so that also F is continuous on U.

For $F_t = F + tG$ write $m_t = T_\alpha(F_t)$ and $M_t = T(F_t)$. Then, in view of (8), we have

$$m_t \in [a', a'']$$

for all $G \in \mathcal{D}(\mathbb{R}) \cap \mathcal{C}(U)$ such that $\|tG\| < \delta$, and

$$F_t(y' + m_t) - F_t(-y' + m_t) < F(a'' + b') - F(a' - b') + 2\delta \leq \beta$$
$$F_t(y'' + m_t) - F_t(-y'' + m_t) > F(a' + b'') - F(b'' + a'') - 2\delta \geq \beta$$

for $y' \leq -b'$ and $y'' \geq b''$, respectively; and therefore

$$b' < M_t < b''$$

which allows a restriction of the increments to the space $\mathcal{C}(U)$.

For every compact $K \subset \mathcal{C}(U)$, and all $G \in K$, it holds that

$$\begin{aligned} 0 &= F_t(m_t + M_t) - F_t(m_t - M_t) - \beta \\ &= (m_t - a)\big(f(a+b) - f(a-b) + \kappa_2(m_t + M_t) - \kappa_3(m_t - M_t)\big) \\ &\quad + (M_t - b)\big(f(a+b) + f(a-b) + \kappa_2(m_t + M_t) + \kappa_3(m_t - M_t)\big) \\ &\quad + t\big(G(m_t + M_t) - G(m_t - M_t)\big) \end{aligned} \tag{47}$$

Using (46) and (11) we conclude that uniformly in $\|G\| \leq m_K$, as $t \to 0$,

$$|M_t - b| = \mathrm{O}(|t|) \tag{48}$$

This order of magnitude, and (11) plugged again into (47), and the uniform modulus of continuity $M_K(.)$ for $G \in K$, imply (41).

(b) Since also $\sqrt{n}\,\|Z_n - \bar{F}_n + F\| \to 0$ in F^n probability, the arguments of the proofs to Proposition 1.5.3 and Corollary 1.5.4 apply. ////

Example 1.5.8 Suppose F satisfies the assumptions of Theorem 1.5.7 and, in addition,

$$f(a - b) = f(a + b) \tag{49}$$

Introduce the relations

$$\begin{aligned} &a_1 = a - b, \quad a_2 = a + b, \quad a = \tfrac{1}{2}(a_1 + a_2), \quad b = \tfrac{1}{2}(a_2 - a_1) \\ &\alpha_1 = F(a_1), \quad \alpha_2 = F(a_2), \quad \alpha = F(a), \qquad\quad \beta = \alpha_2 - \alpha_1 \end{aligned} \tag{50}$$

Then, in view of (6), (35), and (41), the scale functional T has the same compact derivative as the functional $(T_{\alpha_2} - T_{\alpha_1})/2$ (one-half the difference between the α_2 and α_1 quantiles). Hence the corresponding estimators have the same influence curves and as. normal distributions. Nevertheless, these procedures may differ globally. For example, their *breakdown points* are

$$\min\{\alpha, \beta, 1 - \alpha, 1 - \beta\}, \qquad \min\{\alpha_1, \alpha_2, 1 - \alpha_1, 1 - \alpha_2\} \tag{51}$$

respectively, which can be different, as in the case of the median absolute deviation and the interquartile range. Such globally divergent behavior of locally equivalent procedures is not uncommon in robust statistics. ////

Theorem 1.5.7 covers only those quantile scale estimates of form (43), which is based on definition (2) of quantile functionals and some approximation (42) of the empirical process that is continuous on U; for example, $Z_n = \bar{F}_{x,n} - F$. Although there remains considerable flexibility in the choice of particular versions of quantile functionals and smoothed empirical, it is not clear if an arbitrary estimator version employing two quantile statistics in the sense of definition (1) can exactly be reproduced this way.

The bounding technique used in the proofs to Theorem 1.5.1 a, Proposition 1.5.3, and Corollary 1.5.4 does not seem to carry over. Given a particular α quantile statistics $S_{\alpha,n}$ satisfying (1), Z_n could be chosen such that $T_\alpha(F + Z_n) = S_{\alpha,n}$. Using the constructions of Section A.3, this may be achieved by an additional random adjustment,

$$Z_n(x_1, \dots, x_n; y) = \bar{F}_{x,n}(y) - F(y) + \alpha - \bar{F}_{x,n}(S_{\alpha,n}) \tag{52}$$

Since, for all $y \in \mathbb{R}$, $\bar{F}_{x,n}(y)$ is Borel measurable in $x \in \mathbb{R}^n$ (and in the randomization variables $v \in [0,1]^n$), and for all $x \in \mathbb{R}^n$ and $v \in [0,1]^n$, is a continuous function of $F(y)$, indeed $Z_n(\,.\,; y)$ becomes Borel measurable for all $y \in \mathbb{R}$. A further adaptation, however, so that also T_β produces a given β quantile statistic in the sense of (1), evaluated on the transformed sample $|x_i - S_{\alpha,n}|$, seems difficult.

1.6 L Estimates

This section treats linear combinations of $x_{n:1} \le \dots \le x_{n:n}$, the order statistics of real-valued i.i.d. observations $x_1, \dots, x_n$ with law $F \in \mathcal{M}_1(\mathbb{B})$. Employing a weight function $w: \mathbb{R} \to \mathbb{R}$ and a scores function $J: [0,1] \to \mathbb{R}$ (subjected to further conditions), the L *functional* $T: \mathcal{D}(\mathbb{R}) \to \mathbb{R}$ is

$$T(F) = \int w(F^{-1}(s)) J(s)\, \lambda_0(ds) \tag{1}$$

where $F^{-1}: [0,1] \to \bar{\mathbb{R}}$ denotes the usual (left continuous) pseudoinverse of the (right continuous) distribution function F,

$$F^{-1}(s) = \inf\{x \in \mathbb{R} \mid F(x) \ge s\} \tag{2}$$

and $\lambda_0 = R[0,1]$ is the rectangular measure on $[0,1]$, often identified with its distribution function $F_0 = \mathrm{id}_{[0,1]}$.

As the integration in (1) already smoothes in some suitable sense, the empirical can be taken in its original $\mathcal{D}(\mathbb{R})$-valued form,

$$\hat{F}_n(x_1, \dots, x_n; t) = \hat{F}_{x,n}(t) = \frac{1}{n} \sum_{i=1}^{n} \mathbf{I}(x_i \le t) \tag{3}$$

Then the corresponding L *estimate* at sample size n is

$$S_n(x_1,\dots,x_n) = T\big(\hat{F}_n(x_1,\dots,x_n)\big) = \sum_{i=1}^{n} J_{n,i}\, w(x_{n:i}) \tag{4}$$

employing the scores

$$J_{n,i} = \int_{\frac{i-1}{n}}^{\frac{i}{n}} J(s)\,\lambda_0(ds)\,, \qquad i = 1,\dots,n \tag{5}$$

Hadamard Differentiability

Compact differentiability will be proved under the following assumptions:

$$F = F_0 = \mathrm{id}_{[0,1]} \tag{6}$$

$$J \in L_2(\lambda_0) \tag{7}$$

$$w \in \mathcal{C}_1(\mathbb{R})\,, \qquad \|w\|_1 = \sup|w| + \sup|w'| < \infty \tag{8}$$

where, by definition of $\mathcal{C}_1(\mathbb{R})$, both $w \in \mathcal{C}_1(\mathbb{R})$ and its derivative w' are bounded continuous. The space $\mathcal{D}[0,1]$ of right continuous functions from $[0,1]$ to $\mathbb{R}$ with left-hand limits will be equipped with the sup norm. On the one hand, it thus becomes nonseparable, so the converse half of Prokhorov's theorem [Proposition A.1.2 b] is not available. And $\hat{F}_n$ becomes non-Borel measurable [Billingsley (1968; §18, p 150)]. Equipped with the Skorokhod topology, on the other hand, $\mathcal{D}[0,1]$ would not be a topological vector space [Billingsley (1968; Problem 3, p 123)].

In view of Theorem 1.3.3, tightness is settled by the following result.

Proposition 1.6.1 *For every $\varepsilon \in (0,\infty)$ there is a compact $K \subset \mathcal{D}[0,1]$ such that*

$$\liminf_{n\to\infty}\, (F_0^n)_*\big(\sqrt{n}\,(\hat{F}_n - F_0) \in K^{1/\sqrt{n}}\big) \ge 1-\varepsilon \tag{9}$$

PROOF The linearly interpolated empirical $\bar{F}_n\colon [0,1]^n \to \mathcal{C}[0,1]$ defined by A.3(2) is Borel measurable, and under F_0^n converges weakly in $\mathcal{C}[0,1]$ to the Brownian Bridge [Proposition A.3.1]. For this polish space, the converse half of Prokhorov's theorem [Proposition A.1.2 b] is available: Given any $\varepsilon \in (0,\infty)$ there exists a compact $K \subset \mathcal{C}[0,1]$ such that

$$\liminf_{n\to\infty} F_0^n\big(\sqrt{n}\,(\bar{F}_n - F_0) \in K\big) \ge 1-\varepsilon \tag{10}$$

By continuity of the inclusion $\mathcal{C}[0,1] \hookrightarrow \mathcal{D}[0,1]$, this K is also a compact subset of the space $\mathcal{D}[0,1]$. Moreover,

$$\{\sqrt{n}\,(\bar{F}_n - F_0) \in K\} \subset \{\sqrt{n}\,(\hat{F}_n - F_0) \in K^{1/\sqrt{n}}\} \tag{11}$$

because $\|\bar{F}_n - \hat{F}_n\| \le 1/n$. ////

To establish the required compact differentiability, the functional T is decomposed into three mappings: $T = K \circ N \circ I$. The first map denotes the (left continuous) pseudoinverse $I: \mathcal{D}[0,1] \to L_2(\lambda_0)$,

$$I(F)(s) = F^{-1}(s) = \inf\{t \in [0,1] \mid F(t) \geq s\} \tag{12}$$

the second one is the Nemytsky operator $N: L_2(\lambda_0) \to L_1(\lambda_0)$,

$$N(u) = (w \circ u)\, J \tag{13}$$

the third is the integral $K: L_1(\lambda_0) \to \mathbb{R}$,

$$K(v) = \int v\, d\lambda_0 \tag{14}$$

Being continuous linear, $K = d_{\mathrm{F}} K$. Also N is boundedly differentiable.

Proposition 1.6.2 *Under conditions* (7), (8), *the operator N is boundedly differentiable with derivative*

$$d_{\mathrm{F}} N(u) v = (w' \circ u) J v \tag{15}$$

PROOF We shall verify the conditions of Reeds (1976; Theorem A 3.4) who generalizes Vainberg (1964; Theorem 20.1) on the bounded differentiability of Nemytsky operators. The N function in our case reads

$$g(u,s) = w(u) J(s) \tag{16}$$

which is measurable in s and continuous in u. Also the first partial bounded derivative

$$d_{\mathrm{F}1}\, g(u,s) = w'(u) J(s) \tag{17}$$

is an N function; that is, measurable in s, continuous in u. Moreover,

$$|g(u,s)| \leq \|w\|\, |J(s)|, \qquad |d_{\mathrm{F}1}\, g(u,s)| \leq \|w'\|\, |J(s)| \tag{18}$$

involving the finite sup norms of w and w', and the square integrable function J. Thus, conditions (1), (2), and (3) of Reeds (1976; Theorem A 3.4) are in fact satisfied. ////

Remark 1.6.3 Despite identical arguments, Reeds (1976) omits the condition that w' be continuous and $J \in L_2(\lambda_0)$, in his Theorem 6.3.1. But these assumptions are obviously needed when he appeals to Theorem A 3.4. His weaker—in our view, insufficient—condition that $J \in L_1(\lambda_0)$ has been taken over by Boos (1979; Section 5). ////

The differentiability of the pseudoinverse I remains.

Theorem 1.6.4 *For $1 \le q < \infty$ the inverse $I\colon \mathcal{D}[0,1] \to L_q(\lambda_0)$ is compactly differentiable at $F_0 = \mathrm{id}_{[0,1]}$ with derivative*

$$d_{\mathrm{H}} I(F_0) = -\,\mathrm{id}_{\mathcal{D}[0,1]} \tag{19}$$

Remark 1.6.5 Similar to Theorem 1.5.1 b, I is not boundedly differentiable at F_0. The right continuous pseudoinverse $I_+\colon \mathcal{D}[0,1] \to \mathcal{D}[0,1]$,

$$I_+(F)(s) = \sup\bigl\{t \in [0,1] \bigm| F(t) \le s\bigr\}$$

is not weakly differentiable at F_0. ////

Invoking the chain rule for compact differentiation, Theorem 1.6.4 and Proposition 1.6.1, we thus obtain the following result via Theorem 1.3.3.

Theorem 1.6.6 *Assume* (7) *and* (8).

(a) *Then the functional T given by* (1) *is compactly differentiable at F_0,*

$$d_{\mathrm{H}} T(F_0)\Delta = -\int w'(s)J(s)\Delta(s)\,\lambda_0(ds) \tag{20}$$

(b) *The* L *estimator sequence* (4) *has the following as. expansion,*

$$\sqrt{n}\,\bigl(S_n - T(F_0)\bigr) = \frac{1}{\sqrt{n}}\sum_{i=1}^{n} \psi(x_i) + \mathrm{o}_{F_0^n}(n^0) \tag{21}$$

with influence curve

$$\psi(x) = \int_0^1 \bigl(s - \mathbf{I}(x \le s)\bigr)w'(s)J(s)\,\lambda_0(ds) \tag{22}$$

Hence

$$\begin{gathered}\sqrt{n}\,\bigl(S_n - T(F_0)\bigr)(F_0^n) \xrightarrow{\mathrm{w}} \mathcal{N}(0,\sigma^2) \\ \sigma^2 = \int_0^1\!\!\int_0^1 \bigl(s \wedge t - st\bigr)\,w'(s)w'(t)\,J(s)J(t)\,\lambda_0(ds)\lambda_0(dt)\end{gathered} \tag{23}$$

PROOF [Theorem 1.6.4] We start by characterizing the compact subsets of the space $\mathcal{D}[0,1]$ equipped with the sup norm: A subset $K \subset \mathcal{D}[0,1]$ is compact iff it is closed and bounded, and for every $\varepsilon \in (0,1)$ there exists a finite partition of $[0,1)$ by $j(\varepsilon)$ intervals $[a_j, b_j) \ni p_j$ such that

$$\sup\Bigl\{\,\bigl|G(s) - G(p_j)\bigr| \Bigm| G \in K\,,\ s \in [a_j,b_j),\ j = 1,\dots,j(\varepsilon)\Bigr\} < \varepsilon \tag{24}$$

Since $\mathcal{D}[0,1]$ consists precisely of the uniform limits of finite linear combinations of indicators of form $\mathbf{I}_{[a,b)}$ or $\mathbf{I}_{\{1\}}$, the characterization follows from Dunford and Schwartz (1957; Vol. I, Theorem IV 5.6).

First, we show continuity of $I\colon \mathcal{D}[0,1] \to L_q(\lambda_0)$. If $G = F_0 + H$ with $\|H\| \in (0, \frac{1}{2})$, then $G(0) < s < G(1)$ for each $s \in (\|H\|, 1 - \|H\|)$, hence

$$y^+(s) = G^{-1}(s)\,, \qquad y^-(s) = \bigl(G^{-1}(s) - \|H\|\bigr)^+ \tag{25}$$

are well defined, and it holds that $G(y^-(s)) \leq s \leq G(y^+(s))$. Therefore,

$$G^{-1}(s) + H(y^+(s)) = y^+(s) + H(y^+(s)) = G(y^+(s)) \geq s$$
$$G^{-1}(s) - \|H\| + H(y^-(s)) \leq y^-(s) + H(y^-(s)) = G(y^-(s)) \leq s$$

and this implies that

$$\left|G^{-1}(s) - s\right| \leq 2\|H\|, \qquad \|H\| < s < 1 - \|H\| \tag{26}$$

For s outside $(\|H\|, 1-\|H\|)$, as $0 \leq G^{-1} \leq 1$, we have $|G^{-1}(s) - s| \leq 1$. Thus we obtain the bound

$$\int |G^{-1} - F_0^{-1}|^q \, d\lambda_0 \leq 2\|H\| + (2\|H\|)^q \tag{27}$$

which entails continuity of $I\colon \mathcal{D}[0,1] \to L_q(F_0)$.

Second, let $K \subset \mathcal{D}[0,1]$ be compact with norm bound $m_K \in (1, \infty)$. For $H \in K$ and $|t|m_K < 1/2$ put $G = F_0 + tH$. Similarly to (25) define

$$x^+(s) = G^{-1}(s), \qquad x^-(s) = \bigl(G^{-1}(s) - t^2\bigr)^+ \tag{28}$$

for $|t|m_K < s < 1 - |t|m_K$. Then $G(x^-(s)) \leq s \leq G(x^+(s))$. Therefore,

$$G^{-1}(s) + tH(x^+(s)) = x^+(s) + tH(x^+(s)) = G(x^+(s)) \geq s$$
$$G^{-1}(s) - t^2 + tH(x^-(s)) \leq x^-(s) + tH(x^-(s))G(x^-(s)) \leq s$$

Introducing the remainder term

$$t\kappa(tH; s) = G^{-1}(s) - s + tH \tag{29}$$

we conclude for $|t|m_K < s < 1 - |t|m_K$ that

$$t\bigl(H(s) - H(x^+(s))\bigr) \leq t\kappa(tH; s) \leq t\bigl(H(s) - H(x^-(s))\bigr) + t^2$$

As $u \mapsto |u|^q$ is convex for $1 \leq q < \infty$, the following bound is obtained for such s that satisfy $|t|m_K < s < 1 - |t|m_K$,

$$|\kappa(tH; s)|^q \leq 3^{q-1}\bigl(|t|^q + |H(s) - H(x^+(s))|^q + |H(s) - H(x^-(s))|^q\bigr) \tag{30}$$

For $s \leq |t|m_K$ we use the bound

$$G(2|t|m_K) = 2|t|m_K + tH(2|t|m_K) \geq |t|m_K$$

which entails that $0 \leq G^{-1}(s) \leq 2|t|m_K$. Hence, if $0 \leq s \leq |t|m_K$, then

$$|\kappa(tH; s)| \leq 3m_K \tag{31}$$

The same bound obtains for $s \geq 1 - |t|m_K$ since $G(u) < 1 - |t|m_K$ is true for every $0 \leq u < 1 - 2|t|m_K$ and so $1 - 2|t|m_K \leq G^{-1}(s) \leq 1$.

Combining (31) and (30), the following bound is proved,

$$\begin{aligned}\int |\kappa(tH;s)|^q\,\lambda_0(ds) &\le 2|t|m_K(3m_K)^q + 3^{q-1}|t|^q \\ &\quad + 3^{q-1}\int_{|t|m_K}^{1-|t|m_K} \left|H(s)-H(x^+(s))\right|^q\lambda_0(ds) \\ &\quad + 3^{q-1}\int_{|t|m_K}^{1-|t|m_K} \left|H(s)-H(x^-(s))\right|^q\lambda_0(ds)\end{aligned} \tag{32}$$

In these integrals we have

$$\max\{|x^+(s)-s|,|x^-(s)-s|\} \le (2m_K+1)|t| \tag{33}$$

because (26) implies $|G^{-1}(s)-s| \le 2|t|m_K$ for $s\in(|t|m_K, 1-|t|m_K)$, hence

$$|x^+(s)-s| \le 2|t|m_K, \qquad |x^-(s)-s| \le 2|t|m_K + t^2$$

Now let $\varepsilon > 0$ be given. Choose a partition $[a_j,b_j)$, $j=1,\ldots,j(\varepsilon)$, satisfying (24). Let $|t| < \min_{j=1,\ldots,j(\varepsilon)}(b_j-a_j)/(4m_K+2)$. Then for

$$s\in\big(|t|m_K,1-|t|m_K\big)\cap\big[a_j+|t|(2m_K+1),b_j-|t|(2m_K+1)\big)$$

bound (33) implies that $x^+(s),x^-(s)\in[a_j,b_j)$ hence

$$\max\{|H(x^+(s))-H(s)|,|H(x^-(s))-H(s)|\} < 2\varepsilon$$

The rest of the interval $(|t|m_K, 1-|t|m_K)$ has a Lebesgue measure smaller than $(4m_K+2)|t|j(\varepsilon)$. Thus bound (32) can be continued to

$$\begin{aligned}\int |\kappa(tH;s)|^q\,\lambda_0(ds) &\le 2|t|m_K(3m_K)^q + 3^{q-1}|t|^q + 2\cdot 3^{q-1}(2\varepsilon)^q \\ &\quad + 2\cdot 3^{q-1}(4m_K+2)|t|j(\varepsilon)(2m_K)^q\end{aligned} \tag{34}$$

Let first $t\to 0$ and then $\varepsilon\to 0$ to conclude the proof. ////

Remark 1.6.7 In their related proofs, Reeds (1976) and Fernholz (1979), do not split the unit interval at the points $|t|m_K$ and $1-|t|m_K$, with the consequence that their crucial inequality $G(x^-(p)) \le p \le G(x^+(p))$ does not always hold and that $G(x^-(p))$ is undefined sometimes. ////

Fréchet Differentiability

Theorem 1.6.6 puts only the weak square integrability condition (7) on J, but restricts the distribution function to the rectangular $F = F_0 = \mathrm{id}_{[0,1]}$. An extension to general F would require the weight $w = F^{-1}$, which may be nonsmooth, unbounded, and so may have no extension to $\mathbb{R}$ that would satisfy condition (8). The assumptions can however be shifted from F to J, or stronger convergence results on the empirical process may be invoked, in order to establish even bounded differentiability.

Suppose a weight function $w: \mathbb{R} \to \mathbb{R}$ that is left continuous and has bounded variation on each compact interval. That is, identifying measures and their distribution functions also notationally, there exists some measure w on $\mathbb{B}$ of finite total variation on each compact interval, such that

$$w(y) - w(x) = w[x,y), \qquad -\infty < x < y < \infty \tag{35}$$

The corresponding total variation measure is denoted by $|w|$. For example, under assumption (8): $w \in \mathcal{C}_1(\mathbb{R})$, we would have

$$w(dy) = w'(y)\,\lambda(dy), \qquad |w|(dy) = |w'(y)|\,\lambda(dy) \tag{36}$$

Moreover, assume a scores function

$$J: [0,1] \to \mathbb{R} \quad \text{bounded, measurable} \tag{37}$$

Part of the arguments will require that J in addition trims the extremes: There exist some $0 < \alpha < \beta < 1$ such that

$$J(s) \neq 0 \implies \alpha \leq s < \beta \tag{38}$$

Then the L functional T is well defined by (1) and finite for all distribution functions F since

$$-\infty < F^{-1}(\alpha) \leq F^{-1}(s) < F^{-1}(\beta) < \infty, \qquad s \in [\alpha, \beta)$$

the weight w is bounded on $[\alpha, \beta)$, and the scores function J is bounded, vanishing outside $[\alpha, \beta)$. Under condition (38) we choose the domain of T to be the set $\mathcal{F}$ of all right continuous distribution functions belonging to probability measures on $(\mathbb{R}, \mathbb{B})$. For fixed $F \in \mathcal{F}$ the tangent space will be

$$\mathcal{D} = \bigl\{ \gamma(G - F) \bigm| \gamma \in \mathbb{R},\ G \in \mathcal{F} \bigr\} \tag{39}$$

on which the sup norm $\|\,.\,\|$ is reasonable, because the empirical process is in this norm uniformly bounded,

$$\lim_{M\to\infty} \limsup_{n\to\infty} \sup_{F\in\mathcal{F}} F^n\bigl(\sqrt{n}\,\|\hat{F}_n - F\| > M\bigr) = 0 \tag{40}$$

According to Proposition A.3.5 the uniform boundedness extends to even stronger sup norms $\|\,.\,\|_{qF}$ that involve division by $q \circ F$ for certain functions $q \in \mathcal{Q}$ approaching the value 0 at the endpoints of the unit interval; consult A.3(12)–(13) for a definition and Section A.3. Such sup norms other than for q constant 1 allow a similar treatment of nontrimming scores, under an additional moment type assumption on F.

For J not of type (38), choose as domain the set $\mathcal{F}'$ of all those distribution functions F such that the L functional $T(F)$ is well defined by (1) and finite. Fix such an $F \in \mathcal{F}'$. Suppose that, for some function $q \in \mathcal{Q}$,

$$\int_{-\infty}^{\infty} q \circ F \, d|w| < \infty \tag{41}$$

Then define

$$\mathcal{D}' = \left\{ \gamma(G-F) \,\middle|\, \gamma \in \mathbb{R},\, G \in \mathcal{F}',\, \|G-F\|_{qF} < \infty \right\} \tag{42}$$

as the tangent 'space'.

Theorem 1.6.8 *Let $F \in \mathcal{F}$. Assume (35) and (37), and that*

$$w\left\{ y \in \mathbb{R} \,\middle|\, J \text{ discontinuous at } F(y) \right\} = 0 \tag{43}$$

(a) *Then, if J satisfies (38), the functional T defined by (1) is boundedly differentiable at F along $\mathcal{D}$ given by (39) with derivative*

$$d_{\mathrm{F}}T(F)\Delta = -\int J(F(y))\Delta(y)\, w(dy) \tag{44}$$

(b) *Under the assumptions $F \in \mathcal{F}'$ and (41), T is boundedly differentiable at F along $\mathcal{D}'$ given by (42) with the same derivative (44).*

Remark 1.6.9

(a) Without condition (43), the differentiability may fail, and the following asymptotic normality (55) may not hold. This was proved for the trimmed mean by Stigler (1973), correcting the omission in Moore (1968).

(b) Similar to Moore (1968) and Boos (1979), our proof uses integration by parts [Fubini's theorem]. However, Boos' (1979) extra condition that the scores function J be continuous a.e. λ_0 is not needed. ////

PROOF Using (35), the equivalence $y < G^{-1}(s) \iff G(y) < s$, and Fubini's theorem, we may for $G \in \mathcal{F}'$ write

$$\begin{aligned}
&\int w(G^{-1}(s))J(s)\,\lambda_0(ds) - w(0)\int J(s)\,\lambda_0(ds) \\
&\quad = \int_{\{G^{-1}(s)\geq 0\}} \int_{[0,G^{-1}(s))} w(dy)\; J(s)\,\lambda_0(ds) \\
&\qquad - \int_{\{G^{-1}(s)<0\}} \int_{[G^{-1}(s),0)} w(dy)\; J(s)\,\lambda_0(ds) \\
&\quad = \int_{[0,\infty)} \int_{(G(y),1)} J(s)\,\lambda_0(ds)\; w(dy) \\
&\qquad - \int_{(-\infty,0)} \int_{(0,G(y)]} J(s)\,\lambda_0(ds)\; w(dy)
\end{aligned}$$

Therefore,

$$T(G) - T(F) = -\int_{-\infty}^{\infty} \int_{F(y)}^{G(y)} J(s)\,\lambda_0(ds)\,w(dy) \tag{45}$$

and

$$\begin{aligned} &T(G) - T(F) - d_F T(F)(G-F) \\ &\quad = -\int_{-\infty}^{\infty} \int_{F(y)}^{G(y)} J(s)\,\lambda_0(ds) - \big(G(y)-F(y)\big)J(F(y))\,w(dy) \end{aligned} \tag{46}$$

where the outer integration can be restricted to $\{\, y \in \mathbb{R} \mid G(y) \neq F(y)\,\}$.

(a) Let $2\varepsilon = \min\{\alpha, 1-\beta\}$ and put $a = F^{-1}(\varepsilon)$, $b = F^{-1}(1-\varepsilon)$. Then as soon as $0 < \|G-F\| < \varepsilon$,

$$\begin{aligned} &\frac{\big|L(G) - L(F) - d_F L(F)(G-F)\big|}{\|G-F\|} \\ &\quad \le \int_{[a,b)} \left| \frac{1}{G(y)-F(y)} \int_{F(y)}^{G(y)} J(s)\,\lambda_0(ds) - J(F(y)) \right| |w|(dy) \end{aligned} \tag{47}$$

If $\|G-F\| \to 0$ then $G(y) \to F(y)$ for all y. By assumption (43), J is continuous at $F(y)$ for w, hence $|w|$, almost all y. So the fundamental theorem of calculus forces the integrand to go to 0 a.e. $|w|(dy)$. At the same time, the integrand is uniformly bounded by $2\,\|J\| < \infty$. The dominated convergence theorem, since $|w|[a,b) < \infty$, now gives the result. The derivative is indeed continuous linear,

$$\|d_F L(F)\| \le \|J\|\,|w|\big[F^{-1}(\alpha), F^{-1}(\beta)\big) < \infty \tag{48}$$

since $J \circ F$ is bounded and vanishes outside $\big[F^{-1}(\alpha), F^{-1}(\beta)\big)$.

(b) For $G - F \in \mathcal{D}'$ it follows from (46) that

$$\begin{aligned} &\frac{\big|L(G) - L(F) - d_F L(F)(G-F)\big|}{\|G-F\|_{qF}} \\ &\quad \le \int_{-\infty}^{\infty} \left| \frac{1}{G(y)-F(y)} \int_{F(y)}^{G(y)} J(s)\,\lambda_0(ds) - J(F(y)) \right| q(F(y))\,|w|(dy) \end{aligned} \tag{49}$$

Now let $\|G-F\|_{qF} \to 0$. Then $G(y) \to F(y)$ for all y, since any $q \in \mathcal{D}[0,1]$ is necessarily bounded. The previous argument using assumption (43), the fundamental theorem, and boundedness of the integrand by $2\,\|J\| < \infty$, applies again. This time, condition (41) allows us to appeal to the dominated convergence theorem, which gives the result. Conditions (41) and (37) also ensure that

$$\|d_F L(F)\| \le \|J\| \int q \circ F\, d|w| < \infty \tag{50}$$

So the derivative is indeed continuous linear in the present setup. ////

Remark 1.6.10

(a) For $0 < u < v < 1$ the left continuous pseudoinverse F^{-1} satisfies

$$\lambda\{ x \in \mathbb{R} \mid u \leq F(x) < v \} = F^{-1}(v) - F^{-1}(u) \tag{51}$$

Therefore, the measure F^{-1} on $[0,1] \cap \mathbb{B}$ with this distribution function is the image of Lebesgue measure λ under the distribution function F. In case $w = \mathrm{id}_{\mathbb{R}}$ condition (43) amounts to continuity of J a.e. this F^{-1}.

(b) According to A.3(14), for $0 < \delta \leq 1/2$, the functions

$$q(s) = \big(s(1-s)\big)^{1/2-\delta}$$

are in $\mathcal{Q}$. If F for some $k \in (0,\infty)$ has a finite k^{th} absolute moment,

$$\int |x|^k \, F(dx) = k \int_0^\infty x^{k-1} \big(F(-x) + 1 - F(x)\big) \, \lambda(dx) < \infty$$

one may expect, under additional regularity assumptions on the variation of tails, that

$$\lim_{x \to \infty} x^{k-1} \big(F(-x) + 1 - F(x)\big) = 0 \tag{52}$$

Then condition (41) for $w = \mathrm{id}_{\mathbb{R}}$ is satisfied if $k > 3$.

(c) The two sets of assumptions in Theorem 1.6.8 serve the same end; namely, to show that the difference (46) is some suitable $\mathrm{o}(\|G - F\|_{qF})$. Not all situations in which this is true are subsumed under these assumptions. For example, (46) vanishes for J constant 1. An extension of Theorem 1.6.8 and Corollary 1.6.11 to unbounded J only requires a justification of the interchange of limit and integration in (49). ////

Let $x_1, x_2, \ldots$ be i.i.d. observations following some arbitrary distribution function F on $\mathbb{R}$, and consider the linear combinations of order statistics (4) with scores (5) generated by some function J.

Corollary 1.6.11 *Under the assumptions of* Theorem 1.6.8, *the* L *estimator sequence* (4) *satisfies*

$$\sqrt{n}\,\big(S_n - T(F)\big) = \frac{1}{\sqrt{n}} \sum_{i=1}^n \psi(x_i) + \mathrm{o}_{F^n}(n^0) \tag{53}$$

with influence curve

$$\psi(x) = \int \big(F(y) - \mathbf{I}(x \leq y)\big) J(F(y)) \, w(dy) \tag{54}$$

In particular,

$$\begin{aligned} &\sqrt{n}\,\big(S_n - T(F)\big)(F^n) \xrightarrow{\mathbf{w}} \mathcal{N}(0,\sigma^2) \\ &\sigma^2 = \iint \big(F(x \wedge y) - F(x)F(y)\big) J(F(x)) J(F(y)) \, w(dx) \, w(dy) \end{aligned} \tag{55}$$

Remark 1.6.12 Invoking more of the classical theory of weak convergence of stochastic processes, Huber (1969; Chapter 3, p 129) proved the as. normality result differently, for identical weight $w = \mathrm{id}_{\mathbb{R}}$ and under the additional condition that J be of bounded variation. ////

PROOF Given $\varepsilon \in (0,1)$ choose $M \in (0,\infty)$ according to (40), respectively Proposition A.3.5, such that

$$\liminf_{n\to\infty} F^n\bigl(\sqrt{n}\,\|\hat{F}_n - F\|_{qF} \le M\bigr) \ge 1-\varepsilon \tag{56}$$

By Theorem 1.6.8, choose $\delta \in (0,1)$ so that $G \in \mathcal{F}$, $\|G-F\|_{qF} < \delta$ entails

$$\bigl|L(G) - L(F) - d_{\mathrm{F}}L(F)(G-F)\bigr| \le \frac{\varepsilon}{M}\|G-F\|_{qF} \tag{57}$$

Now choose $n_0 > (M/\delta)^2$. Then $n > n_0$ and $\sqrt{n}\,\|\hat{F}_n - F\|_{qF} \le M$ implies that $\|\hat{F}_n - F\|_{qF} < \delta$, and therefore

$$\bigl|\sqrt{n}\,(S_n - L(F)) - d_{\mathrm{F}}L(F)\sqrt{n}\,(\hat{F}_n - F)\bigr| \le \frac{\varepsilon}{M}\sqrt{n}\,\|\hat{F}_n - F\|_{qF} \le \varepsilon \tag{58}$$

Altogether, we have shown that for all $\varepsilon \in (0,1)$,

$$\liminf_{n\to\infty} F^n\bigl(\bigl|\sqrt{n}\,(S_n - L(F)) - d_{\mathrm{F}}L(F)\sqrt{n}\,(\hat{F}_n - F)\bigr| \le \varepsilon\bigr) \ge 1-\varepsilon \tag{59}$$

in analogy to Theorem 1.3.3. ////

Example 1.6.13 Let us calculate the influence curve (54) for L functionals T with general weight w satisfying (35) and the simple *trimming scores* function

$$J(s) = \mathbf{I}(\alpha \le s < \beta) \tag{60}$$

for some $0 < \alpha < \beta < 1$. Define $a = F^{-1}(\alpha)$ and $b = F^{-1}(\beta)$. Then, since $J(F(y)) = 1$ iff $a \le y < b$, we have

$$\psi_T(x) = \int_{[a,b)} \bigl(F(y) - \mathbf{I}(y \ge x)\bigr)\, w(dy) \tag{61}$$

Under condition (43), that is $w(F = \alpha, \beta) = 0$, (61) shows that it does not matter which version of the interval with endpoints α, β is employed in the definition of J. For $x \le a$, we obtain

$$\begin{aligned}\psi_T(x) &= \int_{[a,b)} (F(y) - 1)\, w(dy) = -\int_{[a,b)} \int_{(y,\infty)} F(dz)\, w(dy)\\ &= -\int_{(a,\infty)} \bigl(w(b \wedge z) - w(a)\bigr)\, F(dz) = w(a) - W(F)\end{aligned}$$

where W is the *Winsorizing* L *functional*

$$W(F) = w(a)F(a-0) + T(F) + w(b)\bigl(1 - F(b-0)\bigr) \tag{62}$$

For $a < x < b$, (61) yields

$$\begin{aligned}\psi_T(x) &= \int_{[a,b)} (F(y) - 1)\, w(dy) + \int_{[a,b)} \mathbf{I}(y < x)\, w(dy) \\ &= w(a) - W(F) + w(x) - w(a) = w(x) - W(F)\end{aligned}$$

And for $x \geq b$, (61) gives

$$\begin{aligned}\psi_T(x) &= \int_{[a,b)} F(y)\, w(dy) = \int_{[a,b)} \int_{(-\infty,y]} F(dz)\, w(dy) \\ &= \int_{(-\infty,b]} \big(w(b) - w(a \vee z)\big)\, F(dz) = w(b) - W(F)\end{aligned}$$

Thus, altogether,

$$\psi_T(x) = w\big(a \vee x \wedge b\big) - W(F) \tag{63}$$

Theorem 1.6.8 and Corollary 1.6.11 hold without F being necessarily continuous at a or b. Therefore, in the definition of W, we generally only have

$$F(a-0) \leq \alpha \leq F(a), \qquad F(b-0) \leq \beta \leq F(b)$$

If F is continuous in some neighborhood of $\{a, b\}$, and at a, b has derivatives $f(a) > 0$, $f(b) > 0$, and in addition w is differentiable at a and b, both Corollaries 1.5.4 and 1.6.11 are in force. Applying the chain rule for compact differentiation to w composed with the two quantiles, and adding up the expansions 1.5(30) and (53), verifies the expansion (53), hence as. normality (55), for the sequence of *Winsorizing* L *estimates*, with the following influence curve

$$\begin{aligned}\psi_W(x) &= \alpha \frac{w'(a)}{f(a)} \big(\alpha - \mathbf{I}(x \leq a)\big) + \psi_T(x) + (1-\beta) \frac{w'(b)}{f(b)} \big(\beta - \mathbf{I}(x \leq b)\big) \\ &= \begin{cases} w(a) - \alpha \dfrac{w'(a)}{f(a)} - \widetilde{W}(F), & \text{if } x \leq a \\ w(x) - \widetilde{W}(F), & \text{if } a < x \leq b \\ w(b) + (1-\beta) \dfrac{w'(b)}{f(b)} - \widetilde{W}(F), & \text{if } x > b \end{cases}\end{aligned} \tag{64}$$

where

$$\widetilde{W}(F) = W(F) - \alpha^2 \frac{w'(a)}{f(a)} + (1-\beta)^2 \frac{w'(b)}{f(b)} \tag{65}$$

The influence curve has jumps at a and b. ////

Remark 1.6.14 In this setup, quantile functionals are not boundedly differentiable either. For example, consider the median functional $T = \mathrm{med}$ under perturbations F_t, parametrized by $t \in (0, \frac{1}{4})$, of the rectangular distribution function $F_0 = \mathrm{id}_{[0,1]}$ that are defined by

$$F_t(0) = 0, \quad F_t(\tfrac{1}{2} - t) = \tfrac{1}{2}, \quad F_t(\tfrac{1}{2}) = \tfrac{1}{2} + t^2, \quad F_t(1) = 1$$

and F_t linear in between. Then

$$\|F_t - F_0\| = t, \quad T(F_t) = \tfrac{1}{2} - t, \quad T(F_0) = \tfrac{1}{2}$$

But for the only possible (weak, compact) derivative $-\pi_{\frac{1}{2}}$ we get

$$\frac{\bigl|T(F_t) - T(F_0) + \pi_{\frac{1}{2}}(F_t - F_0)\bigr|}{\|F_t - F_0\|} = |1 - t| > \frac{3}{4} \tag{66}$$

which disproves bounded differentiability. ////

Remark 1.6.15 In statistics, one generally has to deal with more than one measure (even when only testing two simple hypotheses). Accordingly, the as. statistical optimality results [Chapters 3 and 4] will require extensions of the estimator constructions given in this chapter that are suitably uniform, at least in some local as. sense [Chapter 6]. ////

Chapter 2

Log Likelihoods

2.1 General Remarks

This chapter develops, in the necessary minimal generality, the basic technical instruments of asymptotic statistics: contiguity of probability measures, (local) asymptotic normality of statistical models, and log likelihood expansions for L_2 differentiable parametric families. All these notions are associated with the names of LeCam and Hájek. The results are illustrated in the linear regression model.

2.2 Contiguity and Asymptotic Normality

Contiguity is the as. version of absolute continuity. Although the results will mainly be applied to probability measures, the presentation becomes clearer at some instances if, more generally, finite measures are considered.

Let $(\Omega, \mathcal{A})$ be some measurable space, with $\mathcal{M}_b(\mathcal{A})$ denoting the set of all finite measures, and $\mathcal{M}_1(\mathcal{A})$ the set of all probability measures, on $\mathcal{A}$. By definition, given any two finite measures $P, Q \in \mathcal{M}_b(\mathcal{A})$, the *log likelihood* $\log dQ/dP$ of Q relative to P is the set of all measurable functions $L\colon (\Omega, \mathcal{A}) \to (\bar{\mathbb{R}}, \bar{\mathbb{B}})$ such that

$$Q(A \cap \{L < \infty\}) = \int_A e^L \, dP, \qquad A \in \mathcal{A} \tag{1}$$

In particular, it follows that $P(L = \infty) = 0$. The definition can cope with singular parts of Q relative to P. The functions $L \in \log dQ/dP$ are called *versions* of the log likelihood, or simply log likelihoods, of Q relative to P. If p and q are densities relative to any dominating measure $\mu \in \mathcal{M}(\mathcal{A})$ such that $dP = p\,d\mu$ and $dQ = q\,d\mu$, (1) implies the representation

$$L = \log \frac{q}{p} \qquad \text{a.e. } \mu \quad \text{on } \Omega \setminus \{p = 0, q = 0\} \tag{2}$$

As a consequence, any two versions of $\log dQ/dP$ coincide a.e. $(P+Q)$. Furthermore, if $L \in \log dQ/dP$ then $-L \in \log dP/dQ$.

More commonly, one considers the *likelihood* or *Radon–Nikodym derivative* of Q relative to P, which is the set

$$\frac{dQ}{dP} = \Bigl\{ e^L \Bigm| L \in \log \frac{dQ}{dP} \Bigr\} \tag{3}$$

The measure P is said to *dominate* Q (notation: $Q \ll P$) if $P(A) = 0$ implies that $Q(A) = 0$. In terms of log likelihoods $L \in \log dQ/dP$, we have

$$Q \ll P \iff Q(L = \infty) = 0 \iff Q(\Omega) = \int e^L \, dP \iff dQ = e^L \, dP$$

For the following considerations, we assume that $Q \ll P$ and choose any version $L \in \log dQ/dP$; that is, $dQ = e^L \, dP$. Then the law $\mathcal{L}(Q)$ of L under Q can be expressed in terms of the law $\mathcal{L}(P)$ of L under P as follows:

$$\mathcal{L}(Q)(du) = e^u \, \mathcal{L}(P)(du) \tag{4}$$

More generally, if $S \colon (\Omega, \mathcal{A}) \to (\Omega', \mathcal{A}')$ is any statistic, the joint law $W(Q)$ of $W = (S, L)$ under Q is determined by the law $W(P)$ under P according to

$$W(Q)(ds, du) = e^u \, W(P)(ds, du) \tag{5}$$

Assuming (4), the representation (5) can equivalently be written in terms of the conditional distributions given L,

$$Q(S \in A' | L) = P(S \in A' | L) \qquad \text{a.e. } Q\,, \quad A' \in \mathcal{A}' \tag{6}$$

which is seen from the following identity (holding for all $A' \in \mathcal{A}'$, $B \in \bar{\mathbb{B}}$):

$$\begin{aligned}
\int_{\{L \in B\}} Q(S \in A' | L) \, dQ &= Q(S \in A', L \in B) \\
= \int_{\{L \in B\}} \mathbf{I}(S \in A') \, e^L \, dP &= \int_{\{L \in B\}} \mathrm{E}_P\bigl(\mathbf{I}(S \in A') \, e^L \big| L\bigr) \, dP \\
= \int_{\{L \in B\}} P(S \in A' | L) \, e^L \, dP &= \int_{\{L \in B\}} P(S \in A' | L) \, dQ
\end{aligned}$$

Thus, in case $Q \ll P$, the conditional distribution given L of S under Q is the same as the conditional distribution given L of S under P. In other words, the log likelihood L is *sufficient* for the set $\{P, Q\}$ of two measures.

Contiguity extends these ideas to an asymptotic setup. In the as. setup, sequences of finite measures $Q_n \in \mathcal{M}_b(\mathcal{A}_n)$ on general measurable spaces $(\Omega_n, \mathcal{A}_n)$ are considered which are *bounded*; that is,

$$\limsup_{n \to \infty} Q_n(\Omega_n) < \infty \tag{7}$$

Some basic notions and facts about weak convergence of measures are listed in Appendix A. Weak convergence is denoted by $Q_n \xrightarrow{w} Q$ where as usual the sequence index n tends to ∞. Weak convergence on the extended real line $(\bar{\mathbb{R}}, \bar{\mathbb{B}})$, relevant to log likelihoods, is also spelled out there [p 333].

Definition 2.2.1 *A bounded sequence $Q_n \in \mathcal{M}_b(\mathcal{A}_n)$ is called* contiguous *to another bounded sequence $P_n \in \mathcal{M}_b(\mathcal{A}_n)$ of measures if for all sequences of events $A_n \in \mathcal{A}_n$,*

$$\lim_{n\to\infty} P_n(A_n) = 0 \implies \lim_{n\to\infty} Q_n(A_n) = 0 \tag{8}$$

Notation: $(Q_n) \ll (P_n)$. Obviously, $(Q_n) \ll (P_n)$ iff for every $\varepsilon \in (0,1)$ there exists some $\delta \in (0,1)$ such that for all sequences $A_n \in \mathcal{A}_n$,

$$\limsup_{n\to\infty} P_n(A_n) < \delta \implies \limsup_{n\to\infty} Q_n(A_n) < \varepsilon \tag{9}$$

Passing from events to random variables, two other characterizations of contiguity follow: $(Q_n) \ll (P_n)$ iff convergence in probability to 0 under (P_n) implies convergence in probability to 0 under (Q_n). The same goes for tightness. (Look at $X_n = \mathbf{I}_{A_n}$ and $X_n = n\,\mathbf{I}_{A_n}$, respectively.)

For a description by means of log likelihoods, a modification of tightness turns out useful: A bounded sequence $R_n \in \mathcal{M}_b(\bar{\mathbb{B}})$ is called *tight to the right* if for every $\varepsilon \in (0,1)$ there exists some $t \in \mathbb{R}$ such that

$$\limsup_{n\to\infty} R_n((t,\infty]) < \varepsilon \tag{10}$$

Contiguity is connected with tightness to the right of log likelihoods by the following bounds,

$$P(L > t) \le e^{-t} \int e^L \, dP \le e^{-t}\, Q(\Omega) \tag{11}$$

$$Q(A) = \int_{A\cap\{L\le t\}} e^L \, dP + Q(A, L > t) \le e^t P(A) + Q(L > t) \tag{12}$$

holding for all $P, Q \in \mathcal{M}_b(\mathcal{A})$, $L \in \log dQ/dP$, $t \in \mathbb{R}$, and $A \in \mathcal{A}$. Thus:

Proposition 2.2.2 *Given two bounded sequences $P_n, Q_n \in \mathcal{M}_b(\mathcal{A}_n)$ with log likelihood versions $L_n \in \log dQ_n/dP_n$. Then*

(a) *$(L_n(P_n))$ is always tight to the right.*

(b) *$(L_n(Q_n))$ is tight to the right iff $(Q_n) \ll (P_n)$.*

Following Hájek and Šidák (1967; VI.1), part of Proposition 2.2.2 up to Corollary 2.2.6 is sometimes quoted as a triple of 'LeCam lemmas 1–3'.

Obviously, the contiguity $(Q_n) \ll (P_n)$ of the entire sequences is equivalent to the contiguity $(Q_m) \ll (P_m)$ for all subsequences (m) of $\mathbb{N}$. By Prokhorov's theorem for the extended real line $\bar{\mathbb{R}}$ (compact), therefore, the convergence assumption of the following theorem is not restrictive.

Theorem 2.2.3 *Let two sequences $P_n, Q_n \in \mathcal{M}_b(\mathcal{A}_n)$ be bounded with log likelihoods $L_n \in \log dQ_n/dP_n$ that converge weakly under P_n in $\bar{\mathbb{R}}$,*

$$L_n(P_n) \longrightarrow_{\mathrm{w}} M \in \mathcal{M}_b(\bar{\mathbb{B}}) \tag{13}$$

Then the following three statements are pairwise equivalent:

$$(Q_n) \ll (P_n) \tag{14}$$

$$\lim_{n\to\infty} Q_n(\Omega_n) = \int e^u \, M(du) \tag{15}$$

$$L_n(Q_n) \longrightarrow_{\mathrm{w}} M_1 \,, \quad M_1(du) = e^u \, M(du) \tag{16}$$

PROOF

(14)$\Rightarrow$(16): Let $L_m(Q_m)$, along a subsequence (m), tend weakly to some cluster point M_1 in $\mathcal{M}_b(\bar{\mathbb{B}})$, and consider any function $f \in \mathcal{C}(\bar{\mathbb{R}})$ that vanishes on $(t,\infty]$ for some finite t. Then (13) implies that

$$\begin{aligned} \int f \, dM_1 &= \lim_{m\to\infty} \int f(L_m) \, dQ_m \\ &= \lim_{m\to\infty} \int f(L_m) \, e^{L_m} \, dP_m = \int f(u) \, e^u \, M(du) \end{aligned}$$

This determines M_1 uniquely on $\bar{\mathbb{B}} \cap [-\infty,\infty)$ since the indicator of any interval $[s,t]$ with $-\infty \le s < t < \infty$ can, as in A.2(2)–A.2(3), be approximated pointwise from above by such functions. In addition, we have

$$M_1(\{\infty\}) = 0 = M(\{\infty\}) \tag{17}$$

since by the contiguity assumption the sequence $L_n(Q_n)$ is tight to the right. Altogether, the representation $M_1(du) = e^u \, M(du)$ follows.

(16)$\Rightarrow$(15): Weak convergence implies convergence of total masses; that is, of $Q_n(L_n \in \bar{\mathbb{R}}) = Q_n(\Omega_n)$ to $M_1(\bar{\mathbb{R}}) = \int e^u \, M(du)$.

(15)$\Rightarrow$(14): For every finite t such that $M(\{t\}) = 0$ the weak convergence assumption (13) yields

$$\int_{[-\infty,t]} e^u \, M(du) = \lim_{n\to\infty} \int_{\{L_n \le t\}} e^{L_n} \, dP_n = \lim_{n\to\infty} Q_n(L_n \le t)$$

As $M(\{\infty\}) = 0$ by tightness to the right of $(L_n(P_n))$, the monotone convergence theorem and assumption (15) imply

$$\lim_{t\to\infty} \liminf_{n\to\infty} Q_n(L_n \le t) = \int_{[-\infty,\infty]} e^u \, M(du) = \lim_{n\to\infty} Q_n(\Omega_n)$$

hence

$$\lim_{t\to\infty} \limsup_{n\to\infty} Q_n(L_n > t) = 0$$

This means $(L_n(Q_n))$ is tight to the right and therefore $(Q_n) \ll (P_n)$. ////

Note that, if $M(\{-\infty\}) = 0$, then the sequence $(-L_n(P_n))$ will be tight to the right, and thus $(P_n) \ll (Q_n)$ since $-L_n \in \log dP_n/dQ_n$.

In the case of probabilities $P_n, Q_n \in \mathcal{M}_1(\mathcal{A}_n)$, the limiting measure M is necessarily a probability (on $\bar{\mathbb{B}}$), and therefore condition (15) reads

$$\int e^u \, M(du) = 1 \tag{18}$$

Now one may easily verify that the only normal distributions $M = \mathcal{N}(\mu, \sigma^2)$ with $\mu \in \mathbb{R}$, $\sigma \in [0, \infty)$ [and extended to $\bar{\mathbb{B}}$ by $M(\{-\infty, \infty\}) = 0$] that fulfill condition (18) are those with $\mu = -\sigma^2/2$. And then the corresponding measure $M_1(du) = e^u \, M(du)$ is readily calculated to be $\mathcal{N}(\sigma^2/2, \sigma^2)$. Thus the following corollary holds in the as. normal situation.

Corollary 2.2.4 *Let $P_n, Q_n \in \mathcal{M}_1(\mathcal{A}_n)$ be two sequences of probabilities whose log likelihoods $L_n \in \log dQ_n/dP_n$ are under P_n as. normal,*

$$L_n(P_n) \xrightarrow{\mathrm{w}} \mathcal{N}(\mu, \sigma^2) \tag{19}$$

for some $\mu \in \mathbb{R}$, $\sigma \in [0, \infty)$. Then

$$(P_n) \ll (Q_n) \tag{20}$$

$$(Q_n) \ll (P_n) \iff \mu = -\sigma^2/2 \tag{21}$$

$$\mu = -\sigma^2/2 \implies L_n(Q_n) \xrightarrow{\mathrm{w}} \mathcal{N}(\sigma^2/2, \sigma^2) \tag{22}$$

The next result concerns joint weak convergence of log likelihoods with other statistics.

Theorem 2.2.5 *Assume two bounded sequences $P_n, Q_n \in \mathcal{M}_b(\mathcal{A}_n)$ with log likelihoods $L_n \in \log dQ_n/dP_n$ and consider a sequence of statistics S_n on $(\Omega_n, \mathcal{A}_n)$ with values in some polish space $(\Xi, \mathcal{B})$ with Borel σ algebra $\mathcal{B}$, such that*

$$(Q_n) \ll (P_n) \tag{23}$$

and

$$(S_n, L_n)(P_n) \xrightarrow{\mathrm{w}} N \in \mathcal{M}_b(\mathcal{B} \otimes \bar{\mathbb{B}}) \tag{24}$$

Then

$$(S_n, L_n)(Q_n) \xrightarrow{\mathrm{w}} N_1, \quad N_1(ds, du) = e^u \, N(ds, du) \tag{25}$$

PROOF The sequence $((S_n, L_n)(P_n))$ is tight [Prokhorov, converse]. By the contiguity assumption, the sequence $((S_n, L_n)(Q_n))$ also must be tight, hence relatively compact [Prokhorov, direct]. Let us consider any of its cluster points N_1 relative to weak convergence in $\mathcal{M}_b(\mathcal{B} \otimes \bar{\mathbb{B}})$.

Suppose a function $f \in \mathcal{C}(\Xi \times \bar{\mathbb{R}})$ such that, for some $t \in \mathbb{R}$, $f(s, u) = 0$ for all $s \in \Xi$ and $t < u \le \infty$. Then, along some subsequence (m),

$$\begin{aligned} \int f \, dN_1 &= \lim_{m \to \infty} \int f(S_m, L_m) \, dQ_m \\ &= \lim_{m \to \infty} \int f(S_m, L_m) \, e^{L_m} \, dP_m = \int f(s, u) \, e^u \, N(ds, du) \end{aligned}$$

Since the indicator of any rectangle $F \times [s,t]$ with $-\infty \le s < t < \infty$ and closed $F \subset \Xi$ can be approximated pointwise from above by such functions in the manner of A.2(2)–A.2(3), and these sets form a measure determining class for $(\mathcal{B} \otimes \bar{\mathbb{B}}) \cap (\Xi \times [-\infty, \infty))$, N_1 is uniquely determined on this sub-σ algebra. In addition we have

$$N_1(\Xi \times \{\infty\}) = M_1(\{\infty\}) = 0 = M(\{\infty\}) = N(\Xi \times \{\infty\}) \tag{26}$$

due to $(Q_n) \ll (P_n)$. ////

In this setup, denote by M and M_1 the second marginals of N and N_1, respectively, and introduce the canonical projections

$$S: \Xi \times \bar{\mathbb{R}} \longrightarrow \Xi, \qquad L: \Xi \times \bar{\mathbb{R}} \longrightarrow \bar{\mathbb{R}}$$

so that in particular $M = L(N)$ and $M_1 = L(N_1)$. Then, under the assumptions of the preceding theorem, Theorem 2.2.3 is enforced and yields that $M_1(du) = e^u\, M(du)$. In view of (4)–(6), and making use of the polish assumption, the form of $N_1(ds, du) = e^u\, N(ds, du)$ stated in the preceding theorem thus turns out equivalent to

$$N_1(ds|u) = N(ds|u) \qquad \text{a.e. } M_1(du) \tag{27}$$

that is, the regular conditional distribution given L of S under N_1 as a Markov kernel coincides with the corresponding kernel under N.

In the following corollary, N is specified to a $(p+1)$-dimensional normal, which is extended to $\bar{\mathbb{B}}^{p+1}$ in the obvious way.

Corollary 2.2.6 *Let $P_n, Q_n \in \mathcal{M}_1(\mathcal{A}_n)$ be two sequences of probabilities with log likelihoods $L_n \in \log dQ_n/dP_n$, and S_n a sequence of statistics on $(\Omega_n, \mathcal{A}_n)$ taking values in some finite-dimensional $(\bar{\mathbb{R}}^p, \bar{\mathbb{B}}^p)$ such that for some $a, c \in \mathbb{R}^p$, $\sigma \in [0, \infty)$, and $A \in \mathbb{R}^{p \times p}$,*

$$\begin{pmatrix} S_n \\ L_n \end{pmatrix}(P_n) \xrightarrow{\mathrm{w}} \mathcal{N}\left(\begin{pmatrix} a \\ -\sigma^2/2 \end{pmatrix}, \begin{pmatrix} A & c \\ c' & \sigma^2 \end{pmatrix}\right) \tag{28}$$

Then

$$\begin{pmatrix} S_n \\ L_n \end{pmatrix}(Q_n) \xrightarrow{\mathrm{w}} \mathcal{N}\left(\begin{pmatrix} a+c \\ \sigma^2/2 \end{pmatrix}, \begin{pmatrix} A & c \\ c' & \sigma^2 \end{pmatrix}\right) \tag{29}$$

PROOF By assumption we have $M = \mathcal{N}(-\sigma^2/2, \sigma^2)$. Hence Corollary 2.2.4 yields $(Q_n) \ll (P_n)$ as well as $M_1(du) = e^u\, M(du)$. Theorem 2.2.5 being in force, it only remains to verify (27) in the case of the two $(p+1)$ variate normals. But according to the following lemma, the equality of $N_1(ds|u)$ and $N(ds|u)$ a.e. $M_1(du)$ then reduces to the identity

$$a + c\sigma^{-2}(u + \sigma^2/2) = a + c + c\sigma^{-2}(u - \sigma^2/2)$$

in case $\sigma > 0$, and to $a + c = a$ in case $\sigma^2 = 0$, respectively. ////

Lemma 2.2.7 *Consider a finite-dimensional normal random vector*

$$\mathcal{L}\binom{T}{X} = \mathcal{N}\left(\binom{a}{b}, \begin{pmatrix} A & C \\ C' & B \end{pmatrix}\right) \tag{30}$$

with mean vectors $a \in \mathbb{R}^p$, $b \in \mathbb{R}^k$, *covariance* $C \in \mathbb{R}^{p \times k}$, *and variances* $A \in \mathbb{R}^{p \times p}$ *and* $B \in \mathbb{R}^{k \times k}$ *such that* $B > 0$. *Then*

$$\mathcal{L}(T|X = x) = \mathcal{N}\big(a + CB^{-1}(x - b), A - CB^{-1}C'\big) \tag{31}$$

is the regular conditional distribution of T *given* $X = x$, a.e. $\lambda^k(dx)$.

PROOF [Lemma 2.2.7] The variable $Z = T - a - CB^{-1}(X - b)$ satisfies

$$\mathrm{E}\, Z(X - b)' = 0, \qquad \mathrm{E}\, ZZ' = A - CB^{-1}C'$$

The transformation $(T, X) \mapsto (Z, X)$ being linear, Z and X are jointly normal, hence independent. Therefore, a.e. $\lambda^k(dx)$,

$$\mathcal{L}\big(T - a - CB^{-1}(x - b)\big|X = x\big) = \mathcal{L}(Z|X = x) = \mathcal{L}(Z)$$

where $Z \sim \mathcal{N}(0, A - CB^{-1}C')$. ////

Example 2.2.8 The following construction will be taken up in Section 3.2 and Subsection 3.3.1, when a prior π is put on the unknown mean t of a normally $\mathcal{N}(t, B)$ distributed observation:

Given any probability $\pi \in \mathcal{M}_1(\mathbb{B}^k)$ and $0 < B = B' \in \mathbb{R}^{k \times k}$, invoke two stochastically independent random vectors

$$T \sim \pi, \qquad Y \sim \mathcal{N}(0, B) \tag{32}$$

Then their sum $X = T + Y$ has the following conditional distribution given T,

$$\mathcal{L}(X|T = t) = \mathcal{L}(t + Y) = \mathcal{N}(t, B) \qquad \text{a.e. } \pi(dt) \tag{33}$$

and thus represents the normal observation given the mean $T = t$.

Specifying the prior, for any $m \in \mathbb{N}$, to the normal

$$\pi = \pi_m = \mathcal{N}(0, mB) \tag{34}$$

the joint law of T and X may readily be computed to

$$\mathcal{L}\binom{T}{X} = \mathcal{N}\left(\binom{0}{0}, \begin{pmatrix} mB & mB \\ mB & [m+1]B \end{pmatrix}\right) \tag{35}$$

since linear transformations preserve normality. Then Lemma 2.2.7 supplies the following posterior distribution of the mean T,

$$\mathcal{L}(T|X = x) = \mathcal{N}(rx, rB) \qquad \text{a.e. } \lambda^k(dx) \tag{36}$$

given the (unconditional) observation $X = x$, where $r = m/(m + 1)$. ////

Asymptotically normal log likelihoods will by definition arise in connection with parametric families of probability measures that are termed asymptotically normal. To introduce this notion, let

$$\mathcal{Q}_n = \{ Q_{n,t} \mid t \in \mathbb{R}^k \} \subset \mathcal{M}_1(\mathcal{A}_n)$$

be a sequence of statistical models—in short, models $(\mathcal{Q}_n)$—that have the same finite-dimensional parameter spaces $\Theta_n = \mathbb{R}^k$ (or at least $\Theta_n \uparrow \mathbb{R}^k$).

Definition 2.2.9 *The models* $(\mathcal{Q}_n)$ *are called* as. normal *if there exist a sequence of random variables* $Z_n\colon (\Omega_n, \mathcal{A}_n) \to (\mathbb{R}^k, \mathbb{B}^k)$ *that are as. normal,*

$$Z_n(Q_{n,0}) \xrightarrow{\text{w}} \mathcal{N}(0, C) \tag{37}$$

with positive definite covariance $C \in \mathbb{R}^{k\times k}$, *and such that for all* $t \in \mathbb{R}^k$ *the log likelihoods* $L_{n,t} \in \log dQ_{n,t}/dQ_{n,0}$ *have the approximation*

$$L_{n,t} = t'Z_n - \tfrac{1}{2}t'Ct + \mathrm{o}_{Q_{n,0}}(n^0) \tag{38}$$

The sequence $Z = (Z_n)$ *is called the* as. sufficient statistic *and* C *the* as. covariance *of the as. normal models* $(\mathcal{Q}_n)$.

Remark 2.2.10

(a) The covariance C is uniquely determined by (37) and (38). And then (38) tells us that another sequence of statistics $W = (W_n)$ is as. sufficient if and only if $W_n = Z_n + \mathrm{o}_{Q_{n,0}}(n^0)$.

(b) Neglecting the approximation, the terminology of 'as. sufficient' may be justified in regard to Neyman's criterion [Proposition C.1.1]. One speaks of *local as. normality* if, as in Subsection 2.3.2, as. normality depends on suitable local reparametrizations. ////

Example 2.2.11 For the normal location family on $(\mathbb{R}^k, \mathbb{B}^k)$ with covariance $C > 0$ and mean $t \in \mathbb{R}^k$ we observe that

$$\begin{aligned} \mathcal{N}(t, C)(dx) &= \exp\bigl(t'C^{-1}x - \tfrac{1}{2}t'C^{-1}t\bigr)\, \mathcal{N}(0, C)(dx) \\ \log \frac{d\mathcal{N}(t, C)}{d\mathcal{N}(0, C)}(x) &= t'\, C^{-1}x - \frac{1}{2}t'\, C^{-1}t \qquad \text{a.e. } \mathcal{N}(0, C)(dx) \end{aligned} \tag{39}$$

where $C^{-1}x \sim \mathcal{N}(0, C^{-1})$ if $x \sim \mathcal{N}(0, C)$. Thus (37) and (38) are exact. ////

As. normality does not depend on the choice of reference point.

Proposition 2.2.12 *Let the models* $(\mathcal{Q}_n)$ *be as. normal with as. sufficient statistic* Z *and as. covariance* C. *Then*

$$(Q_{n,s}) \ll (Q_{n,t}) \tag{40}$$

$$Z_n(Q_{n,t}) \xrightarrow{\text{w}} \mathcal{N}(Ct, C) \tag{41}$$

$$L_{n,t,s} = (t-s)'(Z_n - Cs) - \tfrac{1}{2}(t-s)'C(t-s) + \mathrm{o}_{Q_{n,s}}(n^0) \tag{42}$$

for all $s, t \in \mathbb{R}^k$ *and log likelihood versions* $L_{n,t,s} \in \log dQ_{n,t}/dQ_{n,s}$.

PROOF It follows from (37) and (38) that for all $t \in \mathbb{R}^k$,

$$L_{n,t}(Q_{n,0}) \xrightarrow{\mathrm{w}} \mathcal{N}(-\tfrac{1}{2}t'Ct, t'Ct)$$

Thus

$$(Q_{n,0}) \ll (Q_{n,t}) \ll (Q_{n,0})$$

by Corollary 2.2.4, which proves (40). Using again (37) and (38), as well as the Cramér–Wold device [Proposition A.1.7], one can show that

$$\begin{pmatrix} Z_n \\ L_{n,t} \end{pmatrix}(Q_{n,0}) \xrightarrow{\mathrm{w}} \mathcal{N}\Bigl(\begin{pmatrix} 0 \\ -\frac{1}{2}t'Ct \end{pmatrix}, \begin{pmatrix} C & Ct \\ t'C & t'Ct \end{pmatrix}\Bigr)$$

thus

$$\begin{pmatrix} Z_n \\ L_{n,t} \end{pmatrix}(Q_{n,t}) \xrightarrow{\mathrm{w}} \mathcal{N}\Bigl(\begin{pmatrix} Ct \\ \frac{1}{2}t'Ct \end{pmatrix}, \begin{pmatrix} C & Ct \\ t'C & t'Ct \end{pmatrix}\Bigr)$$

by Corollary 2.2.6. Hence (41) follows.

To see that as. normality does not depend on the choice of reference point, observe that $L_{n,t,s} = L_{n,t} - L_{n,s}$ a.e. $Q_{n,0} + Q_{n,s} + Q_{n,t}$ on the event where all densities are positive. But the $Q_{n,0}$, $Q_{n,t}$, $Q_{n,s}$ probabilities of the complementary event, by mutual contiguity, go to 0. So (38) implies

$$\begin{aligned} L_{n,t,s} &= (t-s)'Z_n - \tfrac{1}{2}(t'Ct - s'Cs) + \mathrm{o}_{Q_{n,0}}(n^0) \\ &= (t-s)'(Z_n - Cs) - \tfrac{1}{2}(t-s)'C(t-s) + \mathrm{o}_{Q_{n,s}}(n^0) \end{aligned}$$

where indeed $(Z_n - Cs)(Q_{n,s}) \xrightarrow{\mathrm{w}} \mathcal{N}(0, C)$ as shown. ////

2.3 Differentiable Families

In this section, the desired log likelihood expansion for as. normality will be derived under suitable differentiability conditions on the underlying probability model.

2.3.1 Differentiable Arrays

We consider the case of independent, not necessarily identically distributed observations, and let two arrays of probability measures on general sample spaces $(\Omega_{n,i}, \mathcal{A}_{n,i})$ be given,

$$P_{n,i},\, Q_{n,i} \in \mathcal{M}_1(\mathcal{A}_{n,i})$$

where $i = 1, \ldots, i_n$ and $n \geq 1$. The corresponding product probabilities

$$P_n^{(n)} = \bigotimes_{i=1}^{i_n} P_{n,i}\,, \qquad Q_n^{(n)} = \bigotimes_{i=1}^{i_n} Q_{n,i}$$

are defined on the product measurable spaces $\bigl(\times_i \Omega_{n,i}, \bigotimes_i \mathcal{A}_{n,i}\bigr)$. Their log likelihood $\log dQ_n^{(n)}/dP_n^{(n)}$ will be expanded for $n \to \infty$.

Square Root Calculus

Definition 2.3.1 *The array* $(Q_{n,i})$ *is called* L_2 differentiable w.r.t. $(P_{n,i})$ *if there exists a corresponding array of random variables*

$$U_{n,i} \in L_2(P_{n,i})$$

such that for all $i = 1, \ldots, i_n$ and $n \geq 1$, and for all $\varepsilon \in (0, \infty)$,

$$\int U_{n,i}\, dP_{n,i} = 0 \tag{1}$$

$$\limsup_{n\to\infty} \sum_{i=1}^{i_n} \int U_{n,i}^2\, dP_{n,i} < \infty \tag{2}$$

$$\lim_{n\to\infty} \sum_{i=1}^{i_n} \int_{\{|U_{n,i}|>\varepsilon\}} U_{n,i}^2\, dP_{n,i} = 0 \tag{3}$$

$$\lim_{n\to\infty} \sum_{i=1}^{i_n} \bigl\| \sqrt{dQ_{n,i}} - \sqrt{dP_{n,i}}\bigl(1 + \tfrac{1}{2} U_{n,i}\bigr) \bigr\|^2 = 0 \tag{4}$$

Then the array $(U_{n,i})$ *is called* L_2 derivative *of* $(Q_{n,i})$ *relative to* $(P_{n,i})$.

In (4) and subsequently, to get around the domination assumption, the following Hilbert space that includes the ordinary $L_2^k(P)$ spaces is employed: For any measurable space $(\Omega, \mathcal{A})$ and $k \in \mathbb{N}$ define the set

$$\mathcal{L}_2^k(\mathcal{A}) = \bigl\{\, \xi\sqrt{dP} \bigm| \xi \in L_2^k(P),\ P \in \mathcal{M}_b(\mathcal{A}) \,\bigr\} \tag{5}$$

On this set $\mathcal{L}_2^k(\mathcal{A})$, an equivalence relation is given by

$$\xi\sqrt{dP} \equiv \eta\sqrt{dQ} \iff \int \bigl|\xi\sqrt{p} - \eta\sqrt{q}\,\bigr|^2\, d\mu = 0 \tag{6}$$

where $|.|$ denotes Euclidean norm on $\mathbb{R}^k$ and $\mu \in \mathcal{M}_b(\mathcal{A})$ may be any measure, depending on P and Q, so that $dP = p\, d\mu$, $dQ = q\, d\mu$. Linear combinations with real coefficients and a scalar product are then defined:

$$\begin{aligned} \alpha\xi\sqrt{dP} + \beta\eta\sqrt{dQ} &= \bigl(\alpha\xi\sqrt{p} + \beta\eta\sqrt{q}\,\bigr)\sqrt{d\mu} \\ \bigl\langle \xi\sqrt{dP} \bigm| \eta\sqrt{dQ}\,\bigr\rangle &= \int \xi'\eta\sqrt{pq}\, d\mu \end{aligned} \tag{7}$$

Thus, going back to (4), we have for example

$$\bigl\| \sqrt{dQ_{n,i}} - \sqrt{dP_{n,i}}\bigl(1 + \tfrac{1}{2}U_{n,i}\bigr) \bigr\|^2 = \int \bigl(\sqrt{q_{n,i}} - \bigl(1 + \tfrac{1}{2}U_{n,i}\bigr)\sqrt{p_{n,i}}\,\bigr)^2\, d\mu_{n,i}$$

whenever $\mu_{n,i} \in \mathcal{M}_b(\mathcal{A}_{n,i})$ and $dP_{n,i} = p_{n,i}\, d\mu_{n,i}$, $dQ_{n,i} = q_{n,i}\, d\mu_{n,i}$. Then, on passing to equivalence classes, $\mathcal{L}_2^k(\mathcal{A})$ becomes a Hilbert space with norm

$$\bigl\|\xi\sqrt{dP}\,\bigr\|^2 = \int |\xi|^2\, dP$$

To prove completeness, suppose a Cauchy sequence $(\xi_n\sqrt{dP_n})$ in $\mathcal{L}_2^k(\mathcal{A})$. Depending on this sequence, choose $\mu \in \mathcal{M}_b(\mathcal{A})$ to dominate the measures P_n, with densities p_n. By definition, $(\xi_n\sqrt{p_n})$ is Cauchy in the product Hilbert space $L_2^k(\mu)$ (complete), hence has a cluster value $\eta \in L_2^k(\mu)$. But then $\eta\sqrt{d\mu}$ is a cluster value of $(\xi_n\sqrt{dP_n})$ in $\mathcal{L}_2^k(\mathcal{A})$.

According to the following auxiliary result, the centering condition (1) is not restrictive.

Lemma 2.3.2 *If an array* $(U_{n,i})$ *satisfies* (2)–(4), *then*

$$\lim_{n\to\infty}\sum_{i=1}^{i_n}\Bigl|\int U_{n,i}\,dP_{n,i}\Bigr|^2 = 0 = \lim_{n\to\infty}\frac{1}{\sqrt{i_n}}\sum_{i=1}^{i_n}\Bigl|\int U_{n,i}\,dP_{n,i}\Bigr| \tag{8}$$

and conditions (1)–(4) *are fulfilled by the centered array*

$$V_{n,i} = U_{n,i} - \mathrm{E}_{n,i}\,U_{n,i} \tag{9}$$

Notation In this and subsequent proofs, expectation under $P_{n,i}$ may be denoted by $\mathrm{E}_{n,i}$, and summation/maximization over $i = 1,\dots,i_n$ may be abbreviated by $\sum_i$ and $\max_i$, respectively.

PROOF Writing

$$D_{n,i} = \sqrt{dQ_{n,i}} - \sqrt{dP_{n,i}}\bigl(1 + \tfrac{1}{2}U_{n,i}\bigr) \tag{10}$$

we have

$$\mathrm{E}_{n,i}\,U_{n,i} = -\bigl\|\sqrt{dQ_{n,i}} - \sqrt{dP_{n,i}}\,\bigr\|^2 - 2\,\bigl\langle D_{n,i}\big|\sqrt{dP_{n,i}}\,\bigr\rangle \tag{11}$$

The triangle and Cauchy–Schwarz inequalities now show that the following difference becomes negligible,

$$\begin{aligned}&\biggl(\Bigl|\sum\nolimits_i \bigl|\mathrm{E}_{n,i}\,U_{n,i}\bigr|^2\Bigr|^{1/2} - \Bigl|\sum\nolimits_i \bigl\|\sqrt{dQ_{n,i}} - \sqrt{dP_{n,i}}\,\bigr\|^4\Bigr|^{1/2}\biggr)^2 \\ &\quad\le \sum\nolimits_i \Bigl|\mathrm{E}_{n,i}\,U_{n,i} + \bigl\|\sqrt{dQ_{n,i}} - \sqrt{dP_{n,i}}\,\bigr\|^2\Bigr|^2 \\ &\quad= 4\sum\nolimits_i \bigl\langle D_{n,i}\big|\sqrt{dP_{n,i}}\,\bigr\rangle^2 \le 4\sum\nolimits_i \|D_{n,i}\|^2 \underset{(4)}{\longrightarrow} 0\end{aligned} \tag{12}$$

Therefore, (8) is proved if the approximating term tends to 0. But

$$\sum\nolimits_i \bigl\|\sqrt{dQ_{n,i}} - \sqrt{dP_{n,i}}\,\bigr\|^2 \le 2\sum\nolimits_i \|D_{n,i}\|^2 + \frac{1}{2}\sum\nolimits_i \bigl\|U_{n,i}\sqrt{dP_{n,i}}\,\bigr\|^2$$

is bounded due to (2), (4), and

$$\max\nolimits_i \bigl\|\sqrt{dQ_{n,i}} - \sqrt{dP_{n,i}}\,\bigr\|^2 \le 2\sum\nolimits_i \|D_{n,i}\|^2 + \frac{1}{2}\max\nolimits_i \bigl\|U_{n,i}\sqrt{dP_{n,i}}\,\bigr\|^2$$

This upper bound tends to 0 by (4), and since the Lindeberg condition (3) entails Feller's condition, for $(U_{n,i})$. Thus

$$\lim_{n\to\infty} \max_i \left\| \sqrt{dQ_{n,i}} - \sqrt{dP_{n,i}} \right\| = 0 \tag{13}$$

Since $\sum_i \|\,.\,\|^4 \le \max_i \|\,.\,\|^2 \sum_i \|\,.\,\|^2$ the first half of assertion (8) is proved. Then the second half is a consequence of the Cauchy–Schwarz inequality.

Concerning $(V_{n,i})$ we have $\mathrm{E}_{n,i} V_{n,i} = 0$ by definition, and also the sum of L_2 norms is bounded since centering at expectations decreases L_2 norms. As for the Lindeberg condition (3), we have

$$\frac{1}{2}\sum_i \int_{\{|V_{n,i}|>\varepsilon\}} V_{n,i}^2\, dP_{n,i} \le \sum_i \int_{\{|U_{n,i}|>\varepsilon-\gamma_n\}} U_{n,i}^2\, dP_{n,i} + \sum_i \left|\mathrm{E}_{n,i} U_{n,i}\right|^2$$

where by (8) the last sum, hence also $\gamma_n^2 = \max_i \left|\mathrm{E}_{n,i} U_{n,i}\right|^2$, tends to 0. Then we can appeal to the Lindeberg condition (3) for $(U_{n,i})$.

To verify the differentiability (4) for $(V_{n,i})$, we observe that

$$\begin{aligned} \Bigl(\Bigl|\sum_i \|\Delta_{n,i}\|^2\Bigr|^{1/2} - \Bigl|\sum_i \|D_{n,i}\|^2\Bigr|^{1/2}\Bigr)^2 &\le \sum_i \bigl(\|\Delta_{n,i}\| - \|D_{n,i}\|\bigr)^2 \\ \le \sum_i \|\Delta_{n,i} - D_{n,i}\|^2 &= \frac{1}{4}\sum_i \left|\mathrm{E}_{n,i} U_{n,i}\right|^2 \underset{(8)}{\longrightarrow} 0 \end{aligned}$$

where

$$\Delta_{n,i} = \sqrt{dQ_{n,i}} - \sqrt{dP_{n,i}}\bigl(1 + \tfrac{1}{2} V_{n,i}\bigr) \tag{14}$$

Thus the proof is complete. ////

The next lemma shows uniqueness of the L_2 derivative $(U_{n,i})$ up to asymptotic equivalence in the product Hilbert spaces $\bigotimes_i L_2(P_{n,i})$.

Lemma 2.3.3 *Assume the array $(U_{n,i})$ satisfies the conditions (2)–(4). Then another array $V_{n,i} \in L_2(P_{n,i})$ fulfills conditions (2)–(4) iff*

$$\lim_{n\to\infty} \sum_{i=1}^{i_n} \int (V_{n,i} - U_{n,i})^2\, dP_{n,i} = 0 \tag{15}$$

PROOF First, let $(V_{n,i})$ fulfill the conditions (2)–(4). Then

$$\frac{1}{8}\sum_i \mathrm{E}_{n,i} (V_{n,i} - U_{n,i})^2 \le \sum_i \|D_{n,i}\|^2 + \sum_i \|\Delta_{n,i}\|^2$$

by the triangle and $(a+b)^2 \le 2(a^2+b^2)$ inequalities, employing the remainders (10) and (14). Thus (15) already follows from (4).

Secondly, to prove the converse, suppose (15) holds. Then again

$$\frac{1}{2}\sum_i \mathrm{E}_{n,i} V_{n,i}^2 \le \sum_i \mathrm{E}_{n,i} (V_{n,i} - U_{n,i})^2 + \sum_i \mathrm{E}_{n,i} U_{n,i}^2$$

So, property (2) is inherited from $(U_{n,i})$ to $(V_{n,i})$. The same is true concerning property (4), because of (15) and the bound

$$4\Big(\Big|\sum_i \|\Delta_{n,i}\|^2\Big|^{1/2} - \Big|\sum_i \|D_{n,i}\|^2\Big|^{1/2}\Big)^2 \le \sum_i \mathrm{E}_{n,i}\,(V_{n,i}-U_{n,i})^2$$

To verify the Lindeberg condition (3) for $(V_{n,i})$, given $\varepsilon \in (0,1)$, write

$$\begin{aligned}\sum_i \int_{\{|V_{n,i}|>2\varepsilon\}} V_{n,i}^2\,dP_{n,i} \\ \le \sum_i \int_{\{|V_{n,i}|>2\varepsilon\}} \big|V_{n,i}^2 - U_{n,i}^2\big|\,dP_{n,i} + \sum_i \int_{\{|V_{n,i}|>2\varepsilon\}} U_{n,i}^2\,dP_{n,i}\end{aligned} \tag{16}$$

It holds that

$$\begin{aligned}&\sum_i \int_{\{|V_{n,i}|>2\varepsilon\}} U_{n,i}^2\,dP_{n,i} \\ &= \sum_i \int_{\{|V_{n,i}|>2\varepsilon,|U_{n,i}|\le\varepsilon\}} U_{n,i}^2\,dP_{n,i} + \sum_i \int_{\{|V_{n,i}|>2\varepsilon,|U_{n,i}|>\varepsilon\}} U_{n,i}^2\,dP_{n,i} \\ &\le \sum_i \int_{\{|V_{n,i}-U_{n,i}|>\varepsilon\}} U_{n,i}^2\,dP_{n,i} + \sum_i \int_{\{|U_{n,i}|>\varepsilon\}} U_{n,i}^2\,dP_{n,i}\end{aligned}$$

where the last term goes to 0 by (3). As for the preceding term, we have

$$\sum_i P_{n,i}\big(|V_{n,i}-U_{n,i}|>\varepsilon\big) \le \frac{1}{\varepsilon^2}\sum_i \mathrm{E}_{n,i}\,(V_{n,i}-U_{n,i})^2 \longrightarrow 0$$

So its convergence to 0 is ensured by the implication:

$$A_{n,i} \in \mathcal{A}_{n,i},\ \sum_i P_{n,i}(A_{n,i}) \longrightarrow 0 \implies \sum_i \int_{A_{n,i}} U_{n,i}^2\,dP_{n,i} \longrightarrow 0 \tag{17}$$

which is a consequence of the Lindeberg condition (3) and the bound

$$\begin{aligned}&\sum_i \int_{A_{n,i}} U_{n,i}^2\,dP_{n,i} \\ &= \sum_i \int_{A_{n,i}\cap\{|U_{n,i}|\le\varepsilon\}} U_{n,i}^2\,dP_{n,i} + \sum_i \int_{A_{n,i}\cap\{|U_{n,i}|>\varepsilon\}} U_{n,i}^2\,dP_{n,i} \\ &\le \varepsilon^2 \sum_i P_{n,i}(A_{n,i}) + \sum_i \int_{\{|U_{n,i}|>\varepsilon\}} U_{n,i}^2\,dP_{n,i}\end{aligned}$$

Hence, it has been shown that

$$\lim_{n\to\infty} \sum_i \int_{\{|V_{n,i}|>2\varepsilon\}} U_{n,i}^2\,dP_{n,i} = 0 \tag{18}$$

In view of (16), all that remains to prove is

$$\lim_{n\to\infty} \sum\nolimits_i \int_{\{|V_{n,i}|>2\varepsilon\}} \left|V_{n,i}^2 - U_{n,i}^2\right| dP_{n,i} = 0 \tag{19}$$

But as

$$\left|V_{n,i}^2 - U_{n,i}^2\right| \le (V_{n,i} - U_{n,i})^2 + 2\,|U_{n,i}|\,|V_{n,i} - U_{n,i}|$$

we have

$$\begin{aligned}
&\sum\nolimits_i \int_{\{|V_{n,i}|>2\varepsilon\}} \left|V_{n,i}^2 - U_{n,i}^2\right| dP_{n,i} \\
&\le \sum\nolimits_i \mathrm{E}_{n,i}\,(V_{n,i} - U_{n,i})^2 + 2\sum\nolimits_i \int_{\{|V_{n,i}|>2\varepsilon\}} |U_{n,i}||V_{n,i} - U_{n,i}|\,dP_{n,i} \\
&\le \mathrm{o}(n^0) + 2\sum\nolimits_i \left|\mathrm{E}_{n,i}\,(V_{n,i} - U_{n,i})^2\right|^{1/2} \left|\int_{\{|V_{n,i}|>2\varepsilon\}} U_{n,i}^2\,dP_{n,i}\right|^{1/2} \\
&\le \mathrm{o}(n^0) + 2\left|\sum\nolimits_i \mathrm{E}_{n,i}\,(V_{n,i} - U_{n,i})^2\right|^{1/2} \left|\sum\nolimits_i \int_{\{|V_{n,i}|>2\varepsilon\}} U_{n,i}^2\,dP_{n,i}\right|^{1/2}
\end{aligned}$$

where repeated use has been made of Cauchy–Schwarz inequality and (15). This bound tends to 0 by (18), and again (15). ////

For later purposes we note that L_2 differentiability of square root densities implies L_1 differentiablity of densities, as follows.

Lemma 2.3.4 *Under the conditions* (2), (3) *and* (4), *it holds that*

$$\lim_{n\to\infty} \sum_{i=1}^{i_n} \left|\int \left|dQ_{n,i} - dP_{n,i}\,(1 + U_{n,i})\right|\right|^2 = 0 \tag{20}$$

hence also

$$\lim_{n\to\infty} \frac{1}{\sqrt{i_n}} \sum_{i=1}^{i_n} \int \left|dQ_{n,i} - dP_{n,i}\,(1 + U_{n,i})\right| = 0 \tag{21}$$

PROOF Employing the remainders $D_{n,i}$ introduced in (10), we obtain:

$$\begin{aligned}
&\sum\nolimits_i \left|\int \left|dQ_{n,i} - dP_{n,i}\,(1 + U_{n,i})\right|\right|^2 \\
&= \sum\nolimits_i \left|\int \left|D_{n,i}\big(\sqrt{dQ_{n,i}} + \sqrt{dP_{n,i}}\,\big) + \tfrac{1}{2}U_{n,i}\sqrt{dP_{n,i}}\,\big(\sqrt{dQ_{n,i}} - \sqrt{dP_{n,i}}\,\big)\right|\right|^2 \\
&\le 2\sum\nolimits_i \left|\int |D_{n,i}|\left|\sqrt{dQ_{n,i}} + \sqrt{dP_{n,i}}\,\right|\right|^2 \\
&\qquad + \frac{1}{2}\sum\nolimits_i \left|\int \left|U_{n,i}\sqrt{dP_{n,i}}\,\right|\left|\sqrt{dQ_{n,i}} - \sqrt{dP_{n,i}}\,\right|\right|^2 \\
&\le 2\sum\nolimits_i \|D_{n,i}\|^2 \left(\big\|\sqrt{dQ_{n,i}}\,\big\| + \big\|\sqrt{dP_{n,i}}\,\big\|\right)^2 \\
&\qquad + \frac{1}{2}\sum\nolimits_i \big\|U_{n,i}\sqrt{dP_{n,i}}\,\big\|^2 \big\|\sqrt{dQ_{n,i}} - \sqrt{dP_{n,i}}\,\big\|^2 \\
&\le 8\sum\nolimits_i \|D_{n,i}\|^2 + \frac{1}{2}\max\nolimits_i \big\|\sqrt{dQ_{n,i}} - \sqrt{dP_{n,i}}\,\big\|^2 \sum\nolimits_i \big\|U_{n,i}\sqrt{dP_{n,i}}\,\big\|^2
\end{aligned}$$

This upper bound tends to 0, by (2), (4), and (13). Thus (20) is proved. Then appeal to the Cauchy–Schwarz inequality to get (21). ////

Log Likelihood Expansion

With $x_{n,i}\colon \Omega_{n,1} \times \cdots \times \Omega_{n,i_n} \to \Omega_{n,i}$ denoting the canonical projections, the following is the main result of this section.

Theorem 2.3.5 *If the array* $(Q_{n,i})$ *has* L_2 *derivative* $(U_{n,i})$ w.r.t. $(P_{n,i})$, *then*

$$\log \frac{dQ_n^{(n)}}{dP_n^{(n)}} = \sum_{i=1}^{i_n} U_{n,i}(x_{n,i}) - \frac{1}{2}\sum_{i=1}^{i_n} \int U_{n,i}^2\, dP_{n,i} + \mathrm{o}_{P_n^{(n)}}(n^0) \tag{22}$$

In particular, $(P_n^{(n)})$ *and* $(Q_n^{(n)})$ *are mutually contiguous.*

Proof By identity 2.2 (1) and Fubini's theorem, the log likelihood of the products $L^{(n)} \in \log dQ_n^{(n)}/dP_n^{(n)}$ is related to the log likelihoods of the factors $L_{n,i} \in \log dQ_{n,i}/dP_{n,i}$ as follows,

$$L^{(n)} = \sum\nolimits_i L_{n,i} \qquad \text{a.e. } \big(P_n^{(n)} + Q_n^{(n)}\big) \tag{23}$$

provided we define $\sum_i L_{n,i} = \infty$ if one $L_{n,i} = \infty$. Then writing

$$L_{n,i} = 2\log(1 + Z_{n,i}), \qquad Z_{n,i} = e^{L_{n,i}/2} - 1 \tag{24}$$

variables $Z_{n,i}$ are introduced that are linked to $\frac{1}{2}U_{n,i}$ by (4). To alleviate notation, we set $Z_{n,i}(x_{n,i}) = Z_{n,i}$ and $U_{n,i}(x_{n,i}) = U_{n,i}$ in this proof. Then

$$L^{(n)} = 2\sum\nolimits_i \big(Z_{n,i} + c(Z_{n,i})Z_{n,i}^2\big) \qquad \text{a.e. } \big(P_n^{(n)} + Q_n^{(n)}\big) \tag{25}$$

according to the Taylor expansion of the logarithm about 1,

$$\log(1+u) = u + c(u)u^2, \qquad \lim_{u\to 0} c(u) = -\tfrac{1}{2}$$

First step: We show that

$$\max\nolimits_i |Z_{n,i}| \xrightarrow{P_n^{(n)}} 0 \tag{26}$$

Indeed, for all $\varepsilon \in (0,1)$,

$$\begin{aligned} P_n^{(n)}\big(\max\nolimits_i |Z_{n,i}| > 2\varepsilon\big) &= P_n^{(n)}\big(\max\nolimits_i |2Z_{n,i} - U_{n,i} + U_{n,i}| > 4\varepsilon\big) \\ &\le P_n^{(n)}\big(\max\nolimits_i |2Z_{n,i} - U_{n,i}| > 2\varepsilon\big) + P_n^{(n)}\big(\max\nolimits_i |U_{n,i}| > \varepsilon\big) \\ &\le \sum\nolimits_i P_{n,i}\big(|2Z_{n,i} - U_{n,i}| > 2\varepsilon\big) + \sum\nolimits_i P_{n,i}\big(|U_{n,i}| > \varepsilon\big) \\ &\le \frac{1}{\varepsilon^2}\sum\nolimits_i \mathrm{E}_{n,i}\big(Z_{n,i} - \tfrac{1}{2}U_{n,i}\big)^2 + \frac{1}{\varepsilon^2}\sum\nolimits_i \int_{\{|U_{n,i}|>\varepsilon\}} U_{n,i}^2\, dP_{n,i} \end{aligned}$$

where the second sum goes to 0 by (3). In terms of the remainders (10),

$$\left\| Z_{n,i}\sqrt{dP_{n,i}} - \tfrac{1}{2}U_{n,i}\sqrt{dP_{n,i}}\,\right\|^2 = \|D_{n,i}\|^2 - Q_{n,i}(L_{n,i} = \infty) \tag{27}$$

so the first sum tends to 0 by (4). This proves (26).

Second step: Let us show that

$$2\sum_i Z_{n,i} = \sum_i U_{n,i} - \frac{1}{4}\sum_i \mathrm{E}_{n,i}\, U_{n,i}^2 + \mathrm{o}_{P_n^{(n)}}(n^0) \tag{28}$$

In fact, by Bienaymé's equality, (27), and (4),

$$\begin{aligned} \operatorname{Var}_{P_n^{(n)}}\Bigl(\sum_i 2Z_{n,i} - U_{n,i}\Bigr) &= \sum_i \operatorname{Var}_{P_{n,i}}(2Z_{n,i} - U_{n,i}) \\ &\le \sum_i \mathrm{E}_{n,i}(2Z_{n,i} - U_{n,i})^2 \le 4\sum_i \|D_{n,i}\|^2 \longrightarrow 0 \end{aligned} \tag{29}$$

Hence, using the centering condition (1), we conclude that

$$2\sum_i Z_{n,i} = \sum_i U_{n,i} + 2\sum_i \mathrm{E}_{n,i}\, Z_{n,i} + \mathrm{o}_{P_n^{(n)}}(n^0) \tag{30}$$

However,

$$\begin{aligned} 2\sum_i \mathrm{E}_{n,i}\, Z_{n,i} &= 2\sum_i \bigl(\langle \sqrt{dQ_{n,i}}\,|\sqrt{dP_{n,i}}\,\rangle - 1\bigr) \\ &= -\sum_i \bigl\|\sqrt{dQ_{n,i}} - \sqrt{dP_{n,i}}\,\bigr\|^2 \end{aligned} \tag{31}$$

where by the triangle inequality,

$$\biggl(\Bigl|\sum_i \bigl\|\sqrt{dQ_{n,i}} - \sqrt{dP_{n,i}}\,\bigr\|^2\Bigr|^{1/2} - \Bigl|\frac{1}{4}\sum_i \mathrm{E}_{n,i}\, U_{n,i}^2\Bigr|^{1/2}\biggr)^2 \le \sum_i \|D_{n,i}\|^2$$

Due to (4), this upper bound tends to 0. So (28) is proved.

Third step: We show that

$$\sum_i Z_{n,i}^2 = \frac{1}{4}\sum_i \mathrm{E}_{n,i}\, U_{n,i}^2 + \mathrm{o}_{P_n^{(n)}}(n^0) \tag{32}$$

In fact, as in (29), the upper bound in

$$\biggl(\Bigl|\sum_i Z_{n,i}^2\Bigr|^{1/2} - \Bigl|\sum_i \tfrac{1}{4}U_{n,i}^2\Bigr|^{1/2}\biggr)^2 \le \sum_i \bigl(Z_{n,i} - \tfrac{1}{2}U_{n,i}\bigr)^2$$

becomes negligible in $P_n^{(n)}$ mean, and hence in $P_n^{(n)}$ probability. So, instead of (32), we prove that

$$\sum_i U_{n,i}^2 = \sum_i \mathrm{E}_{n,i}\, U_{n,i}^2 + \mathrm{o}_{P_n^{(n)}}(n^0) \tag{33}$$

For any $\eta \in (0,\infty)$, let us put $W_{n,i} = U_{n,i}\,\mathbf{I}(|U_{n,i}| \le \eta)$. Then

$$P_n^{(n)}\Bigl(\sum_i W_{n,i}^2 \ne \sum_i U_{n,i}^2\Bigr) \le \sum_i P_{n,i}(|U_{n,i}| > \eta) \le \frac{1}{\eta^2}\sum_i \int_{\{|U_{n,i}|>\eta\}} U_{n,i}^2\,dP_{n,i} \underset{(3)}{\longrightarrow} 0$$

as well as

$$\sum_i \mathrm{E}_{n,i}\,W_{n,i}^2 - \sum_i \mathrm{E}_{n,i}\,U_{n,i}^2 = -\sum_i \int_{\{|U_{n,i}|>\eta\}} U_{n,i}^2\,dP_{n,i} \underset{(3)}{\longrightarrow} 0$$

Given $\varepsilon \in (0,1)$ choose $\eta = \varepsilon^2/\sigma$ where $\sigma^2 = \limsup_{n\to\infty}\sum_i \mathrm{E}_{n,i}\,U_{n,i}^2$. Then, as soon as $(1+1/\eta^2)\sum_i \int_{\{|U_{n,i}|>\eta\}} U_{n,i}^2\,dP_{n,i} < \varepsilon$,

$$\begin{aligned} &P_n^{(n)}\Bigl(\Bigl|\sum_i (U_{n,i}^2 - \mathrm{E}_{n,i}\,U_{n,i}^2)\Bigr| > 2\varepsilon\Bigr) \\ &\le \varepsilon + P_n^{(n)}\Bigl(\Bigl|\sum_i (W_{n,i}^2 - \mathrm{E}_{n,i}\,W_{n,i}^2)\Bigr| > \varepsilon\Bigr) \\ &\le \varepsilon + \frac{1}{\varepsilon^2}\sum_i \mathrm{E}_{n,i}\,W_{n,i}^4 \le \varepsilon + \frac{\eta^2}{\varepsilon^2}\sum_i \mathrm{E}_{n,i}\,U_{n,i}^2 \le 2\varepsilon + \mathrm{o}(n^0) \end{aligned}$$

which proves (33), hence (32).

Fourth step: The upper bound in

$$\Bigl|\sum_i (2c(Z_{n,i})+1)Z_{n,i}^2\Bigr| \le \max_i \bigl|2c(Z_{n,i})+1\bigr|\sum_i Z_{n,i}^2 \tag{34}$$

converges to 0 in $P_n^{(n)}$ probability. This follows from tightness of the laws of $\sum_i Z_{n,i}^2$ under $P_n^{(n)}$, to be read off from (32) and (2), and (26) plugged into $\lim_{u\to 0} c(u) = -1/2$. Combining this with (25), (28), and (32) completes the proof of the log likelihood expansion.

The mutual contiguity of $(P_n^{(n)})$ and $(Q_n^{(n)})$ is a consequence of this expansion and Corollary 2.2.4. Indeed, in view of the boundedness (2), we may without restriction consider a subsequence (m) along which

$$\lim_{m\to\infty}\sum_i \mathrm{E}_{m,i}\,U_{m,i}^2 = \sigma^2 \tag{35}$$

Suppose $\sigma \ne 0$. Then, by (1) and (3), the Lindeberg–Feller theorem applies, so that

$$\Bigl(\log\frac{dQ^{(m)}}{dP^{(m)}}\Bigr)(P^{(m)}) \xrightarrow{\;w\;} \mathcal{N}(-\sigma^2/2,\sigma^2) \tag{36}$$

The same is true in case $\sigma = 0$ since then the approximating sums converge to 0 in $L_2(P^{(m)})$, and hence in $P^{(m)}$ probability. Finally, argue by subsequences [Lemma A.1.8]. ////

2.3.2 Smooth Parametric Families

The results are now applied to verify as. normality of smooth parametric families. We first consider situations where at least under the null hypothesis the observations are i.i.d.—as in the classical one-sample, two-sample, and correlation cases. Thus let

$$\mathcal{P} = \{\, P_\theta \mid \theta \in \Theta \,\} \subset \mathcal{M}_1(\mathcal{A}) \tag{37}$$

be a family of probability measures on some sample space $(\Omega, \mathcal{A})$, with an open parameter set $\Theta \subset \mathbb{R}^k$ of finite dimension k. Fix $\theta \in \Theta$. (Notation: Expectation under P_* is also denoted by E_*, whatever the subscript.)

Definition 2.3.6 *Model $\mathcal{P}$ is called L_2 differentiable at θ if there exists some function $\Lambda_\theta \in L_2^k(P_\theta)$ such that, as $t \to 0$,*

$$\big\| \sqrt{dP_{\theta+t}} - \sqrt{dP_\theta}\big(1 + \tfrac{1}{2} t'\Lambda_\theta\big) \big\| = \mathrm{o}(|t|) \tag{38}$$

and

$$\mathcal{I}_\theta = \mathrm{E}_\theta\, \Lambda_\theta {\Lambda_\theta}' > 0 \tag{39}$$

The function Λ_θ is called the L_2 derivative *and the $k \times k$ matrix $\mathcal{I}_\theta$* Fisher information *of $\mathcal{P}$ at θ.*

We consider local parameter arrays $(\theta_{n,i})$ about θ of the form

$$\theta_{n,i} = \theta + \tau_{n,i} t_n\,, \qquad t_n \longrightarrow t \in \mathbb{R}^k \tag{40}$$

with coefficients $(\tau_{n,i})$ that fulfill the Noether condition,

$$\lim_{n\to\infty} \max_{i=1,\dots,i_n} |\tau_{n,i}| = 0\,, \qquad \lim_{n\to\infty} \sum_{i=1}^{i_n} \tau_{n,i}^2 = \tau^2 \in (0,\infty) \tag{41}$$

In the one-sample and correlation cases, typically

$$\tau_{n,i} = \frac{1}{\sqrt{n}}\,, \qquad i = 1, \dots, n \tag{42}$$

while in the two-sample case,

$$\tau_{n,i} = \begin{cases} \dfrac{1}{n_1}\sqrt{\dfrac{n_1 n_2}{n}}\,, & i = 1, \dots, n_1 \\[2ex] -\dfrac{1}{n_2}\sqrt{\dfrac{n_1 n_2}{n}}\,, & i = n_1 + 1, \dots, n_1 + n_2 = n \end{cases} \tag{43}$$

Corresponding to such parametric alternatives $(\theta_{n,i})$ of form (40) and to $\theta_{n,i} = \theta$ fixed, respectively, two sequences of product measures are defined on the i_n-fold product measurable spaces $(\Omega^{i_n}, \mathcal{A}^{i_n})$,

$$P_\theta^{(n)} = P_\theta^{i_n} = \bigotimes_{i=1}^{i_n} P_\theta\,, \qquad P_{\theta_n}^{(n)} = \bigotimes_{i=1}^{i_n} P_{\theta_{n,i}} \tag{44}$$

As before, $x_{n,i}$ denotes the ith canonical projection from Ω^{i_n} to Ω.

Theorem 2.3.7 *If $\mathcal{P}$ is L_2 differentiable at θ, its L_2 derivative Λ_θ is uniquely determined in $L_2^k(P_\theta)$. Moreover,*

$$\mathrm{E}_\theta\,\Lambda_\theta = 0 \tag{45}$$

and the alternatives given by (40), (41) and (44) have the log likelihood expansion

$$\log\frac{dP^{(n)}_{\theta_n}}{dP^{(n)}_\theta} = t'\sum_{i=1}^{i_n}\tau_{n,i}\Lambda_\theta(x_{n,i}) - \frac{\tau^2}{2}t'\mathcal{I}_\theta t + \mathrm{o}_{P_\theta^{(n)}}(n^0) \tag{46}$$

where

$$\Bigl(\sum_{i=1}^{i_n}\tau_{n,i}\Lambda_\theta(x_{n,i})\Bigr)(P_\theta^{(n)}) \xrightarrow{\mathrm{w}} \mathcal{N}(0,\tau^2\mathcal{I}_\theta) \tag{47}$$

Proof Given $t_n \to t$ in $\mathbb{R}^k$. For the application of Theorem 2.3.5, identify

$$P_{n,i} = P_\theta\,, \qquad Q_{n,i} = P_{\theta_{n,i}}\,, \qquad U_{n,i} = \tau_{n,i}t'\Lambda_\theta \tag{48}$$

Then the boundedness condition (2) is fulfilled as

$$\lim_{n\to\infty}\sum\nolimits_i \mathrm{E}_{n,i}\,U_{n,i}^2 = \tau^2 t'\mathcal{I}_\theta t$$

Since, with $\bar\tau_n = \max_i|\tau_{n,i}|$,

$$\sum\nolimits_i\int_{\{|U_{n,i}|>\varepsilon\}} U_{n,i}^2\,dP_{n,i} \le \sum\nolimits_i\tau_{n,i}^2\int_{\{|t'\Lambda_\theta|>\varepsilon/\bar\tau_n\}}(t'\Lambda_\theta)^2\,dP_\theta$$

condition (3) is satisfied due to Noether coefficients $(\tau_{n,i})$.

To verify condition (4), we explicitly introduce the remainder $r(.)$,

$$\bigl\|\sqrt{dP_{\theta+s}} - \sqrt{dP_\theta}\bigl(1+\tfrac{1}{2}s'\Lambda_\theta\bigr)\bigr\| = |s|r(s) \tag{49}$$

For later purposes, using triangle inequality and mean zero, we record that

$$|s|r(s) \ge \bigl\|\sqrt{dP_\theta}\bigl(1+\tfrac{1}{2}s'\Lambda_\theta\bigr)\bigr\| - \bigl\|\sqrt{dP_{\theta+s}}\,\bigr\| = \sqrt{1+\tfrac{1}{4}s'\mathcal{I}_\theta s}\;-1 \tag{50}$$

Right now we only make use of $\lim_{s\to 0}r(s)=0$. Then, from the triangle inequality, we obtain the bound

$$\begin{aligned}&\Bigl|\sum\nolimits_i\bigl\|\sqrt{dP_{\theta+\tau_{n,i}t_n}} - \sqrt{dP_\theta}\bigl(1+\tfrac{1}{2}\tau_{n,i}t'\Lambda_\theta\bigr)\bigr\|^2\Bigr|^{1/2}\\ &\qquad\le \Bigl|\sum\nolimits_i\tau_{n,i}^2\Bigr|^{1/2}\Bigl|\frac{1}{2}\sqrt{(t_n-t)'\mathcal{I}_\theta(t_n-t)} + |t_n|\max\nolimits_i r(\tau_{n,i}t_n)\Bigr|\end{aligned}$$

which goes to 0 due to Noether coefficients and $t_n\to t$.

Having verified (2)–(4), Lemma 2.3.2 then tells us that for all $t \in \mathbb{R}^k$,

$$\lim_{n\to\infty} \sum\nolimits_i \tau_{n,i}^2 \big|\mathrm{E}_\theta\, t'\Lambda_\theta\big|^2 = 0$$

Hence Λ_θ has mean 0 under P_θ, which also ensures the centering condition (1). Likewise, it follows from Lemma 2.3.3 that every other L_2 derivative $\Lambda_\theta^\sharp$ of $\mathcal{P}$ at θ must satisfy (15); that is, for all $t \in \mathbb{R}^k$,

$$t'\Lambda_\theta^\sharp = t'\Lambda_\theta \qquad \text{a.e. } P_\theta \tag{51}$$

Uniqueness and mean 0 of Λ_θ can also be shown directly by means of the triangle inequality and L_2 differentiability,

$$\begin{aligned} \big\|\tfrac{1}{2}t'(\Lambda_\theta^\sharp - \Lambda_\theta)\sqrt{dP_\theta}\,\big\| &\le \big\|\sqrt{dP_{\theta+t}} - \sqrt{dP_\theta}\big(1+\tfrac{1}{2}t'\Lambda_\theta^\sharp\big)\big\| \\ &\quad + \big\|\sqrt{dP_{\theta+t}} - \sqrt{dP_\theta}\big(1+\tfrac{1}{2}t'\Lambda_\theta\big)\big\| \\ &= \mathrm{o}(|t|) \end{aligned} \tag{52}$$

respectively, by writing

$$\begin{aligned} -\big\langle \tfrac{1}{2}t'\Lambda_\theta\sqrt{dP_\theta}\,\big|\sqrt{dP_\theta}\,\big\rangle &= \big\langle \sqrt{dP_{\theta+t}} - \sqrt{dP_\theta}\,\big|\sqrt{dP_\theta}\,\big\rangle + \mathrm{o}(|t|) \\ &= \big\langle \sqrt{dP_{\theta+t}}\,\big|\sqrt{dP_\theta}\,\big\rangle - 1 + \mathrm{o}(|t|) \\ &= -\frac{1}{2}\big\|\sqrt{dP_{\theta+t}} - \sqrt{dP_\theta}\,\big\|^2 + \mathrm{o}(|t|) \\ &= \mathrm{O}(|t|^2) + \mathrm{o}(|t|) = \mathrm{o}(|t|) \end{aligned} \tag{53}$$

As the Lindeberg condition (3) has already been verified, the as. normality of $\sum_i \tau_{n,i}\Lambda_\theta(x_{n,i})$ can be concluded from the Lindeberg–Feller theorem, using the Cramér–Wold device. ////

We now turn to situations when, even under the null hypothesis, the stochastically independent observations are not identically distributed. Let

$$\mathcal{P}_{n,i} = \{\, P_{n,i,\theta} \mid \theta \in \Theta \,\} \subset \mathcal{M}_1(\mathcal{A}_{n,i}) \tag{54}$$

be an array of parametric families of probability measures on general sample spaces $(\Omega_{n,i}, \mathcal{A}_{n,i})$, with a common parameter set Θ that is an open subset of some finite-dimensional $\mathbb{R}^k$. Fix $\theta \in \Theta$.

Definition 2.3.8 *The parametric array* $(\mathcal{P}_{n,i})$ *is called* L_2 differentiable *at* θ *if there exist an array of functions* $\Lambda_{n,i,\theta} \in L_2^k(P_{n,i,\theta})$ *such that*

$$\mathrm{E}_{n,i,\theta}\,\Lambda_{n,i,\theta} = 0 \tag{55}$$

for all $i = 1, \dots, i_n$ *and* $n \ge 1$, *and eventually*

$$\mathcal{I}_{n,\theta} = \sum_{i=1}^{i_n} \mathrm{E}_{n,i,\theta}\,\Lambda_{n,i,\theta}\Lambda_{n,i,\theta}{}' > 0 \tag{56}$$

and, for all $\varepsilon \in (0,\infty)$ *and all* $t \in \mathbb{R}^k$,

$$\lim_{n\to\infty} \sum_{i=1}^{i_n} \int_{\{|t'\mathcal{I}_{n,\theta}^{-1/2}\Lambda_{n,i,\theta}|>\varepsilon\}} \left|t'\mathcal{I}_{n,\theta}^{-1/2}\Lambda_{n,i,\theta}\right|^2 dP_{n,i,\theta} = 0 \tag{57}$$

and, for all $b \in (0,\infty)$,

$$\lim_{n\to\infty} \sup_{|t|\le b} \sum_{i=1}^{i_n} \left\| \sqrt{dP_{n,i,\theta_n(t)}} - \sqrt{dP_{n,i,\theta}}\bigl(1 + \tfrac{1}{2}t'\mathcal{I}_{n,\theta}^{-1/2}\Lambda_{n,i,\theta}\bigr) \right\|^2 = 0 \tag{58}$$

The array $(\Lambda_{n,i,\theta})$ *is called* L_2 *derivative of* $(\mathcal{P}_{n,i})$ *at* θ, *and the* $k\times k$ *matrix* $\mathcal{I}_{n,\theta}$ *is called the* Fisher information *of* $(\mathcal{P}_{n,i})$ *at* θ *and time* n.

Since the Fisher informations $\mathcal{I}_{n,\theta}$ need not converge in any sense, the local parameter alternatives $\theta_n(t_n)$ about θ employed in (58) as well as in the following theorem must be suitably standardized,

$$\theta_n(t_n) = \theta + \mathcal{I}_{n,\theta}^{-1/2} t_n, \qquad t_n \longrightarrow t \in \mathbb{R}^k \tag{59}$$

Corresponding product measures on $\bigl(\times_i \Omega_{n,i}, \bigotimes_i \mathcal{A}_{n,i}\bigr)$ are

$$P_\theta^{(n)} = \bigotimes_{i=1}^{i_n} P_{n,i,\theta}, \qquad P_{\theta_n}^{(n)} = \bigotimes_{i=1}^{i_n} P_{n,i,\theta_n(t_n)} \tag{60}$$

Let $x_{n,i}\colon \Omega_{n,1}\times\cdots\times\Omega_{n,i_n} \to \Omega_{n,i}$ denote the ith canonical projection.

Theorem 2.3.9 *If the parametric array* $(\mathcal{P}_{n,i})$ *is* L_2 *differentiable at* θ *with* L_2 *derivative* $(\Lambda_{n,i,\theta})$, *then the log likelihoods of the alternatives defined by* (59) *and* (60) *have the as. expansion*

$$\log \frac{dP_{\theta_n}^{(n)}}{dP_\theta^{(n)}} = t'\mathcal{I}_{n,\theta}^{-1/2} \sum_{i=1}^{i_n} \Lambda_{n,i,\theta}(x_{n,i}) - \frac{1}{2}|t|^2 + o_{P_\theta^{(n)}}(n^0) \tag{61}$$

where

$$\Bigl(\mathcal{I}_{n,\theta}^{-1/2} \sum_{i=1}^{i_n} \Lambda_{n,i,\theta}(x_{n,i})\Bigr)(P_\theta^{(n)}) \xrightarrow{\;\mathrm{w}\;} \mathcal{N}(0,\mathbb{I}_k) \tag{62}$$

PROOF Identifying

$$P_{n,i} = P_{n,i,\theta}, \qquad Q_{n,i} = P_{n,i,\theta_n(t_n)}, \qquad U_{n,i} = t'\mathcal{I}_{n,\theta}^{-1/2}\Lambda_{n,i,\theta}$$

we have again

$$\sum\nolimits_i \mathrm{E}_{n,i}\, U_{n,i}^2 = |t|^2 < \infty$$

hence (2). The Lindeberg condition (3) holds by assumption (57). Since

$$\Bigl|\sum\nolimits_i \bigl\|\sqrt{dP_{n,i,\theta_n(t_n)}} - \sqrt{dP_{n,i,\theta}}\bigl(1 + \tfrac{1}{2}t'\mathcal{I}_{n,\theta}^{-1/2}\Lambda_{n,i,\theta}\bigr)\bigr\|^2\Bigr|^{1/2} - \frac{1}{2}|t_n - t|$$
$$\le \Bigl|\sum\nolimits_i \bigl\|\sqrt{dP_{n,i,\theta_n(t_n)}} - \sqrt{dP_{n,i,\theta}}\bigl(1 + \tfrac{1}{2}t_n'\mathcal{I}_{n,\theta}^{-1/2}\Lambda_{n,i,\theta}\bigr)\bigr\|^2\Bigr|^{1/2}$$

the differentiability (4) is ensured by assumption (58). The mean zero condition (1) is guaranteed by assumption (55). The asymptotic normality (62) follows from assumption (57) and the Lindeberg–Feller theorem. ////

2.3.3 Other Differentiability Notions

Not exactly in the context of log likelihood approximations, two other notions of differentiability are occasionally useful for the parametric family $\mathcal{P}$ given in (37). L_1 differentiability plays a role in the evaluation of expectations of bounded random variables [Sections 5.4 and 6.2].

Definition 2.3.10 *Model $\mathcal{P}$ is called* L_1 differentiable *at θ if there exists some function $\Lambda_\theta \in L_1^k(P_\theta)$, necessarily $\int \Lambda_\theta \, dP_\theta = 0$, such that, as $t \to 0$,*

$$\int \bigl|dP_{\theta+t} - dP_\theta(1 + t'\Lambda_\theta)\bigr| = \mathrm{o}(|t|) \tag{63}$$

and for all $t \in \mathbb{R}^k$,

$$t'\Lambda_\theta = 0 \quad \text{a.e. } P_\theta \implies t = 0 \tag{64}$$

The function Λ_θ is called the L_1 derivative *of $\mathcal{P}$ at θ.*

Cramér–von Mises differentiability leads to a minimum distance estimate with controlled behavior on fairly large neighborhoods [Subsection 5.3.2, Section 6.3]. For this notion, which refers to distribution functions rather than measures, the sample space is supposed to be finite-dimensional Euclidean, $(\Omega, \mathcal{A}) = (\mathbb{R}^m, \mathbb{B}^m)$. Let $\mu \in \mathcal{M}(\mathbb{B}^m)$ be a possibly infinite weight.

Definition 2.3.11 *Model $\mathcal{P}$ is called* CvM (Cramér–von Mises) differentiable *at θ if there exists some function $\Delta_\theta \in L_2^k(\mu)$ such that the distribution functions, as $t \to 0$, satisfy*

$$\int \bigl|P_{\theta+t}(y) - P_\theta(y) - t'\Delta_\theta(y)\bigr|^2 \mu(dy) = \mathrm{o}(|t|^2) \tag{65}$$

and

$$\mathcal{J}_\theta = \int \Delta_\theta \Delta_\theta{}' \, d\mu > 0 \tag{66}$$

The function Δ_θ is called the CvM derivative *and the $k \times k$ matrix $\mathcal{J}_\theta$ the* CvM information *of $\mathcal{P}$ at θ.*

L_2 differentiability of $\mathcal{P}$ at θ implies L_1 differentiability with the same derivative Λ_θ. Indeed, setting $D_{\theta,t} = \sqrt{dP_{\theta+t}} - \sqrt{dP_\theta}\bigl(1 + \frac{1}{2}t'\Lambda_\theta\bigr)$ we conclude essentially as in Lemma 2.3.4 that

$$\begin{aligned}
&\int \bigl|dP_{\theta+t} - dP_\theta(1 + t'\Lambda_\theta)\bigr| \qquad\qquad (67)\\
&\quad = \int \Bigl|D_{\theta,t}\bigl(\sqrt{dP_{\theta+t}} + \sqrt{dP_\theta}\,\bigr) + \tfrac{1}{2}t'\Lambda_\theta\sqrt{dP_\theta}\bigl(\sqrt{dP_{\theta+t}} - \sqrt{dP_\theta}\,\bigr)\Bigr| \\
&\quad \le 2\,\|D_{\theta,t}\| + \frac{1}{2}\sqrt{t'\mathcal{I}_\theta t}\,\bigl\|\sqrt{dP_{\theta+t}} - \sqrt{dP_\theta}\,\bigr\| = \mathrm{o}(|t|)
\end{aligned}$$

And for $\Lambda_\theta \in L_2^k(P_\theta)$ such that $\mathrm{E}_\theta \Lambda_\theta = 0$, (64) is the same as $\mathcal{I}_\theta > 0$.

Let $\mu \in \mathcal{M}_b(\mathbb{B}^m)$ be finite. Then CvM differentiability of $\mathcal{P}$ at θ is implied by L_1 differentiability. This is clear from the following bound, which is uniform in $y \in \mathbb{R}^m$,

$$\bigl|P_{\theta+t}(y) - P_\theta(y) - t'\Delta_\theta(y)\bigr| \le \int \bigl|dP_{\theta+t} - dP_\theta(1 + t'\Lambda_\theta)\bigr| \tag{68}$$

Therefore, the CvM derivative $\Delta_\theta \colon \mathbb{R}^m \to \mathbb{R}^k$ is in this case given by

$$\begin{aligned} \Delta_\theta(y) &= \int \mathbf{I}(x \le y)\,\Lambda_\theta(x)\,P_\theta(dx) \\ &= \int \bigl(\mathbf{I}(x \le y) - P_\theta(y)\bigr)\Lambda_\theta(x)\,P_\theta(dx) \end{aligned} \tag{69}$$

The CvM information $\mathcal{J}_\theta$ based on (69) can have full rank only under (64). Conversely, in view of (69) and the uniqueness theorem for distribution functions, (64) implies $\mathcal{J}_\theta > 0$ if μ has full support. A measure μ of full support also ensures identifiability of $\mathcal{P}$ in $L_2(\mu)$ if $\mathcal{P}$ is identifiable.

Example 2.3.12

(a) The normal location family $P_\theta = \mathcal{N}(\theta, 1)$ under Dirac weight $\mu = \mathbf{I}_0$ is CvM differentiable at every $\theta \in \mathbb{R}$ with derivative $\Delta_\theta(0) = -\varphi(\theta)$ and CvM information $\mathcal{J}_\theta = \varphi^2(\theta)$, where $\varphi = \Phi'$. The measures P_θ are identified with the values $\Phi(-\theta)$ of the standard normal distribution function Φ. Thus, neither for identifiability in $L_2(\mu)$ nor for $\mathcal{J}_\theta > 0$, a weight of full support is necessary.

(b) The rectangular family $P_\theta = R(0, \theta)$, under any weight $\mu \in \mathcal{M}_1(\mathbb{B})$ such that $\mu(\{\theta\}) = 0$ for all $\theta \in (0, \infty)$, is CvM differentiable but not L_1 differentiable, at every $\theta \in (0, \infty)$. Indeed, (65) is fulfilled with the CvM derivative

$$\Delta_\theta(y) = -\frac{y}{\theta^2}\,\mathbf{I}_{(0,\theta)}(y) \qquad \text{a.e. } \mu(dy) \tag{70}$$

But (63) is not, since $P_{\theta+t}(dP_\theta = 0) \ne \mathrm{o}(|t|)$. If $\mu(\{0\}) < 1$ then $\mathcal{J}_\theta > 0$.

For i.i.d. $x_1, \dots, x_n \sim R(0, \theta)$ the largest observation $x_{n:n}$ is sufficient and converges weakly according to

$$\lim_{n\to\infty} P^n_{\theta+t/n}\bigl(n\,(\theta - x_{n:n}) \le s\bigr) = 1 - e^{-(s+t)/\theta}\,\mathbf{I}_{(-t,\infty)}(s) \tag{71}$$

for all $s, t \in \mathbb{R}$. A reduction by sufficiency thus leads to the convergence rate $1/n$ and the exponential translation family in the limit. ////

2.4 Linear Regression

As an example of a structured model we consider the linear model, in which a regression line or plane is observed with real-valued errors.

The error distribution F is required to have a finite Fisher information of location [Huber (1981; Theorem 4.2)]. This means, F is dominated by Lebesgue measure λ and has an absolutely continuous density f with derivative f' such that

$$dF = f\,d\lambda, \qquad \mathcal{I}_f = \mathrm{E}_F\,\Lambda_f^2 < \infty, \qquad \Lambda_f = -f'/f \tag{1}$$

where E_F denotes expectation under F. This assumption guarantees L_2 differentiability of the location model induced by F.

Proposition 2.4.1 *Assume* (1). *Then, as* $s \to 0$,

$$\int \Bigl| \sqrt{f}(u-s) - \sqrt{f}(u)\bigl(1 + \tfrac{1}{2}s\Lambda_f(u)\bigr)\Bigr|^2 \lambda(du) = \mathrm{o}(s^2) \tag{2}$$

PROOF With the derivative $\partial\sqrt{f} = f'/(2\sqrt{f}\,)\,\mathbf{I}(f \neq 0)$ given by C.2(3), it holds that

$$2\int_a^b |\partial\sqrt{f}\,|\,d\lambda = \int_a^b |\Lambda_f|\sqrt{f}\,d\lambda \le \sqrt{b-a}\,\Bigl|\int_a^b \Lambda_f^2\,dF\Bigr|^{1/2} < \infty$$

for all $-\infty < a < b < \infty$. Hence condition C.2(2) of Lemma C.2.2 is verified. Thus $\sqrt{f}$ is absolutely continuous on every bounded interval and has derivative $\partial\sqrt{f}$. In particular, $\sqrt{f}$ is differentiable a.e.,

$$\lim_{s\to 0} \frac{\sqrt{f}(u-s) - \sqrt{f}(u)}{s} = -\partial\sqrt{f} = \frac{1}{2}\Lambda_f(u)\sqrt{f}(u) \qquad \text{a.e. } \lambda(du) \tag{3}$$

By the fundamental theorem of calculus and Cauchy–Schwarz,

$$\begin{aligned} \int \bigl(2\sqrt{f}(u-s) - 2\sqrt{f}(u)\bigr)^2 \lambda(du) &= \int \Bigl|\int_u^{u-s} -\Lambda_f(v)\sqrt{f}(v)\,\lambda(dv)\Bigr|^2 \lambda(du) \\ &\le \int |s| \int_{u-s^-}^{u+s^+} \Lambda_f^2(v) f(v)\,\lambda(dv)\,\lambda(du) = s^2\mathcal{I}_f = \int s^2\Lambda_f^2 f\,d\lambda \end{aligned} \tag{4}$$

Thus the difference quotients, which converge a.e., are uniformly integrable. By Vitali's theorem [Proposition A.2.2], therefore, they converge in L_2 (as $s = s_n \to 0$). ////

Deterministic Carriers

In this setup, at time $n \ge 1$, we make $i_n \ge 1$ real-valued observations

$$y_{n,i} = x_{n,i}{}'\theta + u_{n,i} \tag{5}$$

with a given array of regressors $x_{n,i} \in \mathbb{R}^k$ and unknown regression parameter θ, which may range over $\Theta = \mathbb{R}^k$. Each row $u_{n,1}, \dots, u_{n,i_n}$ of errors are i.i.d. $\sim F$. Then the corresponding probabilities read

$$P_{n,i,\theta}(dy) = f(y - x_{n,i}{}'\theta)\,\lambda(dy) \tag{6}$$

Employing $\Lambda_f = -f'/f$, our candidate L_2 derivative is

$$\Lambda_{n,i,\theta}(y) = \Lambda_f(y - x_{n,i}{}'\theta)\, x_{n,i} \tag{7}$$

The corresponding Fisher information at time n would read

$$\mathcal{I}_{n,\theta} = \sum_{i=1}^{i_n} \mathrm{E}_{n,i,\theta}\, \Lambda_{n,i,\theta}\Lambda_{n,i,\theta}{}' = \mathcal{I}_f\, X_n{}'X_n \tag{8}$$

where X_n denotes the $i_n \times k$ design matrix at time n,

$$X_n{}' = (x_{n,1}, \dots, x_{n,i_n}) \tag{9}$$

Writing $Y_n{}' = (y_{n,1}, \dots, y_{n,i_n})$ and $U_n{}' = (u_{n,1}, \dots, u_{n,i_n})$, we thus obtain the vector formulation of the linear model (5),

$$Y_n = X_n\theta + U_n \tag{10}$$

It will be assumed that, for every $n \geq 1$, the design matrices have full rank,

$$\operatorname{rk} X_n = k \tag{11}$$

so that $X_n{}'X_n > 0$. Then the least squares estimator (LSE) $\hat{\theta}_n$ of θ is

$$\hat{\theta}_n = \bigl(X_n{}'X_n\bigr)^{-1} X_n{}'Y_n = \theta + \bigl(X_n{}'X_n\bigr)^{-1} X_n{}'U_n \tag{12}$$

The corresponding vector of fitted values $\widehat{Y}_n' = (\widehat{y}_{n,1}, \dots, \widehat{y}_{n,i_n})$,

$$\widehat{Y}_n = X_n\hat{\theta}_n = H_nY_n = X_n\theta + H_nU_n \tag{13}$$

is obtained by projecting the observation vector Y_n onto the column space of X_n. The $i_n \times i_n$ symmetric idempotent matrix to achieve this projection,

$$H_n = X_n\bigl(X_n{}'X_n\bigr)^{-1} X_n{}' = \mathcal{I}_f\, X_n \mathcal{I}_{n,\theta}^{-1} X_n{}' = (H_{n;i,j}) \tag{14}$$

is called a *hat matrix*. Since its diagonal bounds the off-diagonal elements,

$$H_{n;i,j}^2 \leq H_{n;i,i} \wedge H_{n;j,j}$$

the hat matrix will get uniformly small along with its diagonal,

$$\lim_{n\to\infty} \max_{i=1,\dots,i_n} H_{n;i,i} = 0 \tag{15}$$

Smallness (15) of the hat matrix is connected with several other properties of the linear regression model.

Theorem 2.4.2 *Assume $\mathcal{I}_f < \infty$ and (11): $\operatorname{rk} X_n = k$. Then, for every $\theta \in \mathbb{R}^k$, the linear model (6) is L_2 differentiable at θ, with derivative $(\Lambda_{n,i,\theta})$ given by (7), iff the hat matrix satisfies (15).*

PROOF First assuming (15), let us verify the L_2 differentiability of the array $(\mathcal{P}_{n,i,\theta})$ given by (6). Since $\mathrm{E}_F\,\Lambda_f = 0$ by Proposition 2.4.1 and Theorem 2.3.7, we have $\mathrm{E}_{n,i,\theta}\,\Lambda_{n,i,\theta} = 0$ for the functions introduced in (7); so 2.3(55) holds. That the Fisher informations (8) have full rank k, hence 2.3(56) is fulfilled, follows from $\operatorname{rk} X_n = k$. To verify the Lindeberg and differentiability conditions 2.3(57) and 2.3(58), write the result of Proposition 2.4.1 more explicitly as

$$\int \Bigl| \sqrt{f}(u-s) - \sqrt{f}(u)\,\bigl(1 + \tfrac{1}{2} s \Lambda_f(u)\bigr)\Bigr|^2 \lambda(du) = s^2\, r^2(s) \tag{16}$$

where $\lim_{s\to 0} r(s) = 0$. Then for all $t \in \mathbb{R}^k$ and $\theta_n(t) = \theta + \mathcal{I}_{n,\theta}^{-1/2} t$,

$$\begin{aligned} &\bigl\| \sqrt{dP_{n,i,\theta_n(t)}} - \sqrt{dP_{n,i,\theta}}\bigl(1 + \tfrac{1}{2} t' \mathcal{I}_{n,\theta}^{-1/2} \Lambda_{n,i,\theta}\bigr)\bigr\|^2 \\ &= \int \Bigl| \sqrt{f}\bigl(y - x_{n,i}{}'(\theta + \mathcal{I}_{n,\theta}^{-1/2} t)\bigr) - \sqrt{f}(y - x_{n,i}{}'\theta) \\ &\qquad - \tfrac{1}{2} \sqrt{f}(y - x_{n,i}{}'\theta)\, t' \mathcal{I}_{n,\theta}^{-1/2} x_{n,i} \Lambda_f(y - x_{n,i}{}'\theta) \Bigr|^2 \lambda(dy) \\ &= \int \Bigl| \sqrt{f}(u - s_{n,i}) - \sqrt{f}(u)\bigl(1 + \tfrac{1}{2} s_{n,i} \Lambda_f(u)\bigr)\Bigr|^2 \lambda(du) = s_{n,i}^2\, r^2(s_{n,i}) \end{aligned} \tag{17}$$

where

$$s_{n,i} = t' \mathcal{I}_{n,\theta}^{-1/2} x_{n,i} \tag{18}$$

These increments satisfy a Noether condition. In fact,

$$\textstyle\sum_i s_{n,i}^2 = t' \mathcal{I}_{n,\theta}^{-1/2} X_n{}' X_n \mathcal{I}_{n,\theta}^{-1/2} t = |t|^2 \mathcal{I}_f^{-1} \le b^2 \mathcal{I}_f^{-1} \tag{19}$$

and

$$\begin{aligned} \max_i s_{n,i}^2 \le\ & |t|^2 \max_i x_{n,i}{}' \mathcal{I}_{n,\theta}^{-1} x_{n,i} \ \le b^2 \max_i e_{n,i}{}' X_n \mathcal{I}_{n,\theta}^{-1} X_n{}' e_{n,i} \\ &= b^2 \mathcal{I}_f^{-1} \max_i e_{n,i}{}' H_n e_{n,i} \underset{(15)}{\longrightarrow} 0 \end{aligned} \tag{20}$$

uniformly in $|t| \le b < \infty$, where $(e_{n,i})$ denotes the canonical basis of $\mathbb{R}^{i_n}$. Now 2.3(58) follows from (17) since $\lim_{s\to 0} r(s) = 0$ and

$$\textstyle\sum_i s_{n,i}^2\, r^2(s_{n,i}) \le \max_i r^2(s_{n,i}) \sum_i s_{n,i}^2$$

The Lindeberg condition 2.3(57) is verified by the following bound,

$$\begin{aligned} &\textstyle\sum_i \displaystyle\int_{\{|t'\mathcal{I}_{n,\theta}^{-1/2} \Lambda_{n,i,\theta}| > \varepsilon\}} \bigl| t' \mathcal{I}_{n,\theta}^{-1/2} \Lambda_{n,i,\theta}\bigr|^2 \, dP_{n,i,\theta} \\ &\quad = \textstyle\sum_i \displaystyle\int_{\{|t'\mathcal{I}_{n,\theta}^{-1/2} x_{n,i} \Lambda_f(u)| > \varepsilon\}} \bigl| t' \mathcal{I}_{n,\theta}^{-1/2} x_{n,i} \Lambda_f(u)\bigr|^2 \, F(du) \\ &\quad \le \textstyle\sum_i s_{n,i}^2 \displaystyle\int_{\{|\Lambda_f(u)| > \varepsilon/\bar{s}_n\}} \Lambda_f^2(u)\, F(du) \end{aligned} \tag{21}$$

where $\bar{s}_n = \max_i |s_{n,i}|$. Due to Noether coefficients $(s_{n,i})$ and finite Fisher information $\mathcal{I}_f < \infty$, this bound goes to 0 uniformly in $|t| \le b < \infty$.

Secondly, to prove the converse, suppose that the hat matrix does not go to 0 in the sense of (15). Then there exist indices $1 \le j_n \le i_n$ such that

$$H_{n;j_n,j_n} \overset{(14)}{=} \mathcal{I}_f x_{n,j_n}{}' \mathcal{I}_{n,\theta}^{-1} x_{n,j_n} = \mathcal{I}_f \bigl| \mathcal{I}_{n,\theta}^{-1/2} x_{n,j_n} \bigr|^2$$

is bounded away from 0. Thus there exist some $\varepsilon \in (0,1)$ and suitable unit vectors $t_n \in \mathbb{R}^k$ such that

$$|s_{n,j_n}| \ge \varepsilon, \qquad s_{n,j_n} = t_n{}' \mathcal{I}_{n,\theta}^{-1/2} x_{n,j_n} \tag{22}$$

Invoking the remainder bound 2.3(50), we obtain that

$$\begin{aligned} \sum\nolimits_i & \bigl\| \sqrt{dP_{n,i,\theta_n(t_n)}} - \sqrt{dP_{n,i,\theta}}\bigl(1 + \tfrac{1}{2} t_n{}' \mathcal{I}_{n,\theta}^{-1/2} \Lambda_{n,i,\theta}\bigr) \bigr\|^2 \\ & \ge \bigl\| \sqrt{dP_{n,j_n,\theta_n(t_n)}} - \sqrt{dP_{n,j_n,\theta}}\bigl(1 + \tfrac{1}{2} t_n{}' \mathcal{I}_{n,\theta}^{-1/2} \Lambda_{n,j_n,\theta}\bigr) \bigr\|^2 \\ & \underset{(17)}{=} s_{n,j_n}^2 \, r^2(s_{n,j_n}) \underset{2.3(50)}{\ge} \Bigl| \sqrt{1 + \tfrac{1}{4}\varepsilon^2 \mathcal{I}_f} - 1 \Bigr|^2 > 0 \end{aligned} \tag{23}$$

Hence 2.3(58) cannot hold. ////

For normal error distribution $F = \mathcal{N}(0, \sigma^2)$, if $\operatorname{rk} X_n = k$, the LSE and the vector of fitted values are, by (12) and (13), exactly normal,

$$\hat{\theta}_n \sim \mathcal{N}\bigl(\theta, \sigma^2 (X_n{}' X_n)^{-1}\bigr), \qquad \widehat{Y}_n \sim \mathcal{N}(X_n \theta, \sigma^2 H_n) \tag{24}$$

For nonnormal error distribution F satisfying

$$\int u \, F(du) = 0, \qquad \sigma^2 = \int u^2 \, F(du) \in (0, \infty) \tag{25}$$

as. normality of the LSE and the fitted values is related to smallness (15) of the hat matrix; conceptually simpler is the characterization by uniform consistency of fitted values. [Huber (1973, 1977, 1981) and Bickel (1976)].

Theorem 2.4.3 *Assume* (11): $\operatorname{rk} X_n = k$, *and* (25) *for* F.

(a) *Then* (15) *holds iff for all* $\varepsilon \in (0, \infty)$,

$$\lim_{n\to\infty} \max_{i=1,\dots,i_n} P_\theta^{(n)}\bigl(|\widehat{y}_{n,i} - x_{n,i}{}'\theta| > \varepsilon\bigr) = 0 \tag{26}$$

(b) *Suppose* F *is not normal. Then* (15) *holds iff*

$$t_n{}' (X_n{}' X_n)^{1/2} (\hat{\theta}_n - \theta)(P_\theta^{(n)}) \xrightarrow{\;\mathrm{w}\;} \mathcal{N}(0, \sigma^2) \tag{27}$$

for every sequence of vectors $t_n \in \mathbb{R}^k$, $|t_n| = 1$.

(c) *Suppose F is not normal. Then (15) holds iff*

$$t_n'(\widehat{Y}_n - X_n\theta)(P_\theta^{(n)}) \xrightarrow{\mathrm{w}} \mathcal{N}(0,\sigma^2) \tag{28}$$

for every sequence of unit vectors $t_n \in \mathbb{R}^{i_n}$ in the column space of X_n.

PROOF

(a) Since $\widehat{Y}_n = X_n\theta + H_nU_n$ we have

$$\mathrm{E}\,\widehat{Y}_n = X_n\theta\,, \qquad \mathrm{Cov}\,\widehat{Y}_n = \sigma^2 H_n$$

Then (15) implies (26) by Chebyshev's inequality,

$$\max_i P_\theta^{(n)}\big(\big|\widehat{y}_{n,i} - x_{n,i}{}'\theta\big| > \varepsilon\big) \le \frac{\sigma^2}{\varepsilon^2}\max_i H_{n;i,i}$$

For the converse, let g_F, g_{U_n}, and $g_{n,j}$ denote the Fourier transforms of F, U_n, $\widehat{y}_{n,j} - x_{n,j}{}'\theta$, respectively, and $(e_{n,i}) \subset \mathbb{R}^{i_n}$ the canonical basis. Then

$$g_{n,j_n}(s) = g_{U_n}(H_n e_{n,j_n}s) = \prod\nolimits_i g_F(H_{n;i,j_n}s) \longrightarrow 1$$

for all $s \in \mathbb{R}$ and $1 \le j_n \le i_n$, since uniform consistency implies uniform weak convergence to 0. As $|g_F| \le 1$, necessarily

$$\lim_{n\to\infty}\big|g_F(H_{n;j_n,j_n}s)\big| = 1$$

Thus, if $h \neq 0$ were any nonzero cluster point of the sequence $H_{n;j_n,j_n}$, it would follow that $|g_F|$ is constant 1, hence $F = \mathbf{I}_0$ (Dirac at 0), by the well-known properties of characteristic functions.

(b) By (12) we can write

$$(X_n'X_n)^{1/2}(\hat\theta_n - \theta) = \Gamma_n U_n \tag{29}$$

where the $k \times i_n$ matrix Γ_n is given by,

$$\Gamma_n = (X_n'X_n)^{-1/2}X_n' \tag{30}$$

and satisfies

$$\Gamma_n\Gamma_n' = \mathbb{I}_k\,, \qquad \Gamma_n'\Gamma_n = H_n \tag{31}$$

Hence, for any unit vectors $t_n \in \mathbb{R}^k$, we obtain the representation

$$t_n'(X_n'X_n)^{1/2}(\hat\theta_n - \theta) = \sum\nolimits_i \gamma_{n,i}u_{n,i} \tag{32}$$

with coefficients

$$\gamma_{n,i} = t_n'\Gamma_n e_{n,i}\,, \qquad \sum\nolimits_i \gamma_{n,i}^2 = 1$$

and i.i.d. random summands $u_{n,i} \sim F$ $(i = 1, \ldots, i_n)$.

Now assume (15). Then the coefficients $(\gamma_{n,i})$ are Noether,

$$\bar{\gamma}_n^2 = \max_i \gamma_{n,i}^2 \le \max_i |\Gamma_n e_{n,i}|^2 = \max_i H_{n:i,i} \longrightarrow 0$$

Because of this and

$$\sum_i \int_{\{|\gamma_{n,i}u_{n,i}|>\varepsilon\}} (\gamma_{n,i}u_{n,i})^2 F(du_{n,i}) \le \sum_i \gamma_{n,i}^2 \int_{\{|u|>\varepsilon/\bar{\gamma}_n\}} u^2 F(du)$$

the Lindeberg–Feller theorem yields the asserted as. normality (27).

Conversely, if (15) is violated, thus

$$\max_i |\Gamma_n e_{n,i}|^2 = \max_i H_{n;i,i} \not\longrightarrow 0$$

there exist indices $1 \le j_n \le i_n$ and unit vectors $t_n \in \mathbb{R}^k$ such that

$$|\gamma_{n,j_n}|^2 = \left|t_n{}'\Gamma_n e_{n,j_n}\right|^2 \not\longrightarrow 0$$

By the Chebyshev inequality, as $\sum_i \gamma_{n,i}^2 = 1$ and $\sigma^2 < \infty$, the two sequences

$$\sum \{\, \gamma_{n,i}u_{n,i} \mid 1 \le i \le i_n \,,\, i \ne j_n \,\}, \qquad \gamma_{n,j_n}u_{n,j_n}$$

are tight in $\mathbb{R}$. By Prokhorov's theorem (direct half) [Proposition A.1.2 a] they converge weakly along some subsequence (m) to limits V and γ_0, respectively,

$$\sum \{\, \gamma_{m,i}u_{m,i} \mid 1 \le i \le i_m \,,\, i \ne j_m \,\} \xrightarrow{\;\mathrm{w}\;} V\,, \qquad \gamma_{m,j_m} \longrightarrow \gamma_0$$

From stochastic independence of $u_{m,i}$ $(i = 1, \ldots, i_m \,,\, i \ne j_m)$ and u_{m,j_m} it follows that

$$\sum_i \gamma_{m,i}u_{m,i} \xrightarrow{\;\mathrm{w}\;} V * \gamma_0 U \tag{33}$$

where the random variable $U \sim F$ is not normal and $\gamma_0 \ne 0$. But then the Cramér–Lévy theorem [Feller (1966, p525)] tells us that the convolution $V * \gamma_0 U$ cannot be normal either.

(c) For vectors $t_n \in \mathbb{R}^{i_n}$, $|t_n| = 1$, $H_n t_n = t_n$, it follows from (13) that

$$t_n{}'(\widehat{Y}_n - X_n\theta) = t_n{}'U_n \tag{34}$$

As shown under (b), the linear combination $t_n{}'U_n$ is as. normal iff the coefficients $(t_{n,i})$ get uniformly small; that is, they are Noether.

Making use of $H_n{}' = H_n = H_n^2$ and the Cauchy–Schwarz inequality, write out $t_n = H_n t_n$ for coordinate number $1 \le j \le i_n$ to obtain that

$$t_{n,j}^2 = \Bigl|\sum_i H_{n;j,i}t_{n,i}\Bigr|^2 \le \sum_i H_{n;j,i}H_{n;i,j} \sum_i t_{n,i}^2 = H_{n;j,j} \tag{35}$$

Hence (15) implies that $\mathcal{L}(t_n{}'U_n) \xrightarrow{\;\mathrm{w}\;} \mathcal{N}(0,\sigma^2)$.

Conversely, if $H_{n;j_n,j_n}$ is bounded away from 0, consider the $j_n{}^{\text{th}}$ column vector $s_n = H_n e_{n,j_n}$ of H_n, which is in the column space of X_n and satisfies $|s_n| \le 1$. Let t_n be s_n rescaled so that $|t_n| = 1$. Then $t_n{}'U_n$ certainly cannot be as. normal since $|t_{n,j_n}| \ge H_{n;j_n,j_n}$. ////

Random Carriers

The linear model may be brought back to the i.i.d. case by treating the regressors as random variables. In this setup, at time $n \geq 1$, we make n i.i.d. observations $z_i = (x_i, y_i) \in \mathbb{R}^{k+1}$ of the form

$$y_i = x_i{}'\theta + u_i \tag{36}$$

where $x_1, \ldots, x_n$ are i.i.d. realizations of the regressor x distributed according to some probability K on $\mathbb{B}^k$, and $u_1, \ldots, u_n$ are i.i.d. copies of the error $u \sim F$. It is assumed that

$$x \sim K \quad \text{and} \quad u \sim F \quad \text{are stochastically independent} \tag{37}$$

The corresponding parametric model with parameter $\theta \in \mathbb{R}^k$ thus reads

$$P_\theta(dx, dy) = F(dy - x'\theta)\, K(dx) = f(y - x'\theta)\, \lambda(dy)\, K(dx) \tag{38}$$

The assumption that $\mathcal{I}_f < \infty$ is taken over. Then define

$$\Lambda_\theta(x, y) = \Lambda_f(y - x'\theta)\, x \tag{39}$$

About the regressor marginal K make the assumption that

$$\mathcal{K} = \int xx'\, K(dx) \in \mathbb{R}^{k \times k}, \qquad \operatorname{rk} \mathcal{K} = k \tag{40}$$

The finiteness of this second moment matrix is related to the previous smallness of the hat matrix, while the full rank condition is similar to the rank k condition on the previous design matrices. In the present setup, the assumption that $\mathcal{K} > 0$ is equivalent to identifiability.

Lemma 2.4.4 *Model* (38) *is identifiable iff for all* $\theta \in \mathbb{R}^k$,

$$x'\theta = 0 \quad \text{a.e. } K(dx) \implies \theta = 0 \tag{41}$$

PROOF We have to show that for all $s, t \in \mathbb{R}^k$,

$$P_t = P_s \iff x't = x's \quad \text{a.e. } K(dx) \tag{42}$$

That the RHS entails equality of the two measures is obvious since

$$P_\theta(x \in A,\, y \in B) = \int_A F(B - x'\theta)\, K(dx) \tag{43}$$

for $\theta = s, t$ and all $A \in \mathbb{B}^k$, $B \in \mathbb{B}$. Conversely, if $P_t = P_s$ and $B \in \mathbb{B}$, insert into (43) the domain

$$A = \{\, x \in \mathbb{R}^k \mid F(B - x't) \neq F(B - x's) \,\}$$

to obtain that $K(A) = 0$. Thus, for every $B \in \mathbb{B}$ there exists some set $A_B \in \mathbb{B}^k$ such that $K(A_B) = 0$ and

$$F(B - x't) = F(B - x's), \qquad x \in \mathbb{R}^k \setminus A_B$$

Employ countably many B_n that generate $\mathbb{B}$ and are closed under finite intersections. Then $K(D) = 0$ still holds for $D = \bigcup_n A_{B_n}$ while

$$F(B_n - x't) = F(B_n - x's), \qquad x \in \mathbb{R}^k \setminus D,\ n \geq 1$$

By the uniqueness theorem, the shifted probabilities coincide,

$$F(dy - x't) = F(dy - x's), \qquad x \in \mathbb{R}^k \setminus D$$

In terms of the Fourier transform $\widehat{F}$ of F this means that

$$e^{iux't}\widehat{F}(u) = e^{iux's}\widehat{F}(u), \qquad u \in \mathbb{R},\ x \in \mathbb{R}^k \setminus D \tag{44}$$

and from this one may conclude that $x't = x's$ for all $x \in \mathbb{R}^k \setminus D$. ////

Remark 2.4.5 More generally, given some matrix $D \in \mathbb{R}^{p \times k}$, identifiability of the linear transform $D\theta$ of the regression parameter θ means that for all $s, t \in \mathbb{R}^k$,

$$P_t = P_s \implies Dt = Ds \tag{45}$$

Equivalently, $D\theta = 0$ is implied by $x'\theta = 0$ a.e. $K(dx)$, that is by $\mathcal{K}\theta = 0$. Thus (45) is realized iff

$$D = A\mathcal{K} \tag{46}$$

for some matrix $A \in \mathbb{R}^{p \times k}$. ////

In view of (39) and (40), Fisher information would read

$$\mathcal{I}_\theta = \mathcal{I}_f \mathcal{K} \tag{47}$$

Theorem 2.4.6 *Assume* $\mathcal{I}_f < \infty$ and (40). *Then the linear model* (38) *is* L_2 *differentiable at every* $\theta \in \mathbb{R}^k$, *with* L_2 *derivative* Λ_θ *given by* (39) *and Fisher information* $\mathcal{I}_\theta$ *given by* (47).

Proof Under the following sample space transformations, so-called *regression translations*,

$$g_\theta(x, y) = (x, x'\theta + y) \tag{48}$$

the model stays invariant, in the sense that, for all $\theta \in \mathbb{R}^k$,

$$P_\theta = g_\theta(P_0), \qquad \Lambda_\theta = \Lambda_0 \circ g_{-\theta} \tag{49}$$

This invariance allows a reduction to the case $\theta = 0$. Invoking the explicit remainder (16) and bound (4) it thus follows that

$$\begin{aligned}
&\big\|\sqrt{dP_{\theta+t}} - \sqrt{dP_\theta}\big(1 + \tfrac{1}{2}t'\Lambda_\theta\big)\big\|^2 \\
&\quad = \iint \Big|\sqrt{f}(u - x't) - \sqrt{f}(u)\big(1 + \tfrac{1}{2}x't\Lambda_f(u)\big)\Big|^2 \lambda(du)\, K(dx) \\
&\quad \le \int_{\{|x|\le M\}} (x't)^2\, r^2(x't)\, K(dx) \\
&\qquad + 2\int_{\{|x|>M\}} \int \big(\sqrt{f}(u - x't) - \sqrt{f}(u)\big)^2 \lambda(du)\, K(dx) \\
&\qquad + \frac{1}{2}\int_{\{|x|>M\}} \int (x't)^2 \Lambda_f^2(u)\, F(du)\, K(dx) \\
&\quad \overset{(4)}{\le} |t|^2 \Big| \operatorname{tr}\mathcal{K} \sup_{|x|\le M} r^2(x't) + \mathcal{I}_f \int_{\{|x|>M\}} |x|^2\, K(dx) \Big|
\end{aligned} \tag{50}$$

for every $M \in (0, \infty)$. By noting that

$$\lim_{t\to 0} \sup_{|x|\le M} r^2(x't) = 0, \qquad \lim_{M\to\infty} \int_{\{|x|>M\}} |x|^2\, K(dx) = 0$$

the proof is concluded. ////

Remark 2.4.7 L_2 differentiablity for generalized linear models has been worked out by Schlather (1994), and for ARMA time series models by Staab (1984) and Hummel (1992), among others. ////

Chapter 3

Asymptotic Statistics

3.1 General Remarks

This chapter derives the asymptotic statistical optimality theorems in the parametric case: Convolution and asymptotic minimax theorems for estimation, and asymptotic maximin theorems for testing hypotheses.

These mathematical results, which are mainly due to LeCam and Hájek, have originally been developed in a finite-dimensional parametric framework. For a review see the introduction to Hájek (1972). We give easier proofs. Subsequent extensions to nonparametric statistics with certain infinite-dimensional aspects rely on the basic parametric versions.

3.2 Convolution Representation

Fisher conjectured that the maximum likelihood estimator minimizes the as. variance among all as. normal estimates. This, however, is not true without further regularity conditions. An early example of an as. normal, but superefficient, estimator is attributed to Hodges by LeCam (1953).

Example 3.2.1 For unknown mean $\theta \in \mathbb{R}^k$ of finite dimension k and i.i.d. observations

$$x_1, \ldots, x_n \sim \mathcal{N}(\theta, \mathbb{I}_k) = P_\theta$$

consider the following modification of the sample mean $\bar{X}_n$,

$$S_n = \begin{cases} \bar{X}_n & \text{if } n\,|\bar{X}_n|^4 > 1 \\ 0 & \text{otherwise} \end{cases} \tag{1}$$

which is obtained by a particular shrinkage of $\bar{X}_n$ towards 0. By the central limit theorem, $\sqrt{n}\,\bar{X}_n$ is under P_0^n as. normal $\mathcal{N}(0, \mathbb{I}_k)$. It follows that, in case $\theta = 0$,

$$P_0^n(S_n = 0) = P_0^n\big(\sqrt{n}\,|\bar{X}_n| \le \sqrt[4]{n}\big) \longrightarrow 1$$

whereas in case $\theta \ne 0$,

$$\begin{aligned} P_\theta^n(S_n = \bar{X}_n) &= P_0^n\big(\sqrt[4]{n}\,|\bar{X}_n + \theta| > 1\big) \\ &\ge P_0^n\big(\sqrt[4]{n}\,|\theta| - \sqrt[4]{n}\,|\bar{X}_n| > 1\big) \\ &= P_0^n\big(\sqrt{n}\,|\bar{X}_n| < \sqrt{n}\,|\theta| - \sqrt[4]{n}\big) \longrightarrow 1 \end{aligned}$$

Therefore,

$$\sqrt{n}\,(S_n - \theta)(P_\theta^n) \xrightarrow{\mathrm{w}} \begin{cases} \mathcal{N}(0, \mathbb{I}_k) & \text{if } \theta \ne 0 \\ 0 & \text{if } \theta = 0 \end{cases} \tag{2}$$

Thus S is superefficient at 0, while seemingly not worse than $\bar{X}$ at other parameter values. Note however that for every $t \in \mathbb{R}^k$,

$$\sqrt{n}\,S_n - t \xrightarrow{P^n_{t/\sqrt{n}}} -t \tag{3}$$

since $\sqrt{n}\,S_n \to 0$ in P_0^n probability and $\big(P^n_{t/\sqrt{n}}\big) \ll (P_0^n)$. This estimator, therefore, will not be regular in the sense of the following definition. ////

Given a sequence of statistical models on sample spaces $(\Omega_n, \mathcal{A}_n)$,

$$\mathcal{Q}_n = \{\, Q_{n,t} \mid t \in \mathbb{R}^k \,\} \subset \mathcal{M}_1(\mathcal{A}_n)$$

with the same finite-dimensional parameter spaces $\Theta_n = \mathbb{R}^k$ (or $\Theta_n \uparrow \mathbb{R}^k$), the parameter of interest is Dt for some $p \times k$ matrix D of full rank $p \le k$. Estimators of Dt in the as. setup in fact arise as sequences of estimates, one for each time or sample size $n \ge 1$,

$$S = (S_n), \qquad S_n \colon (\Omega_n, \mathcal{A}_n) \longrightarrow (\mathbb{R}^p, \mathbb{B}^p)$$

When centered at the estimand, the as. estimators are required to converge weakly to the same limit, for every parameter value. This notion of regularity, which amounts to a certain equivariance of the limit law, and thus expresses some kind of stability in the parametric framework, will effectively rule out superefficiency. The regularity definition of course depends on the statistical models $(\mathcal{Q}_n)$ and the estimand matrix D. [For the subsequent achievement of lower risk and upper power bounds in Sections 3.3 and 3.4, condition (4) may even be required to hold uniformly on t-compacts of R^k.]

Definition 3.2.2 *An as. estimator S is called* regular *for the parameter transform D, with limit law $M \in \mathcal{M}_1(\mathbb{B}^p)$, if for all $t \in \mathbb{R}^k$,*

$$(S_n - Dt)(Q_{n,t}) \xrightarrow{\mathrm{w}} M \tag{4}$$

that is, $S_n(Q_{n,t}) \xrightarrow{\mathrm{w}} M * \mathbf{I}_{Dt}$ as $n \to \infty$, for every $t \in \mathbb{R}^k$.

In the as. normal setup [Definition 2.2.9], the as. distribution of a regular estimator will be more spread out than a certain most concentrated limit law. That no decision-theoretic concepts, like loss functions, are needed, may be considered an advantage of this efficiency concept.

Theorem 3.2.3 *Assume models $(\mathcal{Q}_n)$ that are as. normal with as. covariance $C > 0$ and as. sufficient statistic Z. Let $D \in \mathbb{R}^{p \times k}$ be a matrix of rank $p \leq k$. Let the as. estimator S be regular for D with limit law M. Then there exists a probability $M_0 \in \mathcal{M}_1(\mathbb{B}^p)$ such that*

$$M = M_0 * \mathcal{N}(0, \Gamma), \qquad \Gamma = DC^{-1}D' \tag{5}$$

and

$$\big(S_n - DC^{-1}Z_n\big)(Q_{n,0}) \xrightarrow{\mathrm{w}} M_0 \tag{6}$$

An as. estimator $S^\star$ is regular for D and achieves limit law $M^\star = \mathcal{N}(0, \Gamma)$ iff

$$S_n^\star = DC^{-1}Z_n + \mathrm{o}_{Q_{n,0}}(n^0) \tag{7}$$

Remark 3.2.4

(a) The specialization of the convolution theorem to the smooth parametric i.i.d. case is spelled out in Proposition 4.2.19 a.

(b) In the case $p = 1$, by use of the Neyman–Pearson lemma, an analogue concentration bound based on two-sided confidence intervals can be derived for estimators that are (in a locally uniform sense) as. *median unbiased* [Pfanzagl and Wefelmeyer (1982; Section 9.2)]. ////

PROOF We present three variants (a), (b), and (c) of the proof. The first appears to be in the spirit of Hájek (1970), with somewhat modified arguments; in particular, we dispense with the approximation by exponential families. Statement (6), however, does not seem to be in the reach of this proof and will therefore be shown by means of Fourier transforms.

Fourier transforms and their analytic properties have been employed by Droste and Wefelmeyer (1982), which is the second variant, and by Peter Bickel according to Roussas (1972), which is the third variant of the proof.

In the abstract theory [Millar (1983; pp 95–99, 137–140)], the convolution theorem is a special case of a representation of invariant kernels, which arise as transitions from the translation family $\mathcal{E}$ of limit laws of a distinguished statistic on one hand, to the translation family $\mathcal{F}$ of limit laws of a regular estimator on the other hand; the distance $\delta(\mathcal{E}, \mathcal{F})$ being zero.

(a) Under $t = 0$, both the regular estimator S and the as. sufficient statistic Z converge weakly. Hence the joint laws of $(S_n - DC^{-1}Z_n)$ and $C^{-1}Z_n$ under $Q_{n,0}$ are in $\mathbb{R}^p \times \mathbb{R}^k$ uniformly tight. Along some subsequence (m), therefore, these laws converge weakly [Prokhorov],

$$\big(S_m - DC^{-1}Z_m, C^{-1}Z_m\big)(Q_{m,0}) \xrightarrow{\mathrm{w}} K \in \mathcal{M}_1(\mathbb{B}^p \otimes \mathbb{B}^k) \tag{8}$$

By 2.2(37), K has second marginal $\mathcal{N}(0, C^{-1})$. Invoking expansion 2.2(38) of the log likelihoods $L_{m,t} \in \log dQ_{m,t}/dQ_{m,0}$, we conclude from (8) that

$$\big(S_m, L_{m,t}\big)(Q_{m,0}) \xrightarrow{\ \mathrm{w}\ } K_t \tag{9}$$

for all $t \in \mathbb{R}^k$, and $K_t = \alpha_t(K)$ the image measure of K under the map

$$\alpha_t(s,x) = \big(s + Dx, t'Cx - \tfrac{1}{2}t'Ct\big), \qquad s \in \mathbb{R}^p,\ x \in \mathbb{R}^k$$

Since $(Q_{m,t}) \ll (Q_{m,0})$ by Proposition 2.2.12, Theorem 2.2.5 yields

$$\big(S_m, L_{m,t}\big)(Q_{m,t}) \xrightarrow{\ \mathrm{w}\ } N_t(ds, du) = e^u\, K_t(ds, du) \tag{10}$$

Concerning the first marginal, the assumed estimator regularity says that we must have $M = \beta_t(N_t)$ for all $t \in \mathbb{R}^k$, where the map β_t is given by

$$\beta_t(s,u) = s - Dt\,, \qquad s \in \mathbb{R}^p,\ u \in \bar{\mathbb{R}}$$

Thus, with π_2 denoting the canonical projection onto the second coordinate, the M integral of any bounded continuous function $f\colon \mathbb{R}^p \to \mathbb{R}$, for all $t \in \mathbb{R}^k$, reads

$$\begin{aligned} \int f\, dM = \int f \circ \beta_t\, dN_t &= \int (f \circ \beta_t \circ \alpha_t) \exp(\pi_2 \circ \alpha_t)\, dK \\ &= \int f(s + Dx - Dt) \exp(t'Cx - \tfrac{1}{2}t'Ct)\, K(ds, dx) \end{aligned} \tag{11}$$

Now introduce the regular conditional distribution of the first coordinate given the second, denoted by the Markov kernel $K(ds|x)$, so that

$$K(ds, dx) = K(ds|x)\, \mathcal{N}(0, C^{-1})(dx)$$

and recall Example 2.2.11. Thus, for any $f \in \mathcal{C}(\mathbb{R}^p)$ and all $t \in \mathbb{R}^k$,

$$\int f\, dM = \int f(s + Dx - Dt)\, K(ds|x)\, \mathcal{N}(t, C^{-1})(dx) \tag{12}$$

We now put a prior $\pi \in \mathcal{M}_1(\mathbb{B}^k)$ on the mean $t \in \mathbb{R}^k$, by invoking two stochastically independent random vectors $T \sim \pi$ and $Y \sim \mathcal{N}(0, C^{-1})$. Thus, their sum $X = T + Y$ has $\mathcal{L}(X|T = t) = \mathcal{N}(t, C^{-1})$ a.e. $\pi(dt)$ as conditional distribution given T. Applying Fubini's theorem for Markov kernels to the joint law $\mathcal{L}(T, X)$, and integrating (12) w.r.t. $\pi(dt)$, yields

$$\begin{aligned} \int f\, dM &= \int f(s + Dx - Dt)\, K(ds|x)\ \mathcal{L}(X|T = t)(dx)\ \mathcal{L}(T)(dt) \\ &= \int f(s + Dx - Dt)\, K(ds|x)\ \mathcal{L}(T, X)(dt, dx) \\ &= \int f(s + Dx - Dt)\ \mathcal{L}(T|X = x)(dt)\, K(ds|x)\ \mathcal{L}(X)(dx) \end{aligned} \tag{13}$$

Specifying normal priors $\pi = \pi_m = \mathcal{N}(0, mC^{-1})$, the posterior distribution of T is $\mathcal{L}(T|X = x) = \mathcal{N}(rx, rC^{-1})$ a.e. $\lambda^k(dx)$, with $r = m/(m+1)$ [Example 2.2.8; $B = C^{-1}$].

Upon these specifications, (13) yields, for any $f \in \mathcal{C}(\mathbb{R}^p)$, the following representation of the M integral $\int f\,dM$; namely,

$$\int f(s + Dx - Dt)\,\mathcal{N}(rx, rC^{-1})(dt)\,K(ds|x)\,\mathcal{N}(0, [m+1]C^{-1})(dx) \tag{14}$$
$$= \int f\big(s + (1-r)Dx - \vartheta\big)\,\mathcal{N}(0, r\Gamma)(d\vartheta)\,K(ds|x)\,\mathcal{N}(0, [m+1]C^{-1})(dx)$$

where $\Gamma = DC^{-1}D'$. Since f is bounded, and $\mathcal{N}(0, r\Gamma) \to \mathcal{N}(0, \Gamma)$ in total variation as $m \to \infty$ [Scheffé's Lemma A.2.4], it follows that

$$\int f\,dM + \mathrm{o}(m^0) \tag{15}$$
$$= \int f\big(s + (1-r)Dx - \vartheta\big)\,\mathcal{N}(0, \Gamma)(d\vartheta)\,K(ds|x)\,\mathcal{N}(0, [m+1]C^{-1})(dx)$$

Now choose constants $c_m \in (0, \infty)$ such that

$$c_m/\sqrt{m} \longrightarrow \infty, \qquad c_m/m \longrightarrow 0$$

Then

$$\mathcal{N}(0, [m+1]C^{-1})\big(|x| \ge c_m\big) \le \frac{m+1}{c_m^2}\,\mathrm{tr}\,C^{-1} \longrightarrow 0$$

while

$$\sup_{|x| \le c_m} \big|(1-r)Dx\big| \le \frac{c_m}{m+1}\|D\| \longrightarrow 0$$

Therefore, if f is bounded and even uniformly continuous, the increment $(1-r)Dx$ in the argument of the integrand in (15) becomes negligible. So

$$\int f\,dM = \int_{\mathbb{R}^p} g(s)\,M_m(ds) + \mathrm{o}(m^0) \tag{16}$$

where

$$g(s) = \int_{\mathbb{R}^p} f(s - \vartheta)\,\mathcal{N}(0, \Gamma)(d\vartheta)$$

and

$$M_m(ds) = \int_{\mathbb{R}^k} K(ds|x)\,\mathcal{N}(0, [m+1]C^{-1})(dx)$$

Now suppose f is continuous and even has compact support. Then on one hand, g is continuous on $\mathbb{R}^p$ and tends to 0 for $|s| \to \infty$; hence g may be extended continuously to $\bar{\mathbb{R}}^p$ by the value 0 for $|s| = \infty$. On the other hand, the probabilities M_m, relative to weak convergence in $\bar{\mathbb{R}}^p$, certainly have a cluster point $M_0 \in \mathcal{M}_1(\bar{\mathbb{B}}^p)$. Thus we arrive at the representation

$$\int f\,dM = \int f(s - \vartheta)\,\mathcal{N}(0, \Gamma)(d\vartheta)\,M_0(ds) \tag{17}$$

which holds for all $f \in \mathcal{C}(\bar{\mathbb{R}}^p)$ of compact supports (in $\mathbb{R}^p$). The LHS, hence also the RHS, tends to $M(\mathbb{R}^p) = 1$ if we let $0 \le f \uparrow \mathbf{I}_{\mathbb{R}^p}$. But since the corresponding functions satisfy $0 \le g \le 1$ and vanish at infinity, this can only happen if $M_0(\bar{\mathbb{R}}^p \setminus \mathbb{R}^p) = 0$. So (5) is proved.

(b) Into (12) plug the trigonometric functions

$$f_h(z) = \exp(ih'z), \qquad z \in \mathbb{R}^p,\ h \in \mathbb{R}^p$$

Thus the Fourier transforms $\widehat{M}$ and $\widehat{K}(.\,|x)$ of M, $K(ds|x)$, respectively, satisfy the following identity for all $t \in \mathbb{R}^k$ and $h \in \mathbb{R}^p$,

$$\widehat{M}(h) = \exp(-ih'Dt) \int \widehat{K}(h|x) \exp(ih'Dx)\, \mathcal{N}(t, C^{-1})(dx) \tag{18}$$

Inserting the known form of the characteristic function $\widehat{\mathcal{N}}$ of the normal,

$$\widehat{\mathcal{N}}(t, C^{-1})(D'h) = \exp(ih'Dt - \tfrac{1}{2} h'\Gamma h), \qquad \Gamma = DC^{-1}D'$$

we obtain that, for all $t \in \mathbb{R}^k$ and $h \in \mathbb{R}^p$,

$$\begin{aligned} \exp(\tfrac{1}{2} h'\Gamma h)\, \widehat{M}(h) \int \exp(ih'Dx)\, \mathcal{N}(t, C^{-1})(dx) \\ = \int \widehat{K}(h|x) \exp(ih'Dx)\, \mathcal{N}(t, C^{-1})(dx) \end{aligned} \tag{19}$$

Now invoke completeness of the normal family $\{\, \mathcal{N}(t, C^{-1}) \mid t \in \mathbb{R}^k \,\}$. Thus

$$\widehat{K}(h|x) = \exp(\tfrac{1}{2} h'\Gamma h)\, \widehat{M}(h) \qquad \text{a.e. } \lambda^k(dx) \tag{20}$$

for all $h \in \mathbb{R}^p$. Inserting a sequence $h = h_n$ which is dense in $\mathbb{R}^p$, and using the continuity of characteristic functions, we obtain a Borel set $A \in \mathbb{B}^k$ of Lebesgue measure zero such that for all $x \in \mathbb{R}^k \setminus A$ and all $h \in \mathbb{R}^p$,

$$\widehat{K}(h|x) = \exp(\tfrac{1}{2} h'\Gamma h)\, \widehat{M}(h) \tag{21}$$

The RHS not depending on x, it follows that

$$K(ds|x) = M_0(ds) \qquad \text{a.e. } \mathcal{N}(0, C^{-1})(dx)$$

where M_0 is the first marginal of K. Therefore K is the product of its marginals,

$$K(ds, dx) = M_0(ds) \otimes \mathcal{N}(0, C^{-1})(dx) \tag{22}$$

Like the RHS of (21), K now turns out to be independent of the particular subsequence (m).

Insert (22) into (8) and take the first component to obtain (6). To get (5), add D times the second component to the first. Then (5) and (6) imply (7) since

$$M = \mathcal{N}(0, \Gamma) \overset{(5)}{\iff} M_0 = \mathbf{I}_0 \overset{(6)}{\iff} Y_n(Q_{n,0}) \xrightarrow{\mathrm{w}} 0 \iff Y_n \xrightarrow{Q_{n,0}} 0$$

for $Y_n = S_n^\star - DC^{-1} Z_n$.

(c) Because the regular estimator converges weakly under $t = 0$ and the as. sufficient statistic is as. normal, the laws $(S_n, Z_n)(Q_{n,0})$ are uniformly tight in $\mathbb{R}^p \times \mathbb{R}^k$, hence converge weakly along a subsequence (m) to some limit $\mathcal{L}(\bar{S}, \bar{Z})$. Necessarily $\mathcal{L}(\bar{S}) = M$ and $\mathcal{L}(\bar{Z}) = \mathcal{N}(0, C)$. In view of the expansion of log likelihoods, it follows that for all $t \in \mathbb{R}^k$,

$$(S_m, L_{m,t})(Q_{m,0}) \xrightarrow{\mathrm{w}} \mathcal{L}\big(\bar{S},\, t'\bar{Z} - \tfrac{1}{2}t'Ct\big) \tag{23}$$

Now fix $t \in \mathbb{R}^k$ and $u \in \mathbb{R}^p$, and introduce the random variables

$$\begin{aligned} X_m &= \exp\big(iu'S_m - iu'Dt + L_{m,t}\big) \\ X &= \exp\big(iu'\bar{S} - iu'Dt + t'\bar{Z} - \tfrac{1}{2}t'Ct\big) \end{aligned} \tag{24}$$

Due to (23), we have

$$X_m(Q_{m,t}) \xrightarrow{\mathrm{w}} \mathcal{L}(X) \tag{25}$$

Moreover, since $Q_{m,0}(L_{m,t} = \infty) = 0$ and $(Q_{m,t}) \ll (Q_{m,0})$, it holds that

$$\begin{aligned} \int |X_m|\, dQ_{m,0} &= \int \exp L_{m,t}\, dQ_{m,0} = Q_{m,t}(L_{m,t} < \infty) \\ &= 1 - Q_{m,t}(L_{m,t} = \infty) \longrightarrow 1 = \mathrm{E}\,|X| \end{aligned} \tag{26}$$

where we use

$$\mathrm{E} \exp t'\bar{Z} = \exp \tfrac{1}{2}t'Ct \tag{27}$$

the well-known Laplace transform of $\bar{Z} \sim \mathcal{N}(0, C)$. From equations (25) and (26), uniform integrability follows, and Corollary A.2.3 [Vitali] applies, so that

$$\int X_m\, dQ_{m,0} \longrightarrow \mathrm{E}\, X \tag{28}$$

But it is also true that

$$\begin{aligned} \int X_m\, dQ_{m,0} &= \int \exp\big(iu'S_m - iu'Dt + L_{m,t}\big)\, dQ_{m,0} \\ &= \int \exp\big(iu'(S_m - Dt)\big)\, dQ_{m,t} + \mathrm{o}(m^0) \longrightarrow \widehat{M}(u) \end{aligned} \tag{29}$$

where regularity of the estimator is expressed in terms of characteristic functions, $\widehat{M}$ denoting the Fourier transform of its limit $M = \mathcal{L}(\bar{S})$. Equality holds at the second instance because $\big|\exp\big(iu'(S_m - Dt)\big)\big| \le 1$ and the contiguity $(Q_{m,t}) \ll (Q_{m,0})$ implies that $Q_{m,t}(L_{m,t} = \infty) \to 0$.

For the limit we thus obtain the following representation

$$\widehat{M}(u) = \exp\big(-iu'Dt - \tfrac{1}{2}t'Ct\big)\ \mathrm{E} \exp(iu'\bar{S} + t'\bar{Z}) \tag{30}$$

holding for all $u \in \mathbb{R}^p$, $t \in \mathbb{R}^k$. Now it is a fact on exponential families that for each $u \in \mathbb{R}^p$, the RHS is coordinatewise holomorphic in $t \in \mathbb{C}^k$. By the uniqueness theorem for holomorphic functions therefore, the representation extends to all $t \in \mathbb{C}^k$. Given $u \in \mathbb{R}^p$, plug in $t = -i\, C^{-1} D' u$ to get

$$\widehat{M}(u) = \widehat{M_0}(u) \exp(-\tfrac{1}{2} u' \Gamma u) \tag{31}$$

where

$$\widehat{M_0}(u) = \mathrm{E} \exp\bigl(iu'(\bar{S} - DC^{-1} \bar{Z}\,)\bigr) \tag{32}$$

is the characteristic function of $M_0 = \mathcal{L}(\bar{S} - DC^{-1}\bar{Z})$. This proves (5).

To prove (6) consider any weak cluster point $M_1 \in \mathcal{M}_1(\mathbb{B}^p)$ of the sequence $\bigl(S_n - DC^{-1} Z_n\bigr)(Q_{n,0})$, which is tight in $\mathbb{R}^p$. Hence these laws converge weakly to M_1 along some subsequence (r). Passing to a further subsequence, it may be achieved that the joint laws $(S_r, Z_r)(Q_{r,0})$ tend weakly to some limit $\mathcal{L}(\tilde{S}, \tilde{Z})$. According to (32) and (31), the characteristic function of $\mathcal{L}(\tilde{S} - DC^{-1}\tilde{Z}) = M_1$ must satisfy the identity

$$\widehat{M_1}(u) = \widehat{M}(u) \exp(\tfrac{1}{2} u' \Gamma u) = \widehat{M_0}(u)$$

which implies that $M_1 = M_0$. Hence there is only one such cluster point. To conclude the proof, argue on (7) as previously. ////

Intuitively speaking, the law of the sum of two independent random variables is more diffuse than the law of each summand. To make this interpretation quantitative, loss functions and risks may be employed.

A function $\ell \colon \mathbb{R}^p \to [0, \infty]$ is called *symmetric subconvex* if

$$\ell(z) = \ell(-z), \qquad \{z \in \mathbb{R}^p \mid \ell(z) \le c\} \ \text{ convex} \tag{33}$$

for all $z \in \mathbb{R}^p$, $c \in \mathbb{R}$. The boundaries of convex sets having Lebesgue measure 0, such functions are necessarily Lebesgue measurable. Symmetric subconvex functions will play a role as loss functions in the next section, generalizing square error loss. A function $f \colon \mathbb{R}^p \to [0, \infty]$ is called symmetric subconcave if $-f$ is symmetric subconvex; for example, the Lebesgue density of a centered normal is symmetric subconcave. Conforming with intuition, risks are increased by convolution. The proof is based on a geometric lemma due to Anderson (1955).

Lemma 3.2.5 *If f is a symmetric subconcave Lebesgue density on $\mathbb{R}^p$, and G is a symmetric convex subset of $\mathbb{R}^p$, then for all $t \in \mathbb{R}^p$,*

$$\int_G f(\eta + t)\, \lambda^p(d\eta) \le \int_G f(\eta)\, \lambda^p(d\eta) \tag{34}$$

Corollary 3.2.6 *Let $\ell \colon \mathbb{R}^p \to [0, \infty]$ be symmetric subconvex, and consider probabilities M, M_0, and N on $\mathbb{B}^p$ such that N is absolutely continuous with a symmetric subconcave Lebesgue density, and*

$$M = M_0 * N$$

Then

$$\int \ell\, dM \ge \int \ell\, dN \tag{35}$$

PROOF We apply Anderson's lemma to the symmetric subconcave Lebesgue density f of N and to the symmetric convex level sets $\{\ell \le c\}$. Thus, for every $w \in \mathbb{R}^p$, we obtain the bound

$$\begin{aligned}\int \ell(w+z)\,N(dz) &= \int_0^\infty \Bigl(1 - \int_{\{\ell\le c\}} f(z-w)\,\lambda^p(dz)\Bigr)\lambda(dc)\\ &\underset{\text{erson}}{\overset{\text{And-}}{\ge}} \int_0^\infty \Bigl(1 - \int_{\{\ell\le c\}} f(z)\,\lambda^p(dz)\Bigr)\lambda(dc) = \int \ell\,dN\end{aligned} \tag{36}$$

and then by an integration relative to M_0,

$$\int \ell\,dM = \iint \ell(w+z)\,N(dz)\,M_0(dw) \ge \int \ell\,dN$$

which is the assertion. ////

Example 3.2.7 Likewise, Fisher information of location ($p = 1$) decreases under convolution,

$$\mathcal{I}_{M_0 * N} \le \mathcal{I}_N \tag{37}$$

[Hájek and Šidák (1967; Theorem I.2.3)]. ////

3.3 Minimax Estimation

While the convolution theorem requires estimators to converge to their limiting distributions in some equivariant sense, this section allows arbitrary estimators but evaluates their risks uniformly: The quality of an estimator is assessed by its maximum risk. In this as. setup, regularity—with weak convergence not only pointwise but uniformly on compacts—implies constant risk, so the passage to the maximum would be without effect. The general approach, however, yields an asymptotic lower bound for the maximum risk of arbitrary estimators.

3.3.1 Normal Mean

The starting point is the minimax estimation, at sample size $n = 1$, and for finite dimension k, of (a linear function of) the mean vector $\theta \in \mathbb{R}^k$ of the normal location family on $(\mathbb{R}^k, \mathbb{B}^k)$ with fixed positive definite covariance $C \in \mathbb{R}^{k\times k}$,

$$\mathcal{N}(\theta, C)(dx) = \varphi_C(x-\theta)\,\lambda^k(dx), \qquad \varphi_C(x) = \frac{\exp(-\frac{1}{2}x'C^{-1}x)}{\sqrt{(2\pi)^k \det C}} \tag{1}$$

With θ ranging over $\mathbb{R}^k$, the estimand is $D\theta \in \mathbb{R}^p$ for some $p \times k$ matrix $D \in \mathbb{R}^{p\times k}$ of full rank $p \le k$, which allows the selection of certain parameter coordinates of interest.

Randomized Estimators

As estimators, Markov kernels K from $\mathbb{R}^k$ to $\bar{\mathbb{B}}^p$ are employed,

$$K: \bar{\mathbb{B}}^p \times \mathbb{R}^k \longrightarrow [0,1] \tag{2}$$

where $K(B,x)$ is interpreted as the probability of estimating $D\theta$ by any element s in the Borel set $B \in \bar{\mathbb{B}}^p$ after $x \in \mathbb{R}^k$ has been observed. Extended-valued estimators will arise in the asymptotic setup upon passage to the limit.

Recall that we view $\bar{\mathbb{R}}^p$ homeomorphic to the cube $[-1,1]^p$ in $\mathbb{R}^p$ via the isometry

$$(z_1, \ldots, z_p) \longmapsto \Bigl(\frac{z_1}{1+|z_1|}, \ldots, \frac{z_p}{1+|z_p|}\Bigr)$$

The error of estimating $D\theta$ by some $s \in \bar{\mathbb{R}}^p$ is assessed by means of any measurable map $\ell: \bar{\mathbb{R}}^p \to [0,\infty]$, which in this context is called *loss function*. Under a loss function ℓ, a Markov kernel K from $\mathbb{R}^k$ to $\bar{\mathbb{B}}^p$ is assigned the following *risk* at θ (the true parameter value),

$$\rho(K,\theta,\ell) = \int \ell\,(s - D\theta)\, K(ds,x)\, \mathcal{N}(\theta,C)(dx) \tag{3}$$

The difference $s - D\theta$ when $|s| = \infty$ but $|D\theta| < \infty$, to make sure, is computed coordinatewise according to the usual arithmetic of $\bar{\mathbb{R}}$.

More generally, the *Bayes risk* of K under some prior $\pi \in \mathcal{M}_1(\mathbb{B}^k)$ is

$$\rho(K,\pi,\ell) = \int \rho(K,\theta,\ell)\, \pi(d\theta) \tag{4}$$

The integral is well defined since $(x,\theta,s) \mapsto \ell\,(s - D\theta)\, \varphi_C(x-\theta)$ is jointly Borel measurable and nonnegative, hence the risk $\rho(K,\theta,\ell)$ is a nonnegative Borel measurable function of $\theta \in \mathbb{R}^k$ [Fubini].

(Almost) Bayes [see (†) below] and constant risk imply minimax.

Proposition 3.3.1 *If the kernel K_0 has constant risk and is almost Bayes, then K_0 is minimax and*

$$\sup_\pi \inf_K \rho(K,\pi,\ell) = \inf_K \sup_\pi \rho(K,\pi,\ell) \tag{5}$$

where $\sup_\pi \rho(K,\pi,\ell) = \sup_\theta \rho(K,\theta,\ell)$.

PROOF By definition we have

$$\sup_\pi \inf_K \rho(K,\pi,\ell) \le \inf_K \sup_\pi \rho(K,\pi,\ell) \le \sup_\pi \rho(K_0,\pi,\ell) = \rho_0$$

where $\rho_0 = \rho(K_0,\theta,\ell)$ [constant]. The kernel K_0 *almost Bayes* (†) means: For every $\varepsilon \in (0,1)$ there exists some prior $\pi_\varepsilon \in \mathcal{M}_1(\mathbb{B}^k)$ such that

$$\rho_0 = \rho(K_0,\pi_\varepsilon,\ell) < \inf_K \rho(K,\pi_\varepsilon,\ell) + \varepsilon \le \sup_\pi \inf_K \rho(K,\pi,\ell) + \varepsilon$$

The statement now follows if $\varepsilon \downarrow 0$. ////

As K_0 the nonrandomized estimator $x \mapsto Dx$ itself will serve,

$$K_0(ds, x) = \mathbf{I}_{Dx}(ds) \tag{6}$$

It certainly has constant risk; namely,

$$\rho(K_0, \theta, \ell) = \int \ell\,(Dx - D\theta)\,\mathcal{N}(\theta, C)(dx) = \int \ell \circ D\, d\mathcal{N}(0, C) = \rho_0 \tag{7}$$

Loss Functions

To prove this K_0 almost Bayes, we need loss functions which resemble square error in the respect that the expectation $\mathrm{E}\,\ell\,(Z - a)$, at least for p variate normal Z, is minimized by the mean $a = \mathrm{E}\,Z$; see (11) below.

Denote by L the set of all Borel measurable functions $\ell\colon \bar{\mathbb{R}}^p \to [0, \infty]$ that, first, are symmetric subconvex on $\mathbb{R}^p$ according to 3.2(33), and, second, are upper semicontinuous at infinity, in the sense that for every sequence $z_n \in \mathbb{R}^p$ converging to some $z \in \bar{\mathbb{R}}^p \setminus \mathbb{R}^p$ in $\bar{\mathbb{R}}^p$,

$$\limsup_{n\to\infty} \ell(z_n) \le \ell(z) \tag{8}$$

For example, if the function $v\colon [0, \infty] \to [0, \infty]$ is increasing, and the matrix $A \in \mathbb{R}^{p\times p}$ is symmetric positive definite, then some function $\ell \in \mathsf{L}$ is given by

$$\ell(z) = \begin{cases} v(z'Az) & \text{if } |z| < \infty \\ \infty & \text{if } |z| = \infty \end{cases} \tag{9}$$

This type of loss function will be called *monotone quadratic*.

Identity Almost Bayes

Theorem 3.3.2 *Given a loss function $\ell \in \mathsf{L}$, the constant risk estimator $K_0 = \mathbf{I}_D$ is almost Bayes, hence minimax, among all Markov kernels from $\mathbb{R}^k$ to $\bar{\mathbb{B}}^p$, for the estimation of $D\theta$ in the normal shift model* (1).

PROOF If $\pi \in \mathcal{M}_1(\mathbb{B}^k)$ is any prior on the mean $\theta \in \mathbb{R}^k$, invoke two stochastically independent random vectors $T \sim \pi$, $Y \sim \mathcal{N}(0, C)$. Thus, $X = T + Y$ has $\mathcal{L}(X|T = t) = \mathcal{N}(t, C)$ a.e. $\pi(dt)$ as conditional distribution given T. Using Fubini's theorem for Markov kernels, the Bayes risk reads

$$\begin{aligned} \rho(K, \pi, \ell) &= \int \ell\,(s - D\theta)\,K(ds, x)\ \mathcal{L}(X|T = \theta)(dx)\ \mathcal{L}(T)(d\theta) \\ &= \int \ell\,(s - D\theta)\,K(ds, x)\ \mathcal{L}(T, X)(d\theta, dx) \qquad (10) \\ &= \int \ell\,(s - D\theta)\ \mathcal{L}(T|X = x)(d\theta)\,K(ds, x)\ \mathcal{L}(X)(dx) \end{aligned}$$

Specifying normal priors $\pi = \pi_m = \mathcal{N}(0, mC)$, the posterior distribution of T is $\mathcal{L}(T|X = x) = \mathcal{N}(rx, rC)$ a.e. $\lambda^k(dx)$, with $r = m/(m+1)$ [Example 2.2.8; $B = C$].

It will be shown in this proof that for all $s \in \bar{\mathbb{R}}^p$, $s \neq rDx$,

$$\int \ell\,(s - D\theta)\,\mathcal{N}(rx, rC)(d\theta) \geq \int \ell \circ D\, d\mathcal{N}(0, rC) \tag{11}$$

If this is in fact true, the posterior risk

$$\int \ell\,(s - D\theta)\ \mathcal{L}(T|X = x)(d\theta)\, K(ds, x) \tag{12}$$

is obviously minimized by the kernel

$$K_r(ds, x) = \mathbf{I}_{rDx}(ds) \tag{13}$$

which is Bayes for the prior π_m; that is, K_r achieves

$$\inf_K \rho(K, \pi_m, \ell) = \int \ell \circ D\, d\mathcal{N}(0, rC) \tag{14}$$

Then, for $m \to \infty$, the Lebesgue densities converge, and by Fatou's lemma,

$$\liminf_{m \to \infty} \int \ell \circ D\, d\mathcal{N}(0, rC) \geq \int \ell \circ D\, d\mathcal{N}(0, C) = \rho_0 \tag{15}$$

thus proving K_0 almost Bayes.

To verify (11), distinguish the cases s finite and s infinite. In the first case, bound (11) is easy to see for a monotone quadratic loss function: Choose H orthogonal so as to achieve the diagonalization

$$H(r\Gamma)^{1/2} A\,(r\Gamma)^{1/2} H' = \operatorname{diag}(\lambda_1, \ldots, \lambda_p), \qquad \Gamma = DCD'$$

Then $\lambda_i > 0$ and

$$\begin{aligned} \int \ell\,(s - D\theta)\,\mathcal{N}(rx, rC)(d\theta) &= \int v\big((\eta - s)' A(\eta - s)\big)\, \mathcal{N}(rDx, r\Gamma)(d\eta) \\ &= \int v\big(\textstyle\sum_i \lambda_i(\eta_i - t_i)^2\,\big)\, \mathcal{N}(0, \mathbb{I}_p)(d\eta) \end{aligned} \tag{16}$$

where $t = H(r\Gamma)^{-1/2}(s - rDx)$. The χ^2 distribution having monotone likelihood ratios, hence being stochastically increasing, in the noncentrality parameter, each of the p independent summands $\lambda_i(\eta_i - t_i)^2$ becomes stochastically smallest if $t_i = 0$. As v is increasing, bound (11) follows.

For general loss function $\ell \in \mathsf{L}$, if $|s| < \infty$, bound (11) is a consequence of ℓ being symmetric subconvex and of Corollary 3.2.6 to Anderson's lemma. In fact, identifying

$$M = \mathcal{N}(t, r\Gamma), \quad M_0 = \mathbf{I}_t, \quad N = \mathcal{N}(0, r\Gamma), \quad t = rDx - s, \quad \Gamma = DCD'$$

we obtain that

$$\int \ell\,(s - D\theta)\,\mathcal{N}(rx, rC)(d\theta) = \int \ell\, dM \geq \int \ell\, dN = \int \ell\, d\mathcal{N}(0, r\Gamma)$$

For $|s| = \infty$ employ the clipped version $b \wedge \ell \in \mathsf{L}$, which inherits the loss function properties from $\ell \in \mathsf{L}$. Choose $s_n \in \mathbb{R}^p$ tending to s in $\bar{\mathbb{R}}^p$. Then, for each $\theta \in \mathbb{R}^k$, the sequence $s_n - D\theta$ tends to $s - D\theta$ in $\bar{\mathbb{R}}^p$, and hence

$$\ell\,(s - D\theta) \geq \limsup_{n\to\infty} b \wedge \ell\,(s_n - D\theta)$$

by (8). Now Fatou's lemma and the already proven part of (11) imply that

$$\begin{aligned} \int \ell\,(s - D\theta)\,\mathcal{N}(rx, rC)(d\theta) &\geq \limsup_{n\to\infty} \int b \wedge \ell\,(s_n - D\theta)\,\mathcal{N}(rx, rC)(d\theta) \\ &\geq \int b \wedge \ell \circ D\, d\mathcal{N}(0, rC) \end{aligned}$$

Because the last integral tends to $\int \ell\, d\mathcal{N}(0, r\Gamma)$ as $b \uparrow \infty$ [monotone convergence], bound (11) is also true in case $|s| = \infty$. ////

For the extension of Theorem 3.3.2 to the as. normal case, we need a modification involving prior probabilities of compact, respectively finite, supports.

$$\begin{aligned} \Pi_c &= \{\, \pi \in \mathcal{M}_1(\mathbb{B}^k) \mid \pi(|\theta| \leq c) = 1 \,\}, \qquad c \in (0, \infty) \\ \Pi_{\mathrm{f}} &= \{\, \pi \in \mathcal{M}_1(\mathbb{B}^k) \mid \text{support}(\pi) \text{ finite} \,\} \end{aligned} \tag{17}$$

Recall definition (7) of the risk ρ_0.

Proposition 3.3.3 *Consider the estimation of $D\theta$ in the normal shift model* (1), *by Markov kernels from $\mathbb{R}^k$ to $\bar{\mathbb{B}}^p$, under a loss function $\ell \in \mathsf{L}$.*

(a) *Then*

$$\lim_{c\to\infty} \sup_{\pi\in\Pi_c} \inf_K \rho(K, \pi, \ell) = \lim_{c\to\infty} \inf_K \sup_{\pi\in\Pi_c} \rho(K, \pi, \ell) = \rho_0 \tag{18}$$

(b) *If ℓ is continuous on $\bar{\mathbb{R}}^p$ then*

$$\sup_{\pi\in\Pi_{\mathrm{f}}} \inf_K \rho(K, \pi, \ell) = \inf_K \sup_{\pi\in\Pi_{\mathrm{f}}} \rho(K, \pi, \ell) = \rho_0 \tag{19}$$

PROOF

(a) Since, by inserting the estimator $K_0 = \mathbf{I}_D$, automatically

$$\rho_0 \geq \limsup_{c\to\infty} \inf_K \sup_{\pi\in\Pi_c} \rho(K, \pi, \ell) \geq \limsup_{c\to\infty} \sup_{\pi\in\Pi_c} \inf_K \rho(K, \pi, \ell)$$

the assertion amounts to

$$\rho_0 \leq \liminf_{c\to\infty} \sup_{\pi\in\Pi_c} \inf_K \rho(K, \pi, \ell) \tag{20}$$

Without restriction $\ell \leq b < \infty$ since the risk of K_0 under $b \wedge \ell \in \mathsf{L}$ tends to ρ_0 for $b \uparrow \infty$ [monotone convergence]. To prove (20) let $\varepsilon \in (0, \infty)$. K_0 being almost Bayes [Theorem 3.3.2], there exists a prior $\pi \in \mathcal{M}_1(\mathbb{B}^k)$ such that

$$\rho_0 = \rho(K_0, \pi, \ell) < \inf_K \rho(K, \pi, \ell) + \varepsilon \tag{21}$$

Choose m so large that $\pi(|\theta| \leq m) > 0$, and construct $\pi_m \in \Pi_m$ by conditioning on $|\theta| \leq m$. We shall show that

$$\lim_{m \to \infty} \sup_K \bigl| \rho(K, \pi_m, \ell) - \rho(K, \pi, \ell) \bigr| = 0 \tag{22}$$

hence

$$\lim_{m \to \infty} \bigl| \inf_K \rho(K, \pi_m, \ell) - \inf_K \rho(K, \pi, \ell) \bigr| = 0 \tag{23}$$

This will imply (20) via (21) if $\varepsilon \to 0$.

According to (10), the Bayes risk of a kernel K under π is

$$\rho(K, \pi, \ell) = \int \ell\,(s - D\theta)\, Q(d\theta|x)\, K(ds, x)\, Q(dx) \tag{24}$$

where $Q(dx) = \mathcal{L}(X)(dx)$ and $Q(d\theta|x) = \mathcal{L}(T|X = x)(d\theta)$. Writing out the definitions leads to the following marginal and posterior densities

$$\begin{aligned} Q_m(dx) = h_m(x)\, \lambda^k(dx), \qquad h_m(x) &= \int \varphi_C(x - \theta)\, \pi_m(d\theta) \\ Q_m(d\theta|x) &= \frac{\varphi_C(x - \theta)}{h_m(x)} \cdot \frac{d\pi_m}{d\pi}(\theta)\, \pi(d\theta) \end{aligned} \tag{25}$$

Since $d\pi_m / d\pi \to 1$ pointwise and bounded, and the normal density is bounded, it follows by dominated convergence that for all $x, \theta \in \mathbb{R}^k$,

$$\lim_{m \to \infty} h_m(x) = h(x), \qquad \lim_{m \to \infty} \frac{\varphi_C(x - \theta)}{h_m(x)} \cdot \frac{d\pi_m}{d\pi}(\theta) = \frac{\varphi_C(x - \theta)}{h(x)} \tag{26}$$

which are the limit marginal and posterior densities induced by π. Then Lemma A.2.4 [Scheffé] implies that

$$\lim_{m \to \infty} \int \bigl| Q_m(dx) - Q(dx) \bigr| = 0 = \lim_{m \to \infty} \int \bigl| Q_m(d\theta|x) - Q(d\theta|x) \bigr| \tag{27}$$

for all $x \in \mathbb{R}^k$. As $\ell \leq b$ and the L_1 distance between any two posteriors is always less than 2, we obtain the following bound,

$$\begin{aligned} &\sup_K \bigl| \rho(K, \pi_m, \ell) - \rho(K, \pi, \ell) \bigr| \\ &\quad \leq \sup_K \Bigl| \int \ell\,(s - D\theta) \bigl(Q_m(d\theta|x) - Q(d\theta|x) \bigr) K(ds, x)\, Q_m(dx) \Bigr| \end{aligned}$$

$$+ \sup_K \Bigl| \int \ell\,(s - D\theta)\, Q(d\theta|x)\, K(ds,x) \bigl(Q_m(dx) - Q(dx)\bigr) \Bigr|$$
$$\leq b \iint \bigl| Q_m(d\theta|x) - Q(d\theta|x) \bigr|\, Q_m(dx) + b \int \bigl| Q_m(dx) - Q(dx) \bigr|$$
$$\leq b \iint \bigl| Q_m(d\theta|x) - Q(d\theta|x) \bigr|\, Q(dx) + 3b \int \bigl| Q_m(dx) - Q(dx) \bigr|$$

This bound goes to 0 as $m \to \infty$ by (27) and dominated convergence. Hence (22) is proved.

(b) Since automatically

$$\rho_0 = \sup_{\pi \in \Pi_{\mathrm{f}}} \rho(K_0, \pi, \ell) \geq \inf_K \sup_{\pi \in \Pi_{\mathrm{f}}} \rho(K, \pi, \ell) \geq \sup_{\pi \in \Pi_{\mathrm{f}}} \inf_K \rho(K, \pi, \ell)$$

we must show that

$$\rho_0 \leq \sup_{\pi \in \Pi_{\mathrm{f}}} \inf_K \rho(K, \pi, \ell) \tag{28}$$

For this purpose, by the previous clipping argument, the loss function $\ell \in \mathsf{L}$ may be assumed continuous and bounded by some $b \in (0, \infty)$.

Given $\varepsilon \in (0, \infty)$, approximate the prior π in (21) by some sequence $\pi_m \in \Pi_{\mathrm{f}}$ weakly; Π_{f} is weakly dense in $\mathcal{M}_1(\mathbb{B}^p)$ [Parthasaraty (1967; II Theorem 6.3, p44)]. We shall verify (22) for such π_m.

Recall (25). As the priors converge weakly and the normal density is bounded and continuous, the marginal densities converge at each point, hence (27) holds. For the same reason, at all $x \in \mathbb{R}^k$,

$$\varphi_C(x - \theta)\, \pi_m(d\theta) \xrightarrow{\mathrm{w}} \varphi_C(x - \theta)\, \pi(d\theta)$$

hence also

$$Q_m(d\theta|x) \xrightarrow{\mathrm{w}} Q(d\theta|x) \tag{29}$$

The loss function being continuous on $\bar{\mathbb{R}}^p$ (compact), it is uniformly continuous. As the operator norm $\|D\|$ is also finite, and the metric of $\bar{\mathbb{R}}^p$ decreases Euclidean distances, the maps κ_s from $\mathbb{R}^k$ to $[0, \infty)$,

$$\kappa_s(\theta) = \ell\,(s - D\theta)\,, \qquad \theta \in \mathbb{R}^k\,,\ s \in \bar{\mathbb{R}}^p$$

are bounded, (uniformly) equicontinuous. By Parthasaraty (1967; II Theorem 6.8, p51), therefore, the weak convergence of the posteriors implies that

$$\lim_{m \to \infty} \sup_s \Bigl| \int \ell\,(s - D\theta) \bigl(Q_m(d\theta|x) - Q(d\theta|x)\bigr) \Bigr| = 0 \tag{30}$$

for all $x \in \mathbb{R}^k$. As $\sup_s \bigl| \int \dots \bigr| \leq b$ we conclude that

$$\sup_K \bigl| \rho(K, \pi_m, \ell) - \rho(K, \pi, \ell) \bigr|$$
$$\leq \sup_K \Bigl| \int \ell\,(s - D\theta) \bigl(Q_m(d\theta|x) - Q(d\theta|x)\bigr) K(ds,x)\, Q_m(dx) \Bigr|$$

$$
\begin{aligned}
&\quad + \sup_K \Big| \int \ell\,(s - D\theta)\, Q(d\theta|x)\, K(ds,x) \big(Q_m(dx) - Q(dx)\big) \Big| \\
&\leq \int \sup_s \Big| \int \ell\,(s - D\theta) \big(Q_m(d\theta|x) - Q(d\theta|x)\big) \Big|\, Q_m(dx) \\
&\qquad + b \int \big|Q_m(dx) - Q(dx)\big| \\
&\leq \int \sup_s \Big| \int \ell\,(s - D\theta) \big(Q_m(d\theta|x) - Q(d\theta|x)\big) \Big|\, Q(dx) \\
&\qquad\qquad + 2b \int \big|Q_m(dx) - Q(dx)\big|
\end{aligned}
$$

which bound goes to 0 as $m \to \infty$, by (27), (30), and dominated convergence, respectively. ////

Admissibility

Admissibility complements the (somewhat pessimistic) minimax criterion. James and Stein (1961) proved admissibility of $K_0 = \mathbf{I}_D$ among nonrandomized estimators, for dimensions 1 and 2, under quadratic loss, when specifically

$$
p = k \leq 2\,, \qquad D = \mathbb{I}_k = C\,, \qquad \ell(z) = |z|^2 \tag{31}
$$

The result extends to the estimation of one or two parameters of interest, in the presence of finitely many nuisance components, for general covariance, randomized estimators, and suitably adapted loss function.

Proposition 3.3.4 *Let D be a $p \times k$ matrix and C a $k \times k$ covariance matrix of full ranks p and k, respectively. Assume that $p \leq 2$. Then $K_0 = \mathbf{I}_D$ is admissible among all Markov kernels from $\mathbb{R}^k$ to $\bar{\mathbb{B}}^p$, for the estimation of $D\theta$ in the normal shift model* (1), *under the loss function*

$$
\ell(z) = z'\big(DCD'\big)^{-1} z \tag{32}
$$

PROOF Transform the observation $x \sim \mathcal{N}(\theta, C)$ by $C^{-1/2}$. This gives us the standard normal model with estimand H,

$$
v = C^{-1/2}x \sim \mathcal{N}(\eta, \mathbb{I}_k)\,, \qquad \eta = C^{-1/2}\theta \in \mathbb{R}^k\,, \qquad H = DC^{1/2}
$$

Complement the p linearly independent rows $h_1, \dots, h_p$ of H by vectors $h_{p+1}, \dots, h_k$ to a basis of $\mathbb{R}^k$ such that $h_i \perp h_j$ for all $i \leq p < j$, and denote by $\bar{H}$ the corresponding $(k-p) \times k$ matrix with these rows. Then the transformation by the $k \times k$ matrix G with blocks H and $\bar{H}$ leads us to the normal model

$$
u = \begin{pmatrix} u_{(1)} \\ u_{(2)} \end{pmatrix} = Gv \sim \mathcal{N}\Big(\begin{pmatrix} \delta_{(1)} \\ \delta_{(2)} \end{pmatrix}, \begin{pmatrix} HH' & 0 \\ 0 & \bar{H}\bar{H}' \end{pmatrix} \Big) \tag{33}
$$

Now the first component $\delta_{(1)} \in \mathbb{R}^p$ of $\delta \in \mathbb{R}^k$ is the parameter of interest, the second component $\delta_{(2)} \in \mathbb{R}^{k-p}$ is nuisance, and $K_0(ds, u) = \mathbf{I}_{u_{(1)}}(ds)$.

For any kernel $K(ds, u)$ from $\mathbb{R}^k$ to $\bar{\mathbb{B}}^p$ and any $\delta \in \mathbb{R}^k$ we can write

$$\begin{aligned}\rho(K, \delta, \ell) &= \int \ell\,(s - \delta_{(1)})\, K_{\delta_{(2)}}(ds, u_{(1)})\, \mathcal{N}(\delta_{(1)}, HH')(du_{(1)}) \\ &= \rho(K_{\delta_{(2)}}, \delta_{(1)}, \ell)\end{aligned} \tag{34}$$

with the kernel $K_{\delta_{(2)}}$ from $\mathbb{R}^p$ to $\bar{\mathbb{B}}^p$ obtained by averaging out $u_{(2)}$,

$$K_{\delta_{(2)}}(ds, u_{(1)}) = \int K(ds, u)\, \mathcal{N}(\delta_{(2)}, \bar{H}\bar{H}')(du_{(2)}) \tag{35}$$

This averaging leaves K_0, not depending on $u_{(2)}$, unchanged.

By James and Stein (1961), K_0 is admissible in the normal model with

$$k = p \le 2, \qquad D = \mathbb{I}_p, \quad C = HH', \qquad \ell(z) = z'C^{-1}z \tag{36}$$

Hence, if a kernel K from $\mathbb{R}^k$ to $\bar{\mathbb{B}}^p$ for some $\delta \in \mathbb{R}^k$ satisfies

$$\rho(K_{\delta_{(2)}}, \delta_{(1)}, \ell) = \rho(K, \delta, \ell) < \rho(K_0, \delta, \ell) = \rho(K_0, \delta_{(1)}, \ell) \tag{37}$$

then there exists a $\gamma_{(1)} \in \mathbb{R}^p$ such that, complemented by $\gamma_{(2)} = \delta_{(2)}$,

$$\rho(K, \gamma, \ell) = \rho(K_{\delta_{(2)}}, \gamma_{(1)}, \ell) > \rho(K_0, \gamma_{(1)}, \ell) = \rho(K_0, \gamma, \ell) \tag{38}$$

This proves the asserted admissibility. ////

The result of James and Stein (1961) is indeed true among randomized and possibly infinite-valued estimators.

Lemma 3.3.5 *Consider the estimation of $D\theta$ in the normal model* (1), *under a loss function ℓ that is convex on $\mathbb{R}^p$ such that*

$$\ell(z) \ge c|z| - b, \qquad z \in \bar{\mathbb{R}}^p \tag{39}$$

for some numbers $c \in (0, \infty)$ and $b \in \mathbb{R}$. Then:

(a) *the nonrandomized, finite-valued estimators are essentially complete.*

(b) *If ℓ moreover is strictly convex on $\mathbb{R}^p$ and K_0 is admissible, then K_0* is uniquely determined by its risk function: *Any kernel K whose risk function $\rho(K, .\,, \ell)$ is everywhere less than or equal to ρ_0 necessarily coincides with K_0,*

$$K(ds, x) = \mathbf{I}_{Dx}(ds) \qquad \text{a.e. } \lambda^k(dx) \tag{40}$$

PROOF

(a) Suppose a kernel K has finite risk at some $\theta_0 \in \mathbb{R}^k$,

$$\int \ell\,(s - D\theta_0)\, K(ds, x)\, \mathcal{N}(\theta_0, C)(dx) < \infty \tag{41}$$

Then (39) implies that

$$\int |s - D\theta_0|\, K(ds,x) < \infty \qquad \text{a.e. } \lambda^k(dx) \tag{42}$$

hence

$$K(\mathbb{R}^p, x) = 1, \quad \int |s|\, K(ds,x) < \infty \qquad \text{a.e. } \lambda^k(dx) \tag{43}$$

Thus the nonrandomized estimator S,

$$S(x) = \int s\, K(ds,x) \tag{44}$$

is defined and finite a.e. $\lambda^k(dx)$. In view of $K(\mathbb{R}^p, x) = 1$ and the convexity of ℓ on $\mathbb{R}^p$, Jensen's inequality applies, so that for all $\theta \in \mathbb{R}^k$, a.e. $\lambda^k(dx)$,

$$\ell\big(S(x) - D\theta\big) \leq \int \ell(s - D\theta)\, K(ds,x) \tag{45}$$

Hence

$$\int \ell\big(S(x) - D\theta\big)\, \mathcal{N}(\theta, C)(dx) \leq \int \ell(s - D\theta)\, K(ds,x)\, \mathcal{N}(\theta, C)(dx) \tag{46}$$

This proves essential completeness, since nothing has to be shown in the case of risk functions that are infinite everywhere.

(b) Suppose K has smaller risk than K_0. By part (a), the nonrandomized estimator S also has smaller risk than K_0. Since K_0 has been assumed admissible, the risks of K, S, and K_0 are all identically ρ_0.

In particular, no strict improvement is possible in (46), hence strict inequality in (45) can hold only on a set of Lebesgue measure zero. Therefore, Jensen's inequality for strictly convex functions on $\mathbb{R}^p$ implies that

$$K\big(\{S(x)\}, x\big) = 1 \qquad \text{a.e. } \lambda^k(dx) \tag{47}$$

As S is finite a.e., and ℓ is convex on $\mathbb{R}^p$, Jensen tells us that, a.e. $\lambda^k(dx)$,

$$\ell\big(\tfrac{1}{2}(S(x) + Dx) - D\theta\big) \leq \frac{1}{2}\ell\big(S(x) - D\theta\big) + \frac{1}{2}\ell(Dx - D\theta) \tag{48}$$

Hence the mean of both estimators has risk

$$\int \ell\big(\tfrac{1}{2}(S(x) + Dx) - D\theta\big)\, \mathcal{N}(\theta, C)(dx) \leq \rho_0 \tag{49}$$

As K_0 is admissible, strict improvement is not possible in (49). Thus strict inequality can hold in (48) only on a set of Lebesgue measure zero. Therefore, strict convexity of ℓ implies that

$$S(x) = Dx \qquad \text{a.e. } \lambda^k(dx) \tag{50}$$

which, linked up with (47), proves (40). ////

Remark 3.3.6 In the case $1 = p \le k$, assuming $D = (1,0,\dots,0)'$ and a (nonconstant) monotone quadratic loss function, the uniqueness of K_0 among randomized estimators was proved by Hájek (1972), using and extending the admissibility among nonrandomized estimators shown by Blyth (1951) in the case $k = p = 1$. ////

In dimension 3 or larger, the identity becomes inadmissible as an estimator of the normal mean under quadratic loss, and it actually pays to combine the information contained in stochastically independent observations, in a non-translation equivariant way: To be specific, in the situation

$$p = k \ge 3, \qquad D = \mathbb{I}_k = C, \qquad \ell(z) = |z|^2 \tag{51}$$

shrinking the identity towards 0 (and beyond), James and Stein (1961) invented the nonrandomized estimator $\widetilde{K}_0 = \mathbf{I}_{\tilde{S}_0}$,

$$\tilde{S}_0(x) = \Bigl(1 - \frac{k-2}{|x|^2}\Bigr)x \tag{52}$$

and calculated its mean square error at $\theta \in \mathbb{R}^k$ to

$$\begin{aligned}\int \bigl|\tilde{S}_0(x) - \theta\bigr|^2 \mathcal{N}(\theta, \mathbb{I}_k)(dx) &= k - \mathrm{E}\left(\frac{(k-2)^2}{k-2+2P\bigl(|\theta|^2/2\bigr)}\right)\\ &< k = \int |x-\theta|^2 \mathcal{N}(\theta,\mathbb{I}_k)(dx)\end{aligned} \tag{53}$$

where $P\bigl(|\theta|^2/2\bigr)$ denotes a Poisson random variable of expectation $|\theta|^2/2$. Thus $\widetilde{K}_0$ is strictly better than K_0 everywhere, achieving the maximum improvement at $\theta = 0$, where the MSE of $\widetilde{K}_0$ is just 2.

We formulate the basic inadmissibility result for the estimation of $p \ge 3$ parameters of interest in the presence of a finite number of nuisance components: Given some $p \times k$ matrix D and $k \times k$ covariance C of full ranks p and k, respectively, the nonrandomized estimator

$$\tilde{S}_0(x) = \Bigl(1 - \frac{p-2}{|Dx|_\Gamma^2}\Bigr)Dx \tag{54}$$

achieves the following risk at $\theta \in \mathbb{R}^k$,

$$\begin{aligned}&\int \bigl|\tilde{S}_0(x) - D\theta\bigr|_\Gamma^2 \mathcal{N}(\theta, C)(dx)\\ &= \int \Bigl|\Bigl(1 - \frac{p-2}{|y|_\Gamma^2}\Bigr)y - \eta\Bigr|_\Gamma^2 \mathcal{N}(\eta,\Gamma)(dy), \qquad \eta = D\theta\\ &= \int \Bigl|\Bigl(1 - \frac{p-2}{|z|^2}\Bigr)z - \delta\Bigr|^2 \mathcal{N}(\delta,\mathbb{I}_p)(dz), \qquad \delta = \Gamma^{-1/2}\eta\\ &\overset{(53)}{=} p - \mathrm{E}\left(\frac{(p-2)^2}{p-2+2P\bigl(|\delta|^2/2\bigr)}\right) < p = \int \bigl|Dx - D\theta\bigr|_\Gamma^2 \mathcal{N}(\theta, C)(dx)\end{aligned} \tag{55}$$

if we employ on $\mathbb{R}^p$ the norm

$$|z|_\Gamma^2 = z'\Gamma^{-1}z\,, \qquad \Gamma = DCD' \tag{56}$$

Therefore, under $\ell = |\,.\,|_\Gamma^2$, the nonrandomized estimator $\widetilde{K}_0 = \mathbf{I}_{\tilde{S}_0}$ has strictly smaller risk than $K_0 = \mathbf{I}_D$ at each $\theta \in \mathbb{R}^k$ (and risk 2 for $D\theta = 0$).

Remark 3.3.7 James and Stein (1961) have extended their inadmissibility result to bounded, monotone quadratic loss functions of form (9), with v bounded, twice differentiable, and concave, using the following modification $\bar{S}_0$ of $\tilde{S}_0$,

$$\bar{S}_0(x) = \Big(1 - \frac{b}{a + |x|^2}\Big)x \tag{57}$$

where b is chosen sufficiently small and a large. In addition, James and Stein (1961) state the superiority of $\bar{S}_0$ over K_0 for the location model induced by k arbitrary uncorrelated random variables with finite fourth moments; thus, the normality assumption is not crucial either.

An illuminating review of the inadmissibility phenomenon, and interpretation from a regression perspective, has been given by Stigler (1990). ////

3.3.2 Asymptotic Minimax Bound

The preceding minimax bound for the estimation of the normal mean extends to statistical models that are as. normal in the sense of Definition 2.2.9,

$$\mathcal{Q}_n = \{\, Q_{n,t} \mid t \in \mathbb{R}^k \,\} \subset \mathcal{M}_1(\mathcal{A}_n)$$

having the same finite-dimensional parameter spaces $\Theta_n = \mathbb{R}^k$, or increasing subsets $\Theta_n \uparrow \mathbb{R}^k$ [Θ_n open, for part (a)]. As. estimators

$$S = (S_n), \qquad S_n\colon (\Omega_n, \mathcal{A}_n) \longrightarrow (\bar{\mathbb{R}}^p, \bar{\mathbb{B}}^p)$$

with extended values can be allowed in this subsection. Recall L, the set of all Borel measurable functions $\ell\colon \bar{\mathbb{R}}^p \to [0,\infty]$ that are symmetric subconvex on $\mathbb{R}^p$ and u.s.c. at infinity [see 3.2(33) and (8)].

Theorem 3.3.8 *Assume models $(\mathcal{Q}_n)$ that are as. normal with as. covariance $C > 0$. Let $D \in \mathbb{R}^{p\times k}$ be a matrix of rank $p \le k$. Put*

$$\rho_0 = \int \ell \, d\mathcal{N}(0,\Gamma)\,, \qquad \Gamma = DC^{-1}D' \tag{58}$$

for any Borel measurable function $\ell\colon \mathbb{R}^p \to [0,\infty]$.

(a) *Then, if $\ell \in \mathsf{L}$ and ℓ is lower semicontinuous on $\bar{\mathbb{R}}^p$,*

$$\lim_{b\to\infty}\lim_{c\to\infty}\liminf_{n\to\infty}\inf_{S}\sup_{|t|\le c} \int b \wedge \ell\,(S_n - Dt)\, dQ_{n,t} \ge \rho_0 \tag{59}$$

(b) *If* $\ell \in \mathsf{L}$ *and* ℓ *is continuous on* $\bar{\mathbb{R}}^p$, *then*

$$\sup_{\pi \in \Pi_{\mathrm{f}}} \liminf_{n\to\infty} \inf_{S} \int \ell\,(S_n - Dt)\,dQ_{n,t}\,\pi(dt) \ge \rho_0 \tag{60}$$

(c) *Suppose* $\ell\colon \mathbb{R}^p \to [0,\infty]$ *is continuous* a.e. λ^p *and the as. estimator* $S^\star$ *is for every* $c \in (0,\infty)$ as. *normal, uniformly in* $|t| \le c$,

$$(S_n^\star - Dt)(Q_{n,t}) \xrightarrow{\mathrm{w}} \mathcal{N}(0,\Gamma) \tag{61}$$

Then for all $c \in (0,\infty)$,

$$\lim_{b\to\infty} \lim_{n\to\infty} \sup_{|t|\le c} \int b \wedge \ell\,(S_n^\star - Dt)\,dQ_{n,t} = \rho_0 \tag{62}$$

Remark 3.3.9

(a) The as. minimax theorem has some relevance for finite sample sizes: Given $\varepsilon \in (0,1)$ there exist some $b(\varepsilon), c(\varepsilon) \in (0,\infty)$ [Theorem 3.3.8 a], some finite set $A(\varepsilon) \subset \mathbb{R}^k$ [Theorem 3.3.8 b], and some $n'(\varepsilon) \in \mathbb{N}$ such that for all $n \ge n'(\varepsilon)$,

$$\inf_{S_n} \sup_{|t|\le c(\varepsilon)} \int b(\varepsilon) \wedge \ell\,(S_n - Dt)\,dQ_{n,t} \ge \rho_0 - \varepsilon \tag{63}$$

respectively,

$$\inf_{S_n} \sup_{t \in A(\varepsilon)} \int b(\varepsilon) \wedge \ell\,(S_n - Dt)\,dQ_{n,t} \ge \rho_0 - \varepsilon \tag{64}$$

where the $\inf_{S_n}$ is taken over all estimates S_n at time n.

If $S^\star$ is regular for the transform D with limit law $\mathcal{N}(0,\Gamma)$, or regular uniformly on compacts as in (61), there exists some $n''(\varepsilon) \ge n'(\varepsilon)$ such that for all $n \ge n''(\varepsilon)$,

$$\sup_{t \in A(\varepsilon)} \int b(\varepsilon) \wedge \ell\,(S_n^\star - Dt)\,dQ_{n,t} \le \rho_0 + \varepsilon \tag{65}$$

respectively,

$$\sup_{|t|\le c(\varepsilon)} \int b(\varepsilon) \wedge \ell\,(S_n^\star - Dt)\,dQ_{n,t} \le \rho_0 + \varepsilon \tag{66}$$

In either case, therefore, $S_n^\star$ comes within 2ε of the corresponding minimax risk at finite times, or sample sizes, $n \ge n''(\varepsilon)$.

(b) In technical respects, the as. minimax bound is based on Fatou's lemma. Accordingly, loss functions are employed in (a) and (b) that are l.s.c. on $\bar{\mathbb{R}}^p$. This assumption in particular entails equality in condition (8).

(c) Let the as. estimator S be regular for D with limit law M, and the function $\ell\colon \mathbb{R}^p \to [0,\infty]$ bounded and continuous a.e. M. This suffices to obtain

$$\sup_{\pi \in \Pi_{\mathrm{f}}} \lim_{n\to\infty} \int \ell\,(S_n - Dt)\,dQ_{n,t}\,\pi(dt) = \int \ell\,dM \tag{67}$$

If, according to 3.2(4), S converges to its limit M even uniformly on compacts, and the function $\ell\colon \mathbb{R}^p \to [0,\infty]$ is continuous a.e. M, then

$$\lim_{b\to\infty}\lim_{c\to\infty}\lim_{n\to\infty}\sup_{|t|\le c}\int b\wedge\ell\,(S_n - Dt)\,dQ_{n,t} = \int \ell\,dM \tag{68}$$

In this sense, the passage to the maximum risk has no effect, and the present approach coincides with the approach taken in Section 3.2, for as. estimators that are regular (uniformly on compacts).

(d) Assumed to be as. normal $\mathcal{N}(0,\Gamma)$ à la (61), uniformly on compacts, the estimator $S^\star$ is regular for the transform D with limit law $\mathcal{N}(0,\Gamma)$. By the uniqueness statement of the convolution theorem [Theorem 3.2.3], then necessarily

$$S_n^\star = DC^{-1}Z_n + \mathrm{o}_{Q_{n,0}}(n^0) \tag{95}$$

Under admissibility and uniqueness conditions on $K_0 = \mathbf{I}_D$ in the normal limit model (71), this as. expansion will be derived for arbitrary as. minimax estimators in Proposition 3.3.13 b below.

(e) The as. minimax bound in a similar setup is due to Hájek (1972). For mean square error in one dimension $p = k = 1$, Huber (1966) proved the as. minimax bound differently, using the finite-sample Cramér–Rao bound of Hodges and Lehmann (1950). Also in one dimension, the bound can be derived for confidence intervals and coverage probabilities using Neyman–Pearson techniques [Rieder (1981)]. In the general setup of converging experiments, the as. minimax bound is due to LeCam (1972); see also Millar (1983). The proof given below reflects the general ideas. Instead of passing to bilinear forms, however, we compactify the sample space.

(f) The specialization of the as. minimax theorem to the smooth parametric i.i.d. case is spelled out in Proposition 4.2.19 b. ////

Example 3.3.10 Consider Hodges' estimator of Example 3.2.1 under a l.s.c. loss function $\ell\colon \mathbb{R}^k \to [0,\infty]$. By Fatou and 3.2(3), for every $t \in \mathbb{R}^k$,

$$\liminf_{n\to\infty}\int \ell\,(\sqrt{n}\,S_n - t\,)\,dP^n_{t/\sqrt{n}} \ge \ell(-t) \tag{69}$$

Hence

$$\lim_{c\to\infty}\liminf_{n\to\infty}\sup_{|t|\le c}\int \ell\,(\sqrt{n}\,S_n - t\,)\,dP^n_{t/\sqrt{n}} = \sup \ell \tag{70}$$

which makes for a rather bad—namely, maximal—estimator risk. ////

Passage to the Normal Limit

The proof of the as. minimax bound is based on a passage in the limit to

$$\{\,\mathcal{N}(t, C^{-1}) \mid t \in \mathbb{R}^k\,\} \tag{71}$$

the normal shift model with unknown mean $t \in \mathbb{R}^k$ and fixed covariance C^{-1}.

Lemma 3.3.11 *Assume some $p \times k$ matrix D of rank $p \leq k$, a lower semicontinuous loss function $\ell \in \mathsf{L}$, and let the models $(\mathcal{Q}_n)$ be as. normal with as. covariance $C > 0$. Then, for every as. estimator S and every subsequence of $\mathbb{N}$, there exist a further subsequence (m) and a Markov kernel K_S from $\mathbb{R}^k$ to $\bar{\mathbb{B}}^p$ such that, along (m), for all $t \in \mathbb{R}^k$,*

$$\lim_{b\to\infty} \liminf_{m\to\infty} \int b \wedge \ell\,(S_m - Dt)\, dQ_{m,t} \geq \rho(K_S, t, \ell) \tag{72}$$

where $\rho(K_S, t, \ell)$ denotes the risk of K_S under loss ℓ at t, for estimating Dt in the normal shift model (71).

PROOF Fix S and $(m) \subset \mathbb{N}$. By as. normality, the as. sufficient statistic Z satisfies $Z_n(Q_{n,0}) \xrightarrow{\mathrm{w}} \mathcal{N}(0, C)$ in $\mathbb{R}^k$. $\bar{\mathbb{R}}^p$ being compact, the sequence of joint laws $(S_n, Z_n)(Q_{n,0})$ is tight in $\bar{\mathbb{R}}^p \times \mathbb{R}^k$. By Prokhorov's theorem [Proposition A.1.2] there is a random vector $(\bar{S}, \bar{Z})$ with values in $\bar{\mathbb{R}}^p \times \mathbb{R}^k$ such that along some further subsequence, still denoted by (m),

$$(S_m, Z_m)(Q_{m,0}) \xrightarrow{\mathrm{w}} \mathcal{L}(\bar{S}, \bar{Z}) \tag{73}$$

where necessarily $\bar{Z} \sim \mathcal{N}(0, C)$. Fix $t \in \mathbb{R}^k$. Then the expansion of the log likelihoods under as. normality (and Proposition A.1.4) imply that

$$(S_m, L_{m,t})(Q_{m,0}) \xrightarrow{\mathrm{w}} \mathcal{L}\bigl(\bar{S}, t'\bar{Z} - \tfrac{1}{2}t'Ct\bigr) \tag{74}$$

in $\bar{\mathbb{R}}^p \times \bar{\mathbb{R}}$. From Theorem 2.2.5 we conclude that

$$(S_m, L_{m,t})(Q_{m,t}) \xrightarrow{\mathrm{w}} e^{\pi_2}\, d\mathcal{L}\bigl(\bar{S}, t'\bar{Z} - \tfrac{1}{2}t'Ct\bigr) \tag{75}$$

where π_2 denotes the canonical projection onto the second coordinate. Now Fatou's lemma applies to the l.s.c. function $s \mapsto b \wedge \ell\,(s - Dt) \geq 0$. Thus,

$$\liminf_{m\to\infty} \int b \wedge \ell\,(S_m - Dt)\, dQ_{m,t} \geq \mathrm{E}\, b \wedge \ell\,(\bar{S} - Dt) \exp\bigl(t'\bar{Z} - \tfrac{1}{2}t'Ct\bigr) \tag{76}$$

and then, by the monotone convergence theorem,

$$\lim_{b\to\infty} \liminf_{m\to\infty} \int b \wedge \ell\,(S_m - Dt)\, dQ_{m,t} \geq \mathrm{E}\, \ell\,(\bar{S} - Dt) \exp\bigl(t'\bar{Z} - \tfrac{1}{2}t'Ct\bigr) \tag{77}$$

Conditioning on $X = C^{-1}\bar{Z} \sim \mathcal{N}(0, C^{-1})$ and recalling Example 2.2.11, we obtain that

$$\begin{aligned}
&\mathrm{E}\, \ell\,(\bar{S} - Dt) \exp\bigl(t'\bar{Z} - \tfrac{1}{2}t'Ct\bigr) \\
&\quad = \mathrm{E}\,\mathrm{E}\Bigl(\ell\,(\bar{S} - Dt) \exp\bigl((Ct)'X - \tfrac{1}{2}t'Ct\bigr) \big| X\Bigr) \\
&\quad = \mathrm{E} \exp\bigl((Ct)'X - \tfrac{1}{2}t'Ct\bigr)\, \mathrm{E}\bigl(\ell\,(\bar{S} - Dt) \big| X\bigr) \\
&\quad = \int \mathrm{E}\bigl(\ell\,(\bar{S} - Dt) \big| X = x\bigr) \exp\bigl((Ct)'x - \tfrac{1}{2}t'Ct\bigr)\, \mathcal{N}(0, C^{-1})(dx) \\
&\quad = \int \mathrm{E}\bigl(\ell\,(\bar{S} - Dt) \big| X = x\bigr)\, \mathcal{N}(t, C^{-1})(dx)
\end{aligned} \tag{78}$$

Now invoke the conditional distribution of $\bar{S}$ given X,

$$K_S(B,x) = \Pr\bigl(\bar{S} \in B \big| X = x\bigr) \tag{79}$$

which can be chosen as a Markov kernel from $\mathbb{R}^k$ to $\bar{\mathbb{B}}^p$. Thus we arrive at

$$\begin{aligned}&\int \mathrm{E}\bigl(\ell\,(\bar{S} - Dt)\big|X = x\bigr)\,\mathcal{N}(t, C^{-1})(dx)\\ &\qquad = \int \ell\,(s - Dt)\,K_S(ds,x)\,\mathcal{N}(t,C^{-1})(dx) = \rho(K_S, t, \ell)\end{aligned} \tag{80}$$

and (72) follows. ////

PROOF [Theorem 3.3.8]

(a) For $\varepsilon \in (0,\infty)$, the finite-sample minimax result [in the form of Proposition 3.3.3 a with C replaced by C^{-1}] gives us a $c \in (0,\infty)$ so large that

$$\inf_K \sup_{\pi \in \Pi_c} \rho(K,\pi,\ell) > \rho_0 - \varepsilon \tag{81}$$

Consider any as. estimator S and any subsequence of $\mathbb{N}$. By Lemma 3.3.11 there exist a kernel K_S and a further subsequence (m) such that

$$\begin{aligned}\liminf_{m\to\infty} \sup_{|t|\le c} \int \ell\,(S_m - Dt)\,dQ_{m,t} &\ge \sup_{|t|\le c} \liminf_{m\to\infty} \int \ell\,(S_m - Dt)\,dQ_{m,t}\\ &\overset{(72)}{\ge} \sup_{|t|\le c} \rho(K_S, t, \ell) > \rho_0 - \varepsilon\end{aligned} \tag{82}$$

Since S and the initial subsequence have been arbitrary this proves that

$$\inf_S \liminf_{n\to\infty} \sup_{|t|\le c} \int \ell\,(S_n - Dt)\,dQ_{n,t} > \rho_0 - \varepsilon \tag{83}$$

As $\inf_S$ and $\liminf_n$ may be interchanged, (59) follows if we let $\varepsilon \downarrow 0$.

(b) For $\varepsilon \in (0,\infty)$, Proposition 3.3.3 b supplies a prior $\pi \in \Pi_{\mathrm{f}}$ such that

$$\inf_K \rho(K,\pi,\ell) > \rho_0 - \varepsilon \tag{84}$$

Given any as. estimator S and subsequence (m) of $\mathbb{N}$ apply Lemma 3.3.11 and integrate both sides of (72) w.r.t. $\pi(dt)$. Thus

$$\begin{aligned}&\liminf_{m\to\infty} \int \ell\,(S_m - Dt)\,dQ_{m,t}\,\pi(dt)\\ &\overset{\text{Fa-}}{\underset{\text{tou}}{\ge}} \int \Bigl(\liminf_{m\to\infty} \int \ell\,(S_m - Dt)\,dQ_{m,t}\Bigr)\,\pi(dt) \overset{(72)}{\ge} \rho(K_S,\pi,\ell) > \rho_0 - \varepsilon\end{aligned} \tag{85}$$

Since S and the initial subsequence have been arbitrary, this shows that

$$\inf_S \liminf_{n\to\infty} \int \ell\,(S_n - Dt)\,dQ_{n,t}\,\pi(dt) > \rho_0 - \varepsilon \tag{86}$$

Take the sup over $\pi \in \Pi_{\mathrm{f}}$ and then let $\varepsilon \downarrow 0$ to obtain (60).

(c) Since $b \wedge \ell$ is bounded and continuous a.e. $\mathcal{N}(0,\Gamma)$, the continuous mapping theorem [Proposition A.1.1] and the compact uniform convergence (61) imply that for all $c \in (0,\infty)$, uniformly in $|t| \leq c$,

$$\lim_{n\to\infty} \int b \wedge \ell\,(S_n^\star - Dt)\,dQ_{n,t} = \int b \wedge \ell\,d\mathcal{N}(0,\Gamma) \tag{87}$$

hence

$$\lim_{n\to\infty} \sup_{|t|\leq c} \int b \wedge \ell\,(S_n^\star - Dt)\,dQ_{n,t} = \int b \wedge \ell\,d\mathcal{N}(0,\Gamma) \tag{88}$$

The RHS does not depend on c and tends to ρ_0 as $b \uparrow \infty$ [monotone convergence]. Hence (62) follows. ////

Loss Functions

A monotone quadratic loss function (9) is l.s.c. on $\bar{\mathbb{R}}^p$ if the increasing function $v\colon [0,\infty] \to [0,\infty]$ is left continuous; it is continuous on $\bar{\mathbb{R}}^p$ if v is continuous. But the additional conditions on v are not really needed since Theorem 3.3.8 a, bound (59), apparently extends to any Borel measurable function $\ell\colon \bar{\mathbb{R}}^p \to [0,\infty]$ that satisfies

$$\int \ell\,d\mathcal{N}(0,\Gamma) = \sup\Bigl\{ \int \tilde{\ell}\,d\mathcal{N}(0,\Gamma) \Bigm| \tilde{\ell} \leq \ell,\ \tilde{\ell} \in \mathsf{L},\ \tilde{\ell} \text{ l.s.c.} \Bigr\} \tag{89}$$

And likewise, Theorem 3.3.8 b, bound (60), extends to any Borel measurable function $\ell\colon \bar{\mathbb{R}}^p \to [0,\infty]$ such that

$$\int \ell\,d\mathcal{N}(0,\Gamma) = \sup\Bigl\{ \int \tilde{\ell}\,d\mathcal{N}(0,\Gamma) \Bigm| \tilde{\ell} \leq \ell,\ \tilde{\ell} \in \mathsf{L},\ \tilde{\ell} \text{ continuous} \Bigr\} \tag{90}$$

Lemma 3.3.12 *Condition* (90), *hence also* (89), *are satisfied by monotone quadratic loss functions of form* (9).

PROOF Approximate the increasing function $v\colon [0,\infty] \to [0,\infty]$ by the elementary functions

$$u_m = m\,\mathbf{I}(v > m) + \sum_{i=1}^{m2^m} \frac{i-1}{2^m}\,\mathbf{I}\bigl(i-1 < 2^m v \leq i\bigr)$$

Every function u_m is bounded, increasing, and a step function on $[0,\infty]$ as the sets $\{i-1 < 2^m v \leq i\}$, $\{v > m\}$ are abutting intervals. By the monotone convergence $u_m \uparrow v$ it suffices to approximate a particular u_m.

Except possibly at the strictly positive endpoints of the above intervals, u_m can be approximated pointwise and monotonically from below by functions v_j that are nonnegative, increasing, continuous, and constant outside $[0,c_j]$ for some $c_j \in (0,\infty)$. Then, if $A \in \mathbb{R}^{p\times p}$ is symmetric positive definite, the monotone quadratic loss function based on A and v_j is bounded, symmetric, subconvex, and uniformly continuous on $\bar{\mathbb{R}}^p$ because the quadratic form $z \mapsto z'Az$ is convex, and uniformly continuous on $\{z \mid z'Az \leq c_j\}$. Moreover $\ell_j(z) \uparrow u_m(z'Az)$ a.e. $\lambda^p(dz)$ as $j \to \infty$. ////

Asymptotic Admissibility

The as. minimax criterion may be complemented by results on as. admissibility and uniqueness. Such results follow from Proposition 3.3.4 and Lemma 3.3.5, making use of Lemma 3.3.11 and its proof.

Proposition 3.3.13 *Assume some $p \times k$ matrix D of rank $p \leq k$, a lower semicontinuous loss function $\ell \in \mathsf{L}$, and let the models $(\mathcal{Q}_n)$ be as. normal with as. covariance $C > 0$ and as. sufficient statistic Z.*

(a) *Suppose $K_0 = \mathbf{I}_D$ is admissible. Then, if an as. estimator S, for some $u \in \mathbb{R}^k$, satisfies*

$$\liminf_{n\to\infty} \int \ell\,(S_n - Du)\, dQ_{n,u} < \rho_0 \tag{91}$$

or

$$\lim_{b\to\infty} \limsup_{n\to\infty} \int b \wedge \ell\,(S_n - Du)\, dQ_{n,u} < \rho_0 \tag{92}$$

there exists another $v \in \mathbb{R}^k$ such that

$$\lim_{b\to\infty} \limsup_{n\to\infty} \int b \wedge \ell\,(S_n - Dv)\, dQ_{n,v} > \rho_0 \tag{93}$$

(b) *Suppose that $K_0 = \mathbf{I}_D$ is admissible and uniquely determined by its risk function. Then an as. estimator $S^\star$ achieving*

$$\lim_{b\to\infty} \limsup_{n\to\infty} \int b \wedge \ell\,(S_n^\star - Dt)\, dQ_{n,t} \leq \rho_0 \tag{94}$$

for all $t \in \mathbb{R}^k$, must have the as. expansion

$$S_n^\star = DC^{-1} Z_n + \mathrm{o}_{Q_{n,0}}(n^0) \tag{95}$$

PROOF The conditions on K_0 refer to the estimation of Dt in the normal shift model (71), by Markov kernels from $\mathbb{R}^k$ to $\bar{\mathbb{B}}^p$, under loss ℓ.

(a) Assuming (91), pass to a subsequence (m) along which the risks at u converge. Lemma 3.3.11 provides a further subsequence, still denoted by (m), and a kernel K_S for the limiting model such that

$$\rho(K_S, u, \ell) \leq \lim_{b\to\infty} \liminf_{m\to\infty} \int b \wedge \ell\,(S_m - Du)\, dQ_{m,u} < \rho(K_0, u, \ell) = \rho_0 \tag{96}$$

Starting with $\mathbb{N}$ this can also be achieved under assumption (92) since the $\liminf$ along any subsequence is less or equal the $\limsup$ along the full sequence. In the present situation (e.g., under the assumptions of Proposition 3.3.4) K_0 is admissible. Hence there exists some $v \in \mathbb{R}^k$ such that

$$\rho_0 = \rho(K_0, v, \ell) < \rho(K_S, v, \ell) \leq \lim_{b\to\infty} \liminf_{m\to\infty} \int b \wedge \ell\,(S_m - Dv)\, dQ_{m,v} \tag{97}$$

This proves (93) on passing to the $\limsup$ along the full sequence.

(b) If an estimator $S^\star$ has as. risk everywhere less than ρ_0, any corresponding limiting kernel $K^\star$ supplied by Lemma 3.3.11 satisfies

$$\rho(K^\star,t,\ell) \le \rho_0 = \rho(K_0,t,\ell)\,, \qquad t \in \mathbb{R}^k \tag{98}$$

K_0 being admissible and uniquely determined by its risk function (so under the assumptions of Proposition 3.3.4 and Lemma 3.3.5 b), it follows that

$$K^\star(ds,x) = K_0(ds,x) = \mathbf{I}_{Dx}(ds) \qquad \text{a.e. } \lambda^k(dx) \tag{99}$$

But, recalling construction (73) and (79) from the proof to Lemma 3.3.11, we have

$$K^\star(ds,x) = \Pr\bigl(\bar S^\star \in ds \big| X = x\bigr) \tag{100}$$

where $X = C^{-1}\bar Z \sim \mathcal{N}(0,C^{-1})$ and $(S^\star_m, Z_m)(Q_{m,0}) \xrightarrow{\mathrm{w}} \mathcal{L}(\bar S^\star, \bar Z)$. Hence

$$\bigl(S^\star_m - DC^{-1}Z_m\bigr)(Q_{m,0}) \xrightarrow{\mathrm{w}} \mathcal{L}(\bar S^\star - DX) \tag{101}$$

Therefore,

$$\begin{aligned}
\Pr\bigl(\bar S^\star = DX\bigr) &= \int \Pr\bigl(\bar S^\star = DX \big| X = x\bigr)\, \mathcal{N}(0,C^{-1})(dx) \\
&= \int \Pr\bigl(\bar S^\star = Dx \big| X = x\bigr)\, \mathcal{N}(0,C^{-1})(dx) \\
&\overset{(100)}{=} \int K^\star\bigl(\{Dx\},x\bigr)\, \mathcal{N}(0,C^{-1})(dx) \\
&\overset{(99)}{=} \int K_0\bigl(\{Dx\},x\bigr)\, \mathcal{N}(0,C^{-1})(dx) = 1
\end{aligned} \tag{102}$$

So $\mathcal{L}(\bar S^\star - DX) = \mathbf{I}_0$ (Dirac at 0). Thus (101) implies (95), for a subsequence (m) which has possibly been thinned out twice in the course of the proof. But as we may have started out with an arbitrary subsequence, (95) has in fact been proved for the full sequence [Lemma A.1.8]. ////

Remark 3.3.14 The unique expansion of an as. minimax estimator is due to Hájek (1972, second part of Theorem 4.1) in case $p = 1$ and for monotone quadratic loss functions. In case $p = k = 1$ and for mean square error, Huber (1966) proved the necessary as. normality of an as. minimax estimator actually implied by this expansion. Also in case $p = k = 1$, as. uniqueness can be derived for confidence intervals and coverage probabilities using Neyman–Pearson techniques [Rieder (1981)]. The proof above reflects LeCam's (1979) abstract argument that the as. minimax procedure is as. unique if the limit experiment has a minimax procedure that is nonrandomized, admissible, and uniquely determined by its risk function. ////

In case $p > 2$, the as. estimator $S^\star$ assumed in Theorem 3.3.8 c to achieve the as. minimax bound may be modified as in James and Stein (1961),

$$\tilde S = (\tilde S_n), \qquad \tilde S_n = \left(1 - \frac{p-2}{|S^\star_n|^2_\Gamma}\right) S^\star_n \tag{103}$$

using the norm induced by $\Gamma = DC^{-1}D'$. This transformation being continuous, the pole at 0 aside, (61) entails that for all $c \in (0, \infty)$, uniformly in $|t| \le c$,

$$(\tilde{S}_n - Dt)(Q_{n,t}) \xrightarrow{\mathrm{w}} \mathcal{L}\left\{\left(1 - \frac{p-2}{|Y + Dt|_\Gamma^2}\right)(Y + Dt) - Dt\right\} \tag{104}$$

where $Y \sim \mathcal{N}(0, \Gamma)$. Employ the loss function $\ell = |\,.\,|_\Gamma^2$ and recall $\tilde{S}_0$ given by (54). Then for all $b, c \in (0, \infty)$, uniformly in $|t| \le c$,

$$\begin{aligned} \lim_{n\to\infty} & \int b \wedge \ell\,(\tilde{S}_n - Dt)\,dQ_{n,t} \\ &= \int b \wedge \ell \left\{\left(1 - \frac{p-2}{|y|_\Gamma^2}\right)y - Dt\right\} \mathcal{N}(Dt, \Gamma)(dy) \\ &= \int b \wedge \ell\,(\tilde{S}_0(x) - Dt)\,\mathcal{N}(t, C^{-1})(dx) \end{aligned} \tag{105}$$

hence also

$$\begin{aligned} \lim_{n\to\infty} \sup_{|t|\le c} & \int b \wedge \ell\,(\tilde{S}_n - Dt)\,dQ_{n,t} \\ &= \sup_{|t|\le c} \int b \wedge \ell\,(\tilde{S}_0(x) - Dt)\,\mathcal{N}(t, C^{-1})(dx) \\ &\le \sup_{|t|\le c} \int \ell\,(\tilde{S}_0(x) - Dt)\,\mathcal{N}(t, C^{-1})(dx) \end{aligned} \tag{106}$$

Invoking inequality (55), this implies as. minimaxity of $\tilde{S}$,

$$\lim_{b\to\infty} \lim_{c\to\infty} \lim_{n\to\infty} \sup_{|t|\le c} \int b \wedge \ell\,(\tilde{S}_n - Dt)\,dQ_{n,t} \le p = \rho_0 \tag{107}$$

But (105) and strict inequality in (55) show that for all $t \in \mathbb{R}^k$,

$$\lim_{b\to\infty} \lim_{n\to\infty} \int b \wedge \ell\,(\tilde{S}_n - Dt)\,dQ_{n,t} < p = \rho_0 \tag{108}$$

which renders $S^\star$ inadmissible. We even have

$$\lim_{b\to\infty} \lim_{n\to\infty} \sup_{|t|\le c} \int b \wedge \ell\,(\tilde{S}_n - Dt)\,dQ_{n,t} < p = \rho_0 \tag{109}$$

for all $c \in (0, \infty)$, in view of (106) and since

$$\sup_{|t|\le c} \int \ell\,(\tilde{S}_0(x) - Dt)\,\mathcal{N}(t, C^{-1})(dx) < p = \rho_0 \tag{110}$$

due to (55) and the continuity of the risk function of $\tilde{S}_0$.

3.4 Testing

This section derives as. power bounds for testing some classical hypotheses.

3.4.1 Simple Hypotheses

We start with the simple as. testing problem between two sequences of probabilities P_n vs. Q_n on sample spaces $(\Omega_n, \mathcal{A}_n)$, in the setup of Theorem 2.2.3. That is, we assume contiguity,

$$(Q_n) \ll (P_n) \tag{1}$$

and just one weak cluster point of the log likelihoods $L_n \in \log dQ_n/dP_n$ under P_n,

$$L_n(P_n) \xrightarrow{\mathrm{w}} M \in \mathcal{M}_1(\bar{\mathbb{B}}) \tag{2}$$

Define $M_1 \in \mathcal{M}_1(\bar{\mathbb{B}})$ by

$$M_1(du) = e^u\, M(du) \tag{3}$$

For testing M vs. M_1, the Neyman–Pearson tests

$$\psi(u) = \begin{cases} 1 & \text{if } u > c \\ 0 & \text{if } u < c \end{cases} \tag{4}$$

with critical values $c \in \bar{\mathbb{R}}$ (and constant randomization) are (essentially) complete under a variety of optimality criteria. In the as. setup, tests arise as sequences $\phi = (\phi_n)$ of finite-sample tests ϕ_n on $(\Omega_n, \mathcal{A}_n)$.

Theorem 3.4.1 *Under assumptions* (1) *and* (2), *let* ψ *be a Neyman–Pearson test for* M *vs.* M_1 *with critical value* $c \in \bar{\mathbb{R}}$.

(a) *Then, if an as. test* ϕ *satisfies*

$$\limsup_{n\to\infty} \int \phi_n\, dP_n \le \int \psi\, dM \tag{5}$$

necessarily

$$\limsup_{n\to\infty} \int \phi_n\, dQ_n \le \int \psi\, dM_1 \tag{6}$$

(b) *In case* $M(\{c\}) = 0$, *an as. test* $\phi^\star$ *subject to* (5) *achieves the upper bound* (6) *iff*

$$\phi_n^\star = \mathbf{I}(L_n > c) + \mathrm{o}_{P_n}(n^0) \tag{7}$$

PROOF

(a) Passing to subsequences (m) we may assume that

$$\int \phi_m\, dQ_m \longrightarrow \beta, \qquad \int \phi_m\, dP_m \longrightarrow \alpha \overset{(5)}{\le} \int \psi\, dM$$

and

$$(\phi_m, L_m)(P_m) \xrightarrow{\mathrm{w}} N \in \mathcal{M}_1(\bar{\mathbb{B}}^2) \tag{8}$$

since the sequence of laws $(\phi_m, L_m)(P_m)$ is certainly tight in $\bar{\mathbb{R}}^2$. Due to (8) and assumption (1), Theorem 2.2.5 is in force and yields

$$(\phi_m, L_m)(Q_m) \xrightarrow{w} N_1(ds, du) = e^u\, N(ds, du) \tag{9}$$

Denoting the canonical projections from $\bar{\mathbb{R}}^2$ to $\bar{\mathbb{R}}$ by

$$S(s,u) = s\,, \qquad L(s,u) = u \tag{10}$$

then $M = L(N)$ by our assumption (2), and $M_1 = L(N_1)$ by Theorem 2.2.3. Now, computing conditional expectation under N, define

$$\varphi(u) = \mathrm{E}(S|L=u) = \int s\, N(ds|u) \qquad \text{a.e. } M(du) \tag{11}$$

But, according to 2.2(27), the regular conditional distribution given L of S under N_1 is the same as that of S under N; hence we also have

$$\varphi(u) = \mathrm{E}_1(S|L=u) = \int s\, N_1(ds|u) \qquad \text{a.e. } M_1(du) \tag{12}$$

Moreover, because of (8) and (9), it holds that

$$\phi_m(P_m) \xrightarrow{w} S(N)\,, \qquad \phi_m(Q_m) \xrightarrow{w} S(N_1)$$

Therefore $0 \le S \le 1$ a.e. N by A.2(4) and, using Proposition A.1.1 b,

$$\int S\, dN = \lim_{m\to\infty} \int \phi_m\, dP_m = \alpha\,, \quad \int S\, dN_1 = \lim_{m\to\infty} \int \phi_m\, dQ_m = \beta \tag{13}$$

In view of (11) and (12), it also follows that $0 \le \varphi \le 1$ a.e. M, and

$$\int \varphi\, dM = \int \mathrm{E}(S|L=u)\, L(N)(du) = \int \mathrm{E}(S|L)\, dN = \int S\, dN = \alpha$$
$$\int \varphi\, dM_1 = \int \mathrm{E}_1(S|L=u)\, L(N_1)(du) = \int \mathrm{E}_1(S|L)\, dN_1 = \int S\, dN_1 = \beta$$

Thus a test φ has been constructed for M vs. M_1, and we conclude that

$$\int \varphi\, dM = \alpha \le \int \psi\, dM \implies \beta = \int \varphi\, dM_1 \le \int \psi\, dM_1$$

since ψ maximizes the power subject to its own level. This proves (6).

(b) If an as. test $\phi^\star$, subject to (5), achieves the power bound (6), the corresponding test $S^\star$, in view of (13), achieves power $\beta = \int \psi\, dM_1$ subject to level $\alpha \le \int \psi\, dM$, in the testing problem N vs. N_1. Since

$$\frac{dN_1}{dN} = \frac{dM_1}{dM} \circ L$$

however, $\psi \circ L = \mathbf{I}(L > c)$ is a Neyman–Pearson test for N vs. N_1. Therefore,

$$0 \le \int \bigl(\mathbf{I}(L > c) - S^\star\bigr)(dN_1 - e^c\, dN) \le 0 \tag{14}$$

from which, since $M(\{c\}) = 0$, it follows that

$$S^\star = \mathbf{I}(L > c) \qquad \text{a.e. } N \tag{15}$$

(The optimum test is nonrandomized and unique.) The continuous mapping theorem [Proposition A.1.1], because of (8) and $M(\{c\}) = 0$, implies that

$$\bigl(\phi_m^\star, \mathbf{I}(L_m > c)\bigr)(P_m) \xrightarrow{\text{w}} \bigl(S^\star, \mathbf{I}(L > c)\bigr)(N) \tag{16}$$

hence

$$\limsup_m P_m\bigl(|\phi_m^\star - \mathbf{I}(L_m > c)| \ge \varepsilon\bigr) \le N\bigl(|S^\star - \mathbf{I}(L > c)| \ge \varepsilon\bigr) \overset{(15)}{=} 0$$

for all $\varepsilon \in (0,1)$, and this proves (7), since we may have started out from an arbitrary subsequence (m) of $\mathbb{N}$.

The converse is a matter of dominated and weak convergence. ////

In the normal case of Corollary 2.2.4, and denoting the upper α point of the standard normal $\mathcal{N}(0,1)$ for some level $\alpha \in [0,1]$ henceforth by u_α, this result may be summarized as follows.

Corollary 3.4.2 *Assume that for some $\sigma \in [0,\infty)$,*

$$L_n(P_n) \xrightarrow{\text{w}} \mathcal{N}(-\sigma^2/2, \sigma^2) \tag{17}$$

(a) *Let ϕ be an as. test such that*

$$\limsup_{n\to\infty} \int \phi_n\, dP_n \le \alpha \tag{18}$$

Then

$$\limsup_{n\to\infty} \int \phi_n\, dQ_n \le 1 - \Phi(u_\alpha - \sigma) \tag{19}$$

(b) *An as. test $\phi^\star$ subject to (18) achieves the upper bound (19) iff*

$$\phi_n^\star = \mathbf{I}\bigl(L_n > \sigma u_\alpha - \sigma^2/2\bigr) + \mathrm{o}_{P_n}(n^0) \tag{20}$$

and then equality also holds in (18).

Example 3.4.3 In the normal case (17), the as. lower bound $\Phi(-\sigma/2)$ holds for the maximum error probability, and $2\,\Phi(-\sigma/2)$ for the sum of error probabilites, since $\psi = \mathbf{I}_{(0,\infty)}$ is the optimum Neyman–Pearson test in the limiting problem $\mathcal{N}(-\sigma^2/2, \sigma^2)$ vs. $\mathcal{N}(\sigma^2/2, \sigma^2)$, for both risks. ////

3.4.2 Passage to the Normal Limit

In the sequel, we assume statistical models on sample spaces $(\Omega_n, \mathcal{A}_n)$,

$$\mathcal{Q}_n = \{\, Q_{n,t} \mid t \in \mathbb{R}^k \,\} \subset \mathcal{M}_1(\mathcal{A}_n)$$

having the same finite-dimensional parameter space $\Theta_n = \mathbb{R}^k$ (or open subsets $\Theta_n \uparrow \mathbb{R}^k$) that are as. normal. The comparison with the limiting normal model is based on the following result.

Lemma 3.4.4 *Assume as. normal models $(\mathcal{Q}_n)$ with as. covariance $C > 0$. Let ϕ be an as. test. Then for every subsequence of $\mathbb{N}$ there exist a further subsequence (m) and a test φ on $(\mathbb{R}^k, \mathbb{B}^k)$ such that for all $t \in \mathbb{R}^k$, along (m),*

$$\int \phi_m \, dQ_{m,t} \longrightarrow \int \varphi \, d\mathcal{N}(C^{1/2}t, \mathbb{I}_k) \tag{21}$$

PROOF By as. normality, if Z denotes the as. sufficient statistic, the laws $Z_n(Q_{n,0})$ tend weakly to $\mathcal{N}(0, C)$; hence this sequence is tight in $\mathbb{R}^k$. Then the sequence of joint laws $(\phi_n, Z_n)(Q_{n,0})$ is tight in $[0,1] \times \mathbb{R}^k$, hence weakly relatively sequentially compact [Prokhorov]. Thus, given any subsequence, there exist a random vector $(\bar\varphi, \bar Z)$ and a further subsequence (m) such that, in $[0,1] \times \mathbb{R}^k$,

$$(\phi_m, Z_m)(Q_{m,0}) \xrightarrow{\mathrm{w}} \mathcal{L}(\bar\varphi, \bar Z) \tag{22}$$

In particular, we have $0 \le \bar\varphi \le 1$ and $\bar Z \sim \mathcal{N}(0, C)$. Then, by Theorem 2.2.5, for every $t \in \mathbb{R}^k$, the log likelihoods $L_{m,t} \in \log dQ_{m,t}/dQ_{m,0}$ satisfy

$$\begin{aligned} (\phi_m, L_{m,t})(Q_{m,0}) &\xrightarrow{\mathrm{w}} \mathcal{L}\big(\bar\varphi, t'\bar Z - \tfrac{1}{2}t'Ct\big) \\ (\phi_m, L_{m,t})(Q_{m,t}) &\xrightarrow{\mathrm{w}} e^{\pi_2}\, d\mathcal{L}\big(\bar\varphi, t'\bar Z - \tfrac{1}{2}t'Ct\big) \end{aligned} \tag{23}$$

From weak convergence [Proposition A.1.1 b] it follows that

$$\int \phi_m \, dQ_{m,t} \longrightarrow \mathrm{E}\, \bar\varphi \exp\big(t'\bar Z - \tfrac{1}{2}t'Ct\big) \tag{24}$$

Condition on $X = C^{-1/2}\bar Z \sim \mathcal{N}(0, \mathbb{I}_k)$ and reparametrize $s = C^{1/2}t$ to obtain

$$\begin{aligned} \mathrm{E}\, \bar\varphi \exp\big(t'\bar Z - \tfrac{1}{2}t'Ct\big) &= \mathrm{E}\, \mathrm{E}(\bar\varphi|X) \exp\big(s'X - \tfrac{1}{2}|s|^2\big) \\ &= \int \mathrm{E}(\bar\varphi|X = x) \exp\big(s'x - \tfrac{1}{2}|s|^2\big)\, \mathcal{N}(0, \mathbb{I}_k)(dx) \\ &= \int \mathrm{E}(\bar\varphi|X = x)\, \mathcal{N}(s, \mathbb{I}_k)(dx) = \int \varphi(x)\, \mathcal{N}(s, \mathbb{I}_k)(dx) \end{aligned} \tag{25}$$

with the test

$$\varphi = \mathrm{E}(\bar\varphi|X = .\,) \tag{26}$$

This proves the result. ////

3.4.3 One- and Two-Sided Hypotheses

Given $e \in \mathbb{R}^k$, $|e| = 1$, and numbers $|b| < a \leq \infty$, one-sided hypotheses about the parameter $t \in \mathbb{R}^k$ of the measures $Q_{n,t} \in \mathcal{Q}_n$ are

$$\mathrm{H}_\mathrm{p}^1\colon\ |C^{1/2}t| \leq a\,,\ e'C^{1/2}t \leq b \qquad \mathrm{K}_\mathrm{p}^1\colon\ |C^{1/2}t| \leq a\,,\ e'C^{1/2}t > b \tag{27}$$

with boundary

$$\mathrm{B}_\mathrm{p}^1\colon\ |C^{1/2}t| \leq a\,,\ e'C^{1/2}t = b \tag{28}$$

The subscript p shall indicate pointwise evaluation of power functions.

Theorem 3.4.5 *Assuming models $(\mathcal{Q}_n)$ that are as. normal with as. covariance $C > 0$ and as. sufficient statistic Z, let us consider the one-sided as. testing hypotheses (27) at some level $\alpha \in [0,1]$.*

(a) *$k = 1$: Then an as. test ϕ which satisfies*

$$\limsup_{n\to\infty} \int \phi_n \, dQ_{n,t} \leq 1 - \Phi\bigl(u_\alpha - (C^{1/2}t - b)\bigr) \tag{29}$$

for some $t \in \mathrm{H}_\mathrm{p}^1$, fulfills bound (29) for all $t \in \mathrm{K}_\mathrm{p}^1$.

An as. test $\phi^\star$ subject to (29) for some $t \in \mathrm{H}_\mathrm{p}^1$ can achieve equality in (29) for some other $t \in \mathrm{K}_\mathrm{p}^1$ only if

$$\phi_n^\star = \mathbf{I}\bigl(C^{-1/2}Z_n > u_\alpha + b\bigr) + \mathrm{o}_{Q_{n,0}}(n^0) \tag{30}$$

And (30) entails that

$$\lim_{n\to\infty} \int \phi_n^\star \, dQ_{n,t} = 1 - \Phi\bigl(u_\alpha - (C^{1/2}t - b)\bigr) \tag{31}$$

for all $t \in \mathrm{H}_\mathrm{p}^1 \cup \mathrm{K}_\mathrm{p}^1$.

(b) *$k \geq 1$: Under the as. similarity condition*

$$\lim_{n\to\infty} \int \phi_n \, dQ_{n,t} = \alpha\,, \qquad t \in \mathrm{B}_\mathrm{p}^1 \tag{32}$$

which is implied by as. unbiasedness,

$$\sup_{t\in \mathrm{H}_\mathrm{p}^1} \limsup_{n\to\infty} \int \phi_n \, dQ_{n,t} \leq \alpha \leq \inf_{t\in \mathrm{K}_\mathrm{p}^1} \liminf_{n\to\infty} \int \phi_n \, dQ_{n,t} \tag{33}$$

an as. test ϕ satisfies

$$\limsup_{n\to\infty} \int \phi_n \, dQ_{n,t} \leq 1 - \Phi\bigl(u_\alpha - (e'C^{1/2}t - b)\bigr)\,, \qquad t \in \mathrm{K}_\mathrm{p}^1 \tag{34}$$

$$\liminf_{n\to\infty} \int \phi_n \, dQ_{n,t} \geq 1 - \Phi\bigl(u_\alpha - (e'C^{1/2}t - b)\bigr)\,, \qquad t \in \mathrm{H}_\mathrm{p}^1 \tag{35}$$

Subject to (32), an as. test $\phi^\star$ can achieve equality in (34) for some $t \in \mathrm{K}^1_\mathrm{p}$ only if

$$\phi^\star_n = \mathbf{I}\big(e'C^{-1/2}Z_n > u_\alpha + b\big) + \mathrm{o}_{Q_{n,0}}(n^0) \tag{36}$$

And (36) entails that

$$\lim_{n\to\infty} \int \phi^\star_n \, dQ_{n,t} = 1 - \Phi\big(u_\alpha - (e'C^{1/2}t - b)\big) \tag{37}$$

for all $t \in \mathrm{H}^1_\mathrm{p} \cup \mathrm{K}^1_\mathrm{p}$; in particular, the as. test $\phi^\star$ is unbiased.

PROOF

(a) Lemma 3.4.4 carries us to the testing problem

$$\bar{\mathrm{H}}^1_\mathrm{p}\colon\ |s| \le a\,,\ s \le b \qquad \bar{\mathrm{K}}^1_\mathrm{p}\colon\ |s| \le a\,,\ s > b \tag{38}$$

about the normals $\{\mathcal{N}(s,1) \mid s \in \mathbb{R}\}$. Due to monotone likelihood ratios, the test

$$\psi = \mathbf{I}_{(u_\alpha + b, \infty)} \tag{39}$$

is Neyman–Pearson for each pair $s' < s''$. Uniqueness (30) and optimality (31) now follow from the uniqueness result in the case of simple hypotheses [Theorem 3.4.1 b] and the as. normality assumption; that is, from the convergence $Z_n(Q_{n,t}) \xrightarrow{\ \mathrm{w}\ } \mathcal{N}(Ct, C)$ and the log likelihood expansion.

(b) Lemma 3.4.4 carries us to the limiting model $\{\mathcal{N}(s, \mathbb{I}_k) \mid s \in \mathbb{R}^k\}$ and the corresponding testing problem

$$\bar{\mathrm{H}}^1_\mathrm{p}\colon\ |s| \le a\,,\ e's \le b \qquad \bar{\mathrm{K}}^1_\mathrm{p}\colon\ |s| \le a\,,\ e's > b \tag{40}$$

A transformation by an orthogonal matrix whose first row vector is e' leads us to the normal model

$$\mathcal{N}(\xi, \mathbb{I}_k) = \mathcal{N}(\eta, 1) \otimes \mathcal{N}(\zeta, \mathbb{I}_{k-1})\,, \qquad \xi = (\eta, \zeta')' \in \mathbb{R} \times \mathbb{R}^{k-1} \tag{41}$$

and the corresponding hypotheses

$$\bar{\mathrm{H}}^1_\mathrm{p}\colon\ |\xi| \le a\,,\ \eta \le b \qquad \bar{\mathrm{K}}^1_\mathrm{p}\colon\ |\xi| \le a\,,\ \eta > b \tag{42}$$

Now suppose that a test φ satisfies

$$\iint \varphi(u,v)\, \mathcal{N}(b,1)(du)\, \mathcal{N}(\zeta, \mathbb{I}_{k-1})(dv) = \alpha\,, \qquad |\zeta|^2 \le a^2 - b^2 \tag{43}$$

Then the completeness of the family $\big\{\mathcal{N}(\zeta, \mathbb{I}_{k-1}) \bigm| |\zeta|^2 \le a^2 - b^2\big\}$ tells us that there exists a Borel set A of Lebesgue measure 0 such that

$$\int \varphi(u,v)\, \mathcal{N}(b,1)(du) = \alpha\,, \qquad v \notin A \tag{44}$$

But due to monotone likelihood ratios, the test

$$\psi = \mathbf{I}_{(u_\alpha + b, \infty)} \tag{45}$$

is Neyman–Pearson for b vs. any $\eta > b$ at level α. Therefore,

$$\int \varphi(u,v)\,\mathcal{N}(\eta,1)(du) \le \int \psi(u)\,\mathcal{N}(\eta,1)(du)\,, \qquad v \notin A \tag{46}$$

Since $\mathcal{N}(\zeta, \mathbb{I}_{k-1})(A) = 0$ it follows that for all $|\xi| \le a$ with $\eta > b$,

$$\begin{aligned} \iint \varphi(u,v)\,\mathcal{N}(\eta,1)(du)\,\mathcal{N}(\zeta,\mathbb{I}_{k-1})(dv) &\le \int \psi(u)\,\mathcal{N}(\eta,1)(du) \\ &= 1 - \Phi\bigl(u_\alpha - (\eta - b)\bigr) \end{aligned} \tag{47}$$

which proves bound (34). Bound (35) follows likewise from the fact that the test $1-\psi$ is Neyman–Pearson for b vs. any $\eta < b$ at level $1-\alpha$.

Uniqueness (36) is a consequence of the uniqueness statement for simple hypotheses [Theorem 3.4.1 b] since the test ψ, for every $(\eta,\zeta) \in \bar{\mathrm{K}}^1_{\mathrm{p}}$, is Neyman–Pearson for (b,ζ) vs. (η,ζ). The uniqueness argument as well as the argument to prove (37) use as. normality, namely, the expansion of log likelihoods and $\bigl(C^{-1/2}Z_n\bigr)(Q_{n,t}) \xrightarrow{\mathrm{w}} \mathcal{N}(C^{1/2}t, \mathbb{I}_k)$.

Unbiasedness (33) implies similarity (32), because the power function $s \mapsto \int \varphi\, d\mathcal{N}(s, \mathbb{I}_k)$ of the limit test φ associated with ϕ via Lemma 3.4.4 is continuous. ////

For the following one-sided hypotheses, maxima and minima of power functions (subscript m) are considered. Given any $e \in \mathbb{R}^k$, $|e| = 1$, and numbers $-\infty < b < c < \infty$ such that $|b| \vee |c| \le a \le \infty$, these hypotheses about the parameter $t \in \mathbb{R}^k$ of the measures $Q_{n,t} \in \mathcal{Q}_n$ are

$$\mathrm{H}^1_{\mathrm{m}}\colon\ |C^{1/2}t| \le a\,,\ e'C^{1/2}t \le b \qquad \mathrm{K}^1_{\mathrm{m}}\colon\ |C^{1/2}t| \le a\,,\ e'C^{1/2}t \ge c \tag{48}$$

Theorem 3.4.6 *Assuming models $(\mathcal{Q}_n)$ that are as. normal with as. covariance $C > 0$ and as. sufficient statistic Z, let us consider the one-sided as. testing hypotheses (48) at some level $\alpha \in [0,1]$.*

(a) *Then, every as. test ϕ such that*

$$\limsup_{n\to\infty} \sup_{\mathrm{H}^1_{\mathrm{m}}} \int \phi_n\, dQ_{n,t} \le \alpha \tag{49}$$

necessarily satisfies

$$\limsup_{n\to\infty} \inf_{\mathrm{K}^1_{\mathrm{m}}} \int \phi_n\, dQ_{n,t} \le 1 - \Phi(u_\alpha - (c-b)) \tag{50}$$

(b) *If a sequence of statistics $Z^\star_n\colon (\Omega_n, \mathcal{A}_n) \to (\mathbb{R}^k, \mathbb{B}^k)$ are as. normal,*

$$Z^\star_n(Q_{n,t}) \xrightarrow{\mathrm{w}} \mathcal{N}(Ct, C) \tag{51}$$

uniformly in $|C^{1/2}t| \le a$, *then the as. test* $\phi^\star$,

$$\phi_n^\star = \mathbf{I}\big(e'C^{-1/2}Z_n^\star > u_\alpha + b\big) \tag{52}$$

achieves equalities in (49) *and* (50).

Remark 3.4.7 In case $k=1$ this result is implied by Theorem 3.4.5 a. In case $k>1$ it is not implied by Theorem 3.4.5 b, because Theorem 3.4.6 dispenses with the as. similarity condition (32). ////

PROOF

(a) Let (n) be a subsequence of $\mathbb{N}$ such that

$$\inf_{\mathrm{K}_\mathrm{m}^1} \int \phi_n\, dQ_{n,t} \longrightarrow \bar\beta, \qquad \sup_{\mathrm{H}_\mathrm{m}^1} \int \phi_n\, dQ_{n,t} \longrightarrow \bar\alpha \overset{(49)}{\le} \alpha \tag{53}$$

Then Lemma 3.4.4 carries us to the testing problem

$$\bar{\mathrm{H}}_\mathrm{m}^1\colon\ |s| \le a,\ e's \le b \qquad \bar{\mathrm{K}}_\mathrm{m}^1\colon\ |s| \le a,\ e's \ge c \tag{54}$$

in the normal model $\{\mathcal{N}(s,\mathbb{I}_k) \mid s \in \mathbb{R}^k\}$, and gives us a test φ such that

$$\sup_{\bar{\mathrm{H}}_\mathrm{m}^1} \int \varphi\, d\mathcal{N}(s,\mathbb{I}_k) \overset{(21)}{\le} \lim_{n\to\infty} \sup_{\mathrm{H}_\mathrm{m}^1} \int \phi_n\, dQ_{n,t} \overset{(53)}{=} \bar\alpha \le \alpha \tag{55}$$

Hence φ is of level α in the limiting problem (54). A straightforward solution to this problem is the Neyman–Pearson test

$$\psi(x) = \mathbf{I}(e'x > u_\alpha + b) \tag{56}$$

between the following pair of parameters,

$$s_0 = be \quad \text{vs.} \quad s_1 = ce \tag{57}$$

which turn out least favorable. The test ψ achieves maximin power

$$\beta = 1 - \Phi(u_\alpha - (c-b)) \tag{58}$$

Thus,

$$\bar\beta \overset{(53)}{=} \lim_{n\to\infty} \inf_{\mathrm{K}_\mathrm{m}^1} \int \phi_n\, dQ_{n,t} \overset{(21)}{\le} \inf_{\bar{\mathrm{K}}_\mathrm{m}^1} \int \varphi\, d\mathcal{N}(s,\mathbb{I}_k) \overset{(58)}{\le} \beta \tag{59}$$

which proves (50).

(b) If $Z_n^\star(Q_{n,t})$ tend weakly to $\mathcal{N}(Ct,C)$, uniformly in $|C^{1/2}t| \le a$, then so do $e'C^{-1/2}Z_n^\star(Q_{n,t})$ converge to $\mathcal{N}(e'C^{1/2}t,1)$. Hence also

$$\int \phi_n^\star\, dQ_{n,t} - 1 + \Phi\big(u_\alpha - (e'C^{1/2}t - b)\big) \longrightarrow 0$$

uniformly in $|C^{1/2}t| \le a$, proving (b). ////

Assuming a unit vector $e \in \mathbb{R}^k$, $|e| = 1$, and numbers $b_1 < b_2$ and b, such that $|b| \vee |b_1| \vee |b_2| < a \le \infty$, two kinds of two-sided hypotheses about the parameter $t \in \mathbb{R}^k$ of the measures $Q_{n,t} \in \mathcal{Q}_n$ are defined by

$$\begin{aligned} \mathrm{H}_\mathrm{p}^{[2]}\colon\ & |C^{1/2}t| \le a\,,\ e'C^{1/2}t \in [b_1, b_2] \\ \mathrm{K}_\mathrm{p}^{[2]}\colon\ & |C^{1/2}t| \le a\,,\ e'C^{1/2}t \notin [b_1, b_2] \end{aligned} \tag{60}$$

with boundary

$$\mathrm{B}_\mathrm{p}^{[2]}\colon\ |C^{1/2}t| \le a\,,\ e'C^{1/2}t \in \{b_1, b_2\} \tag{61}$$

respectively, by

$$\mathrm{H}_\mathrm{p}^{\cdot 2}\colon\ |C^{1/2}t| \le a\,,\ e'C^{1/2}t = b \qquad \mathrm{K}_\mathrm{p}^{\cdot 2}\colon\ |C^{1/2}t| \le a\,,\ e'C^{1/2}t \ne b \tag{62}$$

Let $\psi^{\cdot 2}$ denote the following two-sided test on $(\mathbb{R}, \mathbb{B})$,

$$\psi^{\cdot 2} = 1 - \mathbf{I}_{[b - u_{\alpha/2}, b + u_{\alpha/2}]} \tag{63}$$

and $\psi^{[2]}$ the two-sided test on $(\mathbb{R}, \mathbb{B})$ given by

$$\psi^{[2]} = 1 - \mathbf{I}_{[c_1, c_2]}\,, \qquad \int \psi^{[2]}\, d\mathcal{N}(b_1, 1) = \alpha = \int \psi^{[2]}\, d\mathcal{N}(b_2, 1) \tag{64}$$

with the two critical values $c_1 < c_2$ thus suitably adjusted.

Theorem 3.4.8 *Assuming models $(\mathcal{Q}_n)$ that are as. normal with as. covariance $C > 0$ and as. sufficient statistic Z, let us consider the two-sided as. testing hypotheses* (60) *and* (62), *respectively, at some level $\alpha \in [0, 1]$.*

(a) *Then, under the as. similarity condition*

$$\lim_{n\to\infty} \int \phi_n dQ_{n,t} = \alpha\,, \qquad t \in \mathrm{B}_\mathrm{p}^{[2]} \tag{65}$$

which is implied by as. unbiasedness,

$$\sup_{t \in \mathrm{H}_\mathrm{p}^{[2]}} \limsup_{n\to\infty} \int \phi_n\, dQ_{n,t} \le \alpha \le \inf_{t \in \mathrm{K}_\mathrm{p}^{[2]}} \liminf_{n\to\infty} \int \phi_n\, dQ_{n,t} \tag{66}$$

an as. test ϕ satisfies

$$\limsup_{n\to\infty} \int \phi_n dQ_{n,t} \le \int \psi^{[2]}\, d\mathcal{N}(e'C^{1/2}t, 1)\,, \qquad t \in \mathrm{K}_\mathrm{p}^{[2]} \tag{67}$$

$$\liminf_{n\to\infty} \int \phi_n dQ_{n,t} \ge \int \psi^{[2]}\, d\mathcal{N}(e'C^{1/2}t, 1)\,, \qquad t \in \mathrm{H}_\mathrm{p}^{[2]} \tag{68}$$

The as. test $\phi^{[2]}$,

$$\phi_n^{[2]} = \psi^{[2]}(e'C^{-1/2}Z_n) \tag{69}$$

is unbiased and achieves equalities in (67) *and* (68).

(b) *Under the as. unbiasedness condition,*

$$\sup_{t\in \mathrm{H}_{\mathrm{p}}^{\cdot 2}} \limsup_{n\to\infty} \int \phi_n \, dQ_{n,t} \le \alpha \le \inf_{t\in \mathrm{K}_{\mathrm{p}}^{\cdot 2}} \liminf_{n\to\infty} \int \phi_n \, dQ_{n,t} \tag{70}$$

an as. test ϕ *satisfies*

$$\limsup_{n\to\infty} \int \phi_n \, dQ_{n,t} \le \int \psi^{\cdot 2} \, d\mathcal{N}(e'C^{1/2}t, 1), \qquad t \in \mathrm{K}_{\mathrm{p}}^{\cdot 2} \tag{71}$$

The as. test $\phi^{\cdot 2}$,

$$\phi_n^{\cdot 2} = \psi^{\cdot 2}\big(e'C^{-1/2}Z_n\big) \tag{72}$$

is unbiased (70) *and achieves equality in* (71).

PROOF

(a) Applying Lemma 3.4.4 and an orthogonal transformation, as in the proof to Theorem 3.4.5 b, we arrive at the normal model (41) and the hypotheses

$$\bar{\mathrm{H}}_{\mathrm{p}}^{[2]}\colon\ |\xi| \le a,\ \eta \in [b_1, b_2] \qquad \bar{\mathrm{K}}_{\mathrm{p}}^{[2]}\colon\ |\xi| \le a,\ \eta \notin [b_1, b_2] \tag{73}$$

The as. similarity (65) translates into the side condition

$$\iint \varphi(u,v)\, \mathcal{N}(\eta,1)(du)\, \mathcal{N}(\zeta, \mathbb{I}_{k-1})(dv) = \alpha, \qquad |\xi| \le a,\ \eta = b_1, b_2 \tag{74}$$

Then the completeness of the family of normals $\mathcal{N}(\zeta, \mathbb{I}_{k-1})$, when ζ ranges over any nonempty open subset of $\mathbb{R}^{k-1}$, implies that

$$\int \varphi(u,v)\, \mathcal{N}(\eta,1)(du) = \alpha \quad \text{a.e. } \lambda^{k-1}(dv), \qquad \eta = b_1, b_2 \tag{75}$$

But we know from the theory of two-sided tests in one parameter exponential families [fundamental lemma] that the test $\psi^{[2]}$ given by (64) is optimal two-sided unbiased in the model $\{\mathcal{N}(\eta,1) \mid \eta \in \mathbb{R}\}$, so that

$$\int \varphi(u,v)\, \mathcal{N}(\eta,1)(du) \le \int \psi^{[2]}(u)\, \mathcal{N}(\eta,1)(du) \quad \text{a.e. } \lambda^{k-1}(dv) \tag{76}$$

for all $\eta \notin [b_1, b_2]$ on one hand, and on the other hand for all $\eta \in [b_1, b_2]$,

$$\int \varphi(u,v)\, \mathcal{N}(\eta,1)(du) \ge \int \psi^{[2]}(u)\, \mathcal{N}(\eta,1)(du) \quad \text{a.e. } \lambda^{k-1}(dv) \tag{77}$$

Since $\mathcal{N}(\zeta, \mathbb{I}_{k-1}) \ll \lambda^{k-1}$, an integration proves (67) and (68).

Unbiasedness and optimality of the as. test $\phi^{[2]}$ are consequences of the weak convergence $\big(e'C^{-1/2}Z_n\big)(Q_{n,t}) \xrightarrow{\;\mathrm{w}\;} \mathcal{N}(e'C^{1/2}t, 1)$ and the corresponding properties of the limiting test $\psi^{[2]}$.

As. unbiasedness implies as. similarity on the boundary, because the power function $\xi \mapsto \int \varphi \, d\mathcal{N}(\xi, \mathbb{I}_k)$ of the limit test φ given in Lemma 3.4.4 is continuous.

(b) As in part (a), we arrive in the limit at the testing problem

$$\bar{\mathrm{H}}_{\mathrm{p}}^{\cdot 2}\colon\ |\xi| \le a\,,\ \eta = b \qquad \bar{\mathrm{K}}_{\mathrm{p}}^{\cdot 2}\colon\ |\xi| \le a\,,\ \eta \ne b \tag{78}$$

concerning the normal model (41).

As. unbiasedness (70) translates into unbiasedness of the limit test φ,

$$\begin{aligned} \int \varphi\, d\mathcal{N}(\xi, \mathbb{I}_k) &\le \alpha\,, \qquad |\xi| \le a\,,\ \eta = b \\ \int \varphi\, d\mathcal{N}(\xi, \mathbb{I}_k) &\ge \alpha\,, \qquad |\xi| \le a\,,\ \eta \ne b \end{aligned} \tag{79}$$

By continuity and differentiability of power functions in the normal model, (79) implies that for all $|\zeta|^2 \le a^2 - b^2$,

$$\begin{aligned} \iint \varphi(u,v)\, \mathcal{N}(b,1)(du)\, \mathcal{N}(\zeta, \mathbb{I}_{k-1})(dv) &= \alpha \\ \iint u\, \varphi(u,v)\, \mathcal{N}(b,1)(du)\, \mathcal{N}(\zeta, \mathbb{I}_{k-1})(dv) &= \alpha\, b \end{aligned} \tag{80}$$

Now the completeness of the family $\bigl\{\mathcal{N}(\zeta, \mathbb{I}_{k-1}) \bigm| |\zeta|^2 \le a^2 - b^2 \bigr\}$ entails that

$$\begin{aligned} \int \varphi(u,v)\, \mathcal{N}(b,1)(du) &= \alpha \qquad \text{a.e. } \lambda^{k-1}(dv) \\ \int u\, \varphi(u,v)\, \mathcal{N}(b,1)(du) &= \alpha\, b \quad \text{a.e. } \lambda^{k-1}(dv) \end{aligned} \tag{81}$$

The test $\psi^{\cdot 2} = 1 - \mathbf{I}_{[b - u_{\alpha/2}, b + u_{\alpha/2}]}$, by the fundamental lemma, is optimal two-sided unbiased in the model $\{\mathcal{N}(\eta, 1) \mid \eta \in \mathbb{R}\}$, so that for all $\eta \ne b$,

$$\int \varphi(u,v)\, \mathcal{N}(\eta,1)(du) \le \int \psi^{\cdot 2}\, \mathcal{N}(\eta,1)(du) \quad \text{a.e. } \lambda^{k-1}(dv) \tag{82}$$

Since $\mathcal{N}(\zeta, \mathbb{I}_{k-1}) \ll \lambda^{k-1}$, an integration proves (71).

Unbiasedness and optimality of the as. test $\phi^{\cdot 2}$ follow from the weak convergence $\bigl(e'C^{-1/2} Z_n\bigr)(Q_{n,t}) \xrightarrow{\mathrm{w}} \mathcal{N}(e'C^{1/2}t, 1)$, and the corresponding properties of the limiting test $\psi^{\cdot 2}$. ////

3.4.4 Multisided Hypotheses

For the definition of multisided as. hypotheses, let $1 \le p \le k$ and assume some matrix $E \in \mathbb{R}^{p \times k}$ whose rows in $\mathbb{R}^k$ are orthonormal,

$$EE' = \mathbb{I}_p \tag{83}$$

The previous use of a unit vector $e \in \mathbb{R}^k$ in the one- and two-sided problems corresponds to the specification $E = e'$.

Fix any numbers $0 \le b < c < \infty$ and $c \le a \le \infty$. Then multisided hypotheses about the parameter $t \in \mathbb{R}^k$ of $Q_{n,t}$ are defined by

$$\mathrm{H}_{\mathrm{m}}^{\mathrm{m}}\colon\ |C^{1/2}t| \le a\,,\ |EC^{1/2}t| \le b \qquad \mathrm{K}_{\mathrm{m}}^{\mathrm{m}}\colon\ |C^{1/2}t| \le a\,,\ |EC^{1/2}t| \ge c \quad (84)$$

In the limit, the following testing problem about the mean $s \in \mathbb{R}^k$ of the standard normals $\mathcal{N}(s, \mathbb{I}_k)$ will arise,

$$\bar{\mathrm{H}}_{\mathrm{m}}^{\mathrm{m}}\colon\ |s| \le a,\ |Es| \le b \qquad \bar{\mathrm{K}}_{\mathrm{m}}^{\mathrm{m}}\colon\ |s| \le a,\ |Es| \ge c \tag{85}$$

By $\chi^2(p; b^2)$ we denote the χ^2 distribution with p degrees of freedom and noncentrality parameter b^2, and by $c_\alpha(p; b^2)$ its upper α point.

Proposition 3.4.9 *Consider the hypotheses* (85) *at level* $\alpha \in [0,1]$.

(a) *Let* ϕ *be a test on* $(\mathbb{R}^k, \mathbb{B}^k)$ *such that*

$$\sup_{\bar{\mathrm{H}}_{\mathrm{m}}^{\mathrm{m}}} \int \phi\, d\mathcal{N}(s, \mathbb{I}_k) \le \alpha \tag{86}$$

Then

$$\inf_{\bar{\mathrm{K}}_{\mathrm{m}}^{\mathrm{m}}} \int \phi\, d\mathcal{N}(s, \mathbb{I}_k) \le \Pr\bigl(\chi^2(p; c^2) > c_\alpha(p; b^2)\bigr) \tag{87}$$

(b) *The* χ^2 *test* $\chi\colon \mathbb{R}^k \to [0,1]$,

$$\chi(x) = \mathbf{I}\bigl(|Ex|^2 > c_\alpha(p; b^2)\bigr) \tag{88}$$

is maximin—achieving equalities in (86) *and* (87).

(c) *The test* χ *of part* (b) *is admissible: If a test* ϕ *satisfies* (86) *and*

$$\int \phi\, d\mathcal{N}(v, \mathbb{I}_k) > \int \chi\, d\mathcal{N}(v, \mathbb{I}_k) \tag{89}$$

holds for some $v \in \bar{\mathrm{K}}_{\mathrm{m}}^{\mathrm{m}}$, *then there exists another point* $w \in \bar{\mathrm{K}}_{\mathrm{m}}^{\mathrm{m}}$ *such that*

$$\int \phi\, d\mathcal{N}(w, \mathbb{I}_k) < \int \chi\, d\mathcal{N}(w, \mathbb{I}_k) \tag{90}$$

PROOF We complement the p row vectors of $E = (e_1, \dots, e_p)'$ to an orthonormal basis $\{e_1, \dots, e_k\}$ of $\mathbb{R}^k$. In this new coordinate system, the observation vector $y = (e_1, \dots, e_k)'x$ is still k variate standard normal, but the hypotheses on the mean vector $t = (e_1, \dots, e_k)'s$ now read

$$\tilde{\mathrm{H}}_{\mathrm{m}}^{\mathrm{m}}\colon\ |t| \le a,\ |t_1| \le b \qquad \tilde{\mathrm{K}}_{\mathrm{m}}^{\mathrm{m}}\colon\ |t| \le a,\ |t_1| \ge c \tag{91}$$

where

$$t = \begin{pmatrix} t_1 \\ t_2 \end{pmatrix}, \qquad t_1 \in \mathbb{R}^p,\ t_2 \in \mathbb{R}^{k-p}$$

Setting the nuisance parameter $t_2 = 0$, one obtains the following subhypotheses concerning the mean $t \in \mathbb{R}^k$ of $y \sim \mathcal{N}(t, \mathbb{I}_k)$,

$$\hat{\mathrm{H}}^{\mathrm{m}}_{\mathrm{m}}\colon\ |t| \le a,\ t_2 = 0,\ |t_1| \le b \qquad \hat{\mathrm{K}}^{\mathrm{m}}_{\mathrm{m}}\colon\ |t| \le a,\ t_2 = 0,\ |t_1| \ge c \tag{92}$$

for which the first component $z = y_1 \sim \mathcal{N}(u, \mathbb{I}_p)$ is sufficient [Neyman's criterion]. A reduction by sufficiency thus leads to the hypotheses

$$\check{\mathrm{H}}^{\mathrm{m}}_{\mathrm{m}}\colon\ |u| \le a,\ |u| \le b \qquad \check{\mathrm{K}}^{\mathrm{m}}_{\mathrm{m}}\colon\ |u| \le a,\ |u| \ge c \tag{93}$$

Now follows a special variant of the Hunt–Stein theorem [Lehmann (1986; Example 6 and Theorem 3; pp 518, 519)]: The hypotheses (93) are *invariant* under the orthogonal transformations,

$$z \longmapsto Gz, \qquad G \in \mathcal{O}_p \tag{94}$$

The topological group $\mathcal{O}_p$ of orthogonal $p \times p$ matrices, viewed as a compact subset of $\mathbb{R}^{p\times p}$ endowed with the Borel σ algebra $\mathbb{B}^{p\times p} \cap \mathcal{O}_p$, has right invariant *Haar measure*, which may be constructed as follows: Consider a random $p \times p$ matrix U with i.i.d. entries $U_{i,j} \sim \mathcal{N}(0,1)$. Then

$$\det U \ne 0 \qquad \text{a.e.} \tag{95}$$

which is obvious in case $p = 1$. To conclude from dimension $p-1$ to p, we expand the determinant by the jth column,

$$\det U = V_{1,j}U_{1,j} + \cdots + V_{p,j}U_{p,j}$$

The vector V_j of cofactors is stochastically independent of the jth column and its coordinates satisfy $V_{i,j} \ne 0$ a.e. by assumption. Thus we obtain

$$\begin{aligned}
\Pr(\det U = 0) &= \mathrm{E}\Pr\Bigl(\sum\nolimits_i V_{i,j}U_{i,j} = 0 \Big| V_j\Bigr) \\
&= \int \Pr\Bigl(\sum\nolimits_i V_{i,j}U_{i,j} = 0 \Big| V_j = v_j\Bigr)\ \Pr(V_j \in dv_j) \\
&= \int_{\{v_j \ne 0\}} \Pr\Bigl(\sum\nolimits_i v_{i,j}U_{i,j} = 0\Bigr)\ \Pr(V_j \in dv_j) = 0
\end{aligned}$$

since $\sum_i v_{i,j}U_{i,j} \sim \mathcal{N}\bigl(0, |v_j|^2\bigr)$. This proves (95). Moreover, the random matrix $U = (U_1, \ldots, U_p)'$ whose row vectors U_j are i.i.d. $\sim \mathcal{N}(0, \mathbb{I}_p)$ is right invariant in law, in the sense that for all $G \in \mathcal{O}_p$,

$$\mathcal{L}(UG) = \mathcal{L}(U) \tag{96}$$

since $UG = (G'U_1, \ldots, G'U_p)'$ still with i.i.d. rows $G'U_j \sim \mathcal{N}(0, \mathbb{I}_p)$.

By Γ denote the Gram–Schmidt orthonormalization. It transforms any nonsingular matrix $A = (a_1, \ldots, a_p)'$ to $\Gamma(A) = B = (b_1, \ldots, b_p)'$,

$$b_j \propto \quad a_j - \sum_{i=1}^{j-1} (a_j{}' b_i) b_i \,, \qquad |b_j| = 1\,, \quad j = 1, \ldots, p \tag{97}$$

If $G \in \mathcal{O}_p$ then $AG = (G'a_1, \ldots, G'a_p)' \stackrel{\Gamma}{\longmapsto} C = (c_1, \ldots, c_p)'$,

$$c_j \propto G'a_j - \sum_{i=1}^{j-1} (a_j{}' G c_i) c_i \,, \quad |c_j| = 1\,, \quad j = 1, \ldots, p$$

Starting with $c_1 = G'b_1$, one inductively verifies that all $c_j = G'b_j$. Thus, for every nonsingular $A \in \mathbb{R}^{p \times p}$ and $G \in \mathcal{O}_p$, Γ is right equivariant,

$$\Gamma(AG) = \Gamma(A)G \tag{98}$$

Now Haar measure $\mathcal{H}$ on $\mathcal{O}_p$ can be defined as the law

$$\mathcal{H} = \mathcal{L}\big(\Gamma(U)\big) \tag{99}$$

$\mathcal{H}$ is right invariant: If $\mu\colon \mathcal{O}_p \to \mathcal{O}_p$, $\mu(G) = GM$, denotes right multiplication by any fixed $M \in \mathcal{O}_p$, and $\mathcal{E} \in \mathbb{B}^{p \times p} \cap \mathcal{O}_p$, then actually

$$\begin{aligned} \mathcal{H}(\,\mu \in \mathcal{E}\,) &= \Pr\big(\,\Gamma(U)M \in \mathcal{E}\,\big) \stackrel{(98)}{=} \Pr\big(\,\Gamma(UM) \in \mathcal{E}\,\big) \\ &\stackrel{(96)}{=} \Pr\big(\,\Gamma(U) \in \mathcal{E}\,\big) \;=\; \mathcal{H}(\mathcal{E}) \end{aligned} \tag{100}$$

If now ϕ is any test on $(\mathbb{R}^p, \mathbb{B}^p)$, pass to the average

$$\varphi(z) = \int_{\mathcal{O}_p} \phi(Gz)\, \mathcal{H}(dG)\,, \qquad z \in \mathbb{R}^p \tag{101}$$

The map $(z, G) \mapsto Gz$ being Borel measurable, φ is Borel measurable [Fubini's theorem], hence a test. Moreover, φ is invariant under composition with any orthogonal matrix $M \in \mathcal{O}_p$:

$$\begin{aligned} \varphi(Mz) &\stackrel{(101)}{=} \int_{\mathcal{O}_p} \phi(GMz)\, \mathcal{H}(dG) \underset{\text{trafo}}{\stackrel{\text{int.}}{=}} \int_{\mathcal{O}_p} \phi(Hz)\, \mu(\mathcal{H})(dH) \\ &\underset{(100)}{=} \int_{\mathcal{O}_p} \phi(Hz)\, \mathcal{H}(dH) \underset{(101)}{=} \varphi(z)\,, \qquad z \in \mathbb{R}^p \end{aligned} \tag{102}$$

As an average, the test φ has smaller maximum size over $\check{\mathrm{H}}^{\mathrm{m}}_{\mathrm{m}}$ and larger minimum power over $\check{\mathrm{K}}^{\mathrm{m}}_{\mathrm{m}}$ than ϕ, since

$$\begin{aligned} \int \varphi \, d\mathcal{N}(u, \mathbb{I}_p) &\stackrel{(101)}{=} \iint_{\mathcal{O}_p} \phi(Gz)\, \mathcal{H}(dG)\, \mathcal{N}(u, \mathbb{I}_p)(dz) \\ &\underset{\text{bini}}{\stackrel{\text{Fu-}}{=}} \int_{\mathcal{O}_p} \int \phi(Gz)\, \mathcal{N}(u, \mathbb{I}_p)(dz)\, \mathcal{H}(dG) \\ &= \int_{\mathcal{O}_p} \int \phi \, d\mathcal{N}(Gu, \mathbb{I}_p)\, \mathcal{H}(dG) \end{aligned} \tag{103}$$

and the hypotheses $\check{\mathrm{H}}^{\mathrm{m}}_{\mathrm{m}}$ and $\check{\mathrm{K}}^{\mathrm{m}}_{\mathrm{m}}$ are invariant,

$$v \in \check{\mathrm{H}}^{\mathrm{m}}_{\mathrm{m}}, \; w \in \check{\mathrm{K}}^{\mathrm{m}}_{\mathrm{m}}, \; G \in \mathcal{O}_p \implies Gv \in \check{\mathrm{H}}^{\mathrm{m}}_{\mathrm{m}}, \; Gw \in \check{\mathrm{K}}^{\mathrm{m}}_{\mathrm{m}} \tag{104}$$

Therefore,

$$\begin{aligned} \int \varphi \, d\mathcal{N}(u, \mathbb{I}_p) &\le \sup_{v \in \check{\mathrm{H}}^{\mathrm{m}}_{\mathrm{m}}} \int \phi \, d\mathcal{N}(v, \mathbb{I}_p), \qquad \text{if } u \in \check{\mathrm{H}}^{\mathrm{m}}_{\mathrm{m}} \\ \int \varphi \, d\mathcal{N}(u, \mathbb{I}_p) &\ge \inf_{w \in \check{\mathrm{K}}^{\mathrm{m}}_{\mathrm{m}}} \int \phi \, d\mathcal{N}(w, \mathbb{I}_p), \qquad \text{if } u \in \check{\mathrm{K}}^{\mathrm{m}}_{\mathrm{m}} \end{aligned} \tag{105}$$

Thus, the class of invariant tests is essentially complete for maximin testing $\check{\mathrm{H}}^{\mathrm{m}}_{\mathrm{m}}$ vs. $\check{\mathrm{K}}^{\mathrm{m}}_{\mathrm{m}}$. But a test φ is invariant under $\mathcal{O}_p$ iff it is a function

$$\varphi = \psi \circ I \tag{106}$$

of the *maximal invariant* I from $\mathbb{R}^p$ to $[0, \infty)$, which is given by

$$I(z) = |z|^2 \tag{107}$$

In (106) the function ψ is again a test. To see its measurability introduce the map

$$J \colon [0, \infty) \to \mathbb{R}^p, \qquad J(r) = \bigl(\sqrt{r}, 0, \ldots, 0\bigr)'$$

which satisfies $I \circ J = \mathrm{id}_{[0,\infty)}$ and is Borel measurable. As also φ is Borel measurable one indeed obtains for all sets $\Sigma \in \mathbb{B}$ that

$$\psi^{-1}(\Sigma) \underset{\mathrm{id}}{\overset{I \circ J}{=}} (I \circ J)^{-1}\bigl(\psi^{-1}(\Sigma)\bigr) \overset{(106)}{=} J^{-1}\bigl(\varphi^{-1}(\Sigma)\bigr) \in \mathbb{B} \tag{108}$$

The maximal invariant I, under the hypotheses (93), is χ^2 distributed,

$$\chi^2(p; \tau^2), \qquad \tau = |u| \in [0, a]$$

So invariance reduces problem (93) to

$$\check{\mathrm{H}}^{\mathrm{m}}_{\mathrm{m}}\colon \; 0 \le \tau \le b \qquad \check{\mathrm{K}}^{\mathrm{m}}_{\mathrm{m}}\colon \; c \le \tau \le a \tag{109}$$

This maximin testing problem concerning $\chi^2(p; \tau^2)$, due to monotone likelihood ratios, is solved by the Neyman–Pearson test $\check{\chi}$ for $\tau = b$ vs. $\tau = c$,

$$\check{\chi}(r) = \mathbf{I}\bigl(r > c_\alpha(p; b^2)\bigr) \tag{110}$$

Its maximin power is given by (87) since the $\chi^2(p; .)$ distribution increases stochastically in the noncentrality parameter. Thus the test $\check{\chi} = \check{\chi} \circ I$,

$$\check{\chi}(z) = \mathbf{I}\bigl(|z|^2 > c_\alpha(p; b)\bigr) \tag{111}$$

is maximin for $\check{\mathrm{H}}^{\mathrm{m}}_{\mathrm{m}}$ vs. $\check{\mathrm{K}}^{\mathrm{m}}_{\mathrm{m}}$. Then the test $\hat{\chi}$,

$$\hat{\chi}(y) = \mathbf{I}\bigl(|y_1|^2 > c_\alpha(p; b)\bigr) \tag{112}$$

is maximin for $\hat{\mathrm{H}}^{\mathrm{m}}_{\mathrm{m}}$ vs. $\hat{\mathrm{K}}^{\mathrm{m}}_{\mathrm{m}}$, the subhypotheses defined by (92). The law of y_1 not depending on t_2 when $y \sim \mathcal{N}(t, \mathbb{I}_k)$, the test $\hat{\chi}$ remains maximin, with the same maximum size and minimum power, for the larger hypotheses $\tilde{\mathrm{H}}^{\mathrm{m}}_{\mathrm{m}}$ vs. $\tilde{\mathrm{K}}^{\mathrm{m}}_{\mathrm{m}}$ defined by (91). Finally, the test χ given by (88),

$$\chi(x) = \hat{\chi}(Ex) = \mathbf{I}\big(|Ex|^2 > c_\alpha(p; b^2)\big)$$

is maximin for $\bar{\mathrm{H}}^{\mathrm{m}}_{\mathrm{m}}$ vs. $\bar{\mathrm{K}}^{\mathrm{m}}_{\mathrm{m}}$, the original hypotheses (85). Thus parts (a) and (b) are proved.

In order to prove part (c), assume a test φ of level α for $\bar{\mathrm{H}}^{\mathrm{m}}_{\mathrm{m}}$ such that

$$\int \varphi \, d\mathcal{N}(s, \mathbb{I}_k) \geq \int \chi \, d\mathcal{N}(s, \mathbb{I}_k) \tag{113}$$

for every $s \in \bar{\mathrm{K}}^{\mathrm{m}}_{\mathrm{m}}$, and strict inequality holds in (113) at some $v \in \bar{\mathrm{K}}^{\mathrm{m}}_{\mathrm{m}}$. In the coordinate system introduced at the beginning of this proof, and for the corresponding hypotheses (91), we thus get a level α test $\tilde{\varphi}$ for $\tilde{\mathrm{H}}^{\mathrm{m}}_{\mathrm{m}}$ such that

$$\int \tilde{\varphi} \, d\mathcal{N}(t, \mathbb{I}_k) \geq \int \hat{\chi} \, d\mathcal{N}(t, \mathbb{I}_k) \tag{114}$$

for all $t \in \tilde{\mathrm{K}}^{\mathrm{m}}_{\mathrm{m}}$, with strict inequality at some $w \in \tilde{\mathrm{K}}^{\mathrm{m}}_{\mathrm{m}}$. This remains so for the subhypotheses

$$\hat{\mathrm{H}}^{\mathrm{m}}_{\mathrm{p}}\colon \ |t_1| = b, \ t_2 = w_2 \qquad \hat{\mathrm{K}}^{\mathrm{m}}_{\mathrm{p}}\colon \ |t_1| = |w_1|, \ t_2 = w_2 \tag{115}$$

for which the first component $z = y_1$ of the observation vector $y \sim \mathcal{N}(t, \mathbb{I}_k)$ is sufficient [Neyman's criterion]. A reduction by sufficiency thus leads to the following hypotheses about $z \sim \mathcal{N}(u, \mathbb{I}_p)$,

$$\check{\mathrm{H}}^{\mathrm{m}}_{\mathrm{p}}\colon \ |u| = b \qquad \check{\mathrm{K}}^{\mathrm{m}}_{\mathrm{p}}\colon \ |u| = |w_1| \tag{116}$$

and gives us a test $\check{\varphi}$ which is of level α for $\check{\mathrm{H}}^{\mathrm{m}}_{\mathrm{p}}$ and satisfies

$$\int \check{\varphi} \, d\mathcal{N}(u, \mathbb{I}_p) \geq \int \check{\chi} \, d\mathcal{N}(u, \mathbb{I}_p) \tag{117}$$

for all $u \in \check{\mathrm{K}}^{\mathrm{m}}_{\mathrm{p}}$, with strict inequality at $w_1 \in \check{\mathrm{K}}^{\mathrm{m}}_{\mathrm{p}}$. By the continuity of power functions in the normal model there exists some $\varepsilon \in (0,1)$ such that the strict inequality in (117) extends to all $u \in \mathbb{R}^p$ such that $|u - w_1| < \varepsilon$.

The hypotheses (116) being invariant under the orthogonal group $\mathcal{O}_p$, averaging $\check{\varphi}$ w.r.t. Haar measure $\mathcal{H}$ in the manner of (101) yields an invariant test $\bar{\varphi}$ which like $\check{\varphi}$ is still of level α for $\check{\mathrm{H}}^{\mathrm{m}}_{\mathrm{p}}$ and achieves power

$$\begin{aligned} \int \bar{\varphi} \, d\mathcal{N}(w_1, \mathbb{I}_p) &\overset{(103)}{=} \int_{\mathcal{O}_p} \int \check{\varphi} \, d\mathcal{N}(Gw_1, \mathbb{I}_p) \, \mathcal{H}(dG) \\ &\underset{(117)}{>} \int_{\mathcal{O}_p} \int \check{\chi} \, d\mathcal{N}(Gw_1, \mathbb{I}_p) \, \mathcal{H}(dG) = \int \check{\chi} \, d\mathcal{N}(w_1, \mathbb{I}_p) \end{aligned} \tag{118}$$

The inequality in (118) is strict indeed: The matrix orthonormalization Γ is continuous at $\mathbb{I}_p = \Gamma(\mathbb{I}_p)$, so there exists some $\delta \in (0,1)$ such that

$$A \in \mathbb{R}^{p \times p}, \ \|A - \mathbb{I}_p\| < \delta \implies \|\Gamma(A) - \mathbb{I}_p\| < \varepsilon/|w_1|$$

Therefore, and by the full support of the i.i.d. standard normals $U_{i,j}$,

$$\begin{aligned} \mathcal{H}\{ G \in \mathcal{O}_p \mid |Gw_1 - w_1| < \varepsilon \} &\geq \Pr\bigl(\|\Gamma(U) - \mathbb{I}_p\| < \varepsilon/|w_1|\bigr) \\ &\geq \Pr\bigl(\|U - \mathbb{I}_p\| < \delta\bigr) > 0 \end{aligned} \tag{119}$$

A subsequent reduction of $\bar{\varphi} = \breve{\varphi} \circ I$ by invariance yields a level α test $\breve{\varphi}$ for $\chi^2(p;b^2)$ vs. $\chi^2(p;|w_1|^2)$ with power

$$\begin{aligned} \int \breve{\varphi}\, d\chi^2(p;|w_1|^2) &= \int \bar{\varphi}\, d\mathcal{N}(w_1, \mathbb{I}_p) \\ &\underset{(118)}{>} \int \breve{\chi}\, d\mathcal{N}(w_1, \mathbb{I}_p) = \int \breve{\chi}\, d\chi^2(p;|w_1|^2) \end{aligned} \tag{120}$$

Due to monotone likelihood ratios, however, $\breve{\chi}$ has been the Neyman–Pearson test for $\chi^2(p;b^2)$ vs. $\chi^2(p;|w_1|^2)$ at level α. ////

Theorem 3.4.10 *Assuming models $(\mathcal{Q}_n)$ that are as. normal with as. covariance $C > 0$, let us consider the multisided as. testing hypotheses (84) at some level $\alpha \in [0,1]$.*

(a) *Let ϕ be an as. test such that*

$$\limsup_{n\to\infty} \sup_{\mathrm{H}_{\mathrm{m}}^{\mathrm{m}}} \int \phi_n \, dQ_{n,t} \leq \alpha \tag{121}$$

Then

$$\limsup_{n\to\infty} \inf_{\mathrm{K}_{\mathrm{m}}^{\mathrm{m}}} \int \phi_n \, dQ_{n,t} \leq \Pr\bigl(\chi^2(p;c^2) > c_\alpha(p;b^2)\bigr) \tag{122}$$

(b) *If a sequence of statistics $Z_n^\star \colon (\Omega_n, \mathcal{A}_n) \to (\mathbb{R}^k, \mathbb{B}^k)$ are as. normal,*

$$Z_n^\star(Q_{n,t}) \xrightarrow{\;\mathrm{w}\;} \mathcal{N}(Ct, C) \tag{123}$$

uniformly in $|C^{1/2}t| \leq a$, then the as. test $\chi^\star$ given by

$$\chi_n^\star = \mathbf{I}\Bigl(\bigl|EC^{-1/2} Z_n^\star\bigr|^2 > c_\alpha(p;b^2) \Bigr) \tag{124}$$

is maximin—achieving equalities in (121) and (122).

(c) *The as. test $\chi^\star$ defined by (124) is admissible: If an as. test ϕ satisfies (121) and*

$$\limsup_{n\to\infty} \int \phi_n \, dQ_{n,v} > \lim_{n\to\infty} \int \chi_n^\star \, dQ_{n,v} \tag{125}$$

for some $v \in \mathrm{K}_{\mathrm{m}}^{\mathrm{m}}$, then there is another $w \in \mathrm{K}_{\mathrm{m}}^{\mathrm{m}}$ such that

$$\liminf_{n\to\infty} \int \phi_n \, dQ_{n,w} < \lim_{n\to\infty} \int \chi_n^\star \, dQ_{n,w} \tag{126}$$

PROOF

(a) Consider a subsequence (m) of $\mathbb{N}$ such that

$$\inf_{\mathrm{K}^{\mathrm{m}}_{\mathrm{m}}} \int \phi_m \, dQ_{m,t} \longrightarrow \bar{\beta}, \qquad \sup_{\mathrm{H}^{\mathrm{m}}_{\mathrm{m}}} \int \phi_m \, dQ_{m,t} \longrightarrow \bar{\alpha} \overset{(121)}{\leq} \alpha \tag{127}$$

Then Lemma 3.4.4 carries us to the limiting testing problem (85) about the mean $s \in \mathbb{R}^k$ of $\mathcal{N}(s, \mathbb{I}_k)$. Because of (21) the test φ given by (26) satisfies

$$\sup_{\bar{\mathrm{H}}^{\mathrm{m}}_{\mathrm{m}}} \int \varphi \, d\mathcal{N}(s, \mathbb{I}_k) \overset{(21)}{\leq} \lim_{m \to \infty} \sup_{\mathrm{H}^{\mathrm{m}}_{\mathrm{m}}} \int \phi_m \, dQ_{m,t} \overset{(127)}{=} \bar{\alpha} \leq \alpha \tag{128}$$

By Proposition 3.4.9 a it follows that

$$\begin{aligned} \bar{\beta} \overset{(127)}{=} \lim_{m \to \infty} \inf_{\mathrm{K}^{\mathrm{m}}_{\mathrm{m}}} \int \phi_m \, dQ_{m,t} &\overset{(21)}{\leq} \inf_{\bar{\mathrm{K}}^{\mathrm{m}}_{\mathrm{m}}} \int \varphi \, d\mathcal{N}(s, \mathbb{I}_k) \\ &\underset{(87)}{\leq} \Pr\bigl(\chi^2(p; c^2) > c_\alpha(p; b^2)\bigr) \end{aligned} \tag{129}$$

which proves (122).

(b) If $Z_n^\star(Q_{n,t})$ tend weakly to $\mathcal{N}(Ct, C)$, uniformly in $|C^{1/2}t| \leq a$, then so do $EC^{-1/2} Z_n^\star(Q_{n,t})$ converge to $\mathcal{N}(EC^{1/2}t, \mathbb{I}_p)$, hence

$$\int \chi_n^\star \, dQ_{n,t} - \Pr\bigl(\chi^2(p; |EC^{1/2}t|^2) > c_\alpha(p; b^2)\bigr) \longrightarrow 0$$

uniformly in $|C^{1/2}t| \leq a$, proving (b).

(c) Lemma 3.4.4 and Proposition 3.4.9 c. ////

Example 3.4.11 Given a family of probabilities

$$\mathcal{P} = \{ P_\theta \mid \theta \in \Theta \} \subset \mathcal{M}_1(\mathcal{A})$$

which are parametrized by an open subset Θ of $\mathbb{R}^k$. Fix some $\theta \in \Theta$. Suppose $\mathcal{P}$ is L_2 differentiable at this θ with L_2 derivative $\Lambda_\theta \in L_2^k(P_\theta)$ and positive definite Fisher information $\mathcal{I}_\theta = \int \Lambda_\theta \Lambda_\theta' \, dP_\theta > 0$. Theorem 2.3.7 (one-sample case) yields the following log likelihood expansion of the n-fold product measures,

$$\log \frac{dP^n_{\theta + t/\sqrt{n}}}{dP^n_\theta} = \frac{t'}{\sqrt{n}} \sum_{i=1}^n \Lambda_\theta(x_i) - \frac{1}{2} t' \mathcal{I}_\theta t + \mathrm{o}_{P^n_\theta}(n^0) \tag{130}$$

hence the models of n-fold product measures $Q_{n,t} = P^n_{\theta + t/\sqrt{n}}$ are as. normal with as. covariance $\mathcal{I}_\theta$ and as. sufficient statistic

$$Z = (Z_n), \qquad Z_n = \frac{1}{\sqrt{n}} \sum_{i=1}^n \Lambda_\theta(x_i) \tag{131}$$

As Theorem 2.3.7 even applies to convergent sequences $t_n \to t$ in $\mathbb{R}^k$, Proposition 2.2.12, 2.2(41), tells us that uniformly on t-compacts of $\mathbb{R}^k$,

$$Z_n(P^n_{\theta+t/\sqrt{n}}) \xrightarrow{\mathrm{w}} \mathcal{N}(\mathcal{I}_\theta t, \mathcal{I}_\theta) \tag{132}$$

Assuming $a < \infty$, this ensures (123) for the sequence $Z^\star_n = Z_n$.

The compact uniform as. normality (123) can also be achieved by an estimator $S^\star$ which has the as. expansion

$$\sqrt{n}\,(S^\star_n - \theta) = \mathcal{I}_\theta^{-1} \frac{1}{\sqrt{n}} \sum_{i=1}^{n} \Lambda_\theta(x_i) + \mathrm{o}_{P^n_\theta}(n^0) \tag{133}$$

as suggested by the parametric as. minimax and convolution theorems in the present case. Again Theorem 2.3.7 for $t_n \to t \in \mathbb{R}^k$, the Cramèr–Wold device and Corollary 2.2.6 imply that uniformly on t-compacts of $\mathbb{R}^k$,

$$\sqrt{n}\,(S^\star_n - \theta)(P^n_{\theta+t/\sqrt{n}}) \xrightarrow{\mathrm{w}} \mathcal{N}(t, \mathcal{I}_\theta^{-1}) \tag{134}$$

If $a < \infty$, this gives (123) for $Z^\star$ obtained from $S^\star$ via

$$Z^\star_n = \mathcal{I}_\theta \sqrt{n}\,(S^\star_n - \theta) \tag{135}$$

In this smooth parametric i.i.d. setup, let us consider the problem of testing hypotheses concerning the $p \le k$-dimensional first component t_1 of the parameter $t \in \mathbb{R}^k$ of the measures $P^n_{\theta+t/\sqrt{n}}$.

For the suitable choice of the matrix E and the explicit determination of the optimal test statistics, some relations between Fisher information and its inverse are useful. Corresponding to $k = p + (k - p)$ we write both matrices in blocks,

$$\mathcal{I}_\theta = \begin{pmatrix} \mathcal{I}_{\theta,11} & \mathcal{I}_{\theta,12} \\ \mathcal{I}_{\theta,21} & \mathcal{I}_{\theta,22} \end{pmatrix}, \qquad \mathcal{I}_\theta^{-1} = \begin{pmatrix} H_{\theta,11} & H_{\theta,12} \\ H_{\theta,21} & H_{\theta,22} \end{pmatrix} \tag{136}$$

Expanding $\mathcal{I}_\theta^{-1} \circ \mathcal{I}_\theta = \mathbb{I}_k$ we get

$$\begin{aligned} H_{\theta,11}^{-1} &= \mathcal{I}_{\theta,11} - \mathcal{I}_{\theta,12}\,\mathcal{I}_{\theta,22}^{-1}\,\mathcal{I}_{\theta,21} = \mathcal{I}_{\theta,11.2} \\ H_{\theta,12} &= -H_{\theta,11}\,\mathcal{I}_{\theta,12}\,\mathcal{I}_{\theta,22}^{-1} \end{aligned} \tag{137}$$

and

$$\begin{aligned} (\mathbb{I}_p, 0)\,\mathcal{I}_\theta^{-1}\Lambda_\theta &= (H_{\theta,11}, H_{\theta,12}) \begin{pmatrix} \Lambda_{\theta,1} \\ \Lambda_{\theta,2} \end{pmatrix} = \mathcal{I}_{\theta,11.2}^{-1}\,\Lambda_{\theta,1.2} \\ \Lambda_{\theta,1.2} &= \Lambda_{\theta,1} - \mathcal{I}_{\theta,12}\,\mathcal{I}_{\theta,22}^{-1}\,\Lambda_{\theta,2} \end{aligned} \tag{138}$$

Then, taking care of the orthogonality condition (83) by a norming matrix, we define

$$E = \mathcal{I}_{\theta,11.2}^{1/2}\,(\mathbb{I}_p, 0)\,\mathcal{I}_\theta^{-1/2} \tag{139}$$

Since the power bounds derived in Theorems 3.4.10, 3.4.5, 3.4.6, and 3.4.8, do not quantitatively depend on the critical number a, we may choose any a such that

$$b^2 \vee c^2 \le a^2 \operatorname{minev} \mathcal{I}_\theta < \infty \tag{140}$$

This condition will ensure the appropriate inclusion of hypotheses and attainability of the power bounds, respectively. Thus, the multisided hypotheses (84) may be chosen as

$$\mathrm{H}_{\mathrm{m}}^{\mathrm{m}}\colon\ |t| \le a\,,\ t_1{}'\mathcal{I}_{\theta,11.2}\, t_1 \le b^2 \qquad \mathrm{K}_{\mathrm{m}}^{\mathrm{m}}\colon\ |t| \le a\,,\ t_1{}'\mathcal{I}_{\theta,11.2}\, t_1 \ge c^2 \tag{141}$$

In the definitions of one- and two-sided hypotheses (27), (48), and (60), (62), where $p = 1$, the condition that $|C^{1/2}t| \le a$ may likewise be replaced by $|t| \le a$. The remaining inequality and equality constraints, by the choice $e = E'$, concern

$$e'C^{1/2}t = \mathcal{I}_{\theta,11.2}^{1/2}\, t_1 \tag{142}$$

that is, the first parameter coordinate t_1 itself, since $\mathcal{I}_{\theta,11.2} \in (0,\infty)$.

In all four problems, in view of these theorems and (132), optimal tests may be based on the sequence of test statistics

$$EC^{-1/2} Z_n^\star = \mathcal{I}_{\theta,11.2}^{1/2}\, (\mathbb{I}_p\,,\, 0\,)\, \mathcal{I}_\theta^{-1} Z_n^\star \tag{143}$$

and their norms, respectively, where $Z^\star$ may be any as. sufficient statistic for the product models.

(a) *Rao's scores* and *Neyman's $C(\alpha)$ tests* by definition employ $Z^\star = Z$ as given by (131). In view of (138), these statistics are

$$EC^{-1/2} Z_n^\star = \mathcal{I}_{\theta,11.2}^{-1/2}\, \frac{1}{\sqrt{n}} \sum_{i=1}^n \Lambda_{\theta,1.2}(x_i) \tag{144}$$

respectively,

$$\bigl|EC^{-1/2} Z_n^\star\bigr|^2 = \frac{1}{n} \Bigl(\sum\nolimits_i \Lambda_{\theta,1.2}(x_i)\Bigr)' \, \mathcal{I}_{\theta,11.2}^{-1} \Bigl(\sum\nolimits_i \Lambda_{\theta,1.2}(x_i)\Bigr) \tag{145}$$

(b) Based on an as. estimator $S^\star$ with expansion (133), *Wald's estimator tests* by definition employ $Z^\star$ given by (135). Then

$$EC^{-1/2} Z_n^\star = \mathcal{I}_{\theta,11.2}^{1/2}\, \sqrt{n}\, (S_{n,1}^\star - \theta_1) \tag{146}$$

and

$$\bigl|EC^{-1/2} Z_n^\star\bigr|^2 = n\, (S_{n,1}^\star - \theta_1)'\, \mathcal{I}_{\theta,11.2}\, (S_{n,1}^\star - \theta_1) \tag{147}$$

In view of (133) and (138), the Wald statistics agree with the Rao–Neyman statistics up to some negligible term $\mathrm{o}_{P_\theta^n}(n^0)$.

(c) Let $\mathcal{P}$ in addition be dominated by some σ finite $\mu \in \mathcal{M}_\sigma(\mathcal{A})$,

$$dP_\zeta = p_\zeta\, d\mu\,, \qquad \zeta \in \Theta \tag{148}$$

[no extra assumption if L_2 differentiability holds at *all* parameter points]. By definition, the MLE $S = (S_n)$ should for each $n \geq 1$ satisfy on Ω^n the identity

$$\prod_{i=1}^{n} p_{S_n}(x_i) = \sup_{\zeta \in \Theta} \prod_{i=1}^{n} p_\zeta(x_i) \tag{149}$$

Passing to logarithms and presuming some further (pointwise, partial) differentiability, the MLE S becomes the prototype M estimate with scores function

$$\varphi(x, \zeta) = \frac{\partial}{\partial \zeta} \log p_\zeta(x)$$

Under additional regularity, the expansion (133) can in principle be derived from the results of Section 1.4, but may also be taken from the many more explicit papers [e.g., Huber (1965)] on the as. normality of the MLE proved by way of this expansion. So (133) can be assumed for the MLE. Moreover, certain construction techniques discussed in Section 6.4 (discretization of estimators, regularization of log likelihoods) enable us to plug the tight random increment $t_n = \sqrt{n}\,(S_n - \theta)$ into the log likelihood expansion (130). Thus we obtain

$$\begin{aligned} 2 \log \prod_{i=1}^{n} \frac{p_{S_n}}{p_\theta}(x_i) \;&=\; 2\sqrt{n}\,(S_n - \theta)' \frac{1}{\sqrt{n}} \sum_{i=1}^{n} \Lambda_\theta(x_i) \\ &\qquad - n\,(S_n - \theta)'\,\mathcal{I}_\theta\,(S_n - \theta) + \mathrm{o}_{P_\theta^n}(n^0) \\ &\overset{(133)}{=} n\,(S_n - \theta)'\,\mathcal{I}_\theta\,(S_n - \theta) + \mathrm{o}_{P_\theta^n}(n^0) \end{aligned} \tag{150}$$

Similar considerations as for the full model are supposed to go through for a null hypothesis

$$\mathcal{P}_0 = \{\, P_\zeta \mid \zeta \in \Theta_0 \,\}$$

that is parametrized by some subset $\Theta_0 \subset \Theta$. More specifically, we assume that $\Theta_0 = \beta(V)$ is the image of some open neighborhood $V \subset \mathbb{R}^{k-p}$ of 0 under a map $\beta\colon V \to \Theta$ such that $\beta(0) = \theta$ and β is differentiable at 0 with derivative $d\beta(0) = B$ of full rank $k - p$. Then the reparametrized submodel

$$\mathcal{Q} = \{\, Q_v \mid Q_v = P_{\beta(v)}\,,\; v \in V \,\} \tag{151}$$

is L_2 differentiable at 0 with L_2 derivative $B'\Lambda_\theta$ and Fisher information $K_\theta = B'\mathcal{I}_\theta B$ since, as $v \to 0$,

$$\begin{aligned} \sqrt{dQ_v} = \sqrt{dP_{\beta(v)}} &= \sqrt{dP_\theta}\,\bigl(1 + \tfrac{1}{2}(\beta(v) - \theta)'\Lambda_\theta\bigr) + \mathrm{o}\bigl(|\beta(v) - \theta|\bigr) \\ &= \sqrt{dQ_0}\,\bigl(1 + \tfrac{1}{2}v'B'\Lambda_\theta\bigr) + \mathrm{o}(|v|) \end{aligned} \tag{152}$$

On the previous grounds, also the MLE $U = (U_n)$ for model $\mathcal{Q}$,

$$\prod_{i=1}^{n} q_{U_n}(x_i) = \sup_{v \in V} \prod_{i=1}^{n} q_v(x_i)\,, \qquad q_v = p_{\beta(v)} \tag{153}$$

can be assumed to have the expansion corresponding to (133), which is

$$\sqrt{n}\, U_n = K_\theta^{-1} \frac{1}{\sqrt{n}} \sum_{i=1}^n B' \Lambda_\theta(x_i) + o_{Q_0^n}(n^0) \tag{154}$$

Writing (153) in terms of $p_{\beta(v)}$ we see that $R_n = \beta(U_n)$ defines the MLE in $\mathcal{P}_0$. Then the finite-dimensional delta method yields the as. expansion

$$\begin{aligned} \sqrt{n}\,(R_n - \theta) &= B\, K_\theta^{-1} \frac{1}{\sqrt{n}} \sum_{i=1}^n B' \Lambda_\theta(x_i) + o_{P_\theta^n}(n^0) \\ &= A \sqrt{n}\,(S_n - \theta) + o_{P_\theta^n}(n^0) \end{aligned} \tag{155}$$

with

$$A = B\, K_\theta^{-1} B' \mathcal{I}_\theta\,, \qquad AA = A\,, \quad A' \mathcal{I}_\theta\, (\mathbb{I}_k - A) = 0 \tag{156}$$

The matrix A, in the metric induced by $\mathcal{I}_\theta^{-1}$, projects onto the column space of B, which is the tangent space of β through θ. Similarly to (150),

$$\begin{aligned} 2 \log \prod_{i=1}^n \frac{p_{R_n}}{p_\theta}(x_i) &= 2n\,(S_n - \theta)' A' \mathcal{I}_\theta\, (S_n - \theta) \\ &\quad - n\,(S_n - \theta)' A' \mathcal{I}_\theta\, A\, (S_n - \theta) + o_{P_\theta^n}(n^0) \\ &= n\,(S_n - \theta)' A' \mathcal{I}_\theta\, A\, (S_n - \theta) + o_{P_\theta^n}(n^0) \end{aligned} \tag{157}$$

Therefore, the sequence of *likelihood ratio statistics* are

$$\begin{aligned} 2 \log \prod_{i=1}^n \frac{p_{S_n}}{p_{R_n}}(x_i) &= 2 \log \biggl(\sup_{\zeta \in \Theta} \prod_{i=1}^n p_\zeta(x_i) \Big/ \sup_{\zeta \in \Theta_0} \prod_{i=1}^n p_\zeta(x_i) \biggr) \\ &= n\,(S_n - \theta)' (\mathcal{I}_\theta - A' \mathcal{I}_\theta\, A)(S_n - \theta) + o_{P_\theta^n}(n^0) \\ &= n\,(S_n - \theta)' (\mathbb{I}_k - A)' \mathcal{I}_\theta\, (\mathbb{I}_k - A)(S_n - \theta) + o_{P_\theta^n}(n^0) \end{aligned} \tag{158}$$

If, keeping the parameter of interest under the null hypothesis equal to θ_1 (corresponding to $b = 0$) while varying the nuisance component, β is further specialized to

$$\beta(v) = \begin{pmatrix} \theta_1 \\ \theta_2 + v \end{pmatrix} \tag{159}$$

then

$$B = \begin{pmatrix} 0 \\ \mathbb{I}_{k-p} \end{pmatrix}, \quad K_\theta = \mathcal{I}_{\theta,22}\,, \quad A = \begin{pmatrix} 0 & 0 \\ \mathcal{I}_{\theta,22}^{-1} \mathcal{I}_{\theta,21} & \mathbb{I}_{k-p} \end{pmatrix}$$

and

$$\begin{aligned} (\mathbb{I}_k - A)' \mathcal{I}_\theta\, (\mathbb{I}_k - A) &= \mathcal{I}_\theta - \mathcal{I}_\theta\, A \\ &= \begin{pmatrix} \mathcal{I}_{\theta,11} & \mathcal{I}_{\theta,12} \\ \mathcal{I}_{\theta,21} & \mathcal{I}_{\theta,22} \end{pmatrix} \begin{pmatrix} \mathbb{I}_p & 0 \\ -\mathcal{I}_{\theta,22}^{-1} \mathcal{I}_{\theta,21} & 0 \end{pmatrix} \\ &= \begin{pmatrix} \mathcal{I}_{\theta,11.2} & 0 \\ 0 & 0 \end{pmatrix} = \begin{pmatrix} \mathbb{I}_p \\ 0 \end{pmatrix} \mathcal{I}_{\theta,11.2}\, (\mathbb{I}_p\,,\, 0) \end{aligned} \tag{160}$$

Thus, the likelihood ratio statistics agree with the Wald quadratic forms, up to some negligible remainder $\mathrm{o}_{P_\theta^n}(n^0)$.

By the contiguity $\big(P^n_{\theta+t_n/\sqrt{n}}\big) \ll (P^n_\theta)$ for convergent $t_n \to t$ in $\mathbb{R}^k$, the as. equivalence of the test statistics extends accordingly. From the continuity of the limiting normal and χ^2 distribution functions, it follows that the power functions under $P^n_{\theta+t/\sqrt{n}}$ of the two one- and two-sided tests, and of the three multisided tests, respectively, are the same up to some negligible difference $\mathrm{o}(n^0)$, uniformly on t-compacts of $\mathbb{R}^k$. ////

Example 3.4.12 Consider n multinomial i.i.d. observations. In this situation, the laws P_θ are defined on the finite sample space

$$\Omega = \Big\{ x = (x_1, \dots, x_{k+1})' \in \{0,1\}^{k+1} \;\Big|\; x_1 + \cdots + x_{k+1} = 1 \Big\} \tag{161}$$

and may be parametrized by the open subset Θ of $\mathbb{R}^k$,

$$\Theta = \Big\{ \theta = (\theta_1, \dots, \theta_k)' \in (0,1)^k \;\Big|\; \theta_1 + \cdots + \theta_k < 1 \Big\} \tag{162}$$

For $\theta = (\theta_1, \dots, \theta_k)' \in \Theta$ write $\theta_{k+1} = 1 - \theta_1 - \cdots - \theta_k$ by definition. Then the laws P_θ have the following densities relative to counting measure on Ω,

$$p_\theta(x) = \theta_1^{x_1} \cdots \theta_k^{x_k} \theta_{k+1}^{x_{k+1}} \tag{163}$$

Since $\#\Omega = k+1 < \infty$, pointwise differentiability of the log densities

$$\log p_\theta(x) = x_1 \log\theta_1 + \cdots + x_k \log\theta_k + x_{k+1}\log\theta_{k+1} \tag{164}$$

w.r.t. θ already implies L_2 differentiability. Thus, the L_2 derivative is

$$\Lambda_\theta = (\Lambda_{\theta,1}, \dots, \Lambda_{\theta,k})', \qquad \Lambda_{\theta,j}(x) = \frac{x_j}{\theta_j} - \frac{x_{k+1}}{\theta_{k+1}} \tag{165}$$

the Fisher information is

$$\mathcal{I}_\theta = \begin{pmatrix} 1/\theta_1 & 0 & \dots & 0 \\ 0 & 1/\theta_2 & \dots & 0 \\ \vdots & \vdots & \ddots & \vdots \\ 0 & 0 & \dots & 1/\theta_k \end{pmatrix} + \frac{1}{\theta_{k+1}} \begin{pmatrix} 1 & 1 & \dots & 1 \\ 1 & 1 & \dots & 1 \\ \vdots & \vdots & \ddots & \vdots \\ 1 & 1 & \dots & 1 \end{pmatrix} \tag{166}$$

and its inverse is

$$\mathcal{I}_\theta^{-1} = \begin{pmatrix} \theta_1 & 0 & \dots & 0 \\ 0 & \theta_2 & \dots & 0 \\ \vdots & \vdots & \ddots & \vdots \\ 0 & 0 & \dots & \theta_k \end{pmatrix} - \begin{pmatrix} \theta_1^2 & \theta_1\theta_2 & \dots & \theta_1\theta_k \\ \theta_2\theta_1 & \theta_2^2 & \dots & \theta_2\theta_k \\ \vdots & \vdots & \ddots & \vdots \\ \theta_k\theta_1 & \theta_k\theta_2 & \dots & \theta_k^2 \end{pmatrix} \tag{167}$$

For $\theta \in \Theta$ and $t_n \to t \in \mathbb{R}^k$, Theorem 2.3.7 gives us the log likelihood expansion (130) for the n-fold product measures, where in the present case, for all $t = (t_1, \ldots, t_k)' \in \mathbb{R}^k$, and $t_{k+1} = -t_1 - \cdots - t_k$,

$$t'\mathcal{I}_\theta t = \frac{t_1^2}{\theta_1} + \cdots + \frac{t_k^2}{\theta_k} + \frac{t_{k+1}^2}{\theta_{k+1}} \tag{168}$$

Now consider the MLE $S = (S_n)$,

$$S_n = (S_{n,1}, \ldots, S_{n,k})', \qquad S_{n,j}(x_1, \ldots, x_n) = \frac{1}{n}\sum_{i=1}^n x_{i,j} \tag{169}$$

The quadratic form (168) evaluated for $t = \sqrt{n}\,(S_n - \theta)$ is

$$n\,(S_n - \theta)'\mathcal{I}_\theta(S_n - \theta) = \sum_{j=1}^{k+1} \frac{\bigl(\sum_{i=1}^n x_{i,j} - n\theta_j\bigr)^2}{n\theta_j} \tag{170}$$

which is the familiar χ^2 test statistic based on cell frequencies. Moreover, it is straightforward to verify that

$$\sqrt{n}\,(S_n - \theta) = \mathcal{I}_\theta^{-1}\frac{1}{\sqrt{n}}\sum_{i=1}^n \Lambda_\theta(x_i) \tag{171}$$

So (133) is fulfilled, hence the uniform compact as. normality (134); besides, the Rao–Neyman scores and Wald's estimator tests coincide in the present case. Thus, if $a < \infty$, Theorem 3.4.10 is in force with $Z_n^\star = \mathcal{I}_\theta\sqrt{n}\,(S_n - \theta)$:

Given $\theta \in \Theta$ and any $a \in (0, \infty)$ suitably large, define hypotheses $\mathrm{H}_{\mathrm{m}}^{\mathrm{m}}$ and $\mathrm{K}_{\mathrm{m}}^{\mathrm{m}}$ about the local parameter $t \in \mathbb{R}^k$ of the multinomial laws $P_{\theta + t/\sqrt{n}}$ of the n i.i.d. observations by

$$\mathrm{H}_{\mathrm{m}}^{\mathrm{m}}\colon\ |t| \le a,\ \sum_{j=1}^{k+1}\frac{t_j^2}{\theta_j} \le b^2 \qquad \mathrm{K}_{\mathrm{m}}^{\mathrm{m}}\colon\ |t| \le a,\ \sum_{j=1}^{k+1}\frac{t_j^2}{\theta_j} \ge c^2 \tag{172}$$

Then the classical multinomial χ^2 test $\chi^\star$,

$$\chi_n^\star = \mathbf{I}\biggl(\sum_{j=1}^{k+1} \frac{\bigl(\sum_{i=1}^n x_{i,j} - n\theta_j\bigr)^2}{n\theta_j} > c_\alpha(k; b^2)\biggr) \tag{173}$$

is as. maximin at level α, achieving as. power $\Pr\bigl(\chi^2(k; c^2) > c_\alpha(k; b^2)\bigr)$ .////

Chapter 4

Nonparametric Statistics

4.1 Introduction

This chapter treats the estimation and testing of differentiable functionals, which goes back to Koshevnik and Levit (1976), and has been extended to robustness by Beran (1977–1984) and Millar (1979–1984). For the full neighborhoods and the usual procedures of robust statistics, however, bias will dominate the nonparametric as. statistical optimality results.

While the terminology 'nonparametric' suggests independence of any parametric model, parameters will be defined by functionals. A parametric center model then rather serves the local as. investigations and the link up with robustness. Thus we assume a smoothly parametrized family of probability measures on some sample space $(\Omega, \mathcal{A})$,

$$\mathcal{P} = \{ P_\theta \mid \theta \in \Theta \} \subset \mathcal{M}_1(\mathcal{A}) \tag{1}$$

whose parameter space Θ is an open subset of some finite-dimensional $\mathbb{R}^k$. We fix some $\theta \in \Theta$. At this θ, the family $\mathcal{P}$ is assumed to be L_2 differentiable with derivative $\Lambda_\theta \in L_2^k(P_\theta)$,

$$\big\| \sqrt{dP_{\theta+t}} - \sqrt{dP_\theta}\big(1 + \tfrac{1}{2} t'\Lambda_\theta\big) \big\| = \mathrm{o}(|t|) \tag{2}$$

and Fisher information of full rank k,

$$\mathcal{I}_\theta = \mathrm{E}_\theta\, \Lambda_\theta \Lambda_\theta{}' = \mathcal{C}_\theta(\Lambda_\theta) > 0 \tag{3}$$

Here and subsequently, expectation and covariance under P_θ are denoted by E_θ and $\mathcal{C}_\theta$, respectively.

The following investigations are local about the fixed P_θ.

4.2 The Nonparametric Setup

The nonparametric setup locally about P_θ consists of certain neighborhood systems, influence curves, as. linear functionals, and as. linear estimators.

4.2.1 Full Neighborhood Systems

Even if the model distribution P_θ may serve in practice as a reasonable description, the real distribution Q will almost inevitably differ from P_θ but may be contained in a suitable neighborhood about P_θ. From the decomposition

$$Q = P_\theta + (Q - P_\theta) \tag{1}$$

involving the nuisance component $Q - P_\theta$, however, the parameter θ is obviously no longer identifiable.

Full Balls

Robust statistics allows the real distribution to be any member of some suitably full neighborhood of P_θ. We denote by

$$\mathcal{U}(\theta) = \bigl\{\, U(\theta, r) \mid r \in [0, \infty) \,\bigr\} \tag{2}$$

any system of 'neighborhoods' $U(\theta, r)$ of 'radius' $r \in [0, \infty)$ about P_θ such that

$$P_\theta \in U(\theta, r') \subset U(\theta, r'') \subset \mathcal{M}_1(\mathcal{A}), \qquad r' < r'' \tag{3}$$

The following basic types of neighborhood systems $\mathcal{U}_*(\theta)$ have been used in nonparametric and robust statistics all along: contamination $(* = c)$, Hellinger $(* = h)$, total variation $(* = v)$, Kolmogorov $(* = \kappa)$, Cramér–von Mises $(* = \mu)$, Prokhorov $(* = \pi)$, Lévy $(* = \lambda)$. In each of these cases, the system $\mathcal{U}_*(\theta)$ consists of the closed balls about P_θ of type $*$:

$$U_*(\theta, r) = B_*(P_\theta, r), \qquad r \in [0, \infty) \tag{4}$$

The contamination balls employ convex combinations,

$$B_c(P_\theta, r) = \bigl\{\, (1 - r)^+ P_\theta + (1 \wedge r) Q \bigm| Q \in \mathcal{M}_1(\mathcal{A}) \,\bigr\} \tag{5}$$

while the other types of balls

$$B_*(P_\theta, r) = \bigl\{\, Q \in \mathcal{M}_1(\mathcal{A}) \bigm| d_*(Q, P_\theta) \le r \,\bigr\} \tag{6}$$

are based on the following metrics d_*,

$$d_v(Q, P_\theta) = \frac{1}{2} \int |dQ - dP_\theta| = \sup_{A \in \mathcal{A}} \bigl| Q(A) - P_\theta(A) \bigr| \tag{7}$$

$$d_h^2(Q,P_\theta) = \frac{1}{2}\int \big|\sqrt{dQ} - \sqrt{dP_\theta}\,\big|^2 \tag{8}$$

$$d_\kappa(Q,P_\theta) = \sup_{y\in\mathbb{R}^m} \big|Q(y) - P_\theta(y)\big| \tag{9}$$

$$d_\mu^2(Q,P_\theta) = \int \big|Q(y) - P_\theta(y)\big|^2 \mu(dy) \tag{10}$$

$$d_{\mu,1}(Q,P_\theta) = \int \big|Q(y) - P_\theta(y)\big|\,\mu(dy) \tag{11}$$

Contamination, Hellinger, and total variation balls are defined for an arbitrary sample space. In the cases $* = \kappa, \mu, \lambda$, a finite-dimensional Euclidean sample space $(\Omega, \mathcal{A}) = (\mathbb{R}^m, \mathbb{B}^m)$ is assumed so that probabilities may also notationally be identified with their distribution functions,

$$P_\theta(y) = P_\theta(\{x\in\mathbb{R}^m \mid x \le y\}), \qquad Q(y) = Q(\{x\in\mathbb{R}^m \mid x \le y\})$$

Cramér–von Mises (CvM) distance d_μ and its L_1 companion $d_{\mu,1}$ are indexed by some weighting measure $\mu \in \mathcal{M}(\mathbb{B}^m)$. If μ is infinite, the balls B_μ need at some instances to be intersected with balls of type $B_{\mu,1}$. Mostly, μ will be assumed σ finite; μ finite entails considerable simplifications.

Lévy distance $d_\lambda(Q,P_\theta)$ denotes the infimum of those $\varepsilon \in [0,\infty)$ such that for all $y \in \mathbb{R}^m$, and the vector $e = (1,\dots,1)' \in \mathbb{R}^m$,

$$P_\theta(y - \varepsilon e) - \varepsilon \le Q(y) \le P_\theta(y + \varepsilon e) + \varepsilon \tag{12}$$

A general separable metric space $(\Omega,\mathcal{A}) = (\Xi,\mathcal{B})$, with Borel σ algebra $\mathcal{B}$ and distance d, may be assumed for Prokhorov distance d_π. Blowing up the events B to $B^\varepsilon = \{y\in\Omega \mid \inf_{x\in B} d(y,x) \le \varepsilon\}$, $d_\pi(Q,P_\theta)$ denotes the infimum of those $\varepsilon \in [0,\infty)$ such that for all $B\in\mathcal{B}$,

$$Q(B) \le P_\theta(B^\varepsilon) + \varepsilon, \qquad P_\theta(B) \le Q(B^\varepsilon) + \varepsilon \tag{13}$$

The distances d_π and d_λ (and d_κ) metrize weak convergence (to a continuous distribution function, respectively), and d_v and d_π can be interpreted in the light of Strassen's theorem [Huber (1981; Section 2.3)]. The early convex combinations—contaminations—are intuitively very appealing.

Simple Perturbations

Infinitesimal perturbations of P_θ can be defined along arbitrary directions.

Definition 4.2.1 *For any dimension $p\in\mathbb{N}$ and exponent $\alpha\in[1,\infty]$, we set*

$$Z_\alpha^p(\theta) = \big\{\,\zeta \in L_\alpha^p(P_\theta) \;\big|\; \mathrm{E}_\theta\,\zeta = 0\,\big\} \tag{14}$$

The elements of $Z_2^p(\theta)$, $Z_\infty^p(\theta)$ and $Z_1^p(\theta)$ are respectively called square integrable, bounded, *and* integrable p-dimensional tangents at P_θ.

Remark 4.2.2

(a) $Z_2^p(\theta)$ is our basic space of p-dimensional tangents at P_θ, and the attribute 'square integrable' will usually be omitted. In the (product) Hilbert space $L_2^p(P_\theta)$, equipped with scalar product $\langle\eta|\xi\rangle = \mathrm{E}_\theta\,\eta'\xi$, the subspace $Z_2^p(\theta)$ is the (closed) orthocomplement of the constants.

(b) The bounded tangents $Z_\infty^p(\theta)$ play a role as a dense subspace of $Z_2^p(\theta)$ in $L_2^p(P_\theta)$. The space $L_\infty^p(P_\theta)$ itself is equipped with the P_θ essential sup norm $\sup_{P_\theta}|\zeta|$; in this space, $Z_\infty^p(\theta)$ is closed.

(c) The integrable (p-dimensional) tangents $Z_1^p(\theta)$ are rather introduced for the sake of completeness; they will seldom be used.

(d) The L_α derivative of $\mathcal{P}$ at θ satisfies $\Lambda_\theta \in Z_\alpha^k(\theta)$, for $\alpha = 2$ and $\alpha = 1$, respectively, if $\mathcal{P}$ is only L_1 differentiable at θ [Definition 2.3.10]. In either case, Λ_θ is called the *parametric tangent* at P_θ. ////

Every k-dimensional tangent $\zeta \in Z_2^k(\theta)$ gives rise to a sequence of *simple perturbations* $Q_n(\zeta,.)$ *of* P_θ along $\zeta \in Z_2^k(\theta)$, according to

$$dQ_n(\zeta,t) = \Bigl(1 + \frac{1}{\sqrt{n}}\,t'\zeta_n\Bigr)\,dP_\theta\,, \qquad |t| \le \frac{\sqrt{n}}{\sup_{P_\theta}|\zeta_n|} \tag{15}$$

where the approximating bounded tangents $\zeta_n \in Z_\infty^k(\theta)$ are chosen such that

$$\lim_{n\to\infty} \mathrm{E}_\theta\,|\zeta_n - \zeta|^2 = 0\,, \qquad \sup_{P_\theta}|\zeta_n| = \mathrm{o}(\sqrt{n}\,) \tag{16}$$

Every $t \in \mathbb{R}^k$ is eventually admitted as a parameter value. Let us agree to choose $\zeta_n = \zeta$ if $\zeta \in Z_\infty^k(\theta)$.

Remark 4.2.3 Approximation (16) can be achieved by truncation: For any clipping constants $c_n \in (0,\infty)$ such that $\mathrm{o}(\sqrt{n}\,) = c_n \to \infty$, put

$$\zeta_n = \bar{\zeta}_n - \mathrm{E}_\theta\,\bar{\zeta}_n\,, \qquad \bar{\zeta}_n = \zeta\,\mathbf{I}\bigl(|\zeta| \le c_n\bigr)$$

This way, if $\zeta \in Z_\infty^k(\theta)$, we in fact obtain $\zeta_n = \zeta$ eventually. ////

The simple perturbations are smoothly parametrized: The n-fold product measures $Q_n^n(\zeta,.)$ are L_2 differentiable, hence as. normal.

Lemma 4.2.4 *For every convergent sequence $t_n \to t$ in $\mathbb{R}^k$, the simple perturbations $Q_n(\zeta,.)$ along $\zeta \in Z_2^k(\theta)$, as defined by (15) and (16), satisfy*

$$\lim_{n\to\infty} \sqrt{n}\,\Bigl\|\sqrt{dQ_n(\zeta,t_n)} - \sqrt{dP_\theta}\Bigl(1 + \frac{1}{2\sqrt{n}}\,t'\zeta\Bigr)\Bigr\| = 0 \tag{17}$$

and

$$\log\frac{dQ_n^n(\zeta,t_n)}{dP_\theta^n} = t'\frac{1}{\sqrt{n}}\sum_{i=1}^n \zeta(x_i) - \frac{1}{2}t'\mathcal{C}_\theta(\zeta)t + \mathrm{o}_{P_\theta^n}(n^0) \tag{18}$$

PROOF If $t_n \to t$ in $\mathbb{R}^k$ then $\sup_{P_\theta}|t_n'\zeta_n|/\sqrt{n} \to 0$. Into the Taylor expansion

$$2\sqrt{1+u} = 2 + u\,c(u)\,, \qquad \lim_{u\to 0} c(u) = 1$$

plug the random increment $u_n = t_n'\zeta_n/\sqrt{n}$ to obtain that

$$\begin{aligned} 2\sqrt{n}\left[\left(1+\frac{1}{\sqrt{n}}\,t_n{}'\zeta_n\right)^{1/2} - \left(1+\frac{1}{2\sqrt{n}}\,t'\zeta\right)\right] \\ = t_n{}'\zeta_n - t'\zeta + t_n{}'\zeta_n\big(c(u_n)-1\big) \end{aligned} \tag{19}$$

Then (17) is proved as $t_n{}'\zeta_n \to t'\zeta$ in $L_2(P_\theta)$ and $\sup_{P_\theta}|c(u_n)-1| \to 0$. The expansion (18) follows from Theorem 2.3.5, identifying

$$P_{n,i} = P_\theta\,, \quad Q_{n,i} = Q_n(\zeta, t_n)\,, \quad \sqrt{n}\,U_{n,i} = t'\zeta\,, \qquad i_n = n$$

as conditions (1)–(4) of Definition 2.3.1 are obviously fulfilled. ////

Remark 4.2.5 Let $(\Omega, \mathcal{A}) = (\mathbb{R}^m, \mathbb{B}^m)$ be a finite-dimensional Euclidean sample space, and $\mu \in \mathcal{M}(\mathbb{B}^m)$ some possibly infinite weight such that

$$\int P_\theta(y)\big(1 - P_\theta(y)\big)\,\mu(dy) < \infty \tag{20}$$

Then, for the simple perturbations $Q_n = Q_n(\zeta, t_n)$, we have

$$\begin{aligned} &\int\Big|\int \mathbf{I}(x \le y)\big(t_n{}'\zeta_n(x) - t'\zeta(x)\big)\,P_\theta(dx)\Big|^2\,\mu(dy) \\ &\quad = \int\Big|\int \big(\mathbf{I}(x \le y) - P_\theta(y)\big)\big(t_n{}'\zeta_n(x) - t'\zeta(x)\big)\,P_\theta(dx)\Big|^2\,\mu(dy) \\ &\quad \le \mathrm{E}_\theta\,|t_n{}'\zeta_n - t'\zeta|^2 \int P_\theta(1 - P_\theta)\,d\mu \longrightarrow 0 \end{aligned}$$

Hence the simple perturbations are CvM differentiable,

$$\lim_{n\to\infty} n \int \Big|Q_n(y) - P_\theta(y) - \frac{1}{\sqrt{n}}\,t'\Delta(y)\Big|^2\,\mu(dy) = 0 \tag{21}$$

with the following CvM derivative

$$\Delta(y) = \int \big(\mathbf{I}(x \le y) - P_\theta(y)\big)\zeta(x)\,P_\theta(dx) \tag{22}$$

which is square integrable, since

$$\int |\Delta|^2\,d\mu \le \mathrm{E}_\theta\,|\zeta|^2 \int P_\theta(1 - P_\theta)\,d\mu \tag{23}$$

by the Cauchy–Schwarz inequality and condition (20). ////

Inclusion

The basic types of neighborhood systems used in robust statistics cover simple perturbations along $Z_2^k(\theta)$, or at least along $Z_\infty^k(\theta)$, on the $1/\sqrt{n}$ scale.

Definition 4.2.6 *The neighborhood system $\mathcal{U}(\theta)$ about P_θ is called* full *if for every $\zeta \in Z_\infty^k(\theta)$ and $c \in (0,\infty)$ there exist some $r \in (0,\infty)$ and $n_0 \in \mathbb{N}$ such that*

$$t \in \mathbb{R}^k,\ |t| \le c,\ n \in \mathbb{N},\ n > n_0 \implies Q_n(\zeta,t) \in U(\theta, r/\sqrt{n}) \tag{24}$$

Remark 4.2.7 With the $1/\sqrt{n}$ scaling, a neighborhood system is also called *infinitesimal*. For sample size $n \to \infty$, neighborhoods and simple perturbations are scaled down so, because, on the one hand, such deviations from the ideal model have nontrivial effects on statistical procedures, while, on the other hand, they cannot be detected surely by goodness-of-fit tests. ////

Lemma 4.2.8 *The systems $\mathcal{U}_*(\theta)$ of type $* = c, v, h, \kappa, \pi, \lambda$, as well as $\mathcal{U}_\mu(\theta)$, $\mathcal{U}_{\mu,1}(\theta)$ with weight $\mu \in \mathcal{M}(\mathbb{B}^m)$ such that (20) is satisfied, are full; except for $\mathcal{U}_c(\theta)$ and $\mathcal{U}_{\mu,1}(\theta)$, they cover simple perturbations along $Z_2^k(\theta)$.*

PROOF We have the following hierarchy of metrics and balls,

$$B_c \subset B_v\,, \qquad d_h^2 \le d_v \le \sqrt{2}\, d_h\,, \qquad d_\mu \le d_\kappa \le d_v \ge d_\pi \ge d_\lambda \tag{25}$$

provided μ is a probability. Thus we shall prove (24) for bounded tangents in case $* = c$ and for unbounded tangents in case $* = h$. We add the argument in case $* = v$. Since the weight μ may be infinite we give own proofs concerning $\mathcal{U}_\mu(\theta)$ and $\mathcal{U}_{\mu,1}(\theta)$.

(*c*) For $\zeta \in Z_\infty^k(\theta)$ we can write

$$dQ_n(\zeta,t) = \Bigl(1 - \frac{r}{\sqrt{n}}\Bigr) dP_\theta + \frac{r}{\sqrt{n}}\Bigl(1 + \frac{t'\zeta}{r}\Bigr) dP_\theta \tag{26}$$

If $|t| \le c < \infty$ then $|t'\zeta| \le c \sup_{P_\theta}|\zeta|$. Hence we can choose

$$r = c \sup\nolimits_{P_\theta}|\zeta|\,, \qquad n_0 \ge r^2 \tag{27}$$

(*v*) If $\zeta \in Z_2^k(\theta)$ is approximated by $\zeta_n \in Z_\infty^k(\theta)$ according to (16), then, for $Q_n = Q_n(\zeta,t)$, and uniformly in $|t| \le c < \infty$,

$$2\sqrt{n}\, d_v(Q_n, P_\theta) = \sqrt{n} \int \bigl|dQ_n(\zeta,t) - dP_\theta\bigr| = \mathrm{E}_\theta\, |t'\zeta_n| \tag{28}$$

Hence put

$$r = c\, \mathrm{E}_\theta\, |\zeta| \tag{29}$$

and choose n_0 sufficiently large.

(***h***) If $\zeta \in Z_2^k(\theta)$ is approximated by $\zeta_n \in Z_\infty^k(\theta)$ as in (16), it follows from (17) that, for $Q_n = Q_n(\zeta, t)$, uniformly in $|t| \le c < \infty$,

$$\sqrt{8n}\, d_h(Q_n, P_\theta) = \sqrt{4n}\, \big\| \sqrt{dQ_n} - \sqrt{dP_\theta} \, \big\| = \sqrt{t'\mathcal{C}_\theta(\zeta)t} \; + \mathrm{o}(n^0) \tag{30}$$

Hence choose

$$4r^2 = c^2 \operatorname{maxev} \mathcal{C}_\theta(\zeta) \tag{31}$$

and then n_0 sufficiently large.

(***μ***) Let $\zeta_n \in Z_\infty^k(\theta)$ tend to $\zeta \in Z_2^k(\theta)$ in $L_2^k(P_\theta)$. Then (21)–(23) imply the following bound, for $Q_n = Q_n(\zeta, t)$, all $|t| \le c < \infty$ and $n \ge 1$,

$$\begin{aligned} n\, d_\mu^2(Q_n, P_\theta) &= \int \Big| \int \big(\mathbf{I}(x \le y) - P_\theta(y)\big) t'\zeta_n(x)\, P_\theta(dx) \Big|^2 \mu(dy) \\ &\le \mathrm{E}_\theta\, |t'\zeta_n|^2 \int P_\theta(1 - P_\theta)\, d\mu \end{aligned} \tag{32}$$

Hence choose

$$r^2 = 2c^2 \operatorname{maxev} \mathcal{C}_\theta(\zeta) \int P_\theta(1 - P_\theta)\, d\mu \tag{33}$$

and n_0 suitably large.

(***μ*,1**) For $\zeta \in Z_\infty^k(\theta)$, $Q_n = Q_n(\zeta, t)$ and all $|t| \le c < \infty$, $n \ge 1$,

$$\begin{aligned} \sqrt{n}\, d_{\mu,1}(Q_n, P_\theta) &= \int \Big| \int \big(\mathbf{I}(x \le y) - P_\theta(y)\big) t'\zeta(x)\, P_\theta(dx) \Big| \mu(dy) \\ &\le \sup\nolimits_{P_\theta} |t'\zeta| \iint \big| \mathbf{I}(x \le y) - P_\theta(y) \big|\, P_\theta(dx)\, \mu(dy) \end{aligned} \tag{34}$$

Hence choose

$$r = 2c \sup\nolimits_{P_\theta} |\zeta| \int P_\theta(1 - P_\theta)\, d\mu \tag{35}$$

In fact, $\int |\mathbf{I}(x \le y) - P_\theta(y)|\, P_\theta(dx)$ equals $2\, P_\theta(y)(1 - P_\theta(y))$. ////

Remark 4.2.9 The full neighborhood condition (24) for the contamination system $\mathcal{U}_c(\theta)$ is the main reason why we use bounded tangents, which have also been used to show that $\mathcal{U}_{\mu,1}(\theta)$ is full. This latter neighborhood system, however, will only play a technical role subordinate to $\mathcal{U}_\mu(\theta)$; its balls need not be scaled down [Theorem 6.3.8 and Remark 6.3.10]. ////

4.2.2 Asymptotically Linear Functionals

Nonparametric statistics is concerned with certain finite-dimensional features of arbitrary probability measures that can be expressed by functionals. In principle, a functional may be any map $T\colon \mathcal{M}_1(\mathcal{A}) \to \mathbb{R}^k$; it assigns each probability a parameter of interest. In our setup, functionals are linked with the given parametric model $\mathcal{P}$, first, by the condition of Fisher consistency at P_θ:

$$T(P_\theta) = \theta \tag{36}$$

Secondly, they are assumed differentiable at P_θ in some sense, and will be studied only locally about the fixed P_θ. This investigation goes along with increasing sample size n. Therefore, more generally, a functional is formally defined to be a sequence of maps,

$$T = (T_n), \qquad T_n: \mathcal{M}_1(\mathcal{A}) \longrightarrow \mathbb{R}^k \tag{37}$$

one for each sample size. The functionals at different times n will be connected by Fisher consistency and the same derivative at P_θ. Rather than the sample size, the dummy variable n may be interpreted as the distance of order $\mathrm{O}(1/\sqrt{n})$ from P_θ, at which functionals are evaluated at time n. A possible dependence on n substantially alleviates the problem of constructing functionals with prescribed properties, independently of the unknown value θ.

Influence Curves

Influence curves may be introduced as derivatives of as. linear functionals, but will also occur as summands in the as. linear estimator expansions.

Definition 4.2.10 *Suppose $\mathcal{P}$ is L_2 differentiable at θ, and assume some matrix $D \in \mathbb{R}^{p\times k}$ of full rank $p \le k$. Let $\alpha = 2, \infty$, respectively.*

(a) *Then the set $\Psi_2(\theta)$ of all* square integrable *and the subset $\Psi_\infty(\theta)$ of all* bounded influence curves *at P_θ, respectively, are*

$$\Psi_\alpha(\theta) = \big\{\, \psi_\theta \in Z_\alpha^k(\theta) \;\big|\; \mathrm{E}_\theta\, \psi_\theta \Lambda_\theta{}' = \mathbb{I}_k \,\big\} \tag{38}$$

(b) *The set $\Psi_2^D(\theta)$ of all* square integrable, *and the subset $\Psi_\infty^D(\theta)$ of all* bounded, partial influence curves *at P_θ, respectively, are*

$$\Psi_\alpha^D(\theta) = \big\{\, \eta_\theta \in Z_\alpha^p(\theta) \;\big|\; \mathrm{E}_\theta\, \eta_\theta \Lambda_\theta{}' = D \,\big\} \tag{39}$$

Remark 4.2.11

(a) The attribute 'square integrable' will usually be omitted.

(b) The *classical scores* and *classical partial scores*,

$$\psi_{h,\theta} = \mathcal{I}_\theta^{-1} \Lambda_\theta \in \Psi_2(\theta) \tag{40}$$

$$\eta_{h,\theta} = D\psi_{h,\theta} = D\mathcal{I}_\theta^{-1} \Lambda_\theta \in \Psi_2^D(\theta) \tag{41}$$

are always an influence curve, respectively, partial influence curve, at P_θ.

(c) The definition of $\Psi_2(\theta)$ and $\Psi_\infty(\theta)$ requires $\mathcal{I}_\theta > 0$, and Λ_θ nondegenerate in the sense of 2.3(64); otherwise, $\det \mathrm{E}_\theta\, \psi_\theta \Lambda_\theta{}' = 0$.

(d) If the parametric model $\mathcal{P}$ is only L_1 differentiable at θ with derivative $\Lambda_\theta \in L_1^k(P_\theta)$ [Definition 2.3.10], the definition of the bounded (partial) influence curves $\Psi_\infty(\theta)$ and $\Psi_\infty^D(\theta)$ still applies.

For CvM differentiable models $\mathcal{P}$ [Definition 2.3.11], CvM influence curves will be introduced in Example 4.2.15.

(e) In the following representation, the inclusion $\supset$ is obvious,

$$\Psi_\alpha^D(\theta) = \{\, D\psi_\theta \mid \psi_\theta \in \Psi_\alpha(\theta) \,\} \tag{42}$$

To verify $\subset$ we extend the p rows of $D = (D_1, \ldots, D_p)'$ to some basis $\{D_1, \ldots, D_k\}$ of $\mathbb{R}^k$, and put $D_x = (D_{p+1}, \ldots, D_k)'$, $D_c = (D_1, \ldots, D_k)'$. Given $\eta_\theta \in \Psi_\alpha^D(\theta)$, we may choose any $\varrho_\theta \in \Psi_\alpha(\theta)$ to achieve that

$$\eta_\theta = D\psi_\theta\,, \qquad \psi_\theta = D_c^{-1}\begin{pmatrix} \eta_\theta \\ D_x\varrho_\theta \end{pmatrix} \in \Psi_\alpha(\theta) \tag{43}$$

(f) Of course $\Psi_\alpha^D(\theta) = \Psi_\alpha(\theta)$ for $D = \mathbb{I}_k$. Partial influence curves with general D occur when there are nuisance components. In robust regression, moreover, conditionally centered (partial) influence curves will occur.

(g) $\Psi_\alpha^D(\theta)$ are closed convex subsets of $L_\alpha^p(P_\theta)$; $\alpha = 2, \infty$. ////

To derive the nonparametric as. statistical optimality results, weak differentiability of functionals along $Z_\infty^k(\theta)$ suffices, although weak differentiability along $Z_2^k(\theta)$ would be more convenient technically [but not always supported by the inclusion condition (24)]. Stronger notions of differentiability, like compact differentiation [Chapter 1], are relevant mainly for the construction of estimators from functionals, and their weak convergence.

Definition 4.2.12 *Let $\mathcal{P}$ be L_2 differentiable at θ. A functional T is called* as. linear *at P_θ if there exists some $\psi_\theta \in \Psi_2(\theta)$ such that for all bounded tangents $\zeta \in Z_\infty^k(\theta)$ and all convergent sequences $t_n \to t$ in $\mathbb{R}^k$,*

$$\sqrt{n}\,\bigl(T_n(Q_n(\zeta, t_n)) - \theta\bigr) = \mathrm{E}_\theta\, \psi_\theta \zeta' t + \mathrm{o}(n^0) \tag{44}$$

where $Q_n(\zeta, .)$ denote the simple perturbations of form (15) *and* (16). *The function ψ_θ is called the* influence curve *of T at P_θ.*

Remark 4.2.13

(a) As $Z_\infty^k(\theta)$ is dense in $Z_2^k(\theta) \subset L_2^k(P_\theta)$ and $\Psi_2(\theta) \subset Z_2^k(\theta)$, the influence curve ψ_θ is uniquely determined by (44).

(b) If $\mathcal{P}$ is only L_1 differentiable at θ, Definition 4.2.12 applies with the bounded influence curves $\Psi_\infty(\theta) \subset Z_\infty^k(\theta)$ in the place of $\Psi_2(\theta)$.

(c) Property $\mathrm{E}_\theta\, \psi_\theta \Lambda_\theta{}' = \mathbb{I}_k$ corresponds to a locally uniform as. version of Fisher consistency of a functional T which is as. linear at P_θ with influence curve ψ_θ; namely that, uniformly over t-compacts of $\mathbb{R}^k$,

$$\sqrt{n}\,\bigl(T_n(P_{\theta + t/\sqrt{n}}) - \theta\bigr) = t + \mathrm{o}(n^0) \tag{45}$$

This would follow from (44) if Λ_θ were bounded and $P_{\theta+t/\sqrt{n}} = Q_n(\Lambda_\theta, t)$. In general, (45) is true only in the approximate sense implied by Definition 4.2.12; namely that, for all $c \in (0, \infty)$,

$$\lim_{\zeta \to \Lambda_\theta}\ \limsup_{n\to\infty}\ \sup_{|t| \le c} \bigl|\sqrt{n}\,\bigl(T_n(Q_n(\zeta, t)) - \theta\bigr) - t\bigr| = 0 \tag{46}$$

where the convergence of $\zeta \in Z_\infty^k(\theta)$ to Λ_θ is in $L_2^k(P_\theta)$. The functionals of the following examples, however, fulfill (45) as it stands.

(d) These functionals will also be weakly differentiable along $Z_2^k(\theta)$. If in addition the neighborhood system covers simple perturbations along $Z_2^k(\theta)$, the proofs of the nonparametric convolution and as. minimax theorems could be shortened by one approximation. But since the stronger inclusion condition is not fulfilled by contamination balls, weak differentiability along $Z_\infty^k(\theta)$ is our basic notion of as. linearity. ////

Example 4.2.14 A functional T is called *Hellinger differentiable at* P_θ if there exists some $\psi_\theta \in \Psi_2(\theta)$ such that for all $r \in (0,\infty)$ and for all sequences $Q_n \in B_h(P_\theta, r/\sqrt{n})$,

$$\sqrt{n}\,\bigl(T_n(Q_n) - \theta\bigr) = 2\sqrt{n}\int \psi_\theta \sqrt{dP_\theta}\sqrt{dQ_n} + \mathrm{o}(n^0) \tag{47}$$

Then T is as. linear at P_θ with influence curve ψ_θ. Indeed, if $\zeta \in Z_2^k(\theta)$ and $Q_n = Q_n(\zeta, t_n)$, with $t_n \to t$ in $\mathbb{R}^k$, are given by (15) and (16), then, by the Cauchy–Schwarz inequality,

$$\begin{aligned}
&\sqrt{n}\,\Bigl|\,2\int \psi_\theta \sqrt{dP_\theta}\sqrt{dQ_n} - \frac{1}{\sqrt{n}}\int \psi_\theta \zeta'\, dP_\theta\, t\,\Bigr| \\
&\quad = \sqrt{n}\,\Bigl|\,2\int \psi_\theta \sqrt{dP_\theta}\sqrt{dQ_n} - 2\int \psi_\theta \sqrt{dP_\theta}\Bigl(1 + \frac{1}{2\sqrt{n}}\,t'\zeta\Bigr)\sqrt{dP_\theta}\,\Bigr| \\
&\quad \le 2\sqrt{n}\,\bigl\|\,|\psi_\theta|\sqrt{dP_\theta}\,\bigr\|\,\Bigl\|\sqrt{dQ_n} - \sqrt{dP_\theta}\Bigl(1 + \frac{1}{2\sqrt{n}}\,t'\zeta\Bigr)\Bigr\| \stackrel{(17)}{\longrightarrow} 0
\end{aligned}$$

Hellinger differentiable functionals have been introduced by Koshevnik and Levit (1976). If $\mathcal{P}$ is identifiable and, at every $\theta \in \Theta$, L_2 differentiable with Fisher information $\mathcal{I}_\theta > 0$, Beran's (1981) Hellinger MD functional T_h can be constructed globally, to be Hellinger differentiable at every P_θ with influence curve $\psi_{h,\theta}$, the classical scores function (40) [Theorem 6.3.4 a]. ////

Example 4.2.15 Assuming a finite-dimensional Euclidean sample space $(\Omega, \mathcal{A}) = (\mathbb{R}^m, \mathbb{B}^m)$ let $\mathcal{P}$, w.r.t. some weight $\mu \in \mathcal{M}(\mathbb{B}^m)$, be CvM differentiable at θ with CvM derivative Δ_θ and CvM information $\mathcal{J}_\theta > 0$ [Definition 2.3.11]. Then the set $\Phi_\mu(\theta)$ of CvM *influence curves* φ_θ *at* P_θ by definition consists of all functions φ_θ such that

$$\varphi_\theta \in L_2^k(\mu), \qquad \int \varphi_\theta \Delta_\theta{}'\, d\mu = \mathbb{I}_k \tag{48}$$

A functional T is called CvM *differentiable at* P_θ if there is a function $\varphi_\theta \in \Phi_\mu(\theta)$ such that for all $r \in (0,\infty)$, all sequences $Q_n \in B_\mu(P_\theta, r/\sqrt{n})$,

$$\sqrt{n}\,\bigl(T_n(Q_n) - \theta\bigr) = \sqrt{n}\int \bigl(Q_n(y) - P_\theta(y)\bigr)\varphi_\theta(y)\,\mu(dy) + \mathrm{o}(n^0) \tag{49}$$

Then φ_θ is called the CvM *influence curve of* T *at* P_θ. Condition (48), in view of 2.3(65) and (49), is equivalent to (locally uniform compact) Fisher consistency (45). If μ is σ finite and (20) holds, condition (59) below ensures uniqueness of φ_θ.

Now suppose $\mu \in \mathcal{M}_\sigma(\mathbb{B}^m)$ and (20). Then bound A.4(17) justifies Fubini for the measures $Q_n \in B_\mu(P_\theta, r/\sqrt{n}) \cap B_{\mu,1}(P_\theta, r)$, $r \in (0,\infty)$, such that

$$\int (Q_n(y) - P_\theta(y))\varphi_\theta(y)\,\mu(dy) \tag{50}$$
$$= \int \varphi_\theta(y) \int \big(\mathbf{I}(x \le y) - P_\theta(y)\big)\,Q_n(dx)\,\mu(dy) = \int \psi_\theta(x)\,Q_n(dx)$$

with the following function $\psi_\theta \colon \mathbb{R}^m \to \mathbb{R}^k$,

$$\psi_\theta(x) = \int \big(\mathbf{I}(x \le y) - P_\theta(y)\big)\varphi_\theta(y)\,\mu(dy) \tag{51}$$

By assumption (20), bound A.4(20) is in force and shows that for all $e \in \mathbb{R}^k$,

$$e'\mathcal{C}_\theta(\psi_\theta)e \le \int (e'\varphi_\theta)^2\,d\mu \int P_\theta(1 - P_\theta)\,d\mu \tag{52}$$

with the covariance

$$\mathcal{C}_\theta(\psi_\theta) = \iint \big(P_\theta(y \wedge z) - P_\theta(y)P_\theta(z)\big)\,\varphi_\theta(y)\varphi_\theta(z)'\,\mu(dy)\,\mu(dz) \tag{53}$$

In particular, by Fubini, it holds that

$$\mathrm{E}_\theta\,|\psi_\theta|^2 < \infty\,, \qquad \mathrm{E}_\theta\,\psi_\theta = 0 \tag{54}$$

If μ is finite, then ψ_θ is bounded, since, by the Cauchy–Schwarz inequality,

$$|\psi_\theta(x)|^2 \le \mu(\mathbb{R}^m)\,\mathrm{maxev} \int \varphi_\theta\varphi_\theta'\,d\mu \tag{55}$$

In the L_2 or L_1 differentiable setups, where for finite weight μ necessarily

$$\Delta_\theta(y) = \int \big(\mathbf{I}(x \le y) - P_\theta(y)\big)\Lambda_\theta(x)\,P_\theta(dx) \tag{56}$$

according to 2.3(69), such a functional is then as. linear at P_θ with this ψ_θ as influence curve. In fact, again by Fubini, it holds that for all $\zeta \in Z_2^k(\theta)$,

$$\int \varphi_\theta(y)\,\mathbf{I}(x \le y)\,\zeta(x)'\,P_\theta(dx)\,\mu(dy) = \int \psi_\theta(x)\,\zeta(x)'\,P_\theta(dx)$$

Especially,

$$\int \psi_\theta\Lambda_\theta'\,dP_\theta = \int \varphi_\theta\Delta_\theta'\,d\mu = \mathbb{I}_k \tag{57}$$

Consequentially,

$$\mathcal{C}_\theta(\psi_\theta) > 0 \tag{58}$$

In the CvM differentiable setup alone, the nonsingularity (58) of the covariance (53) seems to require some extra conditions. [Example by E. Tabakis: Let $k = m = 1$ and $P = R(-1,0)$, $\mu = \lambda$. Then the function ψ corresponding to $\varphi = \mathbf{I}_{(0,1)}$ via (51) vanishes a.e. P since $\mathbf{I}(x \le y) = P(y) = 1$ for $x \in (-1,0)$ and $y \in (0,\infty)$, while $\int \varphi^2 \, d\mu = 1$.] However, assuming that

$$\operatorname{support} P_\theta = \mathbb{R}^m \tag{59}$$

we can prove (58) by the uniqueness theorem for distribution functions. Or, under the assumption that for all $s \in \mathbb{R}^k$ in some neighborhood of 0,

$$\int \big| P_{\theta+s}(y) - P_\theta(y) \big| \, \mu(dy) < \infty, \qquad P_{\theta+s} \ll P_\theta \tag{60}$$

one may insert $Q_n = P_{\theta+t/\sqrt{n}}$ into the expansion (49), use (50), and exploit the correlation condition (48), and thus obtain (58) again.

The influence curves (51) resemble the influence curves 1.6(54) of L functionals; both coincide if we identify

$$m = 1 = k, \quad P_\theta(y) = F(y), \quad \varphi_\theta(y) = -J(F(y)), \quad \mu(dy) = w(dy) \tag{61}$$

If $\mathcal{P}$ for some weight $\mu \in \mathcal{M}(\mathbb{B}^m)$ is identifiable in $L_2(\mu)$ and CvM differentiable at every $\theta \in \mathbb{R}^k$ with derivative Δ_θ and CvM information $\mathcal{J}_\theta > 0$, the Cramér–von Mises MD functional T_μ studied by Millar (1981, 1983) and Beran (1982,1984) is CvM differentiable at every P_θ, having the expansion (49) with the function $\varphi_{\mu,\theta} = \mathcal{J}_\theta^{-1} \Delta_\theta$ [Theorem 6.3.4 b]. ////

4.2.3 Asymptotically Linear Estimators

Corresponding to as. linear functionals, as. linear estimators are introduced. Here is the definition.

Definition 4.2.16 *An* as. *estimator*

$$S = (S_n) \qquad S_n \colon (\Omega^n, \mathcal{A}^n) \longrightarrow (\mathbb{R}^k, \mathbb{B}^k)$$

is called as. linear *at* P_θ *if there is an* influence curve $\varrho_\theta \in \Psi_2(\theta)$ *such that*

$$R_n = \sqrt{n}\,(S_n - \theta) = \frac{1}{\sqrt{n}} \sum_{i=1}^n \varrho_\theta(x_i) + \mathrm{o}_{P_\theta^n}(n^0) \tag{62}$$

We call $R = (R_n)$ standardization, *and* ϱ_θ *the* influence curve, *of* S *at* P_θ.

Remark 4.2.17 Some remarks on this elementary type of estimators:

(a) The expansion (62) determines the influence curve ϱ_θ uniquely, because $1/\sqrt{n}\,\sum_i \eta(x_i)$ with $\eta \in L_2^k(P_\theta)$, $\mathrm{E}_\theta\,\eta = 0$, can tend to 0 in P_θ^n probability only if $\mathrm{E}_\theta\,|\eta|^2 = 0$; that is, $\eta = 0$ a.e. P_θ.

(b) If S is as. linear at P_θ with influence curve $\varrho_\theta \in \Psi_2(\theta)$, then

$$\sqrt{n}\,(S_n - \theta)(P_\theta^n) \xrightarrow{\mathrm{w}} \mathcal{N}(0, \mathcal{C}_\theta(\varrho_\theta))$$

because of $\varrho_\theta \in L_2^k(P_\theta)$, $\mathrm{E}_\theta\, \varrho_\theta = 0$, and the Lindeberg–Lévy theorem. The third condition $\mathrm{E}_\theta\, \varrho_\theta \Lambda_\theta{}' = \mathbb{I}_k$, as already noted by Rieder (1980; Remarks, p 108), is equivalent to the locally uniform extension of this as. normality [see Lemma 4.2.18 below].

(c) Remarks a) and b) and Definition 4.2.16 are valid also if $\mathcal{P}$ is only L_1 differentiable at θ and if the bounded influence curves $\Psi_\infty(\theta)$ are employed. But then the log likelihood expansion (65) is not available in the proof to Lemma 4.2.18, and it is not clear whether the remainder $\mathrm{o}_{P_\theta^n}(n^0)$ still tends to 0 in $P^n_{\theta+t_n/\sqrt{n}}$ probability. However, setting $\psi_n = \varrho_\theta$ and $Q_{n,i} = P_{\theta+t_n/\sqrt{n}}$ in Proposition 6.2.1 c, d, Lemma 4.2.18 holds under these modified assumptions at least for exact sums $S_n = \theta + 1/n\sum_i \varrho_\theta(x_i)$.

(d) Extending general M estimates, the class of as. linear estimators has in the case $k = 1$ been introduced by Rieder (1980). Bickel (1981) defined the related notion CULAN, employing however compact subsets of Θ instead of compacts in the local parameter space.

(e) The class of as. linear estimators contains the common as. normal M, L, R, and MD (minimum distance) estimates [Chapters 1 and 6]. In fact, most proofs of as. normality in the i.i.d. case end up with an expansion (62); the corresponding conditions need to be verified only under the ideal model.

(f) The nonparametric convolution and as. minimax theorems derived in Section 4.3 may deserve some elementary statistical interpretation but are mathematically valid results. In particular, they prove the as. linear estimators and as. linear estimator tests, respectively, to constitute essentially complete classes, for the estimation and testing of as. linear functionals on full infinitesimal neighborhood systems.

(g) The previous robustness theories of Huber (1964), Hampel (1974), Rieder (1980), and Bickel (1981) have been formulated but for as. linear estimators or, even more specialized, for M estimates. ////

Lemma 4.2.18 *Let the as. estimator S have the as. expansion (62) involving some function $\varrho_\theta \in Z_2^k(\theta)$. Then*

$$\mathrm{E}_\theta\, \varrho_\theta \Lambda_\theta{}' = \mathbb{I}_k \tag{63}$$

holds iff

$$\sqrt{n}\,(S_n - \theta)(P^n_{\theta+t_n/\sqrt{n}}) \xrightarrow{\mathrm{w}} \mathcal{N}(t, \mathcal{C}_\theta(\varrho_\theta)) \tag{64}$$

for all convergent sequences $t_n \to t$ in $\mathbb{R}^k$.

PROOF Because of the assumed L_2 differentiability, Theorem 2.3.7 holds with coefficients $\tau_{n,i} = 1/\sqrt{n}$ so that the log likelihoods of the n-fold product measures have the following as. expansion,

$$\log \frac{dP^n_{\theta+t_n/\sqrt{n}}}{dP^n_\theta} = t' \frac{1}{\sqrt{n}} \sum_{i=1}^n \Lambda_\theta(x_i) - \frac{1}{2} t' \mathcal{I}_\theta t + \mathrm{o}_{P^n_\theta}(n^0) \tag{65}$$

Applying the Cramér–Wold device to (65) and (62), we obtain joint as. normality of $\sqrt{n}\,(S_n - \theta)$ and L_n under P^n_θ. Hence Corollary 2.2.6 is enforced with the identifications

$$a = 0\,, \quad \sigma^2 = t'\mathcal{I}_\theta t\,, \quad A = \mathcal{C}_\theta(\varrho_\theta)\,, \quad c(t) = \mathrm{E}_\theta\, \varrho_\theta \Lambda_\theta{}'\, t$$

Therefore,

$$\sqrt{n}\,(S_n - \theta)(P^n_{\theta+t_n/\sqrt{n}}) \xrightarrow{\mathrm{w}} \mathcal{N}\big(c(t), \mathcal{C}_\theta(\varrho_\theta)\big) \tag{66}$$

Since $t \in \mathbb{R}^k$ has been arbitrary, and the mean of the normal distribution is unique, the asserted equivalence is proved. ////

Parametric Optimality

In the parametric setup of estimating the parameter t of the product models

$$\mathcal{Q}_n = \big\{\, P^n_{\theta+t/\sqrt{n}} \;\big|\; t \in \mathbb{R}^k \,\big\} \subset \mathcal{M}_1(\mathcal{A}^n) \tag{67}$$

the as. linear estimator with influence curve $\psi_{h,\theta}$ given by (40) is distinguished by both the convolution and the local as. minimax theorems [Theorems 3.2.3 and 3.3.8]. In these theorems, an as. estimator R may be any sequence $R_n\colon (\Omega^n, \mathcal{A}^n) \to (\mathbb{R}^k, \mathbb{B}^k)$ [convolution theorem], respectively, any sequence $R_n\colon (\Omega^n, \mathcal{A}^n) \to (\bar{\mathbb{R}}^k, \bar{\mathbb{B}}^k)$ [as. minimax theorem].

Proposition 4.2.19 *Consider the estimation of t in the models (67).*

(a) *Let an as. estimator R be regular for t with limit law $M \in \mathcal{M}_1(\mathbb{B}^k)$. Then there is a probability $M_0 \in \mathcal{M}_1(\mathbb{B}^k)$ such that*

$$M = M_0 * \mathcal{N}(0, \mathcal{I}_\theta^{-1}) \tag{68}$$

A regular estimator $R^\star$ achieves the limit law $M^\star = \mathcal{N}(0, \mathcal{I}_\theta^{-1})$ iff $R^\star$ is the standardization of an estimator $S^\star$ that is as. linear at P_θ with influence curve $\psi_{h,\theta}$.

(b) *Let the loss function $\ell \in \mathsf{L}$ be lower semicontinuous on $\bar{\mathbb{R}}^k$. Then*

$$\lim_{b\to\infty} \lim_{c\to\infty} \liminf_{n\to\infty} \inf_R \sup_{|t|\le c} \int b \wedge \ell\,(R_n - t)\, dP^n_{\theta+t/\sqrt{n}} \ge \rho_0 \tag{69}$$

where

$$\rho_0 = \int \ell\, d\mathcal{N}(0, \mathcal{I}_\theta^{-1}) \tag{70}$$

If the function $\ell\colon \mathbb{R}^k \to [0,\infty]$ is continuous a.e. λ^k, *and the estimator $S^\star$ is as. linear at P_θ with influence curve $\psi_{h,\theta}$, then*

$$\lim_{b\to\infty} \lim_{c\to\infty} \lim_{n\to\infty} \sup_{|t|\le c} \int b \wedge \ell\,\big(\sqrt{n}\,(S^\star_n - \theta) - t\big)\, dP^n_{\theta+t/\sqrt{n}} = \rho_0 \tag{71}$$

PROOF The product models are as. normal with as. covariance $\mathcal{I}_\theta$ and as. sufficient statistics $Z_n = 1/\sqrt{n}\sum_i \Lambda_\theta(x_i)$, by L_2 differentiability and Theorem 2.3.7. Thus Theorems 3.2.3 and 3.3.8 apply. Lemma 4.2.18 ensures the uniformity on t-compacts of weak convergence needed for (71).////

Cramér–Rao Bound

In the parametric setup, and restricted to the class of as. linear estimators, these as. decision-theoretic results coincide with the Cramér–Rao bound. Here is their elementary version.

Proposition 4.2.20 *Consider an estimator S that is as. linear at P_θ with influence curve $\varrho_\theta \in \Psi_2(\theta)$.*

(a) *Then its standardization R is regular with normal limit law*

$$\mathcal{N}(0, \mathcal{C}_\theta(\varrho_\theta)) = \mathcal{N}\big(0, \mathcal{C}_\theta(\varrho_\theta) - \mathcal{I}_\theta^{-1}\big) * \mathcal{N}(0, \mathcal{I}_\theta^{-1}) \tag{72}$$

The limit law $\mathcal{N}(0, \mathcal{I}_\theta^{-1})$ is achieved iff $\varrho_\theta = \psi_{h,\theta}$.

(b) *Assume a loss function $\ell \in \mathrm{L}$ that is continuous* a.e. λ^k. *Then*

$$\begin{aligned} &\lim_{b\to\infty}\lim_{c\to\infty}\lim_{n\to\infty}\sup_{|t|\le c}\int b\wedge\ell\big(\sqrt{n}\,(S_n-\theta)-t\big)\,dP^n_{\theta+t/\sqrt{n}} \\ &\qquad = \int \ell\,d\mathcal{N}(0,\mathcal{C}_\theta(\varrho_\theta)) \ge \int \ell\,d\mathcal{N}(0,\mathcal{I}_\theta^{-1}) \end{aligned} \tag{73}$$

The lower bound is achieved by $\varrho_\theta = \psi_{h,\theta}$. If ℓ is monotone quadratic and not constant a.e. λ^k, *the lower bound can be achieved only by $\varrho_\theta = \psi_{h,\theta}$.*

PROOF The argument is simple.

(a) That R is regular for t with limit $\mathcal{N}(0, \mathcal{C}_\theta(\varrho_\theta))$ [Definition 3.2.2] has been shown in Lemma 4.2.18. The convolution representation (72) actually amounts to the Cramér–Rao bound for the as. variance:

$$\mathcal{C}_\theta(\varrho_\theta) \ge \mathcal{I}_\theta^{-1} \tag{74}$$

with equality iff $\varrho_\theta = \psi_{h,\theta}$. This is an immediate consequence of the bound

$$0 \le \mathrm{E}_\theta\,(\varrho_\theta - \psi_{h,\theta})(\varrho_\theta - \psi_{h,\theta})' = \mathcal{C}_\theta(\varrho_\theta) - \mathcal{I}_\theta^{-1}$$

and the addition of mean, variance of normals under convolution.

(b) Lemma 4.2.18 and Proposition A.1.1 b, since $b \wedge \ell$ is bounded, and continuous a.e. $\mathcal{N}(0, \mathcal{C}_\theta(\varrho_\theta))$, imply that for all $b, c \in (0, \infty)$,

$$\lim_{n\to\infty}\sup_{|t|\le c}\int b\wedge\ell\,(R_n - t)\,dP^n_{\theta+t/\sqrt{n}} = \int b\wedge\ell\,d\mathcal{N}(0,\mathcal{C}_\theta(\varrho_\theta)) \tag{75}$$

For $b \uparrow \infty$ the last integral tends monotonically to

$$\int \ell\,d\mathcal{N}(0,\mathcal{C}_\theta(\varrho_\theta)) \ge \int \ell\,d\mathcal{N}(0,\mathcal{I}_\theta^{-1}) \tag{76}$$

The lower bound follows from Cramér–Rao (74) and Corollary 3.2.6.

For $\ell(z) = v(z'Az)$ of monotone quadratic form 3.3(9), write

$$\begin{aligned}\int \ell \, d\mathcal{N}(0, \mathcal{C}_\theta(\varrho_\theta)) &= \int v(|z|^2)\, \mathcal{N}(0, A^{1/2}\mathcal{C}_\theta(\varrho_\theta)A^{1/2})(dz) \\ &= \int v(z'A^{1/2}\mathcal{C}_\theta(\varrho_\theta)A^{1/2}z)\, \mathcal{N}(0, \mathbb{I}_k)(dz)\end{aligned} \tag{77}$$

Then Cramér–Rao (74) and v increasing imply that for all $z \in \mathbb{R}^k$,

$$v(z'A^{1/2}\mathcal{C}_\theta(\varrho_\theta)A^{1/2}z) \geq v(z'A^{1/2}\mathcal{I}_\theta^{-1}A^{1/2}z) \tag{78}$$

As ℓ is not constant a.e. λ^k, there exists some $u_0 \in (0, \infty)$ such that

$$0 \leq u' < u_0 < u'' < \infty \implies v(u') < v(u'')$$

Thus, if $\varrho_\theta \neq \psi_{h,\theta}$, hence $\mathcal{C}_\theta(\varrho_\theta) \geq \mathcal{I}_\theta^{-1}$ and $\mathcal{C}_\theta(\varrho_\theta) \neq \mathcal{I}_\theta^{-1}$, upon suitable rescaling, continuity gives us a nonempty open subset G of $\mathbb{R}^k$ such that

$$z'A^{1/2}\mathcal{C}_\theta(\varrho_\theta)A^{1/2}z > u_0 > z'A^{1/2}\mathcal{I}_\theta^{-1}A^{1/2}z$$

for all $z \in G$. In view of (77) and (78), this entails that

$$\int \ell \, d\mathcal{N}(0, \mathcal{C}_\theta(\varrho_\theta)) > \int \ell \, d\mathcal{N}(0, \mathcal{I}_\theta^{-1}) \tag{79}$$

since $\mathcal{N}(0, \mathbb{I}_k)(G) > 0$. ////

4.3 Statistics of Functionals

In this nonparametric setup, we shall derive the convolution and as. minimax theorems for the estimation and testing of as. linear functionals. For as. linear estimators, and for tests based on as. linear estimators, the results will be interpreted more elementarily in terms of bias and variance.

The idea of the proofs of the nonparametric convolution and as. minimax theorems below is to apply the parametric versions [Chapter 3] to single tangents—in fact, to just one least favorable tangent: If $\psi_\theta \in \Psi_2(\theta)$ denotes the influence curve of T at P_θ, the least favorable tangents will be

$$\zeta^\star = A\psi_\theta \in Z_2^k(\theta), \qquad A \in \mathbb{R}^{k \times k}, \det A \neq 0 \tag{1}$$

More generally than $T = (T_n)$ itself, let us estimate and test differentiable transforms

$$\tau \circ T = (\tau \circ T_n), \qquad \tau \colon \mathbb{R}^k \to \mathbb{R}^p$$

where $p \leq k$ and τ has a bounded derivative at θ of full rank p,

$$\tau(\theta + t) = \tau(\theta) + d\tau(\theta)t + \mathrm{o}(|t|), \qquad \operatorname{rk} d\tau(\theta) = p \tag{2}$$

for example, p out of k coordinates.

Here is the list of assumptions made throughout this chapter:

- The observations $x_1, \ldots, x_n$ at time n are i.i.d.;
- the parametric family $\mathcal{P}$ is L_2 differentiable at θ with derivative Λ_θ and Fisher information $\mathcal{I}_\theta > 0$;
- a full neighborhood system $\mathcal{U}(\theta)$ about P_θ;
- a differentiable transformation $\tau\colon \mathbb{R}^k \to \mathbb{R}^p$ of kind (2);
- a functional T as. linear at P_θ with influence curve $\psi_\theta \in \Psi_2(\theta)$.

In addition, let $\Gamma_\theta(\psi_\theta)$ denote the covariance of $d\tau(\theta)\psi_\theta$ under P_θ ,

$$\Gamma_\theta(\psi_\theta) = \mathcal{C}_\theta\bigl(d\tau(\theta)\psi_\theta\bigr) = d\tau(\theta)\mathcal{C}_\theta(\psi_\theta)d\tau(\theta)' \tag{3}$$

4.3.1 Unbiased Estimation

Extending Definition 3.2.2 to the nonparametric setup, estimator regularity is defined as follows.

Definition 4.3.1 *An as. estimator*

$$S = (S_n), \qquad S_n\colon (\Omega^n, \mathcal{A}^n) \longrightarrow (\mathbb{R}^p, \mathbb{B}^p)$$

is called regular *for $\tau \circ T$ on $\mathcal{U}(\theta)$ with limit law $M \in \mathcal{M}_1(\mathbb{B}^p)$, if*

$$\sqrt{n}\,\bigl(S_n - \tau \circ T_n(Q_n)\bigr)(Q_n^n) \xrightarrow{\;\mathrm{w}\;} M \tag{4}$$

for all $r \in (0,\infty)$ and all sequences $Q_n \in U(\theta, r/\sqrt{n})$.

Here is the nonparametric convolution theorem.

Theorem 4.3.2 *Let the as. estimator S be regular for $\tau\circ T$ on $\mathcal{U}(\theta)$ with limit law $M \in \mathcal{M}_1(\mathbb{B}^p)$. Then there exists a probability $M_0 \in \mathcal{M}_1(\mathbb{B}^p)$ such that*

$$M = M_0 * \mathcal{N}(0, \Gamma_\theta(\psi_\theta)) \tag{5}$$

and

$$\Bigl(\sqrt{n}\,\bigl(S_n - \tau(\theta)\bigr) - d\tau(\theta)\frac{1}{\sqrt{n}}\sum_{i=1}^{n}\psi_\theta(x_i)\Bigr)(P_\theta^n) \xrightarrow{\;\mathrm{w}\;} M_0 \tag{6}$$

If an as. estimator $S^\star$ is regular for $\tau \circ T$ on $\mathcal{U}(\theta)$ and achieves the limit law $M^\star = \mathcal{N}(0, \Gamma_\theta(\psi_\theta))$, then

$$\sqrt{n}\,\bigl(S_n^\star - \tau(\theta)\bigr) = d\tau(\theta)\frac{1}{\sqrt{n}}\sum_{i=1}^{n}\psi_\theta(x_i) + \mathrm{o}_{P_\theta^n}(n^0) \tag{7}$$

PROOF For any $\zeta \in Z_\infty^k(\theta)$ satisfying

$$\det D(\zeta) \neq 0, \qquad D(\zeta) = \mathrm{E}_\theta\, \psi_\theta \zeta' \tag{8}$$

hence $\mathcal{C}_\theta(\zeta) > 0$, consider the simple perturbations along ζ. The product measures $Q_n^n(\zeta,.)$ are as. normal with as. covariance $\mathcal{C}_\theta(\zeta)$ and as. sufficient statistics $Z_n = 1/\sqrt{n}\,\sum_i \zeta(x_i)$ [Lemma 4.2.4]. The expansion 4.2(44) of T and the differentiability of τ entail that uniformly on t-compacts of $\mathbb{R}^k$,

$$\sqrt{n}\,\bigl(\tau \circ T_n(Q_n(\zeta,t)) - \tau(\theta)\bigr) = d\tau(\theta)D(\zeta)\,t + \mathrm{o}(n^0) \tag{9}$$

Now fix $t \in \mathbb{R}^k$. By the inclusion 4.2(24) and the regularity of S, (4) holds for $Q_n = Q_n(\zeta,t)$, and we may insert (9) into (4). Thus, the standardization

$$R = (R_n), \qquad R_n = \sqrt{n}\,\bigl(S_n - \tau(\theta)\bigr) \tag{10}$$

is regular with limit law M, for the as. normal models $Q_n^n(\zeta,.)$ and the parameter $d\tau(\theta)D(\zeta)\,t$. So Theorem 3.2.3 applies: There exists some probability $M_0(\zeta) \in \mathcal{M}_1(\mathbb{B}^p)$ such that

$$M = M_0(\zeta) * \mathcal{N}(0,\Xi(\zeta)) \tag{11}$$

where

$$\Xi(\zeta) = d\tau(\theta)E(\zeta)d\tau(\theta)', \qquad E(\zeta) = D(\zeta)\mathcal{C}_\theta(\zeta)^{-1}D(\zeta)' \tag{12}$$

and

$$\Bigl(R_n - d\tau(\theta)D(\zeta)\mathcal{C}_\theta(\zeta)^{-1}\frac{1}{\sqrt{n}}\sum\nolimits_i \zeta(x_i)\Bigr)(P_\theta^n) \xrightarrow{\mathrm{w}} M_0(\zeta) \tag{13}$$

To clarify least favorability, compare ψ_θ with $\xi = D(\zeta)\mathcal{C}_\theta(\zeta)^{-1}\zeta$,

$$0 \le \mathcal{C}_\theta(\psi_\theta - \xi) = \mathcal{C}_\theta(\psi_\theta) - E(\zeta) = H(\zeta) \tag{14}$$

where $H(\zeta) = 0$ iff ζ is of least favorable form (1). Hence, these (possibly unbounded) least favorable tangents $\zeta^\star$ would in (11) produce the normal factor $\mathcal{N}(0,\Gamma_\theta(\psi_\theta))$ having largest covariance $\Gamma_\theta(\psi_\theta) = \Xi(\zeta^\star)$.

To prove (5) using only bounded tangents, write (11) as identity on $\mathbb{R}^p$ for characteristic functions,

$$\widehat{M}(u)\exp\bigl(\tfrac{1}{2}u'\Xi(\zeta)u\bigr) = \widehat{M_0}(\zeta,u) \tag{15}$$

and let $\zeta \in Z_\infty^k(\theta)$ approach $\zeta^\star = A\psi_\theta$ of form (1) in $L_2^k(P_\theta)$. Then the matrices $\mathcal{C}_\theta(\zeta)$, $D(\zeta)$, and $E(\zeta)$ converge to those of $\zeta^\star$, and hence also

$$\Xi(\zeta) \longrightarrow \Gamma_\theta(\psi_\theta) \tag{16}$$

and the continuity theorem tells us that $M_0(\zeta)$ tends weakly to the probability $M_0 \in \mathcal{M}_1(\mathbb{B}^p)$ with Fourier transform

$$\widehat{M_0}(u) = \widehat{M}(u)\exp\bigl(\tfrac{1}{2}u'\Gamma_\theta(\psi_\theta)u\bigr) \tag{17}$$

Thus (5) is proved. (6) follows from (13) since for all $\varepsilon \in (0,\infty)$, $n \ge 1$,

$$\begin{aligned} &P_\theta^n\Bigl(\Bigl|d\tau(\theta)D(\zeta)\mathcal{C}_\theta(\zeta)^{-1}\frac{1}{\sqrt{n}}\sum\nolimits_i \zeta(x_i) - d\tau(\theta)\frac{1}{\sqrt{n}}\sum\nolimits_i \psi_\theta(x_i)\Bigr| > \varepsilon\Bigr) \\ &\quad\le \frac{1}{\varepsilon^2}\,\|d\tau(\theta)\|^2 \int \bigl|D(\zeta)\mathcal{C}_\theta(\zeta)^{-1}\zeta - \psi_\theta\bigr|^2\,dP_\theta \end{aligned}$$

and this upper bound goes to 0 as $\zeta \in Z_\infty^k(\theta)$ tends to $\zeta^\star = A\psi_\theta$, hence also $\xi = D(\zeta)\mathcal{C}_\theta(\zeta)^{-1}\zeta \to \psi_\theta$, in $L_2^k(P_\theta)$ [Lemma A.1.5].

Obviously, (5) and (6) imply (7). ////

Remark 4.3.3 As the proof shows, the inclusion 4.2(24) is needed only for single t values, and the expansion 4.2(44) of T only for constant $t_n = t$.

Given a particular functional, the estimator regularity (4) has been used only for the simple perturbations $Q_n(\zeta,.)$ with $\zeta \in Z_\infty^k(\theta)$ arbitrarily close to some least favorable $\zeta^\star$ of form (1); if ψ_θ is bounded, regularity (4) restricted to $Q_n(\zeta^\star,.)$ formally suffices. If ψ_θ is unbounded, and $\mathcal{U}(\theta)$ covers simple perturbations, and T is weakly differentiable, along $Z_2^k(\theta)$, the estimator regularity (4) may have been restricted to the simple perturbations $Q_n(\zeta^\star,.)$ along some least favorable $\zeta^\star$. ////

Recall L, the set of loss functions on $\bar{\mathbb{R}}^p$ satisfying 3.2(33), 3.3(8), and definition (3) of the matrix $\Gamma_\theta(\psi_\theta)$. For any $\ell\colon \mathbb{R}^p \to [0,\infty]$ Borel measurable put

$$\rho_0 = \int \ell \, d\mathcal{N}(0, \Gamma_\theta(\psi_\theta)) \tag{18}$$

Here is the nonparametric as. minimax theorem, in which estimators

$$S = (S_n), \qquad S_n\colon (\Omega^n, \mathcal{A}^n) \longrightarrow (\bar{\mathbb{R}}^p, \bar{\mathbb{B}}^p)$$

may be allowed extended-valued.

Theorem 4.3.4

(a) *If $\ell \in \mathsf{L}$ is a loss function that is continuous on $\bar{\mathbb{R}}^p$, then*

$$\lim_{r\to\infty} \liminf_{n\to\infty} \inf_S \sup_{Q\in U(\theta, r/\sqrt{n})} \int \ell\left(\sqrt{n}\left(S_n - \tau \circ T_n(Q)\right)\right) dQ^n \ge \rho_0 \tag{19}$$

(b) *If $\ell\colon \mathbb{R}^q \to [0,\infty]$ is continuous* a.e. λ^p *and $S^\star$ is regular for $\tau \circ T$ on $\mathcal{U}(\theta)$ with limit law $\mathcal{N}(0, \Gamma_\theta(\psi_\theta))$, then, for all $r \in (0,\infty)$,*

$$\lim_{b\to\infty} \lim_{n\to\infty} \sup_{Q\in U(\theta, r/\sqrt{n})} \int b \wedge \ell\left(\sqrt{n}\left(S_n^\star - \tau \circ T_n(Q)\right)\right) dQ^n = \rho_0 \tag{20}$$

PROOF The proof is easy sailing.

(a) Consider simple perturbations along any $\zeta \in Z_\infty^k(\theta)$ satisfying (8). Then the products $Q_n^n(\zeta,.)$ are as. normal with as. covariance $\mathcal{C}_\theta(\zeta)$ and as. sufficient statistics $Z_n = 1/\sqrt{n}\sum_i \zeta(x_i)$.

Because ℓ is (uniformly) continuous on $\bar{\mathbb{R}}^p$ (compact), the expansion (9) of $\tau \circ T$ carries over so that for all $c \in (0,\infty)$,

$$\ell\left(\sqrt{n}\left[S_n - \tau \circ T_n(Q_n(\zeta, t))\right]\right) - \ell\left(R_n - d\tau(\theta)D(\zeta)\, t\right) \longrightarrow 0 \tag{21}$$

uniformly in $|t| \le c$, on Ω^n, and w.r.t. all as. estimators S with standardization R given by (10). From (21) it follows that for all $c \in (0,\infty)$,

$$\left.\begin{aligned} &\inf_S \sup_{|t|\le c} \int \ell\left(\sqrt{n}\left[S_n - \tau\circ T_n(Q_n(\zeta,t))\right]\right) dQ_n^n(\zeta,t) \\ &\quad - \inf_R \sup_{|t|\le c} \int \ell\left(R_n - d\tau(\theta)D(\zeta)\,t\right) dQ_n^n(\zeta,t) \end{aligned}\right\} \longrightarrow 0 \tag{22}$$

where R, independently of S, ranges over all as. estimators. Therefore, Theorem 3.3.8 applies with the matrix $\Xi(\zeta)$ introduced in (12):

$$\begin{aligned} &\lim_{c\to\infty} \liminf_{n\to\infty} \inf_S \sup_{|t|\le c} \int \ell\left(\sqrt{n}\left[S_n - \tau\circ T_n(Q_n(\zeta,t))\right]\right) dQ_n^n(\zeta,t) \\ &\quad \ge \int \ell\, d\mathcal{N}(0,\Xi(\zeta)) \end{aligned} \tag{23}$$

To prove least favorability of the tangents (1), invoke bound (14), which says

$$\Gamma_\theta(\psi_\theta) = \Xi(\zeta) + \Pi(\zeta), \qquad \Pi(\zeta) = d\tau(\theta)H(\zeta)d\tau(\theta)' \tag{24}$$

hence

$$\mathcal{N}(0,\Gamma_\theta(\psi_\theta)) = \mathcal{N}(0,\Xi(\zeta)) * \mathcal{N}(0,\Pi(\zeta))$$

Then, since ℓ is symmetric subconvex on $\mathbb{R}^p$, Corollary 3.2.6 tells us that

$$\int \ell\, d\mathcal{N}(0,\Gamma_\theta(\psi_\theta)) \ge \int \ell\, d\mathcal{N}(0,\Xi(\zeta)) \tag{25}$$

To show that this upper bound for the as. minimax risk (23) can be approximated by bounded tangents, let $\zeta \in Z_\infty^k(\theta)$ tend to $\zeta^\star = A\psi_\theta$ in $L_2^k(P_\theta)$. Then, in view of (16), the normal densities converge at each $x \in \mathbb{R}^p$,

$$\varphi_{\Xi(\zeta)}(x) \longrightarrow \varphi_{\Gamma_\theta(\psi_\theta)}(x) \tag{26}$$

As $\ell \ge 0$ it follows from Fatou that

$$\int \ell\, d\mathcal{N}(0,\Gamma_\theta(\psi_\theta)) \le \liminf_{\zeta\to\psi_\theta} \int \ell\, d\mathcal{N}(0,\Xi(\zeta)) \tag{27}$$

Altogether, (23), (25), and (27) imply the lower bound in (19). The proof of (a) is concluded by noting that, due to the properties of $\mathcal{U}(\theta)$,

$$\begin{aligned} &\lim_{r\to\infty} \liminf_{n\to\infty} \inf_S \sup_{Q\in U(\theta,r/\sqrt{n})} \int \ell\left(\sqrt{n}\left(S_n - \tau\circ T_n(Q)\right)\right) dQ^n \ge \\ &\sup_{\zeta\in Z_\infty^k(\theta)} \lim_{c\to\infty} \liminf_{n\to\infty} \inf_S \sup_{|t|\le c} \int \ell\left(\sqrt{n}\left[S_n - \tau\circ T_n(Q_n(\zeta,t))\right]\right) dQ_n^n(\zeta,t) \end{aligned} \tag{28}$$

(b) The uniform weak convergence (4) to $M = \mathcal{N}(0, \Gamma_\theta(\psi_\theta))$, as ℓ is continuous a.e. $\mathcal{N}(0, \Gamma_\theta(\psi_\theta))$, implies that for all $b, r \in (0, \infty)$,

$$\lim_{n\to\infty} \sup_{Q \in U(\theta, r/\sqrt{n})} \int b \wedge \ell\bigl(\sqrt{n}\,(S_n^\star - \tau \circ T_n(Q))\bigr)\, dQ^n = \int b \wedge \ell \, d\mathcal{N}(0, \Gamma_\theta(\psi_\theta))$$

[Proposition A.1.1 b] and hence (20) if $b \uparrow \infty$. ////

Remark 4.3.5 More general Borel measurable functions $\ell \colon \bar{\mathbb{R}}^p \to [0, \infty]$ satisfying condition 3.3(90) for the matrix $\Gamma = \Gamma_\theta(\psi_\theta)$ are allowed in Theorem 4.3.4 a; for example, monotone quadratic loss functions. ////

Asymptotic Uniqueness and Inadmissibility

In case $p \le 2$, under some further conditions, an as. minimax estimator $S^\star$ necessarily has expansion (7). To show this uniqueness, we want to employ one possibly unbounded least favorable tangent $\zeta^\star$ directly, and therefore assume that T is weakly differentiable, and $\mathcal{U}(\theta)$ covers simple perturbations, along $Z_2^k(\theta)$. So the following argument applies to the contamination system $\mathcal{U}_c(\theta)$ only in connection with a functional T whose influence curve ψ_θ at P_θ is bounded. Choose the loss function $\ell = |\,.\,|^2_{\Gamma_\theta(\psi_\theta)}$.

In this setup, suppose that an as. estimator $S^\star$, along some least favorable tangent $\zeta^\star = A\psi_\theta$ of form (1) and all $t \in \mathbb{R}^k$, satisfies

$$\begin{aligned} &\lim_{b\to\infty} \limsup_{n\to\infty} \int b \wedge \ell\bigl(\sqrt{n}\,[S_n^\star - \tau \circ T_n(Q_n(\zeta_n^\star, t))]\bigr)\, dQ_n^n(\zeta_n^\star, t) \\ &\qquad \le p = \int |y|^2_{\Gamma_\theta(\psi_\theta)}\, \mathcal{N}(0, \Gamma_\theta(\psi_\theta))(dy) = \rho_0 \end{aligned} \tag{29}$$

Then the assumptions of Proposition 3.3.4 and Lemma 3.3.5 [strict convex loss, and $K_0(ds, y) = \mathbf{I}_y(ds)$ admissible] are fulfilled for $Q_n^n(\zeta_n^\star, .)$, hence Proposition 3.3.13 b applies with the following identifications

$$D = d\tau(\theta)\mathcal{C}_\theta(\psi_\theta)A', \quad C = A\mathcal{C}_\theta(\psi_\theta)A', \quad Z_n = \frac{1}{\sqrt{n}} \sum_{i=1}^n A\psi_\theta(x_i)$$

Then

$$DC^{-1}A = d\tau(\theta)\mathcal{C}_\theta(\psi_\theta)A'\bigl(A\mathcal{C}_\theta(\psi_\theta)A'\bigr)^{-1}A = d\tau(\theta)$$

So 3.3(95) indeed yields

$$\sqrt{n}\,\bigl(S_n^\star - \tau(\theta)\bigr) = d\tau(\theta)\frac{1}{\sqrt{n}} \sum_{i=1}^n \psi_\theta(x_i) + \mathrm{o}_{P_\theta^n}(n^0) \tag{30}$$

If for all $r \in (0, \infty)$ and all sequences $Q_n \in U(\theta, r/\sqrt{n})$,

$$\lim_{b\to\infty} \limsup_{n\to\infty} \int b \wedge \ell\bigl(\sqrt{n}\,(S_n^\star - \tau \circ T_n(Q_n))\bigr)\, dQ_n^n \le p = \rho_0 \tag{31}$$

then (29) is obviously fulfilled. An admissibility result, however, referring to such sequences, and not only restricted to a least favorable tangent, is not available.

In case $p > 2$, the minimax as. estimator $S^\star$ of Theorem 4.3.4 b is not admissible, at least under the loss function $\ell = |\,.\,|^2_{\Gamma_\theta(\psi_\theta)}$ and in the subsystem $\mathcal{U}'(\theta)$ given by

$$U'_n(\theta, r/\sqrt{n}\,) = \bigl\{\, Q \in U(\theta, r/\sqrt{n}\,) \mid \sqrt{n}\,\bigl|\tau \circ T_n(Q) - \tau(\theta)\bigr| \le r \,\bigr\} \tag{32}$$

As the additional boundedness condition is ensured by the functional's expansion (9) along the simple perturbations actually employed in the proofs, both Theorem 4.3.2 (with the correspondingly restricted notion of regularity) and Theorem 4.3.4 hold with $U'_n(\theta, r/\sqrt{n}\,)$ in the place of $U(\theta, r/\sqrt{n}\,)$. Now define the as. estimator $\tilde{S}$ by

$$\tilde{S}_n = \tau(\theta) + \left(1 - \frac{p-2}{n\bigl|S^\star_n - \tau(\theta)\bigr|^2_{\Gamma_\theta(\psi_\theta)}} \right) \bigl(S^\star_n - \tau(\theta)\bigr) \tag{33}$$

Given $r \in (0, \infty)$ and a sequence $Q_n \in U'_n(\theta, r/\sqrt{n}\,)$, put

$$Y_n = \sqrt{n}\,\bigl(S^\star_n - \tau \circ T_n(Q_n)\bigr), \qquad t_n = \sqrt{n}\,\bigl(\tau \circ T_n(Q_n) - \tau(\theta)\bigr)$$

Thus

$$\sqrt{n}\,\bigl(\tilde{S}_n - \tau \circ T_n(Q_n)\bigr) = \left(1 - \frac{p-2}{|Y_n + t_n|^2_{\Gamma_\theta(\psi_\theta)}} \right) (Y_n + t_n) - t_n \tag{34}$$

where $Y_n(Q_n^n) \xrightarrow{\mathrm{w}} \mathcal{N}(0, \Gamma_\theta(\psi_\theta))$ by the regularity of $S^\star$. Hence

$$\begin{aligned} &\limsup_{n\to\infty} \sup_{Q_n \in U'_n(\theta, r/\sqrt{n}\,)} \int b \wedge \ell\bigl(\sqrt{n}\,(\tilde{S}_n - \tau \circ T_n(Q_n))\bigr)\, dQ_n^n \\ &\quad \le \sup_{|t| \le r} \int b \wedge \ell\Bigl(\Bigl(1 - \frac{p-2}{|y|^2_{\Gamma_\theta(\psi_\theta)}}\Bigr) y - t\Bigr)\, \mathcal{N}(t, \Gamma_\theta(\psi_\theta))(dy) \\ &\quad \le \sup_{|t| \le r} \int \Bigl|\Bigl(1 - \frac{p-2}{|y|^2_{\Gamma_\theta(\psi_\theta)}}\Bigr) y - t\Bigr|^2_{\Gamma_\theta(\psi_\theta)} \mathcal{N}(t, \Gamma_\theta(\psi_\theta))(dy) \end{aligned} \tag{35}$$

for all $r, b \in (0, \infty)$. In view of 3.3(55) and 3.3(110), this implies that

$$\begin{aligned} &\lim_{b\to\infty} \limsup_{n\to\infty} \sup_{Q_n \in U'_n(\theta, r/\sqrt{n}\,)} \int b \wedge \ell\bigl(\sqrt{n}\,(\tilde{S}_n - \tau \circ T_n(Q_n))\bigr)\, dQ_n^n \\ &\quad < p = \int |y|^2_{\Gamma_\theta(\psi_\theta)}\, \mathcal{N}(0, \Gamma_\theta(\psi_\theta))(dy) = \rho_0 \end{aligned} \tag{36}$$

which proves $S^\star$ inadmissible, in model $\mathcal{U}'(\theta)$, under loss $\ell = |\,.\,|^2_{\Gamma_\theta(\psi_\theta)}$.

Bias Infinity

In the remainder of this subsection we make the assumption (to be taken up later) that

$$\tau = \mathrm{id}_{\mathbb{R}^k} \tag{82}$$

To achieve the as. minimax bound, in this full parameter case, Theorem 4.3.4 b invokes an as. estimator $S^\star$ that is regular for T on $\mathcal{U}(\theta)$ with limit law $\mathcal{N}(0, \mathcal{C}_\theta(\psi_\theta))$. Among all as. estimators which are regular for T on $\mathcal{U}(\theta)$, such an estimator achieves the most concentrated limit law given by the nonparametric convolution theorem [Theorem 4.3.2]; in view of (7), $S^\star$ is necessarily as. linear at P_θ with the same influence curve ψ_θ as the given functional T.

Without too much respect for abstraction per se, let us shed some light on this optimality by a restriction to as. linear estimators.

Proposition 4.3.6 *Suppose the estimator S is as. linear at P_θ with influence curve*

$$\varrho_\theta \in \Psi_2(\theta), \qquad \varrho_\theta \neq \psi_\theta \tag{91}$$

(a) *Then S is not regular for T on $\mathcal{U}(\theta)$.*
(b) *If $\ell \in \mathsf{L}$ is monotone quadratic, then*

$$\lim_{r\to\infty} \liminf_{n\to\infty} \sup_{Q \in U(\theta, r/\sqrt{n})} \int \ell\left(\sqrt{n}\left(S_n - T_n(Q)\right)\right) dQ^n = \sup \ell \tag{37}$$

Remark 4.3.7

(a) The extension of this result to general differentiable transformations τ is mentioned in Section 4.4; the necessary modifications are given by 4.4(15) below.

(b) In the case $p = 1$, the analogue statement to (a) holds with as. median unbiasedness [Pfanzagl and Wefelmeyer (1982; Definition 9.2.1)] in the place of regularity [Definition 4.3.1] (easy to see). ////

Proof Consider the simple perturbations $Q_n(\zeta, .)$ along some $\zeta \in Z_\infty^k(\theta)$. By the as. linearity of T at P_θ, we have uniformly on t-compacts of $\mathbb{R}^k$,

$$\sqrt{n}\left(T_n(Q_n(\zeta, t)) - \theta\right) = \mathrm{E}_\theta\, \psi_\theta \zeta' t + \mathrm{o}(n^0) \tag{38}$$

Using the log likelihood expansion [Lemma 4.2.4] and the estimator expansion 4.2(62), as in Lemma 4.2.18, we obtain from Corollary 2.2.6 that uniformly on t-compacts of $\mathbb{R}^k$,

$$\sqrt{n}\,(S_n - \theta)(Q_n^n(\zeta, t)) \xrightarrow{\mathrm{w}} \mathcal{N}\left(\mathrm{E}_\theta\, \varrho_\theta \zeta' t, \mathcal{C}_\theta(\varrho_\theta)\right) \tag{39}$$

Combining (38) and (39) yields the following as. normality,

$$\sqrt{n}\left(S_n - T_n(Q_n(\zeta, t))\right)(Q_n^n(\zeta, t)) \xrightarrow{\mathrm{w}} \mathcal{N}\left(B(\zeta, t), \mathcal{C}_\theta(\varrho_\theta)\right) \tag{40}$$

uniformly on t-compacts of $\mathbb{R}^k$. The following bias arises,

$$B(\zeta,t) = \mathrm{E}_\theta(\varrho_\theta - \psi_\theta)\zeta' t \tag{41}$$

Approximating $\varrho_\theta - \psi_\theta$ in $L_2^k(P_\theta)$ by $\zeta \in Z_\infty^k(\theta)$ shows that

$$\mathrm{E}_\theta\, \varrho_\theta \zeta' = \mathrm{E}_\theta\, \psi_\theta \zeta' \quad \forall\, \zeta \in Z_\infty^k(\theta) \implies \varrho_\theta = \psi_\theta \tag{42}$$

If $\varrho_\theta \neq \psi_\theta$, there exist a tangent $\xi \in Z_\infty^k(\theta)$ and a unit vector $e \in \mathbb{R}^k$ such that $b = B(\xi, e) \neq 0$. As a consequence, bias explodes:

$$\lim_{c\to\infty} \sup_{|t|\le c} |B(\xi,t)| \ge \lim_{c\to\infty} |B(\xi, ce)| = \lim_{c\to\infty} |cb| = \infty \tag{43}$$

(a) Because of (43), the limits occurring in (40) for $\zeta = \xi$ and $t \in \mathbb{R}^k$ are not tight in $\mathbb{R}^k$, let alone identical. In view of 4.2(24), therefore, S cannot be regular for T on $\mathcal{U}(\theta)$.

(b) By Fatou, it follows from (40) that

$$\begin{aligned}\liminf_{n\to\infty} \int \ell\bigl(\sqrt{n}\,[S_n - T_n(Q_n(\xi, ce))]\bigr)\, dQ_n^n(\xi, ce) &\ge \int \ell\, d\mathcal{N}(cb, \mathcal{C}_\theta(\varrho_\theta)) \\ = \int v\bigl((z+cb)'A(z+cb)\bigr)\, \mathcal{N}(0, \mathcal{C}_\theta(\varrho_\theta))(dz) &\end{aligned} \tag{44}$$

where the function v is increasing by assumption and the matrix A symmetric positive definite. By another appeal to Fatou we get

$$\liminf_{c\to\infty} \int v\bigl((z+cb)'A(z+cb)\bigr)\, \mathcal{N}(0, \mathcal{C}_\theta(\varrho_\theta))(dz) \ge v(\infty - 0) \tag{45}$$

In view of the inclusion 4.2(24) for ξ and $t = ce$, therefore,

$$\begin{aligned} v(\infty - 0) &\ge \lim_{r\to\infty} \liminf_{n\to\infty} \sup_{Q \in U(\theta, r/\sqrt{n})} \int \ell\bigl(\sqrt{n}\,(S_n - T(Q))\bigr)\, dQ^n \\ &\ge \liminf_{c\to\infty} \liminf_{n\to\infty} \int \ell\bigl(\sqrt{n}\,[S_n - T(Q_n(\xi, ce))]\bigr)\, dQ_n^n(\xi, ce) \\ &\underset{(44)}{\ge} \liminf_{c\to\infty} \int v\bigl((z+cb)'A(z+cb)\bigr)\, \mathcal{N}(0, \mathcal{C}_\theta(\varrho_\theta))(dz) \underset{(45)}{\ge} v(\infty - 0) \end{aligned}$$

and this proves (37). ////

4.3.2 Unbiased Testing

Multisided hypotheses about the transformed functional $\tau \circ T$ on the neighborhood system $\mathcal{U}(\theta)$ of P_θ are set up as follows: For some critical numbers $0 < b < c < \infty$, some radius $r \in (0, \infty)$, and employing on $\mathbb{R}^p$ the norm

$$|y|^2_{\Gamma_\theta(\psi_\theta)} = y'\Gamma_\theta(\psi_\theta)^{-1} y, \qquad \Gamma_\theta(\psi_\theta) = d\tau(\theta)\mathcal{C}_\theta(\psi_\theta)d\tau(\theta)'$$

the law Q of the i.i.d. observations at sample size n is subjected to the following multisided nonparametric hypotheses,

$$\begin{aligned} \mathrm{H}_n^{\mathrm{m}}(T,\theta,r)\colon \quad & Q \in U(\theta, r/\sqrt{n}\,),\; n\left|\tau\circ T_n(Q)-\tau(\theta)\right|^2_{\Gamma_\theta(\psi_\theta)} \le b^2 \\ \mathrm{K}_n^{\mathrm{m}}(T,\theta,r)\colon \quad & Q \in U(\theta, r/\sqrt{n}\,),\; n\left|\tau\circ T_n(Q)-\tau(\theta)\right|^2_{\Gamma_\theta(\psi_\theta)} \ge c^2 \end{aligned} \tag{46}$$

In case $p=1$, given some critical numbers $-\infty < b < c < \infty$, corresponding one-sided nonparametric hypotheses are

$$\begin{aligned} \mathrm{H}_n^{1}(T,\theta,r)\colon \quad & Q \in U(\theta, r/\sqrt{n}\,),\; \sqrt{n}\,\bigl(\tau\circ T_n(Q)-\tau(\theta)\bigr) \le b\,\sqrt{\Gamma_\theta(\psi_\theta)} \\ \mathrm{K}_n^{1}(T,\theta,r)\colon \quad & Q \in U(\theta, r/\sqrt{n}\,),\; \sqrt{n}\,\bigl(\tau\circ T_n(Q)-\tau(\theta)\bigr) \ge c\,\sqrt{\Gamma_\theta(\psi_\theta)} \end{aligned} \tag{47}$$

Two-sided hypotheses in the one-dimensional case $p=1$ can be formulated and treated likewise; the symmetric version $(-b_1 = b_2,\; -c_1 = c_2)$ is covered as a special case of the multisided problem.

In the nonparametric setup, not pointwise but only as. maximum size and as. minimum power are feasible. Then the as. maximin power at as. level $\alpha \in [0,1]$ will be given by the probability that a $\chi^2(p;c^2)$ random variable of noncentrality parameter c^2 exceeds the upper α point $c_\alpha(p;b^2)$ of a $\chi^2(p;b^2)$ distribution, respectively, by the tail probability of a standard normal $\mathcal{N}(0,1)$, where u_α denotes its upper α point.

To achieve the bounds, we use an estimator $S^\star$ which is regular for $\tau\circ T$ on $\mathcal{U}(\theta)$ with limit law $\mathcal{N}(0,\Gamma_\theta(\psi_\theta))$ [Definition 4.3.1]. In addition, since only $\tau(\theta)$ but not θ itself may be known, an as. estimator

$$\Gamma = (\Gamma_n), \qquad \Gamma_n\colon (\Omega^n,\mathcal{A}^n) \longrightarrow (\mathbb{R}^{p\times p},\mathbb{B}^{p\times p})$$

is required such that, for all $r\in(0,\infty)$, all sequences $Q_n \in U(\theta, r/\sqrt{n}\,)$,

$$\Gamma_n = \Gamma_\theta(\psi_\theta) + \mathrm{o}_{Q_n^n}(n^0) \tag{48}$$

Here are the nonparametric as. maximin power bounds.

Theorem 4.3.8 *Consider the nonparametric one- and multisided testing hypotheses* (47) *and* (46), *respectively, and let* ϕ *be any as. test.*

(a) *Then, in the one-sided case,*

$$\lim_{r\to\infty}\limsup_{n\to\infty}\sup_{Q\in \mathrm{H}_n^1(T,\theta,r)}\int \phi_n\,dQ \le \alpha \tag{49}$$

implies that

$$\lim_{r\to\infty}\limsup_{n\to\infty}\inf_{Q\in \mathrm{K}_n^1(T,\theta,r)}\int \phi_n\,dQ \le 1-\Phi\bigl(u_\alpha-(c-b)\bigr) \tag{50}$$

(b) *In the multisided case, if*

$$\lim_{r\to\infty} \limsup_{n\to\infty} \sup_{Q\in \mathrm{H}_n^{\mathrm{m}}(T,\theta,r)} \int \phi_n \, dQ \le \alpha \tag{51}$$

then

$$\lim_{r\to\infty} \limsup_{n\to\infty} \inf_{Q\in \mathrm{K}_n^{\mathrm{m}}(T,\theta,r)} \int \phi_n \, dQ \le \Pr\bigl(\chi^2(p;c^2) > c_\alpha(p;b^2)\bigr) \tag{52}$$

(c) *Assume $S^\star$ is regular for $\tau\circ T$ on $\mathcal{U}(\theta)$ with limit law $\mathcal{N}(0,\Gamma_\theta(\psi_\theta))$, and Γ is consistent for $\Gamma_\theta(\psi_\theta)$ on $\mathcal{U}(\theta)$ satisfying (48). Then the two as. tests $\phi^\star$ and $\chi^\star$ given by*

$$\phi_n^\star = \mathbf{I}\Bigl(\sqrt{n}\,\bigl(S_n^\star - \tau(\theta)\bigr) > (u_\alpha + b)\sqrt{\Gamma_n}\,\Bigr) \tag{53}$$

$$\chi_n^\star = \mathbf{I}\Bigl(n\,\bigl|S_n^\star - \tau(\theta)\bigr|_{\Gamma_n}^2 > c_\alpha(p;b^2)\Bigr) \tag{54}$$

satisfy

$$\begin{aligned} \limsup_{n\to\infty} \sup_{Q\in \mathrm{H}_n^{1}(T,\theta,r)} \int \phi_n^\star \, dQ &= \alpha \\ \limsup_{n\to\infty} \inf_{Q\in \mathrm{K}_n^{1}(T,\theta,r)} \int \phi_n^\star \, dQ &= 1-\Phi\bigl(u_\alpha-(c-b)\bigr) \end{aligned} \tag{55}$$

and

$$\begin{aligned} \limsup_{n\to\infty} \sup_{Q\in \mathrm{H}_n^{\mathrm{m}}(T,\theta,r)} \int \chi_n^\star \, dQ &= \alpha \\ \limsup_{n\to\infty} \inf_{Q\in \mathrm{K}_n^{\mathrm{m}}(T,\theta,r)} \int \chi_n^\star \, dQ &= \Pr\bigl(\chi^2(p;c^2) > c_\alpha(p;b^2)\bigr) \end{aligned} \tag{56}$$

for all $r \in (0,\infty)$, hence are as. maximin—achieving equalities in (a) *and* (b), *respectively.*

PROOF We consider the simple perturbations along some $\zeta \in Z_\infty^k(\theta)$. The product measures $Q_n^n(\zeta,.)$ are as. normal with as. covariance $\mathcal{C}_\theta(\zeta)$ [Lemma 4.2.4]. Recall the compact uniform expansion 4.3(9) of $\tau\circ T$,

$$\sqrt{n}\,\bigl(\tau\circ T_n(Q_n(\zeta,t)) - \tau(\theta)\bigr) = d\tau(\theta)D(\zeta)\,t + \mathrm{o}(n^0)$$

with $D(\zeta) = \mathrm{E}_\theta\,\psi_\theta\zeta'$. Let $A\in\mathbb{R}^{k\times k}$ be nonsingular. Then we have

$$\mathcal{C}_\theta(\zeta) \longrightarrow A\mathcal{C}_\theta(\psi_\theta)A', \qquad D(\zeta) \longrightarrow \mathcal{C}_\theta(\psi_\theta)A' \tag{57}$$

as $\zeta \in Z_\infty^k(\theta)$ tends to $\zeta^\star = A\psi_\theta$ in $L_2^k(P_\theta)$.

(a) Fix any $a\in(0,\infty)$ and $\delta\in(0,1)$. Define parametric hypotheses about the parameter $t\in\mathbb{R}^k$ of the measures $Q_n(\zeta,t)$,

$$\begin{aligned} \mathrm{H}_{\mathrm{m}}^1(\zeta,a)\colon \quad & \bigl|\mathcal{C}_\theta(\zeta)^{1/2}t\bigr| \le a-\delta,\ d\tau(\theta)D(\zeta)\,t \le (b-\delta)\sqrt{\Gamma_\theta(\psi_\theta)} \\ \mathrm{K}_{\mathrm{m}}^1(\zeta,a)\colon \quad & \bigl|\mathcal{C}_\theta(\zeta)^{1/2}t\bigr| \le a-\delta,\ d\tau(\theta)D(\zeta)\,t \ge (c+\delta)\sqrt{\Gamma_\theta(\psi_\theta)} \end{aligned} \tag{58}$$

These hypotheses are of form 3.4(48) with unit vector e proportional to

$$e(\zeta) = \mathcal{C}_\theta(\zeta)^{-1/2} D(\zeta)' d\tau(\theta)' \tag{59}$$

If a is so large that

$$a \geq \delta + \frac{\max\{|b|, |c|\} + \delta}{|e(\zeta)|_G}$$

where $G = \Gamma_\theta(\psi_\theta)\mathbb{I}_k$ here and in (60), Theorem 3.4.6 applies to the as. testing of the hypotheses (58) at level α, and yields the as. maximin power

$$1 - \Phi\left(u_\alpha - \frac{c - b + 2\delta}{|e(\zeta)|_G}\right) \tag{60}$$

By choosing $r = r(\zeta, a) \in (0, \infty)$ according to 4.2(24) and using the expansion of $\tau \circ T$, we achieve that eventually,

$$\begin{aligned} t \in \mathrm{H}^1_{\mathrm{m}}(\zeta, a) &\implies Q_n(\zeta, t) \in \mathrm{H}^1_n(T, \theta, r) \\ t \in \mathrm{K}^1_{\mathrm{m}}(\zeta, a) &\implies Q_n(\zeta, t) \in \mathrm{K}^1_n(T, \theta, r) \end{aligned} \tag{61}$$

Now let $\zeta \in Z^k_\infty(\theta)$ approach $\zeta^\star = A\psi_\theta$ in $L^k_2(P_\theta)$ while keeping δ fixed. Then (57) and (59) imply that

$$\begin{aligned} e(\zeta) \longrightarrow \quad e(\zeta^\star) &= \big(A\mathcal{C}_\theta(\psi_\theta)A'\big)^{-1/2} A\mathcal{C}_\theta(\psi_\theta) d\tau(\theta)' \\ \big|e(\zeta^\star)\big|^2 &= \Gamma_\theta(\psi_\theta) \end{aligned} \tag{62}$$

so bound (60) tends to $1 - \Phi\big(u_\alpha - (c - b + 2\delta)\big)$. Letting $\delta \downarrow 0$ concludes the proof of a).

(b) Fix $a, \delta \in (0, \infty)$ such that $a^2 - \delta^2 \geq c^2 + \delta^2$, and introduce the matrix

$$E = \Gamma_\theta(\psi_\theta)^{-1/2} d\tau(\theta) \mathcal{C}_\theta(\psi_\theta) A' \big(A\mathcal{C}_\theta(\psi_\theta)A'\big)^{-1/2} \tag{63}$$

which has p orthonormal rows in $\mathbb{R}^k$ and thus satisfies 3.4(83) as

$$\begin{aligned} &EE' = \\ &\Gamma_\theta(\psi_\theta)^{-1/2} d\tau(\theta)\mathcal{C}_\theta(\psi_\theta)A'\big(A\mathcal{C}_\theta(\psi_\theta)A'\big)^{-1} A\mathcal{C}_\theta(\psi_\theta) d\tau(\theta)'\Gamma_\theta(\psi_\theta)^{-1/2} \\ &\quad = \Gamma_\theta(\psi_\theta)^{-1/2} d\tau(\theta)\mathcal{C}_\theta(\psi_\theta) d\tau(\theta)'\Gamma_\theta(\psi_\theta)^{-1/2} \\ &\quad = \Gamma_\theta(\psi_\theta)^{-1/2}\Gamma_\theta(\psi_\theta)\Gamma_\theta(\psi_\theta)^{-1/2} = \mathbb{I}_p \end{aligned}$$

Define the following hypotheses about the parameter $t \in \mathbb{R}^k$ of $Q_n(\zeta, t)$,

$$\begin{aligned} \mathrm{H}^{\mathrm{m}}_{\mathrm{m}}(\zeta, a)&: \big|\mathcal{C}_\theta(\zeta)^{1/2} t\big|^2 \leq a^2 - \delta^2, \; t'\mathcal{C}_\theta(\zeta)^{1/2} E'E\,\mathcal{C}_\theta(\zeta)^{1/2} t \leq b^2 - \delta^2 \\ \mathrm{K}^{\mathrm{m}}_{\mathrm{m}}(\zeta, a)&: \big|\mathcal{C}_\theta(\zeta)^{1/2} t\big|^2 \leq a^2 - \delta^2, \; t'\mathcal{C}_\theta(\zeta)^{1/2} E'E\,\mathcal{C}_\theta(\zeta)^{1/2} t \geq c^2 + \delta^2 \end{aligned} \tag{64}$$

For the as. testing of the hypotheses (64) at level α, Theorem 3.4.10 yields the as. maximin power

$$\Pr\bigl(\chi^2(p;c^2+\delta^2) > c_\alpha(p;b^2-\delta^2)\bigr) \tag{65}$$

To eventually include the hypotheses (64) into the hypotheses (46) defined by T, let $\zeta \in Z^k_\infty(\theta)$ approach $\zeta^\star = A\psi_\theta$ in $L^k_2(P_\theta)$, while keeping δ fixed. Then, because of (57), if ζ is sufficiently close to $\zeta^\star$, it holds that

$$t'\mathcal{C}_\theta(\zeta)\,t \le a^2-\delta^2 \implies t'A\mathcal{C}_\theta(\psi_\theta)A't \le a^2 \tag{66}$$

and, with the matrix E given by (63),

$$\begin{aligned}\mathcal{C}_\theta(\zeta)^{1/2}E'E\,\mathcal{C}_\theta(\zeta)^{1/2} &\approx A\mathcal{C}_\theta(\psi_\theta)d\tau(\theta)'\Gamma_\theta(\psi_\theta)^{-1}d\tau(\theta)\mathcal{C}_\theta(\psi_\theta)A' \\ &\approx D(\zeta)'d\tau(\theta)'\Gamma_\theta(\psi_\theta)^{-1}d\tau(\theta)D(\zeta)\end{aligned} \tag{67}$$

where, in this proof, we write $M_1 \approx M_2$ for two $k\times k$ matrices iff

$$t'A\mathcal{C}_\theta(\psi_\theta)A't \le a^2 \implies 2\,\bigl|t'(M_2-M_1)t\bigr| < \delta^2$$

Using the expansion of $\tau\circ T$ this ensures that eventually

$$\begin{aligned}t\in \mathrm{H}^{\mathrm{m}}_{\mathrm{m}}(\zeta,a) &\implies n\,\bigl|\tau\circ T_n(Q_n(\zeta,t))-\tau(\theta)\bigr|^2_{\Gamma_\theta(\psi_\theta)} \le b^2 \\ t\in \mathrm{K}^{\mathrm{m}}_{\mathrm{m}}(\zeta,a) &\implies n\,\bigl|\tau\circ T_n(Q_n(\zeta,t))-\tau(\theta)\bigr|^2_{\Gamma_\theta(\psi_\theta)} \ge c^2\end{aligned}$$

If now $r=r(\zeta,a)\in(0,\infty)$ is chosen according to 4.2(24), then eventually

$$\begin{aligned}t\in \mathrm{H}^{\mathrm{m}}_{\mathrm{m}}(\zeta,a) &\implies Q_n(\zeta,t)\in \mathrm{H}^{\mathrm{m}}_n(T,\theta,r) \\ t\in \mathrm{K}^{\mathrm{m}}_{\mathrm{m}}(\zeta,a) &\implies Q_n(\zeta,t)\in \mathrm{K}^{\mathrm{m}}_n(T,\theta,r)\end{aligned} \tag{68}$$

Since the power (65) converges to $\Pr\bigl(\chi^2(p;c^2) > c_\alpha(p;b^2)\bigr)$ as $\delta\downarrow 0$, bound (52) follows.

(c) Fix $r\in(0,\infty)$ and choose any sequence $Q_n\in U(\theta,r/\sqrt{n})$. Write

$$\sqrt{n}\,\Gamma_n^{-1/2}\bigl(S^\star_n-\tau(\theta)\bigr) = \sqrt{n}\,\Gamma_n^{-1/2}\bigl(S^\star_n-\tau\circ T_n(Q_n)\bigr)+\tilde v_n \tag{69}$$

where

$$\tilde v_n = \sqrt{n}\,\Gamma_n^{-1/2}\bigl(\tau\circ T_n(Q_n)-\tau(\theta)\bigr) \tag{70}$$

The assumptions on Γ and $S^\star$ ensure that

$$\tilde v_n = v_n + \mathrm{o}_{Q^n_n}(n^0)\,,\qquad v_n = \sqrt{n}\,\Gamma_\theta(\psi_\theta)^{-1/2}\bigl(\tau\circ T_n(Q_n)-\tau(\theta)\bigr) \tag{71}$$

and

$$\sqrt{n}\,\Gamma_n^{-1/2}\bigl(S^\star_n-\tau\circ T_n(Q_n)\bigr)(Q^n_n) \xrightarrow{\;\mathrm{w}\;} \mathcal{N}(0,\mathbb{I}_p) \tag{72}$$

The distribution function of the limiting normal being continuous, the latter convergence even holds in Kolmogorov distance. Therefore, up to some terms which tend to 0 in d_κ,

$$\sqrt{n}\,\Gamma_n^{-1/2}\bigl(S_n^\star - \tau(\theta)\bigr)(Q_n^n) = \mathcal{N}(v_n, \mathbb{I}_p) + \mathrm{o}_\kappa(n^0) \tag{73}$$

hence

$$n\,\bigl|S_n^\star - \tau(\theta)\bigr|_{\Gamma_n}^2 (Q_n^n) = \chi^2(p; |v_n|^2) + \mathrm{o}_\kappa(n^0) \tag{74}$$

Since the $\chi^2(p;.)$ distribution is continuous in the noncentrality parameter w.r.t. weak convergence, it follows that

$$Q_m^m\bigl(m\,\bigl|S_m^\star - \tau(\theta)\bigr|_{\Gamma_m}^2 > c_\alpha(p; b^2)\bigr) \longrightarrow \Pr\bigl(\chi^2(p; v^2) > c_\alpha(p; b^2)\bigr) \tag{75}$$

if along the subsequence (m),

$$|v_m|^2 = m\,\bigl|\tau \circ T_m(Q_m) - \tau(\theta)\bigr|_{\Gamma_\theta(\psi_\theta)}^2 \longrightarrow v^2 \tag{76}$$

Under (46), the possible cluster values are $v^2 \le b^2$ and $v^2 \ge c^2$, respectively. Since the $\chi^2(p;.)$ distribution has monotone likelihood ratios, hence is stochastically increasing, in the noncentrality parameter, it follows that

$$\limsup_{n\to\infty} \sup_{Q \in \mathrm{H}_n^{\mathrm{m}}(T,\theta,r)} \int \chi_n^\star \, dQ \le \alpha \tag{77}$$

and

$$\limsup_{n\to\infty} \inf_{Q \in \mathrm{K}_n^{\mathrm{m}}(T,\theta,r)} \int \chi_n^\star \, dQ \ge \Pr\bigl(\chi^2(p; c^2) > c_\alpha(p; b^2)\bigr) \tag{78}$$

In fact, equality holds for r sufficiently large, since then v^2 can approach b^2 under the null hypothesis, and c^2 under the alternative; for example, if the parameter $t \in \mathbb{R}^k$ of simple perturbations $Q_n = Q_n(\zeta, t)$ is chosen according to the restrictions in (46).

This proves (c) in the multisided case. The same arguments are good for the one-sided problem. ////

Remark 4.3.9 We record some technical points.

(a) The proof shows that Theorem 4.3.8 is true without the $\lim_{r\to\infty}$ but with $r \in (0,\infty)$ fixed though sufficiently large, if in the approximation of $\zeta^\star = A\psi_\theta$ by bounded tangents $\zeta_m \in Z_\infty^k(\theta)$ the radius $r_m = r(\zeta_m, a)$ from condition 4.2(24) can be chosen independently of m. This is trivially true if the influence curve ψ_θ itself is bounded.

Depending on the neighborhood system $\mathcal{U}(\theta)$, the fixed radius version of Theorem 4.3.8 may hold if ψ_θ is unbounded. In connection with total variation, Hellinger, Kolmogorov, Lévy, Prokhorov, and CvM balls, Theorem 4.3.8 always holds in the fixed radius version [Lemma 4.2.8 and Remark 6.3.10]. However, for contamination balls and ψ_θ unbounded, we need the $\lim_{r\to\infty}$.

(b) Theorem 4.3.8 is based on Theorems 3.4.6 and 3.4.10 applied to the L_2 differentiable simple perturbations along least favorable tangents, in the i.i.d. case. The parametric versions for the product models 4.2(67) may again be recovered from the preceding proof, by a restriction to the purely parametric neighborhoods of the type

$$U_0(\theta, r/\sqrt{n}) = \{ P_{\theta+t/\sqrt{n}} \mid |t| \le r \} \tag{79}$$

and the formal identifications

$$\begin{array}{ll} \zeta = A\Lambda_\theta\,, \quad \psi_\theta = \psi_{h,\theta} = \mathcal{I}_\theta^{-1}\Lambda_\theta\,, & \mathcal{C}_\theta(\psi_\theta) = \mathcal{I}_\theta^{-1} \\ Q_n(\zeta, t) = P_{\theta+A't/\sqrt{n}}\,, & \sqrt{n}\,(T_n(Q_n(\zeta,t)) - \theta) = A't \end{array} \tag{80}$$

Then, for τ equal to the canonical projection onto the $p \le k$ -dimensional first component, we are back in Example 3.4.11. For general τ, that example is generalized by the preceding proof, if specialized via (79) and (80).

(c) For the one-sided nonparametric testing problem (47) solved by Theorem 4.3.8 a the following simple as. subhypotheses are least favorable,

$$\begin{aligned} dR_{0,n} &= \left(1 + \frac{b}{\sqrt{n\,\Gamma_\theta(\psi_\theta)}}\, d\tau(\theta)\,\psi_\theta\right) dP_\theta \\ dR_{1,n} &= \left(1 + \frac{c}{\sqrt{n\,\Gamma_\theta(\psi_\theta)}}\, d\tau(\theta)\,\psi_\theta\right) dP_\theta \end{aligned} \tag{81}$$

To see this, evaluate the least favorable tangent $\zeta^\star = A\psi_\theta$ at the former least favorable parameter pair 3.4(57) and, using (62), check that

$$\mathcal{C}_\theta(\zeta^\star)^{-1/2}\,\frac{e(\zeta^\star)}{|e(\zeta^\star)|} = \frac{1}{\sqrt{\Gamma_\theta(\psi_\theta)}}\,\bigl(d\tau(\theta)A^{-1}\bigr)'$$

hence

$$\frac{e(\zeta^\star)'}{|e(\zeta^\star)|}\,\mathcal{C}_\theta(\zeta^\star)^{-1/2}\zeta^\star = \frac{1}{\sqrt{\Gamma_\theta(\psi_\theta)}}\,d\tau(\theta)\,\psi_\theta$$

To ensure the eventual inclusion into the one-sided nonparametric hypotheses (47), definition (81) obviously requires some modifications: The numbers b and c have to be replaced by suitable approximations $b_n \uparrow b$ from below and $c_n \downarrow c$ from above, respectively, and an unbounded influence curve ψ_θ by a bounded approximation $\zeta \in Z_\infty^k(\theta)$. This, of course, results in an approximate notion of least favorability (in a setup that is asymptotic anyway). A bounded ψ_θ can be kept in (81). In case ψ_θ is unbounded, if $\mathcal{U}(\theta)$ covers simple perturbations, and T is weakly differentiable, along $Z_2^k(\theta)$, ψ_θ can in (81) be substituted by $\zeta_n \in Z_\infty^k(\theta)$ that approximate ψ_θ in the manner of 4.2(16). ////

Elementary Version

Let us scrutinize the nonparametric as. maximin power bound by an undogmatic evaluation for a restricted class of tests based on as. linear estimators [Definition 4.2.16]. We confine ourselves to the multisided problem, as the one-sided case can be treated the same way, and exclude nuisance parameters for simplicity by setting

$$\tau = \mathrm{id}_{\mathbb{R}^k} \tag{82}$$

Any estimator S which is as. linear at P_θ with influence curve $\varrho_\theta \in \Psi_2(\theta)$ serves as test statistic in the corresponding as. *linear estimator test*

$$\chi_S = (\chi_n), \qquad \chi_n = \mathbf{I}\Bigl(n\,|S_n - \theta|^2_{\mathcal{C}_\theta(\varrho_\theta)} > c_n\Bigr) \tag{83}$$

where the choice of the critical values $c_n \in [0,\infty]$ is left open. Because of (82), the covariance $\mathcal{C}_\theta(\varrho_\theta)$ needs not to be estimated. Thus the maximin test (54) attains form (83) with $\varrho_\theta = \psi_\theta$, in view of the uniqueness statement (7) of Theorem 4.3.2.

It seems natural to base a decision about the difference between the true, unknown parameter and the known, hypothetical parameter on estimates of this difference. Then the arguments for a restriction to as. linear estimators that have been given in Remark 4.2.17 e, f carry over.

We first consider the parametric problem, which concerns the parameter $t \in \mathbb{R}^k$ of the product measures $P^n_{\theta+t/\sqrt{n}}$. If the critical numbers are chosen as $b = 0 < c \le a < \infty$, the hypotheses 3.4(141) read:

$$\mathrm{H}^{\mathrm{m}}_{\mathrm{m}}(\theta,a)\colon\quad t = 0 \qquad \mathrm{K}^{\mathrm{m}}_{\mathrm{m}}(\theta,a)\colon\quad c^2 \le t'\mathcal{I}_\theta t \le a^2 \tag{84}$$

By Theorem 3.4.10, a test that is as. maximin among all tests for the parametric hypotheses (84) can, as in (83), be based on an estimator which is as. linear at P_θ with influence curve $\psi_{h,\theta} = \mathcal{I}_\theta^{-1}\Lambda_\theta$ [Example 3.4.11 a, b].

This optimality and the parametric as. maximin power bound itself [Theorem 3.4.10] are illuminated by the performance of other tests based on as. linear estimators. Here is the elementary result.

Proposition 4.3.10 *Assume the estimator S as. linear at P_θ with influence curve $\varrho_\theta \in \Psi_2(\theta)$, and let $c_n \to c_\alpha(k;0)$. Then the as. linear estimator test χ_S of form (83) has as. minimum power*

$$\lim_{n\to\infty} \inf_{\mathrm{K}^{\mathrm{m}}_{\mathrm{m}}(\theta,a)} \int \chi_n \, dP^n_{\theta+t/\sqrt{n}} = \Pr\bigl(\chi^2(k;\gamma^2_\theta(\varrho_\theta)) > c_\alpha(k;0)\bigr) \tag{85}$$

with noncentrality

$$\gamma^2_\theta(\varrho_\theta) = \frac{c^2}{\max\mathrm{ev}\,\mathcal{I}_\theta^{1/2}\mathcal{C}_\theta(\varrho_\theta)\mathcal{I}_\theta^{1/2}} \le c^2 \tag{86}$$

The upper bound in (86) is achieved iff $\varrho_\theta = \psi_{h,\theta}$.

PROOF By Lemma 4.2.18, uniformly in $t'\mathcal{I}_\theta t \le a^2$,

$$\sqrt{n}\,\big(S_n - \theta - t/\sqrt{n}\big)(P^n_{\theta+t/\sqrt{n}}) \xrightarrow{\;\mathrm{w}\;} \mathcal{N}\big(0, \mathcal{C}_\theta(\varrho_\theta)\big)$$

hence

$$n\,\big|S_n - \theta\big|^2_{\mathcal{C}_\theta(\varrho_\theta)}(P^n_{\theta+t/\sqrt{n}}) \xrightarrow{\;\mathrm{w}\;} \chi^2\big(k; |t|^2_{\mathcal{C}_\theta(\varrho_\theta)}\big) \tag{87}$$

The $\chi^2(k;.)$ distribution being stochastically increasing in the noncentrality parameter, it suffices to bound $|t|^2_{\mathcal{C}_\theta(\varrho_\theta)}$. On the one hand,

$$\inf\{\, t'\mathcal{C}_\theta(\varrho_\theta)^{-1}t \mid c^2 \le t'\mathcal{I}_\theta t \le a^2 \,\} = \frac{c^2}{\operatorname{maxev} \mathcal{I}_\theta^{1/2}\mathcal{C}_\theta(\varrho_\theta)\mathcal{I}_\theta^{1/2}} \tag{88}$$

On the other hand, the Cramér–Rao bound 4.2(74) implies the uniform pointwise bound

$$t'\mathcal{C}_\theta(\varrho_\theta)^{-1}t \le t'\mathcal{I}_\theta t = t'\mathcal{C}_\theta(\psi_{h,\theta})^{-1}t$$

which, in view of (87), translates into a uniform upper bound for the pointwise as. power. Rewriting the Cramér–Rao bound as:

$$\mathcal{I}_\theta^{1/2}\mathcal{C}_\theta(\varrho_\theta)\mathcal{I}_\theta^{1/2} \ge \mathbb{I}_p \tag{89}$$

we see that all eigenvalues of $\mathcal{I}_\theta^{1/2}\mathcal{C}_\theta(\varrho_\theta)\mathcal{I}_\theta^{1/2}$ are greater than 1, and the largest is 1 iff the Cramér–Rao bound is achieved; that is, iff $\varrho_\theta = \psi_{h,\theta}$. ////

This result corresponds to Proposition 4.2.20 and once more demonstrates that the parametric as. maximin power bound is related to the Cramér–Rao bound for the as. variance of as. linear estimators. The as. minimum power of tests based on as. linear estimators is nontrivial; for example, it depends continuously on $\varrho_\theta \in \Psi_2(\theta) \subset L_2^k(P_\theta)$.

The nonparametric hypotheses (46), of radius $r \in (0,\infty)$ and with critical numbers $0 < b < c < \infty$, in the present case (82) simplify to

$$\begin{aligned} \mathrm{H}^{\mathrm{m}}_n(T,\theta,r)\colon\quad & Q \in U(\theta, r/\sqrt{n}),\ n\,\big|T_n(Q) - \theta\big|^2_{\mathcal{C}_\theta(\psi_\theta)} \le b^2 \\ \mathrm{K}^{\mathrm{m}}_n(T,\theta,r)\colon\quad & Q \in U(\theta, r/\sqrt{n}),\ n\,\big|T_n(Q) - \theta\big|^2_{\mathcal{C}_\theta(\psi_\theta)} \ge c^2 \end{aligned} \tag{90}$$

The following analogue to Proposition 4.3.6 now reveals the quite different behavior, in the robust setup of testing functionals, of as. linear estimator tests: All but one are biased; the only one unbiased is already maximin. Here is this bias degeneration.

Proposition 4.3.11 *Let χ_S be an as. linear estimator test of form (83) with influence curve*

$$\varrho_\theta \in \Psi_2(\theta), \qquad \varrho_\theta \ne \psi_\theta \tag{91}$$

and such that

$$\lim_{r\to\infty}\liminf_{n\to\infty}\sup_{Q\in\mathrm{H}^{\mathrm{m}}_n(T,\theta,r)} \int \chi_n\,dQ^n < 1 \tag{92}$$

Then

$$\lim_{n\to\infty} \inf_{Q\in \mathrm{K}_n^{\mathrm{m}}(T,\theta,r)} \int \chi_n \, dQ^n = 0 \tag{93}$$

for all $r \in (0,\infty)$ *sufficiently large.*

PROOF We use the following variant of the 'completeness' implication (42),

$$\varrho_\theta \neq \psi_\theta \implies \exists\, \zeta \in Z_2^k(\theta)\colon\ \mathrm{E}_\theta\, \psi_\theta \zeta' = 0\,,\ \mathrm{E}_\theta\, \varrho_\theta \zeta' \neq 0 \tag{94}$$

Indeed, consider the closed linear subspace $M = \{\, A\psi_\theta \mid A \in \mathbb{R}^{k\times k}\,\}$ of the Hilbert space $L_2^k(P_\theta)$ and decompose ϱ_θ orthogonally relative to M, so that $\varrho_\theta = A\psi_\theta + \zeta$ for some $A \in \mathbb{R}^{k\times k}$ and $\zeta \perp M$. Then $\zeta \in Z_2^k(\theta)$ and $\mathrm{E}_\theta\, \psi_\theta \zeta' = 0$. Therefore $\mathrm{E}_\theta\, \varrho_\theta \zeta' = \mathrm{E}_\theta\, \zeta\zeta'$, which can be the zero matrix only if $\zeta = 0$; that is, if $\varrho_\theta = \psi_\theta$.

For the proof, assume (91), let $\delta \in (0,1)$ and fix any $a \in (0,\infty)$. Choose $\zeta \in Z_2^k(\theta)$ according to (94). Approximate ζ in $L_2^k(P_\theta)$ so closely by some $\xi = \xi(a,\delta) \in Z_\infty^k(\theta)$ that

$$\operatorname{maxev} \mathrm{E}_\theta\, \xi\psi_\theta{}'\, \mathcal{C}_\theta(\psi_\theta)^{-1}\, \mathrm{E}_\theta\, \psi_\theta\xi' < \frac{b^2}{a^2} \tag{95}$$

$$\frac{\operatorname{maxev} \mathrm{E}_\theta\, \xi\varrho_\theta{}'\, \mathcal{C}_\theta(\varrho_\theta)^{-1}\, \mathrm{E}_\theta\, \varrho_\theta\xi'}{\operatorname{maxev} \mathrm{E}_\theta\, \zeta\varrho_\theta{}'\, \mathcal{C}_\theta(\varrho_\theta)^{-1}\, \mathrm{E}_\theta\, \varrho_\theta\zeta'} > 1-\delta \tag{96}$$

Consider the simple perturbations $Q_n(\xi,.)$ along ξ. Then (95) and the expansion 4.2(44) of T ensure that, uniformly in $|t| \le a$,

$$\begin{aligned} n\,\bigl|T_n(Q_n(\xi,t)) - \theta\bigr|^2_{\mathcal{C}_\theta(\psi_\theta)} - \mathrm{o}(n^0) &= \bigl|\mathrm{E}_\theta\, \psi_\theta\xi' t\bigr|^2_{\mathcal{C}_\theta(\psi_\theta)} \\ &\le a^2 \operatorname{maxev} \mathrm{E}_\theta\, \xi\psi_\theta{}'\, \mathcal{C}_\theta(\psi_\theta)^{-1}\, \mathrm{E}_\theta\, \psi_\theta\xi' < b^2 \end{aligned}$$

Because of full neighborhoods, there exist an $n_0 = n_0(\xi,a) \in \mathbb{N}$ and a radius $r = r(\xi,a) \in (0,\infty)$ such that

$$Q_n(\xi,t) \in \mathrm{H}_n^{\mathrm{m}}(T,\theta,r), \qquad |t| \le a,\ n > n_0 \tag{97}$$

By Lemma 4.2.18, the as. linear estimator S with influence curve ϱ_θ at P_θ is as. normal, uniformly in $|t| \le a$,

$$\sqrt{n}\,(S_n - \theta)(Q_n^n(\xi,t)) \xrightarrow{\ \mathrm{w}\ } \mathcal{N}\bigl(\mathrm{E}_\theta\, \varrho_\theta\xi' t\,, \mathcal{C}_\theta(\varrho_\theta)\bigr)$$

Hence, uniformly in $|t| \le a$,

$$n\,\bigl|\, S_n - \theta\,\bigr|^2_{\mathcal{C}_\theta(\varrho_\theta)}(Q_n^n(\xi,t)) \xrightarrow{\ \mathrm{w}\ } \chi^2\bigl(k; \bigl|\mathrm{E}_\theta\, \varrho_\theta\xi' t\bigr|^2_{\mathcal{C}_\theta(\varrho_\theta)}\bigr) \tag{98}$$

By (96), the noncentrality parameter under (97) may be as large as

$$\begin{aligned} &\sup\bigl\{\, t'\, \mathrm{E}_\theta\, \xi\varrho_\theta{}'\, \mathcal{C}_\theta(\varrho_\theta)^{-1}\, \mathrm{E}_\theta\, \varrho_\theta\xi' t \bigm| |t| \le a \,\bigr\} \\ &\qquad = a^2 \operatorname{maxev} \mathrm{E}_\theta\, \xi\varrho_\theta{}'\, \mathcal{C}_\theta(\varrho_\theta)^{-1}\, \mathrm{E}_\theta\, \varrho_\theta\xi' \\ &\qquad\qquad \ge a^2(1-\delta) \operatorname{maxev} \mathrm{E}_\theta\, \zeta\varrho_\theta{}'\, \mathcal{C}_\theta(\varrho_\theta)^{-1}\, \mathrm{E}_\theta\, \varrho_\theta\zeta' \end{aligned} \tag{99}$$

If $r \to \infty$ in (92), $a \in (0, \infty)$ may be chosen arbitrarily large. Thus the noncentrality parameter generated under the null hypotheses $\mathrm{H}_n^{\mathrm{m}}(T, \theta, r)$ is unbounded. Therefore, (92) forces the critical values $c_n \to \infty$.

Exchanging the roles of ψ_θ and ϱ_θ in (94) there is some other $\zeta \in Z_2^k(\theta)$ such that $\mathrm{E}_\theta\, \varrho_\theta \zeta' = 0$ and $\mathrm{E}_\theta\, \psi_\theta \zeta' \neq 0$. Choose $a \in (0, \infty)$ so large and then approximate ζ so closely by some other $\xi \in Z_\infty^k(\theta)$ that

$$a^2 \operatorname{maxev} \mathrm{E}_\theta\, \xi \psi_\theta' \, \mathcal{C}_\theta(\psi_\theta)^{-1} \mathrm{E}_\theta\, \psi_\theta \xi' > c^2 \tag{100}$$

$$\operatorname{maxev} \mathrm{E}_\theta\, \xi \varrho_\theta' \, \mathcal{C}_\theta(\varrho_\theta)^{-1} \mathrm{E}_\theta\, \varrho_\theta \xi' < \frac{\delta}{a^2} \tag{101}$$

Let $e \in \mathbb{R}^k$, $|e| = 1$, be an eigenvector to the maximum eigenvalue of the matrix in (100). Then there exist $n_0 = n_0(\xi, a) \in \mathbb{N}$, $r = r(\xi, a) \in (0, \infty)$ such that the simple perturbations along this ξ satisfy

$$Q_n(\xi, ae) \in \mathrm{K}_n^{\mathrm{m}}(T, \theta, r), \qquad n > n_0 \tag{102}$$

due to full neighborhoods and inequality (100) which, in combination with the expansion of T, ensures that

$$\begin{aligned} n \left| T_n(Q_n(\xi, ae)) - \theta \right|^2_{\mathcal{C}_\theta(\psi_\theta)} - \mathrm{o}(n^0) &= a^2 \left| \mathrm{E}_\theta\, \psi_\theta \xi' e \right|^2_{\mathcal{C}_\theta(\psi_\theta)} \\ &= a^2 \operatorname{maxev} \mathrm{E}_\theta\, \xi \psi_\theta' \, \mathcal{C}_\theta(\psi_\theta)^{-1} \mathrm{E}_\theta\, \psi_\theta \xi' > c^2 \end{aligned}$$

Under the alternatives (102), in view of (98) and (101), however, the test statistic is asymptotically $\chi^2(k; .)$ with noncentrality parameter

$$a^2 \left| \mathrm{E}_\theta\, \varrho_\theta \xi' e \right|^2_{\mathcal{C}_\theta(\varrho_\theta)} < \delta \tag{103}$$

Since c_n tend to ∞, (93) has been proved for this particular $r(\xi, a)$, and hence for all $r \geq r(\xi, a)$. ////

4.3.3 Remarks and Criticisms

Let us summarize some aspects of the nonparametric optimality results.

Approximately Least Favorable Tangents: Theorems 4.3.2, 4.3.4 and 4.3.8 have been derived essentially using only one of the least favorable tangents (1). Such a least favorable $\zeta^\star = A\psi_\theta$ can be inserted directly if ψ_θ is bounded. If ψ_θ is unbounded, some approximation is needed: Whereas the proofs separate this approximation from the passage $n \to \infty$, it is already built into the simple perturbations $Q_n(\zeta^\star, .)$ along the unbounded $\zeta^\star$ itself. The use of $Q_n(\zeta^\star, .)$ just requires the neighborhood system to cover simple perturbations along $Z_2^k(\theta)$ and the functional to be weakly differentiable along $Z_2^k(\theta)$ [Remark 4.2.13 d]. The neighborhood systems $\mathcal{U}_*(\theta)$ of type $* = h, v, \kappa, \pi, \lambda, \mu$ except $* = c$ [Lemma 4.2.8] and the Hellinger and CvM differentiable functionals [Examples 4.2.14 and 4.2.15, Lemma 4.2.4] fulfill these requirements.

Propositions 4.3.6, 4.3.11, contrary to Theorems 4.3.2, 4.3.4, and 4.3.8, cannot be derived using essentially only the least favorable tangents (1).

Smooth Parametric Model: The assumption of a parametric model $\mathcal{P}$ that is L_2 differentiable at θ with Fisher information $\mathcal{I}_\theta > 0$ makes the parametric optimality results in the i.i.d. case, based on $\mathcal{I}_\theta^{-1} = \mathcal{C}_\theta(\psi_{h,\theta})$, available as some standard for the variety of nonparametric optimality results obtained for different functionals. In a technical sense, however, this assumption has been used only implicitly; namely, in the definition of influence curves to express local Fisher consistency.

The proofs to Theorems 4.3.2, 4.3.4, and 4.3.8 go through unchanged without the assumption of such a smooth parametric center model $\mathcal{P}$. In fact, if the functional T has the expansion 4.2(44) with any function

$$\psi_\theta \in L_2^k(P_\theta), \qquad \mathrm{E}_\theta\, \psi_\theta = 0\,, \qquad \mathcal{C}_\theta(\psi_\theta) > 0 \tag{104}$$

where P_θ may stand for any fixed probability, the nonparametric optimality Theorems 4.3.2, 4.3.4, and 4.3.8 obtain. If there is a center parametric model $\mathcal{P}$, it need not be smooth. Intuitively speaking, the inclusion of contamination enforces as. normality (along least favorable tangents) even if the ideal model $\mathcal{P}$ itself is not as. normal at θ [Millar (1981, p85)].

For example, let $\mathcal{P}$ w.r.t. some weight $\mu \in \mathcal{M}_\sigma(\mathbb{B}^m)$ satisfy 6.3(40), be identifiable in $L_2(\mu)$, and CvM differentiable at every $\theta \in \Theta$ with derivative Δ_θ and CvM information $\mathcal{J}_\theta > 0$. Then the Cramér–von Mises MD functional T_μ is CvM differentiable at all P_θ having expansion 4.2(49) with CvM influence curve $\varphi_{\mu,\theta} = \mathcal{J}_\theta^{-1}\Delta_\theta$ [Theorem 6.3.4 b]. Provided that $\psi_{\mu,\theta}$ related to $\varphi_{\mu,\theta}$ via 4.2(51) has covariance $\mathcal{C}_\theta(\psi_{\mu,\theta}) > 0$, the nonparametric optimality results apply to T_μ [Remark 6.3.10].

In particular, they apply to T_μ on a full neighborhood system about the as. nonnormal rectangular shift model treated in Example 2.3.12 b; the tangents least favorable to T_μ point away from the rectangular family.

Hellinger MD Functional: If the parametric model $\mathcal{P}$ is identifiable and L_2 differentiable with Fisher information $\mathcal{I}_\theta > 0$ at all $\theta \in \Theta$, the Hellinger MD functional T_h can be constructed to be Hellinger differentiable at every P_θ with influence curve $\psi_{h,\theta} = \mathcal{I}_\theta^{-1}\Lambda_\theta$ [Theorem 6.3.4 a]. Thus the nonparametric optimality theorems for T_h hold true with $\mathcal{C}_\theta(\psi_{h,\theta}) = \mathcal{I}_\theta^{-1}$, and the tangents $\zeta^\star = A\Lambda_\theta$ are least favorable for T_h.

Because of this and since T_h is Hellinger differentiable, the nonparametric optimality results for the Hellinger MD functional can be derived staying within the parametric family [that is, employing purely parametric neighborhoods of type (79), on which any Fisher consistent functional as. equals the original parameter]. Thus, for T_h, the parametric versions of the convolution and as. minimax theorems are reproduced—as if no contamination was present. The only difference concerns the constructions, to ensure the nonparametric regularity (106) for T_h with influence curve $\psi_{h,\theta}$ on any full neighborhood system $\mathcal{U}(\theta)$; for example, on $\mathcal{U}_h(\theta)$.

Locally Uniform Convergence: For a general as. linear functional T, the optimality of an as. linear estimator S having the same influence curve ψ_θ at P_θ as T depends on suitable constructions. In view of (40) and (41), such S is as. normal, for each $\zeta \in Z_\infty^k(\theta)$, uniformly on t-compacts,

$$\sqrt{n}\,\bigl(S_n - T_n(Q_n(\zeta,t))\bigr)(Q_n^n(\zeta,t)) \xrightarrow{\;\mathrm{w}\;} \mathcal{N}(0, \mathcal{C}_\theta(\psi_\theta)) \tag{105}$$

But additional uniformity is needed in order to make S regular for T on $\mathcal{U}(\theta)$, so that for all $r \in (0,\infty)$, uniformly in $Q_n \in U(\theta, r/\sqrt{n})$,

$$\sqrt{n}\,\bigl(S_n - T_n(Q_n)\bigr)(Q_n^n) \xrightarrow{\;\mathrm{w}\;} \mathcal{N}(0, \mathcal{C}_\theta(\psi_\theta)) \tag{106}$$

This nonparametric regularity will require special, usually awkward, construction efforts [Chapter 6], but at the end turns out to be the essence of nonparametric optimality (that is, unbiasedness) over full neighborhoods.

Qualitative Infinitesimal Robustness? Some qualitative robustness seems to be built into the as. minimax theorems since estimator risk, testing level and power, are evaluated over infinitesimal, though full, neighborhoods of the parametric family. Thus one might consider the nonparametric regularity (4) assumed for the convolution theorem, especially the locally uniform as. normality (106) used to achieve the as. minimax bound, an infinitesimal counterpart to Hampel's (1971) qualitative robustness [Beran (1984; §2.3)].

However, an estimator S can hardly be called robust if it is regular relative to a functional T whose oscillation is out of control. Therefore, (4) per se does not necessarily imply robustness of S. The condition simply ties estimator S and functional T closely—also in robustness respects.

No Effect of Contamination: Equating robustness with local as. minimaxity for functionals, Beran (1984; §2.3, §2.4) and Millar (1981, p75; 1982, p731; 1983, p220) seemed to give robustness a precise decision-theoretic, quantitative meaning. But we must observe that the most concentrated limit law in the convolution representation as well as the as. minimax risk and as. maximin power do not depend on the particular type and radius of full neighborhoods, hence are unable to reflect any given amount of contamination.

An interpretation of the as. maximin power in quantitative terms would be even more difficult since, in addition, the normal and χ^2 upper tail probabilities, respectively, are the same for all functionals. The dependence on the particular functional has been buried in the definition of nonparametric hypotheses.

Therefore, the nonparametric optimality theorems over full neighborhoods cannot be called quantitative results. Only condition (4) of estimator regularity for $\tau\circ T$ on $\mathcal{U}(\theta)$ with limit law $\mathcal{N}(0,\Gamma_\theta(\psi_\theta))$, which is also used to achieve the as. minimax bounds, does depend on the particular type and size of neighborhoods—in some qualitative manner.

Lehmann–Scheffé Analogy: In view of Propositions 4.3.6 and 4.3.11, the nonparametric as. minimax Theorems 4.3.4 and 4.3.8 assign maximum risk, and minimum power, to all as. linear estimators, and as. linear estimator tests, whose influence curves differ from the influence curve of the functional, no matter how large the difference. And without distinction, the nonparametric convolution Theorem 4.3.2 cuts out all as. linear estimators whose influence curves differ from the influence curve of the functional, already by the regularity condition.

In view of (41), (43), and (99), the nonparametric optimality results specialized to as. linear estimators boil down to a nonquantitative $0/\infty$ bias alternative: Bias is unbounded unless $\varrho_\theta = \psi_\theta$, in which case it is zero [or 'swept under the carpet', as one referee of Rieder (1985) put it]. Thus the situation is quite different from the Cramér–Rao bound for as. variance, which the parametric versions extend [Propositions 4.3.6 and 4.3.10].

An analogy emerges to the classical Lehmann–Scheffé theorem on the optimality of the UMVU estimator in the presence of a sufficient statistic that is complete: First, the implications (42) and (94) establish 'completeness' of the simple perturbations, with $\zeta \in Z_\infty^k(\theta)$ acting as (infinite-dimensional) parameter. Secondly, the reduction by 'sufficiency' has its counterpart in the restriction to as. linear estimators and as. linear estimator tests, which form essentially complete classes (for the i.i.d. estimation and testing of as. linear functionals on full neighborhoods). Thirdly, within the respective class, the optimum estimator and estimator test are in both setups distinguished merely by 'unbiasedness'.

On the Lehmann–Scheffé type of optimality, Ferguson (1967, p135) judges: "Being best in a class consisting of one element is no optimum property at all. Unbiasedness itself is no optimum property. ... " By analogy, the nonparametric optimality results for full neighborhoods should rather be interpreted as 'ultimate unbiasedness criteria' than the 'absolute optimality standards' they have been presented as in leading literature [Beran (1981 a, b; 1984, §2.4, pp 747–748), Millar (1981; 1983; 1984)].

Depending on suitable constructions to ensure the required uniformity, every as. linear estimator is optimum for the as. linear functional with the same influence curve—merely on unbiasedness grounds. But this kind of optimality is commonly not deemed very exciting.

Robust Asymptotic Minimaxity: A different as. minimax bound employing the following definition of as. maximum risk would be much more interesting and better suited to robust statistics:

$$\lim_{c\to\infty} \liminf_{n\to\infty} \sup_{|t|\le c} \sup_{Q\in U(\theta+t/\sqrt{n},r/\sqrt{n})} \int \ell\bigl(\sqrt{n}\,[S_n - T_n(Q)]\bigr)\,dQ^n \tag{107}$$

In the spirit of robustness, the measures $P_{\theta+t/\sqrt{n}}$ at sample size n are blown up to full neighborhoods $U(\theta+t/\sqrt{n},r/\sqrt{n})$ with fixed $r\in(0,\infty)$. Keeping the radius r fixed will prevent that the majority of as. linear estimators are ruled out by bias ∞. The corresponding as. minimax bound and as. minimax estimator will depend on the particular type and radius r

of balls as well as on the functional T in a quantitative way [as one may check by looking again at as. linear estimators].

So far, the only as. minimax result of this type has been obtained by Rieder (1981; Theorems 4.1, 4.1 A) for dimension $k = 1$ and a restricted pseudo loss function (over/undershoot probabilities of confidence intervals). In that result, the estimand $T_n(Q)$ has in (107) been replaced by the as. value $\theta + t/\sqrt{n}$ of T_n at the center of $U(\theta + t/\sqrt{n}, r/\sqrt{n})$ [which creates an identifiability problem]. But since r is fixed, no essential difference should arise. An extension to higher dimensions and general loss functions, in either one of the two formulations, remains a challenging open problem.

4.4 Restricted Tangent Space

In structured parametric models $\mathcal{P}$, neighborhood systems $\mathcal{U}(\theta)$ about P_θ may arise that are not full. Then the nonparametric notions and optimality results need some modification, essentially by orthogonal projection of influence curves onto the associated tangent subspace. Otherwise, the technical arguments remain fairly the same as for the nonparametric optimality results for unrestricted tangent space.

Tangent Subspaces

Consider a neighborhood system $\mathcal{U}(\theta)$ that covers simple perturbations of P_θ along some subset $\mathcal{Z} \subset Z_2^k(\theta)$. Since one can rewrite

$$dQ_n(A\zeta, t) = \Bigl(1 + \frac{1}{\sqrt{n}}\, t'A\zeta\Bigr)\, dP_\theta = dQ_n(\zeta, A't)$$

also the tangent $A\zeta$ is admitted if $\zeta \in \mathcal{Z}$ and $A \in \mathbb{R}^{k\times k}$. Moreover, as for any two influence curves $\psi_\theta, \varrho_\theta \in \Psi_2(\theta)$ the identity

$$\mathrm{E}_\theta\, \psi_\theta \zeta' = \mathrm{E}_\theta\, \varrho_\theta \zeta'$$

extends from all $\zeta \in \mathcal{Z}$ to all tangents ζ in the closed linear span of $\mathcal{Z}$, and again to $A\zeta$, one should enlarge $\mathcal{Z}$ to the smallest closed linear subspace $\boldsymbol{\mathcal{Z}} \subset Z_2^k(\theta)$ in the product Hilbert space $L_2^k(P_\theta)$ such that

$$\zeta \in \mathcal{Z},\ A \in \mathbb{R}^{k\times k} \implies A\zeta \in \boldsymbol{\mathcal{Z}} \tag{1}$$

Denoting by $V_2(\theta)$ the closed linear span of $\bigl\{\, q = t'\zeta \bigm| \zeta \in \mathcal{Z},\, t \in \mathbb{R}^k \,\bigr\}$ in $L_2(P_\theta)$, we have

$$\boldsymbol{\mathcal{Z}} = V_2^k(\theta) \tag{2}$$

Thus we may equally have started out with any closed linear subspace of one-dimensional tangents $V_2(\theta) \subset Z_2(\theta)$, a so-called *tangent subspace*.

However, in the setup of an assumed central ideal model $\mathcal{P}$, the parametric tangent Λ has been made part of the definition of influence curves and as. linearity; namely, by the requirement $\mathrm{E}\,\psi\Lambda' = \mathbb{I}_k$ that ensures local Fisher consistency of functionals and estimators at the ideal model [Remarks 4.2.13 c and 4.2.17 b]. For this reason, contrary to the standard literature on this subject [e.g., Pfanzagl and Wefelmeyer (1982)], we also assume that

$$\Lambda_\theta \in V_2^k(\theta) \tag{3}$$

Then $\psi_{h,\theta} = \mathcal{I}_\theta^{-1}\Lambda_\theta \in V_2^k(\theta)$ always. In addition, we require that

$$V_\infty(\theta) = V_2(\theta) \cap Z_\infty(\theta) \quad \text{dense in} \quad V_2(\theta) \subset L_2(P_\theta) \tag{4}$$

but only for the sake of such neighborhood systems $\mathcal{U}(\theta)$ and functionals T that may cover simple perturbations, respectively may be weakly differentiable, just along the bounded tangents $V_\infty^k(\theta)$ in $V_2^k(\theta)$.

Remark 4.4.1

(a) Merely for the proofs of the statistical optimality results on estimation and testing of functionals along tangent subspaces, neither condition (3), nor L_2 differentiability of $\mathcal{P}$ at θ, nor some central model $\mathcal{P}$ at all, are needed.

(b) Assuming (3) we implicitly pass to the (closed) linear span $W_2(\theta)$ of a tangent subspace $V_2(\theta)$ and the k coordinates of Λ_θ in $Z_2(\theta) \subset L_2(P_\theta)$.

(c) If condition (4) is violated, as by $W_{2;*,\varepsilon}$ determined in 7.2(32), we need instead to assume that the neighborhoods $\mathcal{U}(\theta)$ cover simple perturbations, and the functionals T are weakly differentiable, along all tangents in $V_2^k(\theta)$ [replacement $Z_\infty(\theta) \rightsquigarrow V_2(\theta)$ in Definitions 4.2.6 and 4.2.12]. ////

Semiparametric Optimality

Thus suppose some tangent space $V_2(\theta)$ with properties (3) and (4), and let T be a functional that is weakly differentiable at P_θ along $V_\infty^k(\theta)$. Then, obviously, the influence curve $\psi_\theta \in \Psi_2(\theta)$ in the expansion 4.2(44) of T is no longer unique if $V_2(\theta) \neq Z_2(\theta)$. If we introduce the orthogonal projection

$$\Pi_\theta^k \colon L_2^k(P_\theta) \longrightarrow V_2^k(\theta) \tag{5}$$

(acting coordinatewise), however, $\Pi_\theta^k \psi_\theta$ is unique and called the *influence curve of T along $V_2^k(\theta)$*. Then for all $\zeta \in V_2^k(\theta)$,

$$\mathrm{E}_\theta(\Pi_\theta^k\psi_\theta)\zeta' = \mathrm{E}_\theta\,\psi_\theta\zeta' \tag{6}$$

Inserting $\zeta = \Lambda_\theta$, which is legal by (3), it follows that

$$\Pi_\theta^k\psi_\theta \in \Psi_2(\theta) \tag{7}$$

Moreover, for $\zeta = \Pi_\theta^k\psi_\theta$ itself one obtains that

$$0 \le \mathrm{E}_\theta\bigl(\psi_\theta - \Pi_\theta^k\psi_\theta\bigr)\bigl(\psi_\theta - \Pi_\theta^k\psi_\theta\bigr)' \underset{(6)}{=} \mathcal{C}_\theta(\psi_\theta) - \mathcal{C}_\theta(\Pi_\theta^k\psi_\theta) \tag{8}$$

with equality iff $\Pi_\theta^k \psi_\theta = \psi_\theta$; that is, $\psi_\theta \in V_2^k(\theta)$.

Then the proofs to Theorems 4.3.2, 4.3.4 and 4.3.8 go through step-by-step with the replacements: $Z_\infty(\theta) \rightsquigarrow V_\infty(\theta)$, $\psi_\theta \rightsquigarrow \Pi_\theta^k \psi_\theta$. Note in particular that the least favorable tangents $\zeta^\star = A\Pi_\theta^k \psi_\theta$ all belong to $V_2^k(\theta)$.

Since for $\mathcal{Z} = \{\Lambda_\theta\}$ one always gets $\Pi_\theta^k \psi_\theta = \psi_{h,\theta}$, and thus obtains the parametric versions, while for $\mathcal{Z} = Z_2^k(\theta)$ the nonparametric versions follow with $\Pi_\theta^k \psi_\theta = \psi_\theta$, the variants in between are termed *semiparametric*.

Remark 4.4.2 It is a consequence of (6) that

$$\Pi_\theta^p\bigl(d\tau(\theta)\psi_\theta\bigr) = d\tau(\theta)\,\Pi_\theta^k \psi_\theta \tag{9}$$

where Π_θ^p is the orthogonal projection from $L_2^p(P_\theta)$ onto $V_2^p(\theta)$. ////

Comparisons

Compared with the nonparametric convolution theorem, the semiparametric version distinguishes the normal $\mathcal{N}(0, \Gamma_\theta(\Pi_\theta^k \psi_\theta))$ with the smaller covariance $\Gamma_\theta(\Pi_\theta^k \psi_\theta) \le \Gamma_\theta(\psi_\theta)$ as best limit law of as. estimators that are regular for $\tau \circ T$ on $\mathcal{U}(\theta)$ in the sense of Definition 4.3.1. The optimal regular estimator $S^\star$ must necessarily have the as. expansion

$$\sqrt{n}\,\bigl(S_n^\star - \tau(\theta)\bigr) = d\tau(\theta)\frac{1}{\sqrt{n}}\sum_{i=1}^n \Pi_\theta^k \psi_\theta(x_i) + \mathrm{o}_{P_\theta^n}(n^0) \tag{10}$$

The semiparametric as. minimax lower bound is

$$\int \ell \, d\mathcal{N}(0, \Gamma_\theta(\Pi_\theta^k \psi_\theta)) \tag{11}$$

which, by (8), for ℓ symmetric subconvex on $\mathbb{R}^p$, is smaller than the nonparametric analogue based on $\Gamma_\theta(\psi_\theta)$ [Corollary 3.2.6].

The comparison of the semiparametric and nonparametric as. maximin power bounds is not so clear as the as. maximin power stays the same. In the multisided case 4.3(46), in view of (8), the null hypothesis becomes smaller, and the alternative larger, since

$$|t|_{\Gamma_\theta(\Pi_\theta^k\psi_\theta)} \le b \implies |t|_{\Gamma_\theta(\psi_\theta)} \le b, \quad |t|_{\Gamma_\theta(\psi_\theta)} \ge c \implies |t|_{\Gamma_\theta(\Pi_\theta^k\psi_\theta)} \ge c$$

At least in the one-dimensional, one-sided problem 4.3(47) the 'separation' of null hypothesis and alternative decreases, as

$$(c-b)\sqrt{\Gamma_\theta(\psi_\theta)} \ge (c-b)\sqrt{\Gamma_\theta(\Pi_\theta \psi_\theta)}$$

A comparison with the parametric versions works the other way round since (7) enforces the Cramér–Rao inequality 4.2(74),

$$\mathcal{C}_\theta(\Pi_\theta^k \psi_\theta) \ge \mathcal{I}_\theta^{-1} \tag{12}$$

Choice of Subspace

Given $\mathcal{U}(\theta)$ and T the considerations so far also apply to any smaller tangent space $\widetilde{V}_2(\theta) \subset V_2(\theta)$. Denoting by $\widetilde{\Pi}_\theta$ the orthogonal projection onto $\widetilde{V}_2(\theta)$, the corresponding semiparametric optimality results are based on the covariance $\Gamma_\theta(\widetilde{\Pi}_\theta^k \psi_\theta)$, and the corresponding optimal regular estimator $\tilde{S}$ would necessarily have the as. expansion

$$\sqrt{n}\,\bigl(\tilde{S}_n - \tau(\theta)\bigr) = d\tau(\theta)\frac{1}{\sqrt{n}}\sum_{i=1}^{n}\widetilde{\Pi}_\theta^k \psi_\theta(x_i) + \mathrm{o}_{P_\theta^n}(n^0) \tag{13}$$

Similarly to (8), upon composition with $d\tau(\theta)$, it follows that

$$\Gamma_\theta(\widetilde{\Pi}_\theta^k \psi_\theta) \le \Gamma_\theta(\Pi_\theta^k \psi_\theta) \tag{14}$$

with equality iff $d\tau(\theta)\widetilde{\Pi}_\theta^k \psi_\theta = d\tau(\theta)\Pi_\theta^k \psi_\theta$. Therefore, to obtain a more concentrated limit normal and a smaller as. minimax risk, on the one hand, one would choose the tangent space small. But the proof to Proposition 4.3.6 with the replacements:

$$\begin{array}{lll} \psi_\theta \rightsquigarrow d\tau(\theta)\Pi_\theta^k \psi_\theta\,, & \varrho_\theta \rightsquigarrow d\tau(\theta)\widetilde{\Pi}_\theta^k \psi_\theta & \\ Z_\infty(\theta) \rightsquigarrow V_\infty(\theta), & k \rightsquigarrow p, & \xi \rightsquigarrow (\xi', 0')' \end{array} \tag{15}$$

shows that $\tilde{S}$ cannot satisfy the regularity condition 4.3(4) under simple perturbations $Q_n = Q_n(\zeta, .)$, along all tangents $\zeta \in V_2^k(\theta) \setminus \widetilde{V}_2^k(\theta)$, unless

$$d\tau(\theta)\widetilde{\Pi}_\theta^k \psi_\theta = d\tau(\theta)\Pi_\theta^k \psi_\theta \tag{16}$$

and then

$$\Gamma_\theta(\widetilde{\Pi}_\theta^k \psi_\theta) = \Gamma_\theta(\Pi_\theta^k \psi_\theta) \tag{17}$$

as well as

$$\sqrt{n}\,(\tilde{S}_n - S_n^\star) = \mathrm{o}_{P_\theta^n}(n^0) \tag{18}$$

Thus the attainability of the limit normal $\mathcal{N}(0, \Gamma_\theta(\Pi_\theta^k \psi_\theta))$ by an as. estimator $S^\star$ regular for $\tau \circ T$ on $\mathcal{U}(\theta)$, which may also be employed to achieve the semiparametric as. minimax risk and as. maximin power on $\mathcal{U}(\theta)$, prohibits a choice of $V_2(\theta)$ too small.

On the other hand, in view of (14), a convolution representation 4.3(5) with factor $\mathcal{N}(0, \Gamma_\theta(\Pi_\theta^k \psi_\theta))$ implies the one with factor $\mathcal{N}(0, \Gamma_\theta(\widetilde{\Pi}_\theta^k \psi_\theta))$. And the as. minimax lower bound based on $\Gamma_\theta(\Pi_\theta^k \psi_\theta)$, for ℓ symmetric subconvex on $\mathbb{R}^p$, entails the smaller one based on $\Gamma_\theta(\widetilde{\Pi}_\theta^k \psi_\theta)$ [Corollary 3.2.6].

Therefore, the tangent space $V_2(\theta)$ must be chosen as large as possible, subject to coverage of the corresponding simple perturbations by $\mathcal{U}(\theta)$.

Optimization

Optimization problems in robustness will concern the minimization of the covariance $\Gamma_\theta(\Pi_\theta^k \psi_\theta)$ subject to bounds on the oscillation of functionals caused by simple perturbations along $V_\infty^k(\theta)$. In such problems, the projection Π_θ^k can equally be ignored. For, in view of (8), the covariance of influence curves $\psi_\theta \in \Psi_2(\theta)$ is decreased by projection, while the oscillation terms, in view of (6), stay the same [which is obvious for the exact oscillation terms, but is also true for the approximate versions; see 7.4(17), p 278].

Therefore, the projection onto the tangent subspace, which is the essence of the semiparametric statistical optimality results, will actually be taken care of automatically—by the kind of optimization problems we are going to solve for optimally robust influence curves.

Chapter 5

Optimal Influence Curves

5.1 Introduction

The nonparametric as. minimax risk, based on the covariance of the most concentrated (normal) limit law in the nonparametric convolution representation, will subsequently be used just to quantify the costs of estimating a given functional (by its unique unbiased as. linear estimator). In addition, robustness aspects of functionals (e.g., their oscillations) will be taken into account. Optimization of the two criteria: nonparametric as. minimax risk (covariance) and robustness, then determines the functional, which has so far been arbitrary. Likewise, that functional will be determined (implying the use of its unique unbiased as. linear estimator test) which maximizes as. maximin power subject to the inclusion of capacity balls into the nonparametric hypotheses formulated by functionals.

Thus, nonparametric statistics for functionals may be linked with other robustness theories that have directly concerned bias and variance of as. linear estimators. These other approaches, therefore, do not contradict nonparametric as. unbiasedness, but rather complement it by some optimality.

As in Chapter 4, a parametric family is employed as ideal center model,

$$\mathcal{P} = \{\, P_\theta \mid \theta \in \Theta \,\} \subset \mathcal{M}_1(\mathcal{A}) \tag{1}$$

whose parameter space Θ is an open subset of some finite-dimensional $\mathbb{R}^k$. Some $\theta \in \Theta$ is fixed. The assumptions of the nonparametric and robust setup locally about this P_θ are taken over from Section 4.3:

- L_2 differentiability of $\mathcal{P}$ at θ with derivative Λ_θ and Fisher information $\mathcal{I}_\theta > 0$;
- a full neighborhood system $\mathcal{U}(\theta)$ about P_θ;
- a transform $\tau\colon \mathbb{R}^k \to \mathbb{R}^p$ differentiable at θ, $\operatorname{rk} d\tau(\theta) = p \le k$;
- functionals T as. linear at P_θ with influence curves $\psi_\theta \in \Psi_2(\theta)$.

Then the covariances $\Gamma_\theta(\psi_\theta)$ determine the most concentrated (normal) limits in the nonparametric convolution representations as well as the nonparametric as. minimax risks,

$$\int \ell \, d\mathcal{N}(0, \Gamma_\theta(\psi_\theta)), \qquad \Gamma_\theta(\psi_\theta) = d\tau(\theta)\mathcal{C}_\theta(\psi_\theta)d\tau(\theta)' \tag{2}$$

In the absence of further constraints, the Cramér–Rao inequality 4.2(74) says that the covariance $\mathcal{C}_\theta(\psi_\theta)$ is minimized among all $\psi_\theta \in \Psi_2(\theta)$ by the classical scores function $\psi_{h,\theta} = \mathcal{I}_\theta^{-1}\Lambda_\theta$ uniquely, in the positive definite sense. For symmetric subconvex ℓ, the classical $\psi_{h,\theta}$ also minimizes the as. minimax risk [Corollary 3.2.6]; uniquely, under further conditions on ℓ [nonconstant monotone quadratic; cf. Proposition 4.2.20 b].

According to Section 4.4, under assumption 4.4(3): $\Lambda_\theta \in V_2^k(\theta)$, on the tangent subspace $V_2(\theta) \subset Z_2(\theta)$, also the semiparametric analogue covariance $\mathcal{C}_\theta(\Pi_\theta\psi_\theta)$ is uniquely minimized by $\psi_{h,\theta}$, in the strong sense.

Remark 5.1.1 If there are finitely many nuisance parameters, and in the absence of further constraints, the classical partial scores $\eta_{h,\theta} = D\psi_{h,\theta}$ introduced in 4.2(41) will be optimum, by the Cramér–Rao bound 5.5(6). Begun et al. (1983) in fact extend this result to an infinite-dimensional nuisance component. These authors [p434] reject the similar suggestion by a referee concerning the optimization of Fisher consistent functionals as being 'too cumbersome', but rather want to stay 'consistent with the philosophical principle' that 'the likelihoods tell the story'. ////

The solution $\psi_{h,\theta}$ may be viewed the influence curve at P_θ of the Hellinger MD functional T_h as well as (under suitable regularity conditions) of the classical maximum likelihood estimator. In this chapter, however, we just argue in terms of influence curves. Constructions of functionals and their as. unbiased estimators, with (optimal) influence curves prescribed, will be the subject of Chapter 6.

5.2 Minimax Risk

The first approach to robustify the estimation and testing of functionals is restricted to the one-dimensional location problem with $\tau = \mathrm{id}_{\mathbb{R}}$. We continue to use the nonparametric as. minimax risk, or variance taken from the nonparametric convolution representation, as the only criterion. However, it will be evaluated in a supermodel that is composed of a collection of ideal location models and their respective full infinitesimal neighborhood systems. For each of these location models, a nonparametric as. minimax bound obtains. The maximum of these bounds is going to be minimized. Thus, at the cost of a certain conceptual complexity, Huber's (1964) saddle point result on the as. variance of location M estimates may be translated to the estimation and testing of functionals.

In more detail, let $\mathcal{F} \subset \mathcal{M}_1(\mathbb{B})$ be a set of distribution functions on the real line. Denote by $\mathcal{F}_2$ the subset of all $dF = f\,d\lambda \in \mathcal{F}$ that have finite Fisher informations $\mathcal{I}_f$ of location [cf. 2.4(1)]. Each $F \in \mathcal{F}_2$ is member number zero of a one-dimensional location model,

$$F_\theta(y) = F(y-\theta), \qquad y \in \mathbb{R},\ \theta \in \mathbb{R} \tag{1}$$

This model, by Proposition 2.4.1, is L_2 differentiable at every $\theta \in \mathbb{R}$ with the following derivative $\Lambda_{f,\theta}$ and Fisher information $\mathcal{I}_f$,

$$\Lambda_{f,\theta}(y) = \Lambda_f(y-\theta), \qquad \mathcal{I}_f = \mathrm{E}_F\,\Lambda_f^2, \qquad \Lambda_f = -f'/f \tag{2}$$

where E_F means expectation under F. We fix $\theta = 0$, and denote by $\Psi_2(F)$ the set of influence curves at $F = F_0$ in model (1). Given any $\psi\colon \mathbb{R} \to \mathbb{R}$ Borel measurable, let $\mathcal{F}_2(\psi)$ be the subset of all $F \in \mathcal{F}_2$ such that

$$\psi \in L_2(F), \qquad \mathrm{E}_F\,\psi = 0, \qquad \mathrm{E}_F\,\psi\Lambda_f \neq 0 \tag{3}$$

Then, by standardization, for every $F \in \mathcal{F}_2(\psi)$,

$$\psi_F = \bigl(\mathrm{E}_F\,\psi\Lambda_f\bigr)^{-1}\psi \in \Psi_2(F) \tag{4}$$

Under certain conditions, the M equation with scores function ψ in fact defines a functional T_ψ that, at every $F \in \mathcal{F}_2(\psi)$, is as. linear in model (1) with influence curve ψ_F [Theorem 6.2.3].

Given some numbers $-\infty < b < c < \infty$ and full neighborhood systems $\mathcal{U}(F) = \{U(F,.)\}$ about every $F \in \mathcal{F}_2(\psi)$, one-sided hypotheses concerning such a functional $T_\psi = (T_n)$ at $F \in \mathcal{F}_2(\psi)$ are in the manner of 4.3(47) defined by

$$\begin{aligned} \mathrm{H}_n^1(T,F,r)\colon&\quad Q \in U(F, r/\sqrt{n}),\ \sqrt{n}\,T_n(Q) \le b \\ \mathrm{K}_n^1(T,F,r)\colon&\quad Q \in U(F, r/\sqrt{n}),\ \sqrt{n}\,T_n(Q) \ge c \end{aligned} \tag{5}$$

According to Theorem 4.3.8 a, the as. maximin power for this testing problem at level $\alpha \in (0,1)$ is

$$1 - \Phi\left(u_\alpha - \frac{c-b}{\sqrt{\mathcal{C}_F(\psi_F)}}\right) \tag{6}$$

where

$$\mathcal{C}_F(\psi_F) = \mathrm{E}_F(\psi_F)^2 = \bigl(\mathrm{E}_F\,\psi\Lambda_f\bigr)^{-2}\,\mathrm{E}_F\,\psi^2 \tag{7}$$

We determine that scores function ψ which maximizes the minimum of the testing powers (6) when F ranges over $\mathcal{F}_2(\psi)$.

The nonparametric as. minimax risk for estimating T_ψ on a full neighborhood system $\mathcal{U}(F)$ about $F \in \mathcal{F}_2(\psi)$ is given by

$$\int \ell\,\bigl(y^2\,\mathcal{C}_F(\psi_F)\bigr)\,\mathcal{N}(0,1)(dy) \tag{8}$$

Likewise, that scores function ψ can be recommended for use (estimation) which minimizes the maximum of risks (8) when F ranges over $\mathcal{F}_2(\psi)$.

In the following saddle point result, the original set $\mathcal{F}$, instead of the smaller $\mathcal{F}_2(\psi^\star)$, would be desirable in the left-hand inequality, but cannot be allowed if things are to be well defined. The right-hand inequality holds for all (nonzero scalar multiples of) influence curves at the least favorable distribution $F_\star$, in the location model generated by $F_\star$.

Theorem 5.2.1 *Suppose the set $\mathcal{F}$ is convex. Assume there exists an element $dF_\star = f_\star \, d\lambda$ of $\mathcal{F}_2$ such that for all other $dF = f \, d\lambda \in \mathcal{F}_2$,*

$$\mathcal{I}_f \geq \mathcal{I}_{f_\star} \tag{9}$$

and

$$F \ll F_\star \tag{10}$$

Set

$$\psi^\star = \Lambda_{f_\star} \tag{11}$$

Then, for all distribution functions $F \in \mathcal{F}_2(\psi^\star)$ and all scores functions ψ such that $F_\star \in \mathcal{F}_2(\psi)$, it holds that

$$\begin{aligned} 1 - \Phi\left(u_\alpha - \frac{c-b}{\sqrt{\mathcal{C}_F(\psi_F^\star)}}\right) &\geq 1 - \Phi\left(u_\alpha - \frac{c-b}{\sqrt{\mathcal{C}_{F_\star}(\psi_{F_\star}^\star)}}\right) \\ &\geq 1 - \Phi\left(u_\alpha - \frac{c-b}{\sqrt{\mathcal{C}_{F_\star}(\psi_{F_\star})}}\right) \end{aligned} \tag{12}$$

and, for increasing function $\ell\colon [0,\infty] \to [0,\infty]$,

$$\begin{aligned} \int \ell\left(y^2 \mathcal{C}_F(\psi_F^\star)\right) \mathcal{N}(0,1)(dy) &\leq \int \ell\left(y^2 \mathcal{C}_{F_\star}(\psi_{F_\star}^\star)\right) \mathcal{N}(0,1)(dy) \\ &\leq \int \ell\left(y^2 \mathcal{C}_{F_\star}(\psi_{F_\star})\right) \mathcal{N}(0,1)(dy) \end{aligned} \tag{13}$$

PROOF If $\mathcal{F}$ is convex, so are $\mathcal{F}_2$ and $\mathcal{F}_2(\psi^\star)$. Invoking Huber's (1981) saddle point result for the as. variance of M estimates of location, we obtain that for all $F \in \mathcal{F}_2(\psi^\star)$ and $F_\star \in \mathcal{F}_2(\psi)$,

$$\mathcal{C}_F(\psi_F^\star) \leq \mathcal{C}_{F_\star}(\psi_{F_\star}^\star) = \mathcal{I}_{f_\star}^{-1} \leq \mathcal{C}_{F_\star}(\psi_{F_\star}) \tag{14}$$

where the right-hand inequality is just Cramér–Rao 4.2(74), in the location model generated by $F_\star$. The left-hand inequality is a consequence of formulas (5.1) and (5.2) of Huber (1981, p 82) combined with his correspondence (6.8) between as. variance and Fisher information [Huber (1981, p 95)]. In his formulas (5.1) and (6.8), however, the domination $F_1 \ll F_0$ is required; hence our condition (10). ////

If $\mathcal{F}$ is compact in the vague topology, an element of minimum Fisher information exists because Fisher information is lower semicontinuous in that topology [Huber (1981; Theorem 4.2, Proposition 4.3)]. And if, in addition to satisfying (9) and (10), $F_\star \in \mathcal{F}_2$ has convex support, then it is unique [Huber (1981, Proposition 4.5), for which (10) is again needed].

Remark 5.2.2 Theorem 5.2.1 remains unaffected by a restriction of the tangent spaces $Z_2(F)$ at $F \in \mathcal{F}_2$ to tangent subspaces $V_2(F) \subset Z_2(F)$ as long as

$$\Lambda_{F_\star} \in V_2(F_\star) \tag{15}$$

Denoting the orthogonal projection from $L_2(F)$ onto $V_2(F)$ by Π_F, and explicitly using the influence curves (4), the semiparametric analogue of the variance $\mathcal{C}_F(\psi_F)$ at $F \in \mathcal{F}_2(\psi)$ is

$$\mathcal{C}_F(\Pi_F\psi_F) = \int (\Pi_F\psi_F)^2\,dF \tag{16}$$

Since $\Pi_{F_\star}\psi^\star_{F_\star} = \psi^\star_{F_\star}$ by (15), the following chain of inequalities is valid for all $F \in \mathcal{F}_2(\psi^\star)$ and $F_\star \in \mathcal{F}_2(\psi)$,

$$\begin{aligned} \mathcal{C}_F(\Pi_F\psi^\star_F) &\overset{4.4(8)}{\leq} \mathcal{C}_F(\psi^\star_F) \overset{(14)}{\leq} \mathcal{C}_{F_\star}(\psi^\star_{F_\star}) \\ &\underset{(15)}{=} \mathcal{C}_{F_\star}(\Pi_{F_\star}\psi^\star_{F_\star}) \underset{(14)}{=} \mathcal{I}_{f_\star}^{-1} \underset{4.4(12)}{\overset{(15)}{\leq}} \mathcal{C}_{F_\star}(\Pi_{F_\star}\psi_{F_\star}) \end{aligned} \tag{17}$$

Thus the semiparametric analogue variance $\mathcal{C}_F(\Pi_F\psi_F)$ has the same saddle point and value of the game as $\mathcal{C}_F(\psi_F)$. ////

In the best-known Example 5.2 of Huber (1981, pp84–86), the set $\mathcal{F}$ denotes a contamination ball of radius $\varepsilon \in (0,1)$,

$$\mathcal{F} = B_c(G,\varepsilon) = \{\, F \mid F = (1-\varepsilon)G + \varepsilon H,\ H \in \mathcal{M}_1(\mathbb{B}) \,\} \tag{18}$$

centered at a distribution function G whose Lebesgue density g is twice differentiable on $\mathbb{R}$ and such that $\log g$ is concave on $\mathbb{R}$. In this example, the least favorable $F_\star$ has full support, is unique, and the minimax scores function is

$$\psi^\star = (-m) \vee \Lambda_g \wedge m \tag{19}$$

with clipping constant $m \in (0,\infty)$ determined by

$$\int 1 \vee \frac{|\Lambda_g|}{m}\,dG = \frac{1}{1-\varepsilon} \tag{20}$$

Such a $\psi^\star$ is bounded increasing. If $\psi^\star$ is also Lipschitz bounded, as in the special case $G = \mathcal{N}(0,1)$, the constructions of a corresponding functional as well as estimator by means of M equations go through for infinitesimal neighborhood systems as large as the total variation systems $\mathcal{U}_v(F)$ about every $F \in \mathcal{F}_2(\psi^\star)$ [Theorems 6.2.3 and 6.2.4].

Although we come up with Huber's (1981) solution $\psi^\star$ again, the functional saddle point for as. maximin power and as. minimax risk now incorporates all achievements of the nonparametric optimality theorems:

(i) There is no restriction to M estimates;

(ii) In the local submodels it is clearly defined what to estimate or test;

(iii) We get uniform as. normality of the as. minimax M estimate in each one of the local submodels.

(iv) The infinitesimal perturbations, which are the fairly arbitrary measures in $U(F, r/\sqrt{n})$ belonging to any full neighborhood system $\mathcal{U}(F)$ about $F \in \mathcal{F}_2(\psi^\star)$, may be asymmetric and of different type than the elements of the fixed neighborhood $\mathcal{F}$. For example, the solution (19) still obtains if infinitesimal perturbations of type Hellinger or total variation are composed with the original fixed contamination ball $\mathcal{F} = B_c(G, \varepsilon)$.

As one referee of Rieder (1985) put it: 'Nonparametric decision theory can have Huber's cake and eat it too'; or the other way round, maybe.

5.3 Oscillation

Functionals whose oscillation get out of control under minor departures from the ideal P_θ (e.g., moments) are certainly not desirable, even though they might be estimable at low as. minimax risk. Therefore, oscillation should be used as a robustness criterion. The notion is understood in an infinitesimal sense; so we actually consider the slope of the oscillation at 0, also called *sensitivity* elsewhere. In view of the connection 6.1(2) between functionals and their unbiased (regular efficient, as. minimax) estimators, the oscillation of functionals translates into as. estimator bias, when $\tau(\theta)$ with the fixed parameter value θ remains the estimand.

In this section, oscillation terms are introduced and investigated technically (preparing the ground for a subsequent combination with other criteria), and a (pure) oscillation saddle point result with the two Hilbert MD functionals is derived.

5.3.1 Oscillation/Bias Terms

We study the basic oscillation terms of as. linear functionals in the general parameter case; further bias variants and similar computations become due in structured models. Matching the definition of as. linear functionals and estimators, oscillation/bias terms are introduced in the neighborhood submodel based on the simple perturbations 4.2(15) of P_θ along $Z^k_\infty(\theta)$. At least for the optimal influence curves it must be made sure later that the passage from the neighborhood submodel to full neighborhoods does not increase these bias terms [Chapter 6; also Lemma 5.3.4 and Section 5.4].

Definitions

Fix $\theta \in \Theta$. Define the following subclasses of one-dimensional bounded tangents,

$$\mathcal{G}_*(\theta) = \big\{ q \in Z_\infty(\theta) \bigm| (*1) \big\} \tag{1}$$

where E_θ denotes expectation under P_θ and condition $(*1)$ for the different choices $* = c, v, h, \kappa, \mu$ is, respectively,

$$q \ge -1 \qquad \text{a.e. } P_\theta \tag{$c1$}$$

$$\mathrm{E}_\theta \, |q| \le 2 \tag{$v1$}$$

$$\mathrm{E}_\theta \, q^2 \le 8 \tag{$h1$}$$

$$\sup_{y \in \mathbb{R}^m} \Big| \int \mathbf{I}(x \le y) q(x) \, P_\theta(dx) \Big| \le 1 \tag{$\kappa1$}$$

$$\int_{\mathbb{R}^m} \Big| \int \mathbf{I}(x \le y) q(x) \, P_\theta(dx) \Big|^2 \mu(dy) \le 1 \tag{$\mu1$}$$

The weight $\mu \in \mathcal{M}_\sigma(\mathbb{B}^m)$ is assumed σ finite. Identifying $t'\zeta = rq$ formally, the simple perturbations along $\zeta \in Z^k_\infty(\theta)$ are, for $\sqrt{n} \ge -r \inf_{P_\theta} q$,

$$dQ_n(q, r) = dQ_n(\zeta, t) = \Big(1 + \frac{r}{\sqrt{n}} q \Big) \, dP_\theta \tag{2}$$

Lemma 5.3.1 *Given $q \in Z_\infty(\theta)$ and $r \in (0, \infty)$. Then, in the cases $* = c, v, \kappa, \mu$, for every $n \in \mathbb{N}$ such that $\sqrt{n} \ge -r \inf_{P_\theta} q$,*

$$Q_n(q, r) \in B_*(P_\theta, r/\sqrt{n}) \iff q \in \mathcal{G}_*(\theta) \tag{3}$$

In case $* = h$, there exist $r_n \to r$ such that $Q_n(q, r) \in B_h(P_\theta, r_n/\sqrt{n})$ for all $n \in \mathbb{N}$, iff $q \in \mathcal{G}_h(\theta)$.

PROOF On identifying $t'\zeta_n = rq$ with $\zeta_n = \zeta$ we can refer to the proof of Lemma 4.2.8; in particular, see 4.2(26) for $* = c$, 4.2(28) for $* = v$, and 4.2(30) for $* = h$. The cases $* = \kappa, \mu$ are similar to prove. ////

Given $q \in Z_\infty(\theta)$ and $r_n \to r \in \mathbb{R}$, the expansion 4.3(9) of a functional T that is as. linear with influence curve $\psi_\theta \in \Psi_2(\theta)$ at P_θ applies to the k-dimensional bounded tangent $\zeta = (q, 0, \ldots, 0)' \in Z^k_\infty(\theta)$ and the convergent vectors $t_n = (r_n, 0, \ldots, 0)' \in \mathbb{R}^k$, so that

$$\sqrt{n} \big(\tau \circ T_n(Q_n(q, r_n)) - \tau(\theta) \big) = r \, d\tau(\theta) \, \mathrm{E}_\theta \, \psi_\theta q + \mathrm{o}(n^0) \tag{4}$$

Therefore, setting $D = d\tau(\theta)$ and $\eta_\theta = D\psi_\theta \in \Psi^D_2(\theta)$, infinitesimal oscillation terms may in a restricted and simplistic manner be defined by

$$\begin{aligned} \omega_{*,\theta}(\eta_\theta) &= \frac{1}{r} \sup \Big\{ \lim_{n \to \infty} \sqrt{n} \, \big| \tau \circ T_n(Q_n(q, r)) - \tau(\theta) \big| \Bigm| q \in \mathcal{G}_*(\theta) \Big\} \\ &= \sup \big\{ |\mathrm{E}_\theta \, \eta_\theta q| \bigm| q \in \mathcal{G}_*(\theta) \big\} = \omega_{*,\theta;0}(\eta_\theta) \end{aligned} \tag{5}$$

where the index $s=0$ is an artifact. Indexed by $s=2,\infty$ are the following variants, which are based on the coordinatewise oscillations,

$$\omega_{*,\theta;2}^2(\eta_\theta)=\sum_{j=1}^p \omega_{*,\theta}^2(\eta_{\theta,j}), \qquad \omega_{*,\theta;\infty}(\eta_\theta)=\max_{j=1,\ldots,p}\omega_{*,\theta}(\eta_{\theta,j}) \tag{6}$$

Definitions (5) and (6) extend to all $\eta_\theta=(\eta_{\theta,1},\ldots,\eta_{\theta,p})'\in L_1^p(P_\theta)$.

Abbreviated Notation

Dropping the fixed parameter θ from notation, let us write $\omega_*=\omega_{*,\theta}$ and $\omega_{*;s}=\omega_{*,\theta;s}$ for the bias terms as well as $\mathcal{G}_*=\mathcal{G}_*(\theta)$ and $Z_\alpha=Z_\alpha(\theta)$ for $\alpha=1,2,\infty$ for the classes of one-dimensional tangents at $P=P_\theta$. Let $\mathrm{E}=\mathrm{E}_\theta$ denote expectation, and $\mathcal{C}=\mathcal{C}_\theta$ covariance, under $P=P_\theta$. The pointwise extrema inf, sup are accompanied by $\inf_P$, $\sup_P$, the essential extrema under $P=P_\theta$.

General Properties

Prior to any explicit bias computations we record some general properties. Recall the weak topology on the Hilbert space $L_2^p(P)$ as the initial topology generated by the scalar products $\eta\mapsto \mathrm{E}\,\eta'\zeta$, for $\zeta\in L_2^p(P)$.

Lemma 5.3.2 *Let $*=c,v,h,\kappa,\mu$, $s=0,2,\infty$, and $\eta\in L_1^p(P)$. Then*

$$\omega_{*;s}(\eta)=\omega_{*;s}(\eta-\mathrm{E}\,\eta) \tag{7}$$

$$\omega_*(\eta)=\sup\big\{\,\omega_*(e'\eta)\bigm| e\in\mathbb{R}^p,\ |e|=1\big\} \tag{8}$$

$$\omega_{*;\infty}(\eta)\le\omega_{*;0}(\eta)\le\omega_{*;2}(\eta)\le\sqrt{p}\,\omega_{*;\infty}(\eta) \tag{9}$$

$$\omega_{c;s}(\eta)\le\omega_{v;s}(\eta)\le 2\,\omega_{c;s}(\eta), \qquad \omega_{v;s}(\eta)\le\omega_{\kappa;s}(\eta) \tag{10}$$

The terms $\omega_{*;s}$ are positively homogeneous, subadditive, hence convex, on $L_1^p(P)$, and weakly lower semicontinuous on $L_2^p(P)$.

PROOF If $q\in\mathcal{G}_*$ then $\mathrm{E}\,q=0$ hence $\mathrm{E}\,\eta q=\mathrm{E}(\eta-\mathrm{E}\,\eta)q$, which proves (7). To prove (8), write out the definition for $s=0$,

$$\begin{aligned}\omega_*(\eta)=\sup_{q\in\mathcal{G}_*}|\mathrm{E}\,\eta q|&=\sup_{q\in\mathcal{G}_*}\sup_{|e|=1}\mathrm{E}\,e'\eta q\\ &=\sup_{|e|=1}\sup_{q\in\mathcal{G}_*}\mathrm{E}\,e'\eta q=\sup_{|e|=1}\omega_*(e'\eta)\end{aligned} \tag{11}$$

where the $\sup_{|e|=1}$ needs to be evaluated only on a countable dense subset of the unit sphere of $\mathbb{R}^p$. If we restrict e to the canonical basis of $\mathbb{R}^p$ it already follows that $\omega_*\ge\omega_{*,\infty}$. Moreover, $|\mathrm{E}\,\eta q|^2=\sum_j|\mathrm{E}\,\eta_j q|^2$, hence we obtain $\omega_*\le\omega_{*,2}$. That $\omega_{*,2}\le\sqrt{p}\,\omega_{*,\infty}$ is the usual bound between Euclidean and sup norms on $\mathbb{R}^p$. Thus (9) is proved.

For $* = c, v$ we have $\mathcal{G}_c \subset \mathcal{G}_v$, thus $\omega_c \le \omega_v$, since $\mathrm{E}\, q^+ = \mathrm{E}\, q^-$ whenever $\mathrm{E}\, q = 0$, and $q^- \le 1$ hence $\mathrm{E}\,|q| = \mathrm{E}\, q^+ + \mathrm{E}\, q^- \le 2$ for $q \in \mathcal{G}_c$. For $q \in \mathcal{G}_v$ both summands in $q = (q^+ - \mathrm{E}\, q^+) - (q^- - \mathrm{E}\, q^-)$ are elements of $\mathcal{G}_c$. Thus $\omega_v \le 2\omega_c$. Bound (10) for ω_* carries over to $\omega_{*,2}$, $\omega_{*,\infty}$. That $\omega_{v;s} \le \omega_{\kappa;s}$ is a consequence of $\mathcal{G}_v \subset \mathcal{G}_\kappa$.

From $|\mathrm{E}\,\gamma\eta q| = |\gamma||\mathrm{E}\,\eta q|$ and $|\mathrm{E}(\eta + \zeta)q| \le |\mathrm{E}\,\eta q| + |\mathrm{E}\,\zeta q|$, by taking the sup over $q \in \mathcal{G}_*$, it follows that for all $\gamma \in \mathbb{R}$ and $\eta, \zeta \in L_1^p(P)$,

$$\omega_{*;s}(\gamma\eta) = |\gamma|\,\omega_{*;s}(\eta), \qquad \omega_{*;s}(\eta + \zeta) \le \omega_{*;s}(\eta) + \omega_{*;s}(\zeta) \tag{12}$$

in case $s = 0$. This carries over to $\omega_{*,2}$, $\omega_{*,\infty}$ as the Euclidean and sup norms on $\mathbb{R}^p$ are positively homogeneous, increasing in each coordinate.

Since $\mathcal{G}_* \subset L_\infty(P) \subset L_2(P)$ each function $\eta \mapsto \mathrm{E}\,\eta q$ with $q \in \mathcal{G}_*$ is weakly continuous on $L_2^p(P)$, hence ω_* is weakly l.s.c. being the sup of these functions. To see that weak l.s.c. is inherited to the coordinatewise versions, let $\eta_n \to \eta$ weakly in $L_2^p(P)$, that is $\eta_{n,j} \to \eta_j$ weakly in $L_2(P)$ for each $j = 1, \dots, p$. Then $\omega_*(\eta_j) \le \liminf_n \omega_*(\eta_{n,j})$ for each j. Since the Euclidean and sup norms on $\mathbb{R}^p$ are continuous, increasing in each coordinate, this implies that $\omega_{*;s}(\eta) \le \liminf_n \omega_{*;s}(\eta_n)$ for $s = 2, \infty$. ////

Explicit Expressions

The oscillation terms ω_* can be calculated more or less explicitly.

Proposition 5.3.3 *Let* $\eta \in Z_1^p$.

(a) *Then*

$$\omega_c(\eta) = \sup_P |\eta| \tag{13}$$

$$\omega_v(\eta) = \sup\{\, \sup_P e'\eta - \inf_P e'\eta \mid e \in \mathbb{R}^p,\ |e| = 1 \,\} \tag{14}$$

$$\omega_h^2(\eta) = 8 \max\mathrm{ev}\, \mathcal{C}(\eta) \tag{15}$$

(b) *If, for some measure* $\nu \colon \mathbb{B}^m \to \mathbb{R}^p$ *of total variation* $\|\nu\|$,

$$\eta(x) = \nu(y \ge x) + \mathrm{const} \qquad \text{a.e. } P(dx) \tag{16}$$

then

$$\omega_\kappa(\eta) \le \|\nu\| \tag{17}$$

For dimension $p = 1$ *and the function* $\eta \colon \mathbb{R} \to \mathbb{R}$ *monotone, we have*

$$\omega_\kappa(\eta) = \omega_v(\eta) = \sup_P \eta - \inf_P \eta \tag{18}$$

(c) *If the measures* $\mu \in \mathcal{M}_\sigma(\mathbb{B}^m)$ *and* P *satisfy*

$$\int P(y)\big(1 - P(y)\big)\, \mu(dy) < \infty \tag{19}$$

and if there exists some function $\varphi \in L_2^p(\mu)$ *such that*

$$\eta(x) = \int \big(\mathbf{I}(x \le y) - P(y)\big)\,\varphi(y)\,\mu(dy) \qquad \text{a.e. } P(dx) \tag{20}$$

then

$$\omega_\mu^2(\eta) \le \operatorname{maxev} \int \varphi\varphi'\,d\mu \tag{21}$$

with equality if, for some $\xi \in Z_2^p$,

$$\varphi(y) = \int \big(\mathbf{I}(x \le y) - P(y)\big)\,\xi(x)\,P(dx) \qquad \text{a.e. } \mu(dy) \tag{22}$$

PROOF First assume $p = 1$. Then, for $q \in \mathcal{G}_c$ we have $1 + q \ge 0$ hence

$$\mathrm{E}\,\eta q = \mathrm{E}\,\eta(1 + q) \le \sup_P \eta \tag{23}$$

where the upper bound is approximated for $\delta \downarrow 0$ by

$$q = \frac{1}{P(A)}\,\mathbf{I}_A - 1\,, \qquad A = \{\,\eta > \sup_P \eta - \delta\,\} \tag{24}$$

The same argument applying to $-\eta$, indeed (13) holds since

$$\omega_c(\eta) = \sup\{\,\mathrm{E}\,\eta q \vee \mathrm{E}\,(-\eta)q \mid q \in \mathcal{G}_c\,\} = \sup_P \eta \vee \sup_P(-\eta) = \sup_P |\eta|$$

To prove (14), decompose any $q \in \mathcal{G}_v$ into positive and negative parts to get

$$\mathrm{E}\,\eta q = \mathrm{E}\,\eta q^+ - \mathrm{E}\,\eta q^- \le \sup_P \eta - \inf_P \eta \tag{25}$$

since $\mathrm{E}\,q^+ = \mathrm{E}\,q^- \le 1$. The upper bound, which stays the same for $-\eta$, can for $\delta \downarrow 0$ be approximated by

$$q = \frac{1}{P(A'')}\,\mathbf{I}_{A''} - \frac{1}{P(A')}\,\mathbf{I}_{A'}\,, \qquad \begin{array}{l} A' = \{\,\eta < \inf_P \eta + \delta\,\} \\ A'' = \{\,\eta > \sup_P \eta - \delta\,\} \end{array} \tag{26}$$

To prove (15), apply the Cauchy–Schwarz inequality to any $q \in \mathcal{G}_h$. Thus

$$|\mathrm{E}\,\eta q|^2 \le \mathrm{E}\,\eta^2\,\mathrm{E}\,q^2 \le 8\,\mathrm{E}\,\eta^2 \tag{27}$$

The upper bound can be approximated as $m \to \infty$ by choosing

$$q_m \propto \eta_m - \mathrm{E}\,\eta_m\,, \qquad \eta_m = \eta\,\mathbf{I}(|\eta| \le m)\,, \qquad \mathrm{E}\,q_m^2 = 8\,\mathbf{I}\big(\mathrm{E}\,\eta^2 \ne 0\big)$$

To prove (18), note that the expectation of any increasing $\eta\colon \mathbb{R} \to \mathbb{R}$,

$$Q \longmapsto \int \eta\,dQ = \int_0^\infty Q(\eta > t)\,dt - \int_{-\infty}^0 Q(\eta < t)\,dt \tag{28}$$

is increasing w.r.t. stochastic ordering of Q. Given $q \in \mathcal{G}_\kappa$ pick $\varepsilon \in (0,1)$ so that $\varepsilon q \geq -1$ and consider $dQ = (1+\varepsilon q)\,dP$, which is in $B_\kappa(P,\varepsilon)$ by Lemma 5.3.1. But the distribution function $t \mapsto \bigl(P(t) - \varepsilon\bigr) \vee 0$ moving mass ε from the left flank of P to ∞ is stochastically larger than any other element of $B_\kappa(P,\varepsilon)$. It follows that

$$\int \eta\, dQ \leq \int \eta\, dP + \varepsilon\bigl(\eta(\infty) - \inf_P \eta\bigr) \tag{29}$$

hence

$$\mathrm{E}\,\eta q \leq \eta(\infty) - \inf_P \eta \tag{30}$$

Under P however, η cannot be distinguished from $\inf_P \eta \vee \eta \wedge \sup_P \eta$, which is still increasing. For decreasing η, use the fact that $-q \in \mathcal{G}_\kappa$ whenever $q \in \mathcal{G}_\kappa$. This proves that $\omega_\kappa(\eta) \leq \omega_v(\eta)$ for monotone η. But then equality holds since always $\omega_\kappa \geq \omega_v$.

If P satisfies (19) and η is of form (20) we invoke Fubini 4.2(50) for $Q \in B_\mu(P, r/\sqrt{n}\,) \cap B_{\mu,1}(P,r)$ and conclude by Cauchy–Schwarz that

$$\sqrt{n}\int \eta\, dQ = \sqrt{n}\int \bigl(Q(y) - P(y)\bigr)\varphi(y)\,\mu(dy) \leq r\,\Bigl|\int \varphi^2\, d\mu\Bigr|^{1/2} \tag{31}$$

For φ of the special form (22) approximate ξ in $Z_2 \subset L_2(P)$ by bounded functions $q \in Z_\infty$. Then

$$\int \Bigl|\int \mathbf{I}(x \leq y)\bigl(\xi(x) - q(x)\bigr)\,P(dx)\Bigr|^2 \mu(dy) \leq \mathrm{E}\,|\xi - q|^2 \int P(1-P)\,d\mu$$

Hence the functions $\mathrm{E}\,q\,\mathbf{I}_{(-\infty,y]}$ approximate φ in $L_2(\mu)$. It follows that the Cauchy–Schwarz inequality in (31) for $Q = Q_n(q,r)$ turns into an equality as $q \to \xi$. Finally rescale q so that $q \in \mathcal{G}_\mu$. This proves (21), with attainment. Hence all the results that are stated under (a)–(c) for one dimension $p = 1$ have been proved.

In case $p \geq 1$, the reduction (8) to $p = 1$ settles the cases $* = v, h, \mu$ immediately. In case $* = c$, the essential (as opposed to pointwise) supremum asks for a little extra argument. Obviously $\omega_c(\eta) \leq \sup_P |\eta|$ holds since $|e'\eta| \leq |\eta|$ for all unit vectors. But choosing (e_n) dense in the unit sphere of $\mathbb{R}^p$, the reverse bound $\omega_c(\eta) \geq \sup_P |\eta|$ is also true because

$$t \in (0,\infty),\quad P({e_n}'\eta > t) = 0 \;\; \forall\, n \geq 1 \implies P(|\eta| > t) = 0 \tag{32}$$

Alternatively, we could bound $|\mathrm{E}\,\eta q|$ for $q \in \mathcal{G}_c$ by $\mathrm{E}\,|\eta|(1+q) \leq \sup_P |\eta|$, and then directly apply the approximation (24) with $|\eta|$ in the place of η.

To prove (17), assume η of bounded variation satisfying (16), and any $Q \in B_\kappa(P, r/\sqrt{n}\,)$. Then making use of $\mathrm{E}\,\eta = 0$ and Fubini's theorem, we obtain

$$\sqrt{n}\int \eta\, dQ = \sqrt{n}\int \bigl(Q(y) - P(y)\bigr)\,\nu(dy) \leq r\,\|\nu\| \tag{33}$$

by means of the polar representation $d\nu = h\,d|\nu|$ with $|h| = 1$. ////

Full Balls B_{cv}

The following exact bias evaluation for full contamination and total variation balls serves as some control, and is relevant for exact sums. These type of balls about P of radii $\varepsilon_n, \delta_n \in [0,1]$, respectively, can be combined to contamination/total variation balls of the form

$$B_{cv}(P;\varepsilon_n,\delta_n) = \bigcup \{\, B_v(Q,\delta_n) \mid Q \in B_c(P,\varepsilon_n) \,\} \tag{34}$$

For $Q \in \mathcal{M}_1(\mathcal{A})$, according to Rieder (1977, proof to Lemma 4.3), membership to these balls can be characterized by either one of the following two bounds holding identically on the σ algebra $\mathcal{A}$,

$$\begin{aligned} Q \in B_{cv}(P;\varepsilon_n,\delta_n) &\iff Q \le (1-\varepsilon_n)P + \varepsilon_n + \delta_n \\ &\iff Q \ge (1-\varepsilon_n)P - \delta_n \end{aligned} \tag{35}$$

Given $\varepsilon, \delta \in [0,\infty)$, the simple perturbations submodel of these neighborhoods with $\varepsilon_n = \varepsilon/\sqrt{n}$ and $\delta_n = \delta/\sqrt{n}$ would consist of the sequences

$$dQ_n = \Big(1 + \frac{1}{\sqrt{n}}\,(\varepsilon q_c + \delta q_v)\Big)\, dP\,, \qquad q_c \in \mathcal{G}_c\,,\ q_v \in \mathcal{G}_v \tag{36}$$

For $\eta \in L_1(P)$ such that $\mathrm{E}\,\eta = 0$, according to Proposition 5.3.3 and its proof, they generate the infinitesimal oscillation term $\omega_{cv}(\eta;\varepsilon,\delta)$,

$$\omega_{cv}(\eta;\varepsilon,\delta) = \varepsilon \sup_P |\eta| + \delta\big(\sup_P \eta - \inf_P \eta\big) \tag{37}$$

Given any $\varepsilon_n, \delta_n \in [0,1]$ and any measurable function $\eta\colon (\mathbb{R},\mathbb{B}) \to (\mathbb{R},\mathbb{B})$, let us introduce the following quantiles of the image law of η under P,

$$\begin{aligned} z_n' &= \big(\eta(P)\big)^{-1}\Big(1 - \frac{\delta_n}{1-\varepsilon_n}\Big) \\ &= \inf\Big\{\, t \in [\inf_P \eta, \sup_P \eta] \;\Big|\; P(\eta > t) \le \frac{\delta_n}{1-\varepsilon_n} \,\Big\} \\ z_n'' &= \big(\eta(P)\big)^{-1}\Big(\frac{\delta_n}{1-\varepsilon_n}\Big) \\ &= \inf\Big\{\, t \in [\inf_P \eta, \sup_P \eta] \;\Big|\; P(\eta \le t) \ge \frac{\delta_n}{1-\varepsilon_n} \,\Big\} \end{aligned} \tag{38}$$

Assuming $\eta \in L_\infty(P)$ even bounded pointwise, define

$$\begin{aligned} w_{cv,n}'(\eta) &= (1-\varepsilon_n) \int z_n' \wedge \eta \, dP - \delta_n z_n' + (\varepsilon_n + \delta_n) \inf \eta \\ w_{cv,n}''(\eta) &= (1-\varepsilon_n) \int z_n'' \vee \eta \, dP - \delta_n z_n'' + (\varepsilon_n + \delta_n) \sup \eta \end{aligned} \tag{39}$$

It turns out that the standardized as. oscillation for the full balls B_{cv} does not exceed ω_{cv} by more than the increase of some P essential to pointwise extrema. In fact, the extension of $\omega_{cv}(\eta;\varepsilon,\delta)$, for $\eta \in L_1(P)$ such that η is bounded and $\mathrm{E}\,\eta = 0$, to full balls B_{cv} is

$$\begin{aligned} w_{cv}(\eta;\varepsilon,\delta) &= (-w'_{cv}(\eta;\varepsilon,\delta)) \vee w''_{cv}(\eta;\varepsilon,\delta) \\ &= \bigl((\varepsilon+\delta)\sup\eta - \delta\inf\nolimits_P\eta\bigr) \vee \bigl(\delta\sup\nolimits_P\eta - (\varepsilon+\delta)\inf\eta\bigr) \end{aligned} \tag{40}$$

This is proved by the following lemma. Seen from the full balls, the discrepancy is unavoidable since the perturbations $Q_n(q,r)$, being dominated by P, cannot distinguish between η and $\inf_P\eta \vee \eta \wedge \sup_P\eta$.

Lemma 5.3.4 *Consider any $\eta \in Z_\infty$ such that $\sup|\eta| < \infty$.*
(a) *Then, for $\varepsilon_n, \delta_n \in [0,1]$, we have*

$$\begin{aligned} \inf\Bigl\{\int \eta\,dQ \Bigm| Q \in B_{cv}(P;\varepsilon_n,\delta_n)\Bigr\} &= w'_{cv,n}(\eta) \\ \sup\Bigl\{\int \eta\,dQ \Bigm| Q \in B_{cv}(P;\varepsilon_n,\delta_n)\Bigr\} &= w''_{cv,n}(\eta) \end{aligned} \tag{41}$$

(b) *For $\varepsilon,\delta \in [0,\infty)$ and $\varepsilon_n = \varepsilon/\sqrt{n}$, $\delta_n = \delta/\sqrt{n}$, we have*

$$\begin{aligned} \lim_{n\to\infty} \sqrt{n}\,w'_{cv,n}(\eta) &= (\varepsilon+\delta)\inf\eta - \delta\sup\nolimits_P\eta = w'_{cv}(\eta;\varepsilon,\delta) \\ \lim_{n\to\infty} \sqrt{n}\,w''_{cv,n}(\eta) &= (\varepsilon+\delta)\sup\eta - \delta\inf\nolimits_P\eta = w''_{cv}(\eta;\varepsilon,\delta) \end{aligned} \tag{42}$$

PROOF Define $W'_{cv,n}, W''_{cv,n}\colon \mathcal{A} \to [0,1]$ (lower/upper probabilities) by

$$\begin{aligned} W'_{cv,n}(A) &= \bigl((1-\varepsilon_n)P(A) - \delta_n\bigr) \vee 0, & A \neq \Omega;\ \ W'_{cv,n}(\Omega) = 1 \\ W''_{cv,n}(A) &= \bigl((1-\varepsilon_n)P(A) + \varepsilon_n + \delta_n\bigr) \wedge 1, & A \neq \emptyset;\ \ W''_{cv,n}(\emptyset) = 0 \end{aligned} \tag{43}$$

so that, in view of (35),

$$B_{cv}(P;\varepsilon_n,\delta_n) = \bigl\{\, Q \in \mathcal{M}_1(\mathcal{A}) \bigm| W'_{cv,n} \le Q \le W''_{cv,n} \,\bigr\} \tag{44}$$

We shall prove the second halves of (41) and (42). The first halves then follow for $-\eta$, or can be shown directly by evaluating $W'_{cv,n}$ like $W''_{cv,n}$.
(a) In this proof, we may drop $\mathrm{E}\,\eta = 0$ and assume $\eta \ge 0$. Then, as

$$z''_n = \inf\bigl\{\, t \in [\inf\nolimits_P\eta, \sup\nolimits_P\eta] \bigm| (1-\varepsilon_n)P(\eta > t) + \varepsilon_n + \delta_n \le 1 \,\bigr\} \tag{45}$$

we obtain that

$$\begin{aligned} &\int_0^\infty W''_{cv,n}(\eta > t)\,dt \\ &\quad = \int_0^{z''_n} 1\,dt + \int_{z''_n}^{\sup\eta} \bigl((1-\varepsilon_n)P(\eta > t) + \varepsilon_n + \delta_n\bigr)\,dt \end{aligned} \tag{46}$$

$$
\begin{aligned}
&= z_n'' + (1-\varepsilon_n) \int_{z_n''}^{\sup \eta} P(\eta > t)\, dt + (\varepsilon_n + \delta_n)(\sup \eta - z_n'') \\
&= (1-\varepsilon_n) \int_0^{\sup \eta} P(z_n'' \vee \eta > t)\, dt - \delta_n z_n'' + (\varepsilon_n + \delta_n) \sup \eta \\
&= (1-\varepsilon_n) \int z_n'' \vee \eta \, dP - \delta_n z_n'' + (\varepsilon_n + \delta_n) \sup \eta \;=\; w_{cv,n}''(\eta)
\end{aligned}
$$

To approximate this upper bound, choose $\gamma_n \in [0,1]$ so that

$$
\gamma_n (1-\varepsilon_n) P(\eta = z_n'') = (1-\varepsilon_n) P(\eta \le z_n'') - \delta_n \tag{47}
$$

and pick any $y \in \Omega$. Then the probability R_n on $\mathcal{A}$ defined by

$$
\begin{aligned}
R_n(A) = \gamma_n & (1-\varepsilon_n) P(A, \eta = z_n'') \\
& + (1-\varepsilon_n) P(A, \eta > z_n'') + (\varepsilon_n + \delta_n)\, \mathbf{I}_y(A)
\end{aligned} \tag{48}
$$

is an element of $B_{cv}(P; \varepsilon_n, \delta_n)$ and achieves

$$
\int \eta \, dR_n = (1-\varepsilon_n) \int z_n'' \vee \eta \, dP - \delta_n z_n'' + (\varepsilon_n + \delta_n) \eta(y) \tag{49}
$$

Letting $\eta(y)$ tend to $\sup \eta$ proves the second half of (41).

(b) Definition (38) ensures that

$$
\inf_P \eta \le z_n'' \longrightarrow \inf_P \eta \tag{50}
$$

hence

$$
\begin{aligned}
0 \le \sqrt{n} \int (z_n'' \vee \eta - \eta)\, dP &= \sqrt{n} \int (z_n'' - \eta)^+ \, dP \\
&\le \sqrt{n}\, \frac{\delta_n}{1-\varepsilon_n} \, (z_n'' - \inf_P \eta) \longrightarrow 0
\end{aligned} \tag{51}
$$

which proves the second half of (42). ////

5.3.2 Minimax Oscillation

A classical result of Huber (1981) says that among all translation equivariant functionals, the median minimaxes oscillation on contamination and Lévy neighborhoods, of a symmetric, strongly unimodal distribution. Concerning the oscillation of two classes of equivariant regression functionals on fixed contamination balls, the result has been extended to the linear regression model by Martin , Yohai, and Zamar (1989).

In the infinitesimal setup, the Hellinger and Cramér–von Mises MD functionals T_h and T_μ minimize the maximum oscillation among all Fisher consistent functionals, on shrinking Hellinger and CvM neighborhoods, respectively. In fact, a saddle point exists with the two minimum Hilbert distance functionals, on the one hand, being minimax and, on the other hand, the ideal model $\mathcal{P}$ itself being least favorable. This result, which is due to Beran (1977) in the Hellinger case and, in more generality, to Donoho and Liu (1988 a), does not play any role in the nonparametric as. optimality theorems. It is therefore not mentioned in Beran's (1981–1984) further work and has not even been recognized by Millar (1981–1984). Fix $\theta \in \Theta$.

Theorem 5.3.5 *Consider any functional $T = (T_n)$ such that*

$$\sqrt{n}\,\bigl(\tau \circ T_n(P_{\theta+t/\sqrt{n}}) - \tau(\theta)\bigr) = d\tau(\theta)t + \mathrm{o}(n^0) \tag{52}$$

for every $t \in \mathbb{R}^k$.

(a) *Assume $\mathcal{P}$ is L_2 differentiable at θ with derivative Λ_θ, and Fisher information $\mathcal{I}_\theta > 0$. Suppose the functional T_h is Hellinger differentiable at P_θ with influence curve $\psi_{h,\theta} = \mathcal{I}_\theta^{-1}\Lambda_\theta$. Then, for all $r \in (0,\infty)$,*

$$\begin{aligned}
&\liminf_{n\to\infty} \sup\Bigl\{\, n\bigl|\tau \circ T_n(Q) - \tau(\theta)\bigr|^2 \Bigm| Q \in B_h(P_\theta, r/\sqrt{n}\,)\Bigr\} \\
&\quad \ge \liminf_{n\to\infty} \sup\Bigl\{\, n\bigl|\tau \circ T_n(Q) - \tau(\theta)\bigr|^2 \Bigm| Q \in \mathcal{P} \cap B_h(P_\theta, r/\sqrt{n}\,)\Bigr\} \\
&\quad \ge 8r^2 \operatorname{maxev} d\tau(\theta)\mathcal{I}_\theta^{-1} d\tau(\theta)' \\
&\quad = \lim_{n\to\infty} \sup\Bigl\{\, n\bigl|\tau \circ T_{h,n}(Q) - \tau(\theta)\bigr|^2 \Bigm| Q \in B_h(P_\theta, r/\sqrt{n}\,)\Bigr\}
\end{aligned} \tag{53}$$

(b) *Assume that $\mathcal{P}$, relative to some weight $\mu \in \mathcal{M}(\mathbb{B}^m)$, is* CvM *differentiable at θ with derivative Δ_θ and* CvM *information $\mathcal{J}_\theta > 0$, and suppose that the functional T_μ is* CvM *differentiable at P_θ with* CvM *influence curve $\varphi_{\mu,\theta} = \mathcal{J}_\theta^{-1}\Delta_\theta$. Then, for all $r \in (0,\infty)$,*

$$\begin{aligned}
&\liminf_{n\to\infty} \sup\Bigl\{\, n\bigl|\tau \circ T_n(Q) - \tau(\theta)\bigr|^2 \Bigm| Q \in B_\mu(P_\theta, r/\sqrt{n}\,)\Bigr\} \\
&\quad \ge \liminf_{n\to\infty} \sup\Bigl\{\, n\bigl|\tau \circ T_n(Q) - \tau(\theta)\bigr|^2 \Bigm| Q \in \mathcal{P} \cap B_\mu(P_\theta, r/\sqrt{n}\,)\Bigr\} \\
&\quad \ge r^2 \operatorname{maxev} d\tau(\theta)\mathcal{J}_\theta^{-1} d\tau(\theta)' \\
&\quad = \lim_{n\to\infty} \sup\Bigl\{\, n\bigl|\tau \circ T_{\mu,n}(Q) - \tau(\theta)\bigr|^2 \Bigm| Q \in B_\mu(P_\theta, r/\sqrt{n}\,)\Bigr\}
\end{aligned} \tag{54}$$

Remark 5.3.6 Some comments to keep track of the technical assumptions:

(a) If $\mathcal{P}$ is identifiable and L_2 differentiable with Fisher information $\mathcal{I}_\theta > 0$ at every $\theta \in \Theta$, the Hellinger MD functional T_h can be constructed to be Hellinger differentiable with influence curve $\psi_{h,\theta} = \mathcal{I}_\theta^{-1}\Lambda_\theta$ at every P_θ [Example 4.2.14, Theorem 6.3.4 a]. Formally, the oscillation saddle point requires the differentiability of T_h only at the fixed P_θ.

(b) Given a finite-dimensional Euclidean sample $(\Omega, \mathcal{A}) = (\mathbb{R}^m, \mathbb{B}^m)$, let $\mathcal{P}$, relative to some measure $\mu \in \mathcal{M}(\mathbb{B}^m)$, be identifiable in $L_2(\mu)$ and at every $\theta \in \mathbb{R}^k$ CvM differentiable with derivative Δ_θ and CvM information $\mathcal{J}_\theta > 0$ [Definition 2.3.11]. Then the CvM MD functional T_μ is

CvM differentiable at every P_θ with CvM influence curve $\varphi_{\mu,\theta} = \mathcal{J}_\theta^{-1}\Delta_\theta$ [Example 4.2.15, Theorem 6.3.4 b]. For the oscillation saddle point, it would formally suffice that $\mathcal{P}$ and T_μ be CvM differentiable at the fixed θ, respectively, P_θ. ////

PROOF In the Hellinger case, by L_2 differentiability of $\mathcal{P}$ at θ and by Fisher consistency (52), respectively, we have for every $t \in \mathbb{R}^k$,

$$\begin{aligned} \lim_{n\to\infty} 8n\, d_h^2(P_{\theta+t/\sqrt{n}}, P_\theta) &= t'\mathcal{I}_\theta t \\ \lim_{n\to\infty} n\left|\tau\circ T_n(P_{\theta+t/\sqrt{n}}) - \tau(\theta)\right|^2 &= \left|d\tau(\theta)t\right|^2 \end{aligned} \tag{55}$$

and, using the identity C.2(24), p 373,

$$\begin{aligned} \sup\Bigl\{ \left|d\tau(\theta)t\right|^2 \Bigm| t'\mathcal{I}_\theta t < 8r^2 \Bigr\} &= 8r^2 \operatorname{maxev} \mathcal{I}_\theta^{-1/2} d\tau(\theta)' d\tau(\theta) \mathcal{I}_\theta^{-1/2} \\ &= 8r^2 \operatorname{maxev} d\tau(\theta)\mathcal{I}_\theta^{-1} d\tau(\theta)' \end{aligned} \tag{56}$$

As for T_h, the expansion 4.2(47) with $\psi_\theta = \psi_{h,\theta}$ and the differentiability of τ at θ ensure that, for all sequences $Q_n \in B_h(P_\theta, r/\sqrt{n})$,

$$\begin{aligned} &\limsup_{n\to\infty} \sqrt{n}\left|\tau\circ T_{h,n}(Q_n) - \tau(\theta)\right| \\ &\quad = \limsup_{n\to\infty} \sup_{|e|=1} 2\sqrt{n} \int e' d\tau(\theta)\psi_{h,\theta}\sqrt{dP_\theta}\,\bigl(\sqrt{dQ_n} - \sqrt{dP_\theta}\,\bigr) \\ &\quad \le r\sqrt{8 \operatorname{maxev} d\tau(\theta)\mathcal{I}_\theta^{-1} d\tau(\theta)'} \end{aligned} \tag{57}$$

by Cauchy–Schwarz. In the CvM case, we argue similarly. ////

The following variant of Theorem 5.3.5 is suited to as. linear functionals, which need to be Fisher consistent only in the approximate sense 4.2(46), and also shows that the simplifications and restrictions do not affect the lower oscillation bounds.

Corollary 5.3.7 *Let $\mathcal{P}$ be L_2 differentiable at θ with derivative Λ_θ and Fisher information $\mathcal{I}_\theta > 0$. Consider functionals $T = (T_n)$ such that*

$$\lim_{\zeta\to\Lambda_\theta} \limsup_{n\to\infty} \left|\sqrt{n}\,\bigl(\tau\circ T_n(Q_n(\zeta,t)) - \tau(\theta)\bigr) - d\tau(\theta)t\,\right| = 0 \tag{58}$$

for every $t \in \mathbb{R}^k$, as $\zeta \in Z_\infty^k(\theta)$ tends to Λ_θ in $L_2^k(P_\theta)$.

(a) *Suppose T_h is as. linear at P_θ with influence curve $\psi_{h,\theta} = \mathcal{I}_\theta^{-1}\Lambda_\theta$. Then, for all $r \in (0,\infty)$,*

$$\begin{aligned} &\sup\Bigl\{ \liminf_{n\to\infty} n\left|\tau\circ T_n(Q_n(q,r)) - \tau(\theta)\right|^2 \Bigm| q \in \mathcal{G}_h(\theta) \Bigr\} \ge \\ &\liminf_{\zeta\to\Lambda_\theta} \sup\Bigl\{ \liminf_{n\to\infty} n\left|\tau\circ T_n(Q_n(\zeta,t)) - \tau(\theta)\right|^2 \Bigm| t \in \mathbb{R}^k,\, t'\zeta \in r\,\mathcal{G}_h(\theta) \Bigr\} \\ &\quad \ge 8r^2 \operatorname{maxev} d\tau(\theta)\mathcal{I}_\theta^{-1} d\tau(\theta)' = r^2\omega_h^2\bigl(d\tau(\theta)\psi_{h,\theta}\bigr) \end{aligned} \tag{59}$$

(b) *Assume $\mu \in \mathcal{M}_\sigma(\mathbb{B}^m)$ and 4.2(20). Suppose that $\mathcal{J}_\theta > 0$ for Δ_θ given by 2.3(69), and let T_μ be as. linear at P_θ with influence curve $\psi_{\mu,\theta}$ derived from $\varphi_{\mu,\theta} = \mathcal{J}_\theta^{-1}\Delta_\theta$ via 4.2(51). Then, for all $r \in (0,\infty)$,*

$$\sup\Bigl\{ \liminf_{n\to\infty} n\bigl|\tau\circ T_n(Q_n(q,r)) - \tau(\theta)\bigr|^2 \Bigm| q \in \mathcal{G}_\mu(\theta) \Bigr\} \ge$$

$$\liminf_{\zeta\to\Lambda_\theta} \sup\Bigl\{ \liminf_{n\to\infty} n\bigl|\tau\circ T_n(Q_n(\zeta,t)) - \tau(\theta)\bigr|^2 \Bigm| t \in \mathbb{R}^k,\, t'\zeta \in r\,\mathcal{G}_\mu(\theta) \Bigr\}$$

$$\ge r^2 \operatorname{maxev} d\tau(\theta)\mathcal{J}_\theta^{-1} d\tau(\theta)' = r^2\omega_\mu^2\bigl(d\tau(\theta)\psi_{\mu,\theta}\bigr) \tag{60}$$

PROOF In the CvM case, pick any $\zeta \in Z_\infty^k(\theta)$ and put

$$\xi(y) = \int \mathbf{I}(x \le y)\zeta(x)\, P_\theta(dx)$$

As $\zeta \to \Lambda_\theta$ in $L_2^k(P_\theta)$ it follows that

$$\int |\xi - \Delta_\theta|^2\, d\mu = \int \Bigl| \int \mathbf{I}(x \le y)\bigl(\zeta(x) - \Lambda_\theta(x)\bigr)\, P_\theta(dx)\Bigr|^2 \mu(dy)$$
$$\le \mathrm{E}_\theta\, |\zeta - \Lambda_\theta|^2 \int P_\theta(1 - P_\theta)\, d\mu \longrightarrow 0$$

so that for all $t \in \mathbb{R}^k$ and all $n \ge 1$,

$$n\, d_\mu^2\bigl(Q_n(\zeta,t), P_\theta\bigr) = \int |t'\xi|^2\, d\mu \longrightarrow t'\mathcal{J}_\theta t \tag{61}$$

Hence eventually $t'\zeta \in r\,\mathcal{G}_\mu(\theta)$ provided that $t'\mathcal{J}_\theta t < r^2$. By (58),

$$\liminf_{\zeta\to\Lambda_\theta} \liminf_{n\to\infty} n\,\bigl|\tau\circ T_n(Q_n(\zeta,t)) - \tau(\theta)\bigr|^2 = \bigl|d\tau(\theta)t\bigr|^2 \tag{62}$$

and similarly to (56),

$$\begin{aligned} \sup\Bigl\{ \bigl|d\tau(\theta)t\bigr|^2 \Bigm| t'\mathcal{J}_\theta t < r^2 \Bigr\} &= r^2 \operatorname{maxev} \mathcal{J}_\theta^{-1/2} d\tau(\theta)' d\tau(\theta) \mathcal{J}_\theta^{-1/2} \\ &= r^2 \operatorname{maxev} d\tau(\theta)\mathcal{J}_\theta^{-1} d\tau(\theta)' \end{aligned} \tag{63}$$

As for T_μ we have

$$\begin{aligned} \psi_{\mu,\theta}(x) &= \int \bigl(\mathbf{I}(x \le y) - P_\theta(y)\bigr)\varphi_{\mu,\theta}(y)\, \mu(dy), \qquad \varphi_{\mu,\theta} = \mathcal{J}_\theta^{-1}\Delta_\theta \\ \Delta_\theta(y) &= \int \mathbf{I}(x \le y)\Lambda_\theta(x)\, P_\theta(dx), \qquad \mathcal{J}_\theta = \int \Delta_\theta \Delta_\theta'\, d\mu \end{aligned} \tag{64}$$

Invoking 4.2(50) for simple perturbations $Q_n = Q_n(q,r)$ along $q \in \mathcal{G}_\mu(\theta)$, and the Cauchy–Schwarz inequality, we thus obtain

$$\bigl|\mathrm{E}_\theta\, d\tau(\theta)\psi_{\mu,\theta}q\bigr| = \sup_{|e|=1} \int e' d\tau(\theta)\varphi_{\mu,\theta}(y) \int \mathbf{I}(x \le y) q(x)\, P_\theta(dx)\, \mu(dy)$$
$$\le \sqrt{\operatorname{maxev} d\tau(\theta)\mathcal{J}_\theta^{-1} d\tau(\theta)'} \tag{65}$$

In the Hellinger case, we argue similarly. ////

The classical scores function $\psi_{h,\theta}$ minimizes the covariance $\mathcal{C}_\theta(\psi_\theta)$ among all influence curves $\psi_\theta \in \Psi_2(\theta)$ in the positive definite sense [Cramér–Rao 4.2(74)]; hence also the maximum eigenvalue,

$$\begin{aligned}\omega_{h,\theta}^2(\psi_{h,\theta}) &= 8\,\mathrm{maxev}\,\mathcal{I}_\theta^{-1} \\ &\le 8\,\mathrm{maxev}\,\mathcal{C}_\theta(\psi_\theta) = \omega_{h,\theta}^2(\psi_\theta)\,, \qquad \psi_\theta \in \Psi_2(\theta)\end{aligned} \tag{66}$$

since $\mathrm{maxev}\,\mathcal{C}_\theta(\psi_\theta) = \sup_{|e|=1} e'\mathcal{C}_\theta(\psi_\theta)e$. Without using the explicit form of $\omega_{h,\theta}$, this bound for the Hellinger oscillation of as. linear functionals also follows from Corollary 5.3.7 a. In case $* = h$, therefore, the classical scores function $\psi_{h,\theta}$ minimizes both criteria simultaneously.

Except for the influence curve $\psi_{\mu,\theta}$ given by (64), which satisfies the extra assumptions made in Proposition 5.3.3, the explicit determination of the CvM oscillation $\omega_{\mu,\theta}$ is difficult. Independently of the explicit form of $\omega_{\mu,\theta}$ however, Corollary 5.3.7 b implies the bound

$$\omega_{\mu,\theta}^2(\psi_{\mu,\theta}) = \mathrm{maxev}\,\mathcal{J}_\theta^{-1} \le \omega_{\mu,\theta}^2(\psi_\theta)\,, \qquad \psi_\theta \in \Psi_2(\theta) \tag{67}$$

In general, the influence curves $\psi_{h,\theta}$ and $\psi_{\mu,\theta}$ are not the only solutions to (66) and (67), respectively. We assume the full parameter case $\tau = \mathrm{id}_{\mathbb{R}^k}$.

Proposition 5.3.8 *Let $\mathcal{P}$ be L_2 differentiable at θ with derivative Λ_θ and Fisher information $\mathcal{I}_\theta > 0$.*

(a) *Then, if $\mathcal{I}_\theta$ and the unit matrix $\mathbb{I}_k$ are linearly dependent,* Hellinger *oscillation $\omega_{h,\theta}$ is over $\Psi_2(\theta)$ minimized uniquely by $\psi_{h,\theta}$.*

If $\mathcal{I}_\theta$, $\mathbb{I}_k$ are linearly independent and $\dim L_2(P_\theta) > 2k$, then $\omega_{h,\theta}$ is minimized over $\Psi_2(\theta)$ not only by $\psi_{h,\theta}$.

(b) *Let $\mu \in \mathcal{M}_\sigma(\mathcal{A})$ and P_θ be such that 4.2(20) and 4.2(59) hold. Suppose that $\mathcal{J}_\theta > 0$ for Δ_θ given by 2.3(69). Then, if $\mathcal{J}_\theta$ and $\mathbb{I}_k$ are linearly independent and $\dim L_2(\mu) > 2k$,* CvM *oscillation $\omega_{\mu,\theta}$ is minimized over $\Psi_2(\theta)$ not only by $\psi_{\mu,\theta}$.*

PROOF Using θ free notation, differences δ of any two influence curves are characterized by

$$\delta \in L_2^k(P), \quad \mathrm{E}\,\delta = 0, \quad \mathrm{E}\,\delta\Lambda' = 0 \tag{68}$$

so that

$$\mathcal{C}(\psi_h + \delta) = \mathcal{I}^{-1} + \mathcal{C}(\delta) \tag{69}$$

First, let $\gamma\mathcal{I} = \mathbb{I}_k$ for some $\gamma \in (0,\infty)$. Then, for all unit vectors e,

$$e'\mathcal{C}(\psi_h + \delta)e = \gamma + \mathrm{E}\,(e'\delta)^2 \tag{70}$$

Thus

$$\delta \neq 0 \implies \mathrm{maxev}\,\mathcal{C}(\psi_h + \delta) > \gamma = \mathrm{maxev}\,\mathcal{I}^{-1} \tag{71}$$

Secondly, let $\mathcal{I}$ and $\mathbb{I}_k$ be linearly independent. Choose $\{e_1,\dots,e_k\}$ an orthonormal basis of $\mathbb{R}^k$ made up by eigenvectors of $\mathcal{I}$ so that

$$\mathcal{I}^{-1} = \lambda_1 e_1 e_1{}' + \cdots + \lambda_k e_k e_k{}' \tag{72}$$

where $\lambda_1 \geq \cdots \geq \lambda_k$ can be arranged; then $\lambda_1 > \lambda_k$ by assumption. Because $\dim L_2(P) > 2k$, there exists some function $g \in L_2^k(P)$ such that

$$\mathrm{E}\,g = 0, \quad \mathrm{E}\,g\Lambda' = 0, \quad \mathrm{E}\,gg' = \mathbb{I}_k \tag{73}$$

Considering the nonzero difference

$$\delta = \sqrt{(\lambda_1 - \lambda_k)}\, e_k {e_k}' g \tag{74}$$

and any unit vector $e = \alpha_1 e_1 + \cdots + \alpha_k e_k$, we obtain that

$$\begin{aligned} e'\mathcal{C}(\psi_h + \delta)e &= e'\mathcal{I}^{-1}e + (\lambda_1 - \lambda_k)e' e_k {e_k}' e \\ &= \sum\nolimits_j \lambda_j \alpha_j^2 + (\lambda_1 - \lambda_k)\alpha_k^2 \leq \lambda_1 \sum\nolimits_j \alpha_j^2 = \lambda_1 \end{aligned} \tag{75}$$

which proves the result in case $* = h$.

For $* = \mu$, the previous arguments apply with Δ, $\mathcal{J}$, $L_2(\mu)$ in the place of Λ, $\mathcal{I}$, $L_2(P)$, respectively. Thus some $g \in L_2^k(\mu)$ exists such that

$$\int g(y)P(y)\,\mu(dy) = 0\,, \quad \int g(y)\Delta(y)'\,\mu(dy) = 0\,, \quad \int g(y)g(y)'\,\mu(dy) = \mathbb{I}_k$$

and so that, with the corresponding $\delta \in L_2^k(\mu)$ given by (74),

$$\operatorname{maxev} \int (\varphi_\mu + \delta)(\varphi_\mu + \delta)'\,d\mu = \operatorname{maxev} \mathcal{J}^{-1} \tag{76}$$

It remains to check that $\varphi_\mu + \delta$ via (20) induces a $\psi \neq \psi_\mu$. But if, on the contrary,

$$\int \bigl(\mathbf{I}(x \leq y) - P(y)\bigr)\delta(y)\,\mu(dy) = 0 \qquad \text{a.e. } P(dx) \tag{77}$$

then

$$\int \mathbf{I}(x \leq y)e'\delta(y)\,\mu(dy) = 0 \tag{78}$$

for every unit vector $e \in \mathbb{R}^k$, on a set of x values that is dense in $\mathbb{R}^m$ (full support). Thus the two finite measures $(e'\delta)^+\,d\mu$, $(e'\delta)^-\,d\mu$ coincide on a determining class, hence must be the same, and $e'\delta = 0$ a.e. $d\mu$. Employing for e the canonical basis of $\mathbb{R}^k$ we conclude that $\delta = 0$ a.e. $d\mu$. This, however, contradicts the choice of g. ////

Remark 5.3.9

(a) In view of the oscillation optimality of T_h and T_μ, the $0/\infty$ bias mechanism of the nonparametric optimality theorems seems particularly inadequate in connection with the two Hilbert MD functionals.

(b) One of the referees of Rieder (1985) raised the suspicion that a similar degeneration might underly the oscillation saddle point. This is certainly not true for the usual Hellinger and CvM balls since (in the θ-free notation of the preceeding proof) Hellinger oscillation ω_h is finite (and continuous) on $\Psi_2^D \subset L_2^p(P)$ [Proposition 5.3.3 a], while CvM oscillation ω_μ has been proved finite at least for (partial) influence curves derived from arbitrary CvM influence curves via 4.2(51) [Proposition 5.3.3 c, Example 4.2.15].

But, as that referee conjectured, it is principally possible to tailor the neighborhoods to your favorite estimator so that it becomes bias minimax: Given any partial influence curve $\eta \in \Psi_2^D$, look at the corresponding polar class $\mathcal{G}_\eta$ of one-dimensional tangents, with any radius or bound $b \in [0, \infty)$,

$$\mathcal{G}_\eta = \{ q \in Z_2 \mid |\mathrm{E}\,\eta q| \le b \} \tag{79}$$

Then, for $g = t'\Lambda$ with $t \in \mathbb{R}^k$ such that $|Dt| = b$, and any other $\zeta \in \Psi_2^D$, you get

$$\sup_{q \in \mathcal{G}_\eta} |\mathrm{E}\,\zeta q| \ge |\mathrm{E}\,\zeta g| = |\mathrm{E}\,\eta g| = |Dt| = b \ge \sup_{q \in \mathcal{G}_\eta} |\mathrm{E}\,\eta q| \tag{80}$$

Thus $(\eta, t'\Lambda)$ is an oscillation saddle point. Moreover, one encounters the bias degeneration

$$\sup_{q \in \mathcal{G}_\eta} |\mathrm{E}\,\zeta q| = \infty \tag{81}$$

for $\zeta \in \Psi_2^D$ any different from η: Decompose $\zeta = A\eta + \xi$ in $L_2^p(P)$ orthogonally relative to η such that $\mathrm{E}\,\eta\xi' = 0$ [cf. proof to Proposition 4.3.11]. Note that $\xi \ne 0$ if $\zeta \ne \eta$, so that $e'\xi \ne 0$ for some unit vector $e \in \mathbb{R}^p$. Then, setting $q = re'\xi$, we obtain that $|\mathrm{E}\,\eta q| = 0$ for all $r \in (0, \infty)$, whereas $|\mathrm{E}\,\zeta q| = r\,|\mathrm{E}\,\xi\xi' e| \to \infty$ as $r \uparrow \infty$.

This polar type of neighborhoods, however, are just a formal construct lacking any statistical interpretation. ////

5.4 Robust Asymptotic Tests

As a measure of how easily a given as. linear functional can be tested on a full infinitesimal neighborhood system, the nonparametric as. maximin power bound [Theorem 4.3.8] is more complicated to handle than the nonparametric as. minimax risk, based on the covariance of the most concentrated (normal) limit law in the nonparametric convolution representation. For the normal and χ^2 upper tail probabilities are not only independent of the particular type and size of neighborhoods but, in addition, coincide for all functionals. Therefore, the hidden dependence on the functional must be made more explicit, and additional robustness ideas are needed, so that the nonparametric as. maximin power bound obtains some relevance for size and power of tests on specific neighborhoods.

One-Sided Robust Hypotheses

We first look into the one-sided testing problem with $k = p = 1$. In this case, the nonparametric as. hypotheses 4.3(47) about a functional T on a full infinitesimal neighborhood system $\mathcal{U}(\theta)$ about P_θ are

$$\begin{aligned} \mathrm{H}_n^1(T,\theta,r)\colon\quad & Q_n \in U(\theta, r/\sqrt{n}\,),\ \sqrt{n}\,\bigl(T_n(Q_n)-\theta\bigr) \le b\,\sqrt{\mathcal{C}_\theta(\psi_\theta)} \\ \mathrm{K}_n^1(T,\theta,r)\colon\quad & Q_n \in U(\theta, r/\sqrt{n}\,),\ \sqrt{n}\,\bigl(T_n(Q_n)-\theta\bigr) \ge c\,\sqrt{\mathcal{C}_\theta(\psi_\theta)} \end{aligned} \tag{1}$$

where the numbers $-\infty < b < c < \infty$ are fixed, and $r \in (0,\infty)$ will have to be chosen suitably large, or even will become arbitrarily large. Assuming T as. linear at P_θ with influence curve ψ_θ according to Theorem 4.3.8 a, the corresponding as. maximin power at level $\alpha \in (0,1)$ is

$$1 - \Phi(u_\alpha - (c-b)) \tag{2}$$

From a robustness point of view, the contamination/total variation balls introduced in 5.3(34) appear more natural than the nonparametric hypotheses (1), which are defined and restricted by the values of some functional. Therefore, given some radii $\varepsilon, \delta \in [0,\infty)$, we blow up the measures $P_{\theta+t/\sqrt{n}}$ at sample size n to shrinking but full neighborhoods

$$B_{cv}\bigl(P_{\theta+t/\sqrt{n}}\,; \varepsilon/\sqrt{n}\,, \delta/\sqrt{n}\,\bigr)$$

The minimax testing problem between two such balls has been solved by Huber and Strassen (1973) who, in the general framework of two-fold alternating Choquet capacities, proved the existence of least favorable pairs. For the special contamination/total variation balls, Huber (1965, 1968) intuitively found some least favorable pairs explicitly.

Fixing some numbers $-a \le \kappa_0 < \kappa_1 \le a < \infty$ and employing shrinking contamination/total variation balls of possibly different radii $\varepsilon_j, \delta_j \in [0,\infty)$ under the null hypothesis ($j=0$) and alternative ($j=1$), we thus formulate the following as. testing hypotheses,

$$\begin{aligned} \mathrm{H}_n^1(B_{cv},\theta,a) &= \bigcup \bigl\{ B_{cv}\bigl(P_{\theta+t/\sqrt{n}}\,; \varepsilon_0/\sqrt{n}\,, \delta_0/\sqrt{n}\,\bigr) \bigm| -a \le t \le \kappa_0 \bigr\} \\ \mathrm{K}_n^1(B_{cv},\theta,a) &= \bigcup \bigl\{ B_{cv}\bigl(P_{\theta+t/\sqrt{n}}\,; \varepsilon_1/\sqrt{n}\,, \delta_1/\sqrt{n}\,\bigr) \bigm| \kappa_1 \le t \le a \bigr\} \end{aligned} \tag{3}$$

In this section, the as. testing problem between these robust hypotheses will be solved using the nonparametric framework of functionals. The idea is to include the robust hypotheses into the nonparametric hypotheses (1),

$$\mathrm{H}_n^1(B_{cv},\theta,a) \subset \mathrm{H}_n^1(T,\theta,r)\,, \qquad \mathrm{K}_n^1(B_{cv},\theta,a) \subset \mathrm{K}_n^1(T,\theta,r) \tag{4}$$

eventually, so as to guarantee as. level α on $\mathrm{H}_n^1(B_{cv},\theta,a)$ and as. minimum power (2) on $\mathrm{K}_n^1(B_{cv},\theta,a)$, by Theorem 4.3.8 a. Subject to this inclusion, that functional will be determined which maximizes as. maximin power (2).

Side condition (4) will be simplified by the choice of the total variation neighborhood system,

$$\mathcal{U}(\theta) = \mathcal{U}_v(\theta) \tag{5}$$

since, by L_1 differentiability of $\mathcal{P}$ at θ, eventually

$$\mathrm{H}_n^1(B_{cv},\theta,a)\cup\mathrm{K}_n^1(B_{cv},\theta,a)\subset B_v(P_\theta,r/\sqrt{n}\,) \tag{6}$$

as soon as

$$r>\varepsilon_0\vee\varepsilon_1+\delta_0\vee\delta_1+\tfrac{1}{2}a\,\mathrm{E}_\theta|\Lambda_\theta| \tag{7}$$

With choice (5), moreover, Theorem 4.3.8 becomes valid in the fixed radius version [Remark 4.3.9 a]. The equivalent conditions

$$\begin{aligned}\frac{\varepsilon_0+\delta_0+\delta_1}{\kappa_1-\kappa_0}&<\int\Bigl(\Lambda_\theta-\frac{\varepsilon_1-\varepsilon_0}{\kappa_1-\kappa_0}\Bigr)^+dP_\theta\\ \frac{\varepsilon_1+\delta_0+\delta_1}{\kappa_1-\kappa_0}&<\int\Bigl(\frac{\varepsilon_1-\varepsilon_0}{\kappa_1-\kappa_0}-\Lambda_\theta\Bigr)^+dP_\theta\end{aligned} \tag{8}$$

which require the radii ε_j,δ_j to be sufficiently small in relation to κ_j and Λ_θ, ensure that the robust hypotheses become eventually disjoint,

$$\mathrm{H}_n^1(B_{cv},\theta,a)\cap\mathrm{K}_n^1(B_{cv},\theta,a)=\emptyset \tag{9}$$

With $<$ alleviated to $\le$, condition (8) is also necessary for eventual disjointness; which is a consequence of Rieder (1977, Lemma 4.3), using once more L_1 differentiability of $\mathcal{P}$ at θ.

The functions $t\mapsto\mathrm{E}_\theta(t-\Lambda_\theta)^+$, $\mathrm{E}_\theta(\Lambda_\theta-t)^+$ are continuous [dominated convergence] due to $\Lambda_\theta\in L_1(P_\theta)$, and increasing, respectively decreasing, the monotony being strict where the values are positive. Therefore, the equations

$$\begin{aligned}\int(\lambda_\theta'-\Lambda_\theta)^+\,dP_\theta&=\frac{\varepsilon_1+\delta_0+\delta_1}{\kappa_1-\kappa_0}\\ \int(\Lambda_\theta-\lambda_\theta'')^+\,dP_\theta&=\frac{\varepsilon_0+\delta_0+\delta_1}{\kappa_1-\kappa_0}\end{aligned} \tag{10}$$

define unique clipping points $\lambda_\theta'\ge\inf_{P_\theta}\Lambda_\theta$ and $\lambda_\theta''\le\sup_{P_\theta}\Lambda_\theta$. Under condition (8), and the additional assumption

$$\varepsilon_0\wedge\varepsilon_1+(\delta_0+\delta_1)>0 \tag{11}$$

we obtain

$$\inf\nolimits_{P_\theta}\Lambda_\theta<\lambda_\theta'<\frac{\varepsilon_1-\varepsilon_0}{\kappa_1-\kappa_0}<\lambda_\theta''<\sup\nolimits_{P_\theta}\Lambda_\theta \tag{12}$$

Thus the influence curve $\psi_\theta^\star$ given by (14) below will be bounded.

Theorem 5.4.1 *Assume* (5), (8), *and* (11).

(a) *Then the as. power problem concerning functionals that are as. linear at P_θ with influence curves $\psi_\theta\in\Psi_2(\theta)$,*

$$\Phi(u_\alpha-(c-b))=\min!\qquad\psi_\theta\in\Psi_2(\theta);\ b,c\in\mathbb{R}\ \textit{subject to (4)} \tag{13}$$

can only have the solution

$$\psi_\theta^\star=\bigl(\mathrm{E}_\theta\,\xi_\theta^\star\Lambda_\theta\bigr)^{-1}\xi_\theta^\star,\qquad\xi_\theta^\star=\lambda_\theta'\vee\Lambda_\theta\wedge\lambda_\theta''-\frac{\varepsilon_1-\varepsilon_0}{\kappa_1-\kappa_0} \tag{14}$$

(b) *If a functional $T^\star = (T^\star_n)$, in addition to being as. linear at P_θ with influence curve $\psi^\star_\theta$ given by (14), has the as. expansion*

$$T^\star_n(Q_n) = \theta + \int \psi^\star_\theta \, dQ_n + \mathrm{o}\Big(\frac{1}{\sqrt{n}}\Big) \tag{15}$$

for all $r \in (0,\infty)$ and all sequences $Q_n \in B_v(P_\theta, r/\sqrt{n})$, then $T^\star$ solves problem (13), *and the as. test $\phi^\star = (\phi^\star_n)$,*

$$\phi^\star_n = \mathbf{I}\Big(\frac{1}{\sqrt{n}}\sum_{i=1}^n \psi^\star_\theta(x_i) > (u_\alpha + b)\sqrt{\mathcal{C}_\theta(\psi^\star_\theta)}\,\Big) \tag{16}$$

is maximin at as. level α for the nonparametric hypotheses (1) *about $T^\star$, achieving as. minimum power*

$$1 - \Phi\big(u_\alpha - (\kappa_1 - \kappa_0)\sqrt{\mathcal{C}_\theta(\xi^\star_\theta)}\,\big) \tag{17}$$

(c) *The as. test $\phi^\star$ is also maximin for the robust subhypotheses* (3), *at the same as. level α, and achieving the same as. minimum power* (17).

PROOF In this proof, we actually relax the inequalities on T in the nonparametric as. hypotheses (1) to hold only in the limit as $n \to \infty$, and also work with the sequence models (19) below, which are somewhat larger than the robust as. hypotheses (3), at least in case $\delta = 0$.

(a) In view of (5)–(7), the inclusions (4) are fulfilled for r suitably large iff eventually

$$\begin{aligned} \sqrt{n}\,\big(T_n(Q) - \theta\big) &\le b\,\sqrt{\mathcal{C}_\theta(\psi_\theta)}\,, \qquad Q \in \mathrm{H}^1_n(B_{cv}, \theta, a) \\ \sqrt{n}\,\big(T_n(Q) - \theta\big) &\ge c\,\sqrt{\mathcal{C}_\theta(\psi_\theta)}\,, \qquad Q \in \mathrm{K}^1_n(B_{cv}, \theta, a) \end{aligned} \tag{18}$$

Using the expansion 5.3(4) of as. linear functionals, we evaluate the upper and lower oscillation in the submodel of sequences

$$\begin{gathered} dQ_n = \Big(1 + \frac{1}{\sqrt{n}}(\varepsilon q_c + \delta q_v + t\zeta)\Big)\, dP_\theta\,, \qquad |t| \le a \\ q_c \in \mathcal{G}_c(\theta),\ q_v \in \mathcal{G}_v(\theta),\ \zeta \in Z_\infty(\theta), \quad v(\zeta) = \mathrm{E}_\theta\,|\zeta - \Lambda_\theta| \longrightarrow 0 \end{gathered} \tag{19}$$

Indeed, if $\zeta \in Z_\infty(\theta)$, and $q_c \in \mathcal{G}_c(\theta)$, $q_v \in \mathcal{G}_v(\theta)$ [the tangent classes introduced in 5.3(1)], these measures are on $\mathcal{A}$ uniformly bounded by

$$Q_n \ge \Big(1 - \frac{\varepsilon}{\sqrt{n}}\Big) P_{\theta + t/\sqrt{n}} - \frac{\delta + a v(\zeta)}{\sqrt{n}} - \mathrm{o}\Big(\frac{1}{\sqrt{n}}\Big) \tag{20}$$

Hence, for every $\delta' > \delta$ there exists an $n_0 \in \mathbb{N}$ such that for all $\zeta \in Z_\infty(\theta)$ satisfying $v(\zeta) < (\delta' - \delta)/a$ and for all $q_c \in \mathcal{G}_c(\theta)$, $q_v \in \mathcal{G}_v(\theta)$,

$$Q_n \in B_{cv}\big(P_{\theta + t/\sqrt{n}}\,; \varepsilon/\sqrt{n}\,, \delta'/\sqrt{n}\,\big)\,, \qquad |t| \le a,\ n > n_0 \tag{21}$$

In this sense, the sequences (19) are part of the robust as. hypotheses (3), hence have to be included into the nonparametric as. hypotheses (1).

T being as. linear at P_θ with influence curve ψ_θ, we have

$$\sqrt{n}\,\bigl(T_n(Q_n)-\theta\bigr)=\mathrm{E}_\theta\,\psi_\theta\bigl(\varepsilon q_c+\delta q_v+t\zeta\bigr)+\mathrm{o}(n^0)$$

Now let q_c vary over $\mathcal{G}_c(\theta)$, q_v over $\mathcal{G}_v(\theta)$, and then let $\zeta\in Z_\infty(\theta)$ approach Λ_θ in $L_1(P_\theta)$. By Proposition 5.3.3 and its proof, we conclude from (18) that

$$\begin{aligned}\kappa_0+\varepsilon_0\sup\nolimits_{P_\theta}\psi_\theta+\delta_0\bigl(\sup\nolimits_{P_\theta}\psi_\theta-\inf\nolimits_{P_\theta}\psi_\theta\bigr)&\le b\sqrt{\mathcal{C}_\theta(\psi_\theta)}\\ \kappa_1+\varepsilon_1\inf\nolimits_{P_\theta}\psi_\theta-\delta_1\bigl(\sup\nolimits_{P_\theta}\psi_\theta-\inf\nolimits_{P_\theta}\psi_\theta\bigr)&\ge c\sqrt{\mathcal{C}_\theta(\psi_\theta)}\end{aligned}\tag{22}$$

To maximize power (2) given T, the most favorable definition of b and c subject to (22) is by

$$\begin{aligned}b(\psi_\theta)\sqrt{\mathcal{C}_\theta(\psi_\theta)}&=\kappa_0+\varepsilon_0\sup\nolimits_{P_\theta}\psi_\theta+\delta_0\bigl(\sup\nolimits_{P_\theta}\psi_\theta-\inf\nolimits_{P_\theta}\psi_\theta\bigr)\\ c(\psi_\theta)\sqrt{\mathcal{C}_\theta(\psi_\theta)}&=\kappa_1+\varepsilon_1\inf\nolimits_{P_\theta}\psi_\theta-\delta_1\bigl(\sup\nolimits_{P_\theta}\psi_\theta-\inf\nolimits_{P_\theta}\psi_\theta\bigr)\end{aligned}\tag{23}$$

Thus we arrive at the problem to maximize, among all $\psi_\theta\in\Psi_2(\theta)$,

$$\frac{(\kappa_1-\kappa_0)+(\varepsilon_1+\delta_0+\delta_1)\inf_{P_\theta}\psi_\theta-(\varepsilon_0+\delta_0+\delta_1)\sup_{P_\theta}\psi_\theta}{\sqrt{\mathcal{C}_\theta(\psi_\theta)}}\tag{24}$$

Eliminating the covariance condition, we obtain the following variant:

$$\begin{aligned}&\frac{(\kappa_1-\kappa_0)\,\mathrm{E}_\theta\,\xi_\theta\Lambda_\theta+(\varepsilon_1+\delta_0+\delta_1)\inf_{P_\theta}\xi_\theta-(\varepsilon_0+\delta_0+\delta_1)\sup_{P_\theta}\xi_\theta}{\sqrt{\mathcal{C}_\theta(\xi_\theta)}}\\ &\qquad=\max!\qquad\text{subject to: }\xi_\theta\in Z_2(\theta),\ \xi_\theta\ne 0\end{aligned}\tag{25}$$

Note that $\xi_\theta^\star$ given by (14) satisfies the side conditions of (25) since

$$\begin{aligned}\mathrm{E}_\theta\,\xi_\theta^\star&=\mathrm{E}_\theta\bigl(\lambda_\theta'\vee\Lambda_\theta\wedge\lambda_\theta''-\Lambda_\theta\bigr)-\frac{\varepsilon_1-\varepsilon_0}{\kappa_1-\kappa_0}\\ &=\mathrm{E}_\theta\,(\lambda_\theta'-\Lambda_\theta)^+-\mathrm{E}_\theta\,(\Lambda_\theta-\lambda_\theta'')^+-\frac{\varepsilon_1-\varepsilon_0}{\kappa_1-\kappa_0}\overset{(10)}{=}0\end{aligned}$$

Moreover, the following bound holds for the functions ξ_θ considered in (25),

$$\begin{aligned}\mathrm{E}_\theta\,\xi_\theta\Lambda_\theta&=\mathrm{E}_\theta\,\xi_\theta\xi_\theta^\star+\mathrm{E}_\theta\,\xi_\theta\bigl(\Lambda_\theta-\lambda_\theta'\vee\Lambda_\theta\wedge\lambda_\theta''\bigr)\\ &\le\sqrt{\mathcal{C}_\theta(\xi_\theta)\,\mathcal{C}_\theta(\xi_\theta^\star)}-\frac{\varepsilon_1+\delta_0+\delta_1}{\kappa_1-\kappa_0}\inf\nolimits_{P_\theta}\xi_\theta+\frac{\varepsilon_0+\delta_0+\delta_1}{\kappa_1-\kappa_0}\sup\nolimits_{P_\theta}\xi_\theta\end{aligned}\tag{26}$$

The bound is achieved iff ξ_θ is a positive multiple of $\xi_\theta^\star$. Therefore, $\xi_\theta^\star$ maximizes the objective function in (25), whose maximum equals

$$\sigma_\star=(\kappa_1-\kappa_0)\sqrt{\mathcal{C}_\theta(\xi_\theta^\star)}=c(\psi_\theta^\star)-b(\psi_\theta^\star)\tag{27}$$

and this is also the maximum of the objective function (24) that is achieved by the influence curve $\psi_\theta^\star$ uniquely.

(b) First, we have to verify the inclusion (4) of full capacity balls into the nonparamatric as. hypotheses (1) about $T^\star$, with $b = b(\psi_\theta^\star)$ and $c = c(\psi_\theta^\star)$ determined by (23). Making use of (5)–(7) and of the expansion (15) of $T^\star$, we show that

$$\limsup_{n\to\infty} \sqrt{n}\,\bigl(T_n^\star(Q_n') - \theta\bigr) \overset{(15)}{=} \limsup_{n\to\infty} \sqrt{n} \int \psi_\theta^\star\, dQ_n' \le b(\psi_\theta^\star)\sqrt{\mathcal{C}_\theta(\psi_\theta^\star)} \tag{28}$$

for all sequences

$$Q_n' \in B_{cv}\bigl(P_{\theta+t_n'/\sqrt{n}}\,; \varepsilon_0/\sqrt{n}\,, \delta_0/\sqrt{n}\,\bigr)\,, \qquad t_n' \longrightarrow t' \in [-a, \kappa_0]$$

and

$$\liminf_{n\to\infty} \sqrt{n}\,\bigl(T_n^\star(Q_n'') - \theta\bigr) \overset{(15)}{=} \liminf_{n\to\infty} \sqrt{n} \int \psi_\theta^\star\, dQ_n'' \ge c(\psi_\theta^\star)\sqrt{\mathcal{C}_\theta(\psi_\theta^\star)} \tag{29}$$

for all sequences

$$Q_n'' \in B_{cv}\bigl(P_{\theta+t_n''/\sqrt{n}}\,; \varepsilon_1/\sqrt{n}\,, \delta_1/\sqrt{n}\,\bigr)\,, \qquad t_n'' \longrightarrow t'' \in [\kappa_1, a]$$

For the proof of this, we invoke Lemma 5.3.4, whose part (a) applies with $P = P_\theta$ replaced by $P_n = P_{\theta+t_n/\sqrt{n}}$. Part (b) of Lemma 5.3.4 requires a suitable extension to moving center measure, however. Thus, as for (28), we must evaluate an upper expectation of the following type:

$$\sqrt{n}\,(1-\varepsilon_n) \int \tilde{z}_n'' \vee \psi_\theta^\star\, dP_n - \delta\, \tilde{z}_n'' + (\varepsilon+\delta) \sup \psi_\theta^\star \tag{30}$$

where

$$\inf \psi_\theta^\star \le \tilde{z}_n'' = \bigl(\psi_\theta^\star(P_n)\bigr)^{-1}\Bigl(\frac{\delta_n}{1-\varepsilon_n}\Bigr) \tag{31}$$

From $d_v(P_n, P_\theta) \to 0$ and $\varepsilon_n, \delta_n \to 0$, we can only conclude that

$$\limsup_n \tilde{z}_n'' \le \inf{}_{P_\theta} \psi_\theta^\star$$

But, since the P_θ essential coincides with the pointwise infimum of $\psi_\theta^\star$, actually

$$\lim_{n\to\infty} \tilde{z}_n'' = \inf \psi_\theta^\star = \inf{}_{P_\theta} \psi_\theta^\star \tag{32}$$

Then, similarly to 5.3(51),

$$\begin{aligned} 0 \le \sqrt{n} \int (\tilde{z}_n'' \vee \psi_\theta^\star - \psi_\theta^\star)\, dP_n &= \sqrt{n} \int (\tilde{z}_n'' - \psi_\theta^\star)^+\, dP_n \\ &\le \sqrt{n}\, \frac{\delta_n}{1-\varepsilon_n}\, (\tilde{z}_n'' - \inf \psi_\theta^\star) \longrightarrow 0 \end{aligned} \tag{33}$$

From L_1 differentiability of $\mathcal{P}$ at θ and boundedness of $\psi_\theta^\star$ again, we get

$$\lim_{n\to\infty} \sqrt{n} \int \psi_\theta^\star\, dP_{\theta+t_n/\sqrt{n}} = t\,, \qquad t_n \longrightarrow t \in \mathbb{R} \tag{34}$$

These arguments yield the upper bound

$$\kappa_0 - \delta_0 \inf \psi_\theta^\star + (\varepsilon_0 + \delta_0) \sup \psi_\theta^\star \tag{35}$$

for the LHS in (28). But as the P_θ essential extrema of $\psi_\theta^\star$ coincide with its respective pointwise extrema, (35) is the same as the upper LHS of (22), which by definition (23) equals the RHS of (28), in the case of $\psi_\theta^\star$.

Secondly, we show that $\phi^\star$ given by (16) is as. maximin for the hypotheses (1) about $T^\star$. Fix $r \in (0,\infty)$ and any sequence $Q_n \in B_v(P_\theta, r/\sqrt{n})$. Then $\sup|\psi_\theta^\star| < \infty$ and $d_v(Q_n, P_\theta) \to 0$ imply that

$$\int \psi_\theta^\star \, dQ_n \longrightarrow \mathrm{E}_\theta \, \psi_\theta^\star = 0\,, \qquad \int (\psi_\theta^\star)^2 \, dQ_n \longrightarrow \mathrm{E}_\theta \, (\psi_\theta^\star)^2 \in (0,\infty)$$

Therefore Proposition 6.2.1 applies with $\psi_n = \psi_\theta^\star$ and $Q_{ni} = Q_n$. Thus the Lindeberg–Feller theorem and the as. expansion (15) of $T^\star$ yield

$$\begin{aligned} \Bigl(\frac{1}{\sqrt{n}} \sum\nolimits_i \psi_\theta^\star(x_i)\Bigr)(Q_n^n) &= \mathcal{N}\bigl(v_n, \mathcal{C}_\theta(\psi_\theta^\star)\bigr) + \mathrm{o}_\kappa(n^0) \\ v_n = \sqrt{n} \int \psi_\theta^\star \, dQ_n &\overset{(15)}{=} \sqrt{n}\,\bigl(T_n^\star(Q_n) - \theta\bigr) + \mathrm{o}(n^0) \end{aligned} \tag{36}$$

Similarly to the proof of Theorem 4.3.8 c, this shows that the as. test $\phi^\star$ is maximin for the hypotheses (1) about $T^\star$. Its as. minimum power, in view of (27) and (2), is given by (17).

(c) The smaller robust as. hypotheses are in fact not easier to test than the larger nonparametric as. hypotheses about the functional $T^\star$. According to Remark 4.3.9 c, the following simple as. hypotheses are least favorable for the nonparametric as. testing problem (1) about $T^\star$,

$$\begin{aligned} dR_{0,n} &= \Bigl(1 + \frac{b(\psi_\theta^\star)}{\sqrt{n\,\mathcal{C}_\theta(\psi_\theta^\star)}}\,\psi_\theta^\star\Bigr)\, dP_\theta \\ dR_{1,n} &= \Bigl(1 + \frac{c(\psi_\theta^\star)}{\sqrt{n\,\mathcal{C}_\theta(\psi_\theta^\star)}}\,\psi_\theta^\star\Bigr)\, dP_\theta \end{aligned} \tag{37}$$

At least if

$$\varepsilon_0 = \varepsilon_1 = 0\,, \quad \delta_0 = \delta_1 = \delta\,, \quad \kappa_1 = -\kappa_0 = \kappa \tag{38}$$

the sequence of pairs $(R_{0,n}, R_{1,n})$ of probabilities already lie in the smaller robust as. hypotheses (3). Indeed, in the special case (38), we have

$$\mathrm{E}_\theta \, \xi_\theta^\star \Lambda_\theta = \mathrm{E}_\theta \, \xi_\theta^\star \xi_\theta^\star + \mathrm{E}_\theta \, \xi_\theta^\star (\Lambda_\theta - \xi_\theta^\star) \underset{(38)}{\overset{(10)}{=}} \mathcal{C}_\theta(\xi_\theta^\star) + \frac{\delta}{\kappa}(\lambda_\theta'' - \lambda_\theta')$$

hence

$$\begin{aligned} \frac{c(\psi_\theta^\star)}{\sqrt{\mathcal{C}_\theta(\psi_\theta^\star)}}\,\psi_\theta^\star &\overset{(23)}{=} \frac{\kappa - \delta\bigl(\sup_{P_\theta} \psi_\theta^\star - \inf_{P_\theta} \psi_\theta^\star\bigr)}{\mathcal{C}_\theta(\psi_\theta^\star)}\,\psi_\theta^\star \\ &= \frac{\kappa \, \mathrm{E}_\theta \, \xi_\theta^\star \Lambda_\theta - \delta(\lambda_\theta'' - \lambda_\theta')}{\mathcal{C}_\theta(\xi_\theta^\star)}\,\xi_\theta^\star = \kappa\, \xi_\theta^\star \end{aligned} \tag{39}$$

By L_1 differentiability of $\mathcal{P}$ at θ, and a similar argument for $R_{0,n}$, thus

$$2\delta \underset{(10)}{=} \kappa \,\mathrm{E}_\theta\, |\xi^\star_\theta - \Lambda_\theta| \underset{(39)}{=} 2\sqrt{n}\, d_v\big(R_{1,n}, P_{\theta+\kappa/\sqrt{n}}\big) + \mathrm{o}(n^0) \\ = 2\sqrt{n}\, d_v\big(R_{0,n}, P_{\theta-\kappa/\sqrt{n}}\big) + \mathrm{o}(n^0) \tag{40}$$

Now we apply Proposition 2.2.12 with the identifications

$$Z_n = \frac{1}{\sqrt{n}\,\mathcal{C}_\theta(\psi^\star_\theta)} \sum\nolimits_i \psi^\star_\theta(x_i)\,, \qquad C = 1, \quad s = b(\psi^\star_\theta), \quad t = c(\psi^\star_\theta)$$

and check that, in the sense of Corollary 3.4.2 b, with $\sigma_\star$ defined by (27), the test $\phi^\star$ given by (16) is as. optimum for $R_{0,n}$ vs. $R_{1,n}$ at as. level α.

It is a pity that in general the pairs $(R_{0,n}, R_{1,n})$ are not in the robust as. hypotheses (3). Instead, we have to invoke Huber's (1965, 1968) least favorable pairs $(Q_{0,n}, Q_{1,n})$ for the boundary balls

$$B_{cv}\big(P_{\theta+\kappa_0/\sqrt{n}}\,; \varepsilon_0/\sqrt{n}\,, \delta_0/\sqrt{n}\big) \quad \text{vs.} \quad B_{cv}\big(P_{\theta+\kappa_1/\sqrt{n}}\,; \varepsilon_1/\sqrt{n}\,, \delta_1/\sqrt{n}\big)$$

Using the shorthand notations

$$\varepsilon_{j,n} = \frac{\varepsilon_j}{\sqrt{n}}\,, \quad \delta_{j,n} = \frac{\delta_j}{\sqrt{n}}\,, \quad \nu_{j,n} = \frac{\varepsilon_{j,n} + \delta_{j,n}}{1 - \varepsilon_{j,n}}\,, \quad \omega_{j,n} = \frac{\delta_{j,n}}{1 - \varepsilon_{j,n}} \\ P_{0,n} = P_{\theta+\kappa_0/\sqrt{n}}\,, \quad P_{1,n} = P_{\theta+\kappa_1/\sqrt{n}}\,, \qquad \Delta_n = \frac{dP_{1,n}}{dP_{0,n}}$$

these pair of measures are

$$\frac{dQ_{0,n}}{1 - \varepsilon_{0,n}} = \begin{cases} \dfrac{\nu_{1,n}\, dP_{0,n} + \omega_{0,n}\, dP_{1,n}}{\nu_{1,n} + \omega_{0,n}\vartheta'_n} & \text{if } \Delta_n < \vartheta'_n \\ dP_{0,n} & \text{if } \vartheta'_n \le \Delta_n \le \vartheta''_n \\ \dfrac{\omega_{1,n}\, dP_{0,n} + \nu_{0,n}\, dP_{1,n}}{\omega_{1,n} + \nu_{0,n}\vartheta''_n} & \text{if } \vartheta''_n < \Delta_n \end{cases} \\ \frac{dQ_{1,n}}{1 - \varepsilon_{1,n}} = \Big(\vartheta'_n \vee \Delta_n \wedge \vartheta''_n\Big) \frac{dQ_{0,n}}{1 - \varepsilon_{0,n}} \tag{41}$$

with clipping constants $\vartheta'_n, \vartheta''_n \in (0, \infty)$ that are the unique solutions of

$$\vartheta'_n P_{0,n}(\Delta_n < \vartheta'_n) - P_{1,n}(\Delta_n < \vartheta'_n) = \nu_{1,n} + \omega_{0,n}\vartheta'_n \\ P_{1,n}(\Delta_n > \vartheta''_n) - \vartheta''_n P_{0,n}(\Delta_n > \vartheta''_n) = \nu_{0,n}\vartheta''_n + \omega_{1,n} \tag{42}$$

All Huber–Strassen least favorable pairs have been characterized, and another one given explicitely, in Rieder (1977; Theorem 5.2). The distribution of their likelihood is unique [Rieder (1977; Proposition 2.2, formula (2.7))]; even more, their likelihood itself is unique [Rieder (1977; Proposition 2.3,

Theorem 5.1)]. Along the proofs of Huber-Carol (1970), Rieder (1978; Theorem 4.1, Lemmas 4.2, 4.3, and 1981; Lemma 3.1), we can show that the log likelihoods of the least favorable products, with $\sigma_\star$ given by (27), satisfy

$$\Bigl(\log \frac{dQ_{1,n}^n}{dQ_{0,n}^n}\Bigr)(Q_{0,n}^n) \xrightarrow{\mathrm{w}} \mathcal{N}(-\sigma_\star^2/2, \sigma_\star^2) \tag{43}$$

The proof only uses the contiguity $(P_{0,n}^n), (P_{1,n}^n) \ll (P_\theta^n)$ and L_1 differentiability of $\mathcal{P}$ at θ. It follows that our test $\phi^\star$, in view of its as. level α and as. minimum power (17) for the larger nonparametric hypotheses (1), achieves equalities in Corollary 3.4.2 a, when testing $Q_{0,n}$ vs. $Q_{1,n}$. ////

Remark 5.4.2 Using the work by Wang (1981), it should be possible to extend the preceding result to the case $p = 1 \le k$; that is, to the testing of one real functional component of interest in the presence of a finite number of nuisance components. ////

Multisided Robust Hypotheses

We consider the full multiparameter case $k = p \ge 1$ and start out with the parametric multisided hypotheses 3.4(141) about the parameter $t \in \mathbb{R}^k$ of the product measures $P_{\theta+t/\sqrt{n}}^n$,

$$\mathrm{H}_{\mathrm{m}}^{\mathrm{m}}(\theta, a)\colon \quad t'\mathcal{I}_\theta t \le b_0^2 \qquad \mathrm{K}_{\mathrm{m}}^{\mathrm{m}}(\theta, a)\colon \quad c_0^2 \le t'\mathcal{I}_\theta t \le a^2 \tag{44}$$

where $0 \le b_0 < c_0 \le a < \infty$ are fixed. If T is a functional, as. linear at P_θ with influence curve $\psi_\theta \in \Psi_2(\theta)$, consider the nonparametric multisided hypotheses 4.3(46) about T on a full neighborhood system $\mathcal{U}(\theta)$ of P_θ,

$$\begin{aligned} \mathrm{H}_n^{\mathrm{m}}(T, \theta, r)\colon \quad Q \in U(\theta, r/\sqrt{n}),\ n\,\bigl|T_n(Q) - \theta\bigr|_{\mathcal{C}_\theta(\psi_\theta)}^2 \le b_T^2 \\ \mathrm{K}_n^{\mathrm{m}}(T, \theta, r)\colon \quad Q \in U(\theta, r/\sqrt{n}),\ n\,\bigl|T_n(Q) - \theta\bigr|_{\mathcal{C}_\theta(\psi_\theta)}^2 \ge c_T^2 \end{aligned} \tag{45}$$

of radius $r \in (0, \infty)$, and with critical numbers $0 < b_T < c_T < \infty$ now to be specified: In order that eventually

$$\mathrm{H}_{\mathrm{m}}^{\mathrm{m}}(\theta, a) \subset \mathrm{H}_n^{\mathrm{m}}(T, \theta, r), \qquad \mathrm{K}_{\mathrm{m}}^{\mathrm{m}}(\theta, a) \subset \mathrm{K}_n^{\mathrm{m}}(T, \theta, r) \tag{46}$$

the radius r must be chosen sufficiently large, and then b_T, c_T so that

$$b_T^2 \ge \sup\Bigl\{ |t|_{\mathcal{C}_\theta(\psi_\theta)}^2 \Bigm| |t|_{\mathcal{I}_\theta^{-1}}^2 \le b_0^2 \Bigr\} = \frac{b_0^2}{\operatorname{minev} \mathcal{I}_\theta^{1/2} \mathcal{C}_\theta(\psi_\theta) \mathcal{I}_\theta^{1/2}} = b_1^2(\psi_\theta) \tag{47}$$

$$c_T^2 \le \inf\Bigl\{ |t|_{\mathcal{C}_\theta(\psi_\theta)}^2 \Bigm| |t|_{\mathcal{I}_\theta^{-1}}^2 \ge c_0^2 \Bigr\} = \frac{c_0^2}{\operatorname{maxev} \mathcal{I}_\theta^{1/2} \mathcal{C}_\theta(\psi_\theta) \mathcal{I}_\theta^{1/2}} = c_1^2(\psi_\theta) \tag{48}$$

Thus, we formulate the hypotheses

$$\mathrm{H}_T^{\mathrm{m}}(\theta, a)\colon\ |t|_{\mathcal{C}_\theta(\psi_\theta)}^2 \le b_1^2(\psi_\theta) \qquad \mathrm{K}_T^{\mathrm{m}}(\theta, a)\colon\ c_1^2(\psi_\theta) \le |t|_{\mathcal{C}_\theta(\psi_\theta)}^2 \le a^2 \tag{49}$$

which enlarge the original parametric hypotheses (44) to achieve (46). [This enlargement is without effect, and not needed, in case $k = 1$.]

Fix some $\varepsilon \in (0, \infty)$. At sample size n, blow up each parametric element $P_{\theta+t/\sqrt{n}}$ of $\mathrm{H}_T^{\mathrm{m}}(\theta, a)$ and $\mathrm{K}_T^{\mathrm{m}}(\theta, a)$ respectively, to a contamination ball of radius $\varepsilon/\sqrt{n}$,

$$\begin{aligned} \mathrm{H}_{n,T}^{\mathrm{m}}(B_c, \theta, a) &= \bigcup \{ B_c(P_{\theta+t/\sqrt{n}}, \varepsilon/\sqrt{n}) \mid t \in \mathrm{H}_T^{\mathrm{m}}(\theta, a) \} \\ \mathrm{K}_{n,T}^{\mathrm{m}}(B_c, \theta, a) &= \bigcup \{ B_c(P_{\theta+t/\sqrt{n}}, \varepsilon/\sqrt{n}) \mid t \in \mathrm{K}_T^{\mathrm{m}}(\theta, a) \} \end{aligned} \tag{50}$$

For the eventual inclusions

$$\mathrm{H}_{n,T}^{\mathrm{m}}(B_c, \theta, a) \subset \mathrm{H}_n^{\mathrm{m}}(T, \theta, r), \qquad \mathrm{K}_{n,T}^{\mathrm{m}}(B_c, \theta, a) \subset \mathrm{K}_n^{\mathrm{m}}(T, \theta, r) \tag{51}$$

we must then choose $b_T \geq b(\psi_\theta)$ and $c_T \leq c(\psi_\theta)$, where these lower and upper bounds are given by

$$\begin{aligned} b_T^2 &\geq \sup_{Q \in \mathcal{M}_1(\mathcal{A})} \left\{ \left| t + \varepsilon \int \psi_\theta \, dQ \right|_{\mathcal{C}_\theta(\psi_\theta)}^2 \,\middle|\, |t|_{\mathcal{C}_\theta(\psi_\theta)}^2 \leq b_1^2(\psi_\theta) \right\} \\ &= \left(\frac{b_0}{\sqrt{\operatorname{minev} \mathcal{I}_\theta^{1/2} \mathcal{C}_\theta(\psi_\theta) \mathcal{I}_\theta^{1/2}}} + \varepsilon \sup |\psi_\theta|_{\mathcal{C}_\theta(\psi_\theta)} \right)^2 = b^2(\psi_\theta) \end{aligned} \tag{52}$$

respectively,

$$\begin{aligned} c_T^2 &\leq \inf_{Q \in \mathcal{M}_1(\mathcal{A})} \left\{ \left| t + \varepsilon \int \psi_\theta \, dQ \right|_{\mathcal{C}_\theta(\psi_\theta)}^2 \,\middle|\, |t|_{\mathcal{C}_\theta(\psi_\theta)}^2 \geq c_1^2(\psi_\theta) \right\} \\ &= \left(\frac{c_0}{\sqrt{\operatorname{maxev} \mathcal{I}_\theta^{1/2} \mathcal{C}_\theta(\psi_\theta) \mathcal{I}_\theta^{1/2}}} - \varepsilon \sup |\psi_\theta|_{\mathcal{C}_\theta(\psi_\theta)} \right)^2 = c^2(\psi_\theta) \end{aligned} \tag{53}$$

At this instance, we have for T assumed an expansion of the kind (15); employing simple perturbations and the class $\mathcal{G}_c(\theta)$ as in the proof to Theorem 5.4.1 a, would result in the term $\sup_{P_\theta} |\psi_\theta|_{\mathcal{C}_\theta(\psi_\theta)}$ instead. Moreover, with

$$s = \mathcal{C}_\theta(\psi_\theta)^{-1/2} t, \qquad \varphi_\theta = \mathcal{C}_\theta(\psi_\theta)^{-1/2} \psi_\theta$$

and by the Cauchy–Schwarz inequality, the following bound holds,

$$\begin{aligned} \left| t + \varepsilon \int \psi_\theta \, dQ \right|_{\mathcal{C}_\theta(\psi_\theta)}^2 &= |s|^2 + 2\varepsilon \, s' \int \varphi_\theta \, dQ + \varepsilon^2 \left| \int \varphi_\theta \, dQ \right|^2 \\ &\leq |s|^2 + 2\varepsilon \, |s| \left| \int \varphi_\theta \, dQ \right| + \varepsilon^2 \left| \int \varphi_\theta \, dQ \right|^2 \\ &= \left(|s| + \varepsilon \left| \int \varphi_\theta \, dQ \right| \right)^2 \end{aligned}$$

The upper bound is achieved for s_Q determined by

$$\left| \int \varphi_\theta \, dQ \right| s_Q = |s| \int \varphi_\theta \, dQ$$

Letting $s \in \mathbb{R}^k$ vary subject to $|s| \le b_1(\psi_\theta)$ and then Q over $\mathcal{M}_1(\mathcal{A})$ proves (52). Switching signs proves (53).

With the most favorable choices $b_T = b(\psi_\theta)$ and $c_T = c(\psi_\theta)$ determined by (52) and (53), respectively, and in view of the corresponding as. maximin power given by Theorem 4.3.8 b, we thus arrive at the problem,

$$\Pr\bigl(\chi^2(k; c^2(\psi_\theta)) > c_\alpha(k; b^2(\psi_\theta))\bigr) = \max! \qquad \psi_\theta \in \Psi_2(\theta) \tag{54}$$

In an intricate fashion it combines the largest summand and, in an opposite way, the smallest summand of the objective function of the information standardized problem with the side condition of the self-standardized problem [cf. pp 214, 217]. The general solution is not known.

To proceed, put $b_0 = 0$. Then a solution $\psi_\theta^\star$ to problem (54), if there exists one, would solve the following problem for $b = \sup_{P_\theta} |\psi_\theta^\star|_{\mathcal{C}_\theta(\psi_\theta^\star)}$,

$$\operatorname{maxev} \mathcal{I}_\theta^{1/2} \mathcal{C}_\theta(\psi_\theta) \mathcal{I}_\theta^{1/2} = \min! \qquad \psi_\theta \in \Psi_2(\theta),\ \sup_{P_\theta} |\psi_\theta|_{\mathcal{C}_\theta(\psi_\theta)} \le b \tag{55}$$

In fact, the set of solutions to the problems (55) with b varying over $(0, \infty]$ is an essentially complete class for problem (54).

Remark 5.4.3 Problem (55) employs the same self-standardized oscillation as problem 5.5(142) discussed in Subsection 5.5.4, whose solution is the influence curve $\hat{\varrho}_\theta$ given by 5.5(140). However, minimum trace does not imply smallest maximum eigenvalue. In addition, the covariance in problem (55) is standardized by $\mathcal{I}_\theta^{-1}$ instead by $\mathcal{C}_\theta(\hat{\varrho}_\theta)$. Not even the admissibility argument for minimum trace solutions [see p 211] carries over to smallest maximum eigenvalue solutions, because the latter may not be unique. ////

Two-Sided Robust Hypotheses

In the one-dimensional $(k = 1)$ two-sided case, problem (55) simplifies as the maximum eigenvalue coincides with the variance and the standardization by the Fisher information can be ignored: Blow up the elements of the parametric hypotheses (44) with $b_0 = 0$ to $\varepsilon/\sqrt{n}$ contamination balls and require the nonparametric hypotheses (45) to include these enlarged hypotheses. Then the as. linear functionals T with influence curves at P_θ of form

$$\psi_\theta^\star = \bigl(A_\theta \Lambda_\theta - a_\theta\bigr) \min\Bigl\{1, \frac{b}{|A_\theta \Lambda_\theta - a_\theta|}\Bigr\} \tag{56}$$

for any $b \in (0, \infty)$ and $a_\theta, A_\theta \in \mathbb{R}$, constitute an essentially complete class for problem (54).

5.5 Minimax Risk and Oscillation

After as. minimax risk and oscillation of functionals have been optimized separately, both criteria are now combined in some feasible way. The basic optimality, speaking estimator terminology, consists in minimum as. variance subject to some upper bound on as. bias.

In solving this type of problems, which date back to Hampel (1968, Lemma 5), some ease and generality are gained by a systematic use of convex optimization (Lagrange multipliers), which is developed in Appendix B. This technique also allows the minimization of a convex combination of both criteria, like as. mean square error (MSE).

Minimization of the as. covariance matrix is w.r.t. trace (which amounts to quadratic loss in the nonparametric as. minimax bound). An influence curve that, subject to some bias bound, would minimize the covariance matrix in the positive definite sense does in general not exist.

The robust influence curves thus obtained for the usual bias terms are not equivariant under reparametrizations. Suitable non-Euclidean norms make the objective function and the bias terms invariant, and thus enforce equivariance of the corresponding solutions.

The optimality is not of a monolithic type that would admit just one solution. The variety of robust solutions, depending on the neighborhoods and loss function, rather constitute a complete class. If the size of balls is unknown, the bias bound, respectively, bias weight in MSE, may be determined such that, for example, the efficiency loss incurred at the ideal model amounts to a certain insurance premium (e.g., 10%).

5.5.1 Minimum Trace Subject to Bias Bound

In view of the nonparametric convolution theorem and as. minimax bound [Theorems 4.3.2 and 4.3.4], assuming the loss function $\ell(z) = |z|^2$, the following type of problems will be studied,

$$\operatorname{tr} \mathcal{C}_\theta(\eta_\theta) = \min! \qquad \psi_\theta \in \Psi_2(\theta), \ \ \omega_{*,\theta;s}(\eta_\theta) \le b \tag{1}$$

where $\eta_\theta = d\tau(\theta)\psi_\theta$, bound $b \in (0,\infty)$ is fixed, and $\omega_{*,\theta;s}$ stands for one of the bias terms introduced in Section 5.3. The general assumptions of Section 5.1 are enforced.

Notation

Recall Definition 4.2.10 b of the set $\Psi_2^D(\theta)$ of partial influence curves at P_θ corresponding to the matrix

$$D = d\tau(\theta) \in \mathbb{R}^{p\times k}, \qquad \operatorname{rk} D = p \le k \tag{2}$$

To lighten notation, the fixed θ is now dropped. Thus we write $\omega_{*;s} = \omega_{*,\theta;s}$ for the bias terms, $\Psi_2 = \Psi_2(\theta)$ for the set of influence curves, $\Psi_2^D = \Psi_2^D(\theta)$ for the set of partial influence curves, at $P = P_\theta$, and $\Lambda = \Lambda_\theta$ for the L_2 derivative, $\mathcal{I} = \mathcal{I}_\theta$ for the Fisher information, of $\mathcal{P}$ at θ, as well as $\mathrm{E} = \mathrm{E}_\theta$ for expectation under $P = P_\theta$. The Hilbert space $L_2^p(P)$

with scalar product $\langle\eta|\zeta\rangle = \mathrm{E}\,\eta'\zeta$, and the Banach space $L^p_\infty(P)$ with P essential sup norm $\sup_P|\eta|$, are subsequently denoted by

$$\mathbb{H} = L^p_2(P)\,, \qquad \mathbb{L} = L^p_\infty(P) \tag{3}$$

Thus, in lightened notation, the problem $O^{\mathrm{tr}}_{*;s}(b)$ formulated by (1) is

$$\mathrm{E}\,|\eta|^2 = \min! \qquad \eta\in\Psi^D_2\,,\ \omega_{*;s}(\eta)\le b \tag{4}$$

Aspects of Problem $O^{\mathrm{tr}}_{*;s}(b)$

We outline some general features of this type of optimization problems.

Minimum Norm: Expectation and scalar products being linear weakly continuous, and the bias terms being convex, weakly l.s.c. by Lemma 5.3.2, problem $O^{\mathrm{tr}}_{*;s}(b)$ is a minimum norm problem over a convex closed subset in Hilbert space, which has a unique solution $\tilde\eta$ provided only there exists some $\eta\in\Psi^D_2$ that satisfies the bias bound $\omega_{*;s}(\eta)\le b$ [Lemma B.1.1].

Restrictions on b: Concerning bound b, we may assume that

$$\omega^{\min}_{*;s} = \inf\{\,\omega_{*;s}(\eta)\mid \eta\in\Psi^D_2\,\}\le b<\omega_{*;s}(\eta_h) \tag{5}$$

where $\eta_h = D\psi_h = D\mathcal{I}^{-1}\Lambda$ is the classical partial scores function introduced in 4.2(41). In case $b<\omega^{\min}_{*;s}$, the bias constraint $\omega_{*;s}(\eta)\le b$, by definition, cannot be met by any $\eta\in\Psi^D_2$. And in case $b\ge\omega_{*,s}(\eta_h)$, always $\tilde\eta=\eta_h$ is the solution. This is a consequence of the Cramèr–Rao bound, which says that for all $\eta\in\Psi^D_2$,

$$0\le \mathrm{E}\,(\eta-\eta_h)(\eta-\eta_h)' = \mathrm{E}\,\eta\eta' - \mathrm{E}\,\eta_h\eta_h{}' \tag{6}$$

Hence, in particular, $\mathrm{E}\,|\eta_h|^2\le\mathrm{E}\,|\eta|^2$, for all $\eta\in\Psi^D_2$.

Convex Optimization: Problem $O^{\mathrm{tr}}_{*;s}(b)$ is convex in the sense of optimization theory. For the objective function f, as well as the functions G and H defining the convex and linear side conditions,

$$f(\eta)=\mathrm{E}\,|\eta|^2\,,\qquad G(\eta)=\omega_{*;s}(\eta)\,,\qquad H\eta=\bigl(\mathrm{E}\,\eta,\mathrm{E}\,\eta\Lambda'\bigr)\in\mathbb{R}^p\times\mathbb{R}^{p\times k} \tag{7}$$

are convex and linear, respectively; sometimes, $G(\eta)=\omega^2_{*;s}(\eta)$ [convex, too] is more convenient, in particular for $s=2$. Thus, the Lagrange multiplier theorems of Section B.2 become available with these identifications and the choices $Z=\mathbb{R}$, $Y=\mathbb{R}^{p+pk}$, and $A=\mathbb{H}$ or $A=\mathbb{L}$ there.

The oscillation terms may be extended to all of $\mathbb{H}$ using the explicit expressions derived under the condition that $\mathrm{E}=0$, which will be taken care of separately by the linear operator H; thus 5.3(7) is given up.

Infinite Values: The Lagrange multiplier theorems require finite-valued functions [see Remark B.2.10 b, however]. Finite values of $\omega_{*;s}$ in the cases $*=c,v$ can be achieved by a restriction onto $\mathbb{L}$, which is a dense subspace of $\mathbb{H}$. Even Ψ_∞^D is dense in Ψ_2^D, which is shown by clipping, centering and standardization as follows: Given any $\eta = D\psi \in \Psi_2^D$ with $\psi \in \Psi_2$, let

$$\psi_m = \psi\,\mathbf{I}(|\psi| \le m), \qquad \psi_m^\natural = \psi_m - \mathrm{E}\,\psi_m \tag{8}$$

Then

$$\psi_m^\flat = \bigl(\mathrm{E}\,\psi_m^\natural \Lambda'\bigr)^{-1} \psi_m^\natural \in \Psi_\infty, \quad \eta_m = D\psi_m^\flat \in \Psi_\infty^D, \quad \eta_m \to \eta \tag{9}$$

as $m \to \infty$. Thus the restriction onto $\mathbb{L}$ will only be a minor handicap to subsequent Lagrange multiplier arguments.

Well-Posedness: For example, the operator H has full range $\mathbb{R}^{p+pk}$ not only when defined on $\mathbb{H}$ but also if restricted onto $\mathbb{L}$: Pick any bounded influence curve $\psi \in \Psi_\infty$. Then, for any $a \in \mathbb{R}^p$ and $A \in \mathbb{R}^{p\times k}$, you achieve

$$\eta = a + A\psi \in \mathbb{L}, \qquad a = \mathrm{E}\,\eta, \quad A = \mathrm{E}\,\eta\Lambda' \tag{10}$$

This way, conditions B.2(18) and B.2(22) on the linear constraints, for problem $O_{*;s}^{\mathrm{tr}}(b)$ to be well-posed in the sense of Definition B.2.9, can be verified. Condition B.2(19) may be checked by letting a and A vary over bounded neighborhoods of 0 in $\mathbb{R}^p$, respectively, of D in $\mathbb{R}^{p\times k}$, but can actually be ignored in this section since $\dim Y < \infty$ [Remarks B.2.4 and B.2.7 a]. Well-posedness condition B.2(23) on the convex constraints will be ensured by choice of the oscillation bound b.

One At a Time Optimality: Similarly to the product spaces $\mathbb{H}$ and $\mathbb{L}$, we have

$$\Psi_\alpha^D = \Psi_\alpha^{D_1} \times \cdots \times \Psi_\alpha^{D_p}, \qquad \alpha = 2, \infty \tag{11}$$

where $D_j \in \mathbb{R}^{1\times k}$ denotes the jth row vector of the matrix D. Thus, for example, the minimal values of the bias variants $\omega_{*;2}$ and $\omega_{*;\infty}$ satisfy the relations

$$\inf\{\, \omega_{*;2}^2(\eta) \mid \eta \in \Psi_\alpha^D \,\} = \sum_{j=1}^p \inf\{\, \omega_*^2(\eta_j) \mid \eta_j \in \Psi_\alpha^{D_j} \,\} \tag{12}$$

$$\inf\{\, \omega_{*;\infty}(\eta) \mid \eta \in \Psi_\alpha^D \,\} = \max_{j=1,\dots,p} \inf\{\, \omega_*(\eta_j) \mid \eta_j \in \Psi_\alpha^{D_j} \,\} \tag{13}$$

And $\omega_{*;s}$ is minimized over Ψ_α^D by $\bar\eta \in \Psi_\alpha^D$ if, and in case $s = 2$ only if, each coordinate $\bar\eta_j$ minimizes ω_* over $\Psi_\alpha^{D_j}$.

As $\mathrm{E}\,|\eta|^2 = \sum_j \mathrm{E}\,\eta_j^2$, the problems $O_{*;s}^{\mathrm{tr}}(b)$ fall apart into p separate such problems, one for each row vector D_j of D with $p = 1$, provided that the coordinates can also be separated in the constraint $\omega_{*;s}(\eta) \le b$. This is obviously possible for the oscillation version $s = \infty$.

The problems $O_{*;\infty}^{\mathrm{tr}}(b)$ are in fact special cases of the corresponding coordinatewise problems with $p = 1$, which even allow possibly different upper bounds $\omega_*(\eta_j) \le b_j$. The solution $\tilde\eta = (\tilde\eta_1, \dots, \tilde\eta_p)'$ thus obtained from the coordinatewise problems minimizes not only the trace but necessarily each single diagonal element of the covariance matrix $\mathcal{C}(\eta)$, among all partial influence curves $\eta \in \Psi_2^D$ such that $\omega_*(\eta_j) \le b_j$ for $j = 1, \dots, p$.

Because the side conditions $\omega_{*;s} \le b$ for $s = 0,2$ and $s = \infty$ are different [we only have the bounds 5.3(9)], the solutions to $O^{\rm tr}_{*;s}(b)$ for $s = 0,2$ are however not rendered inadmissible by the solution to problem $O^{\rm tr}_{*;\infty}(b)$.

Full and Partial Solutions: Contrary to $\eta_h = D\psi_h$ (the classical case), we have $\tilde\eta \ne D\tilde\varrho$ for the solution $\tilde\eta$ to the general problem $O^{\rm tr}_{*;s}(b)$, and the solution $\tilde\varrho$ to $O^{\rm tr}_{*;s}(b)$ in the special case $p = k$, $D = \mathbb{I}_k$. The reason is that the ordering of $\operatorname{tr}\mathcal{C}(\psi)$ for $\psi \in \Psi_2$ may not be inherited to $\eta = D\psi \in \Psi_2^D$ [though positive definite ordering of $\mathcal{C}(\psi)$ would be], and the oscillation terms $\omega_{*;s}(\psi)$ are not invariant under linear transforms.

Hellinger Solution: The classical partial scores η_h is the universal solution to all the Hellinger problems $O^{\rm tr}_{h;s}(b)$ with $s = 0,2,\infty$. For the Cramèr–Rao bound (6) implies that η_h minimizes maxev and each one of the p diagonal elements of the covariance $\mathcal{C}(\eta)$, among all $\eta \in \Psi_2^D$. In particular,

$$\omega^{\rm min}_{h;s} = \omega_{h;s}(\eta_h), \qquad s = 0,2,\infty \tag{14}$$

Minimal Bias: In the cases $* = c, v$ the lower bias bounds $\omega^{\rm min}_{*;s}$ are always attained. For $\Psi^D_\infty \ne \emptyset$ guarantees that $\omega^{\rm min}_{*;s} < \infty$. If $\omega_{*;s}(\eta_r)$ tends to $\omega^{\rm min}_{*;s}$ along some sequence $\eta_r \in \Psi_2^D$ then the norms are in these cases bounded by $\mathrm{E}\,|\eta_r|^2 \le p\,\omega^2_{*;s}(\eta_r)$. Hence there exists a weak cluster point $\bar\eta$ [Lemma B.1.2 c]. Necessarily $\bar\eta \in \Psi_2^D$ since Ψ_2^D is weakly closed. And $\omega_{*;s}(\bar\eta) \le \omega^{\rm min}_{*;s}$ by weak l.s.c. of $\omega_{*;s}$. Thus $\omega^{\rm min}_{*;s} = \omega_{*;s}(\bar\eta)$.

Neglect of Centering: The constraint $\mathrm{E}\,\eta = 0$ may be removed from the total variation problems $O^{\rm tr}_{v;s}(b)$ with $s = 0,2,\infty$: If $\eta \in \mathbb{H}$ is recentered at its expectation, then $\mathrm{E}\,\eta\Lambda'$ and the explicit expression derived for $\omega_v(\eta)$ under the assumption $\mathrm{E}\,\eta = 0$ stay the same, whereas the matrix $\mathrm{E}\,\eta\eta'$ decreases (in the positive definite sense), and so does its trace $\mathrm{E}\,|\eta|^2$.

Kolmogorov, Cramèr–von Mises Problems: Problems $O^{\rm tr}_{*;s}(b)$ for $* = \kappa$ and $* = \mu$ remain unsolved in general. In the case $p = 1$, if the solution $\tilde\eta$ to the problem $O^{\rm tr}_v(b)$ turns out monotone, $\tilde\eta$ also solves problem $O^{\rm tr}_\kappa(b)$, because then $\omega_\kappa(\tilde\eta) = \omega_v(\tilde\eta)$ [Proposition 5.3.3 b], while $\omega_v \le \omega_\kappa$ always.

General Parameter Optimality

We make a restriction to $* = c, v$ (contamination, total variation).

The vector a and matrix A below generalize $a = 0$ and $A = D\mathcal{I}^{-1}$ in $\eta_h = D\mathcal{I}^{-1}\Lambda$. In the *main case* $\omega^{\rm min}_{*;s} < b < \omega_{*;s}(\eta_h)$, the robust solutions $\tilde\eta$ thus generalize the classical η_h. Mere achievement of the minimal oscillation will enforce sign type functions in the *lower case* $b = \omega^{\rm min}_{*;s}$.

Theorem 5.5.1 [Problem $O^{\rm tr}_c(b)$]

(a) *In case $\omega^{\rm min}_c < b \le \omega_c(\eta_h)$, there exist some $a \in \mathbb{R}^p$ and $A \in \mathbb{R}^{p\times k}$ such that the solution is of the form*

$$\tilde\eta = (A\Lambda - a)w, \qquad w = \min\Bigl\{1, \frac{b}{|A\Lambda - a|}\Bigr\} \tag{15}$$

and

$$\omega_c(\tilde\eta) = b \tag{16}$$

Conversely, if some $\tilde\eta \in \Psi_2^D$ *is of form (15) for any* $b \in (0,\infty)$, $a \in \mathbb{R}^p$, *and* $A \in \mathbb{R}^{p\times k}$, *then* $\tilde\eta$ *is the solution, and the following representations hold,*

$$a = Az, \quad z = \frac{\mathrm{E}\,\Lambda w}{\mathrm{E}\,w}, \qquad D = A\,\mathrm{E}\,(\Lambda - z)(\Lambda - z)'w \tag{17}$$

where $AD' = DA' > 0$.

(b) *It holds that*

$$\omega_c^{\min} = \max\Bigl\{ \frac{\operatorname{tr} AD'}{\mathrm{E}\,|A\Lambda - a|} \Bigm| a \in \mathbb{R}^p,\ A \in \mathbb{R}^{p\times k}\setminus\{0\} \Bigr\} \tag{18}$$

There exist $a \in \mathbb{R}^p$, $A \in \mathbb{R}^{p\times k}\setminus\{0\}$ *and* $\bar\eta \in \Psi_2^D$ *achieving* $\omega_c^{\min} = b$, *respectively. And then necessarily*

$$\bar\eta = b\,\frac{A\Lambda - a}{|A\Lambda - a|} \qquad \text{on } \{A\Lambda \neq a\} \tag{19}$$

Moreover, $a = Az$ *for some* $z \in \mathbb{R}^k$, *and* $AD' = DA' \ge 0$.

If $\bar\eta$ *in addition is constant on* $\{A\Lambda = a\}$, *then it is the solution.*

PROOF Recall that $\omega_c(\eta) = \sup_P |\eta|$.

(a) If $\omega_c^{\min} < b < \omega_c(\eta_h)$, then $\tilde\eta$ solves the convex, well-posed problem

$$\mathrm{E}\,|\eta|^2 = \min! \qquad \eta \in \mathbb{L},\ \mathrm{E}\,\eta = 0,\ \mathrm{E}\,\eta\Lambda' = D,\ \omega_c(\eta) \le b \tag{20}$$

Theorem B.2.6 provides some multipliers $a \in \mathbb{R}^p$, $A \in \mathbb{R}^{p\times k}$, $\beta \in [0,\infty)$ such that for all $\eta \in \mathbb{L}$,

$$\begin{aligned} L(\eta) &= \mathrm{E}\,|\eta|^2 + 2a'\,\mathrm{E}\,\eta - 2\,\mathrm{E}\,\eta' A\Lambda + \beta\omega_c^2(\eta) \\ &\ge \mathrm{E}\,|\tilde\eta|^2 - 2\operatorname{tr} AD' + \beta b^2 = L(\tilde\eta) \end{aligned} \tag{21}$$

In case $\beta = 0$ we get $\mathrm{E}\,|\eta|^2 \ge \mathrm{E}\,|\tilde\eta|^2$ for all $\eta \in \Psi_\infty^D$ hence all $\eta \in \Psi_2^D$, and thus $\eta = \eta_h$, contradicting the choice of $b < \omega_c(\eta_h)$. Thus $\beta > 0$, and (16) follows. Moreover, $\tilde\eta$ must also be the solution to

$$L_1(\eta) = \mathrm{E}\,|\eta - (A\Lambda - a)|^2 = \min! \qquad \eta \in \mathbb{L},\ \omega_c(\eta) \le b \tag{22}$$

which problem can be solved by pointwise minimization of the integrand, subject to $|\eta| \le b$. Thus (15) is proved.

Conversely, if some $\tilde\eta \in \Psi_2^D$ is of form (15) for any $a \in \mathbb{R}^p$, $A \in \mathbb{R}^{p\times k}$, then $\tilde\eta$ minimizes the corresponding Lagrangian L_1 among all $\eta \in \mathbb{H}$ subject to $\sup_P |\eta| \le b$. But for $\eta \in \Psi_2^D$ we have $L_1(\eta) = \mathrm{E}\,|\eta|^2 + \text{const}$.

To prove (17) write out

$$\begin{aligned} 0 = \mathrm{E}\,\tilde{\eta} = \mathrm{E}\,(A\Lambda - a)w = A\,\mathrm{E}\,\Lambda w - a\,\mathrm{E}\,w \\ D = \mathrm{E}\,\tilde{\eta}\Lambda' = \mathrm{E}\,A(\Lambda - z)\Lambda' w = A\,\mathrm{E}\,(\Lambda - z)(\Lambda - z)' w \end{aligned} \tag{23}$$

The representation $D = A\widetilde{\mathcal{I}}$ with any matrix $\widetilde{\mathcal{I}} = \widetilde{\mathcal{I}}' \geq 0$, and $\operatorname{rk} D = p$, imply that $0 < DD' = A\widetilde{\mathcal{I}}\widetilde{\mathcal{I}}A'$, and hence also $DA' = AD' = A\widetilde{\mathcal{I}}A' > 0$; in particular, $\operatorname{rk} A = p$. Of course, $z = \mathrm{E}\,\Lambda w / \mathrm{E}\,w$ in $a = Az$ is unique only modulo $\ker A$, and the second line of (23) is true for any such z.

(b) If $\bar{\eta} \in \Psi_2^D$ achieves $\omega_c^{\min} = b$ then $\bar{\eta}$ solves the convex, well-posed problem

$$\omega_c(\eta) = \min! \qquad \eta \in \mathbb{L},\ \mathrm{E}\,\eta = 0,\ \mathrm{E}\,\eta\Lambda' = D \tag{24}$$

Theorem B.2.3 supplies multipliers $\bar{a} \in \mathbb{R}^p$, $\bar{A} \in \mathbb{R}^{p \times k}$ so that for all $\eta \in \mathbb{L}$,

$$L_2(\eta) = \omega_c(\eta) + \bar{a}'\,\mathrm{E}\,\eta - \mathrm{E}\,\eta'\bar{A}\Lambda \geq b - \operatorname{tr}\bar{A}D' = L_2(\bar{\eta}) \tag{25}$$

Plugging in $\eta = 0$ we get $\operatorname{tr}\bar{A}D' \geq b$, where $b > 0$ as $\bar{\eta} \neq 0$; hence in particular $\bar{A} \neq 0$. Moreover, it follows that $\bar{\eta}$ is a solution to the problem

$$L_3(\eta) = \mathrm{E}\,\eta'(\bar{A}\Lambda - \bar{a}) = \max! \qquad \eta \in \mathbb{H},\ \sup_P |\eta| \leq b \tag{26}$$

which is solved by pointwise maximization of the integrand. Thus form (19) is obtained. Furthermore, by this form,

$$\operatorname{tr}\bar{A}D' = \mathrm{E}\,\bar{\eta}'(\bar{A}\Lambda - \bar{a}) = b\,\mathrm{E}\,|\bar{A}\Lambda - \bar{a}| \tag{27}$$

while for arbitrary $a \in \mathbb{R}^p$, $A \in \mathbb{R}^{p \times k}$, and $\eta \in \Psi_2^D$,

$$\operatorname{tr} AD' = \mathrm{E}\,\eta'(A\Lambda - a) \leq \omega_c(\eta)\,\mathrm{E}\,|A\Lambda - a| \tag{28}$$

Hence (18) is proved, and the pair of multipliers $\bar{a}$, $\bar{A}$ achieve the max.

If any $a \in \mathbb{R}^p$, $A \in \mathbb{R}^{p \times k} \setminus \{0\}$, and $\eta \in \Psi_2^D$ achieve $\omega_c^{\min}$, respectively, the orthogonal decomposition of $a = Az + \hat{a}$ with $\hat{a}'A = 0$ shows that $\mathrm{E}\,|A\Lambda - a| > \mathrm{E}\,|A(\Lambda - z)|$ unless $\hat{a} = 0$; that is, $a = Az$ for some $z \in \mathbb{R}^k$. Moreover, equality can hold in (28) only if η is of form (19). Then

$$DA' = \mathrm{E}\,\eta(A\Lambda - a)' = b \int_{\{A\Lambda \neq a\}} \frac{(A\Lambda - a)(A\Lambda - a)'}{|A\Lambda - a|}\,dP = AD' \geq 0 \tag{29}$$

If among these functions, $\tilde{\eta}$ is constant on the event $\Xi = \{A\Lambda = a\}$, which may be assumed of positive probability, then

$$\frac{\mathrm{E}\,|\eta|^2 - \mathrm{E}\,|\tilde{\eta}|^2}{P(\Xi)} = \mathrm{E}_\Xi\,|\eta|^2 - \left|\mathrm{E}_\Xi\,\eta\right|^2 = \operatorname{tr}\operatorname{Cov}_\Xi\,\eta \geq 0 \tag{30}$$

where the subscript indicates (elementary) conditioning on the event Ξ. Hence $\tilde{\eta}$ is the minimum norm solution. ////

Remark 5.5.2 The relations (17) suggest the following algorithm to compute the vector a and matrix A of the solution (15):

$$\begin{aligned} w_n &= \min\Bigl\{1, \frac{b}{|A_n\Lambda - a_n|}\Bigr\}, & z_{n+1} &= \frac{\mathrm{E}\,\Lambda w_n}{\mathrm{E}\,w_n} \\ A_{n+1} &= D\bigl(\mathrm{E}\,(\Lambda - z_n)(\Lambda - z_n)'w_n\bigr)^{-1}, & a_{n+1} &= A_{n+1}z_{n+1} \end{aligned} \tag{31}$$

starting with $a_0 = 0$ and $A_0 = D\mathcal{I}^{-1}$. ////

Problem $O^{\mathrm{tr}}_{c;2}(b)$ is included for the sake of completeness and comparability with $O^{\mathrm{tr}}_{v;2}(b)$. In the solutions to these problems, the multiplier β may be interpreted as a neighborhood radius [Theorem 5.5.7 c below].

Theorem 5.5.3 [Problem $O^{\mathrm{tr}}_{c;2}(b)$]

(a) *In case* $\omega^{\min}_{c;2} < b < \omega_{c;2}(\eta_h)$, *there exist* $\beta, b_j \in (0,\infty)$, $a_j \in \mathbb{R}$, *and rows* $A_j \in \mathbb{R}^{1\times k}\setminus\{0\}$, *such that the solution* $\tilde\eta$ *has the coordinates*

$$\tilde\eta_j = (A_j\Lambda - a_j)\min\Bigl\{1, \frac{b_j}{|A_j\Lambda - a_j|}\Bigr\} \tag{32}$$

where

$$\beta b_j = \mathrm{E}\,\bigl(|A_j\Lambda - a_j| - b_j\bigr)^+ \tag{33}$$

and

$$b_1^2 + \cdots + b_p^2 = b^2 \tag{69}$$

Conversely, if some $\tilde\eta \in \Psi_2^D$, *for any numbers* $\beta, b_j \in [0,\infty)$, $a_j \in \mathbb{R}$, *and rows* $A_j \in \mathbb{R}^{1\times k}\setminus\{0\}$, *is of form (32), (33), and (69), then* $\tilde\eta$ *is the solution.*

(b) *It holds that*

$$\bigl(\omega^{\min}_{c;2}\bigr)^2 = \sum_{j=1}^p \max\Bigl\{\Bigl|\frac{A_jD_j{}'}{\mathrm{E}\,|A_j\Lambda - a_j|}\Bigr|^2 \Bigm| a_j \in \mathbb{R},\ A_j \in \mathbb{R}^{p\times k}\setminus\{0\}\Bigr\} \tag{34}$$

There exist $a_j \in \mathbb{R}$, $A_j \in \mathbb{R}^{p\times k}\setminus\{0\}$, *and* $\bar\eta \in \Psi_2^D$ *attaining* $\omega^{\min}_{c;2}$, *respectively. And then, with* b_j^2 *denoting the* j^{th} max *in (34), necessarily*

$$\bar\eta_j = b_j\frac{A_j\Lambda - a_j}{|A_j\Lambda - a_j|} \qquad \text{on } \{A_j\Lambda \ne a_j\} \tag{35}$$

Proof In view of (12) and Theorem 5.5.1 b for $p = 1$, only (a) needs to be proved. Thus let $\omega^{\min}_{c;2} < b < \omega_{c;2}(\eta_h)$. Then problem $O^{\mathrm{tr}}_{c;2}(b)$,

$$\mathrm{E}\,|\eta|^2 = \min! \qquad \eta \in \mathbb{L},\ \mathrm{E}\,\eta = 0,\ \mathrm{E}\,\eta\Lambda' = D,\ \omega_{c;2}(\eta) \le b \tag{36}$$

is convex and well-posed. Theorem B.2.6 supplies some multipliers $a \in \mathbb{R}^p$, $A \in \mathbb{R}^{p\times k}$, and $\beta \in [0,\infty)$ such that, for the solution $\tilde\eta$ and all $\eta \in \mathbb{L}$,

$$\begin{aligned} L(\eta) &= \mathrm{E}\,|\eta|^2 + 2a'\,\mathrm{E}\,\eta - 2\,\mathrm{E}\,\eta' A\Lambda + \beta\omega^2_{c;2}(\eta) \\ &\ge \mathrm{E}\,|\tilde\eta|^2 - 2\,\mathrm{tr}\,AD' + \beta b^2 = L(\tilde\eta) \end{aligned} \tag{37}$$

Since $\tilde\eta \neq \eta_h$ as $b < \omega_{c;2}(\eta_h)$ necessarily $\beta > 0$, hence $\omega_{c;2}(\tilde\eta) = b$. Moreover, $\tilde\eta$ solves the problem

$$L_1(\eta) = \mathrm{E}\,|\eta - (A\Lambda - a)|^2 = \min! \qquad \eta \in \mathbb{H},\ \omega_{c;2}(\eta) \le b \tag{38}$$

Denote by a_j and A_j the $j^{\rm th}$ coordinate of a, respectively, the $j^{\rm th}$ row of A, and set $b_j = \omega_c(\eta_j)$. Then any such function η is improved by modifying its coordinates to form (32); so $\tilde\eta$ itself must have this form. Its b_j values minimize the convex Lagrangian

$$L_2(\boldsymbol{b}) = \sum_{j=1}^{p} \int \Bigl| \bigl(|A_j\Lambda - a_j| - b_j\bigr)^+ \Bigr|^2 dP + \beta b_j^2 \tag{39}$$

Condition (33) follows by differentiation w.r.t. b_j [Lemma C.2.3].

Conversely, any $\tilde\eta \in \Psi_2^D$ of this form minimizes the corresponding Lagrangian L_2, and then L, and hence $\mathrm{E}\,|\eta|^2$ among all $\eta \in \Psi_2^D$ subject to $\omega_{c;2}(\eta) \le b$. ////

Remark 5.5.4 [Problem $O^{\rm tr}_{c;\infty}(b)$] General, possibly asymmetric, bounds may be imposed on the coordinates of $\eta = (\eta_1, \dots, \eta_p)' \in \Psi_2^D$,

$$b_j' \le \eta_j \le b_j'' \tag{40}$$

The corresponding minimum trace problem can be solved for each coordinate separately: If the bounds (40) are verified by at least one $\eta \in \Psi_2^D$ (so necessarily $b_j' < 0 < b_j''$), there exist multipliers $a_j \in \mathbb{R}$ and $A_j \in \mathbb{R}^{1\times k}$ such that the $j^{\rm th}$ coordinate of the solution $\tilde\eta$ attains form

$$\tilde\eta_j = b_j' \vee (A_j\Lambda - a_j) \wedge b_j'' \tag{41}$$

or

$$\tilde\eta_j\, \mathbf{I}(A_j\Lambda \neq a_j) = b_j'\, \mathbf{I}(A_j\Lambda < a_j) + b_j''\, \mathbf{I}(A_j\Lambda > a_j) \tag{42}$$

This follows from Theorem B.2.3 if we keep bound (40) in the domain A there, and set $H\eta_j = (\mathrm{E}\,\eta_j, \mathrm{E}\,\eta_j\Lambda')'$. Then condition B.2(18) is fulfilled [with interior points $(0, \delta D_j)'$ for $\delta \in [0,1)$]. However, the well-posedness condition B.2(22) with $y_0 = (0, D_j)'$ itself is possibly not fulfilled; and then the second form (42) may occur.

Conversely, form (41) of some $\tilde\eta_j \in \Psi_2^{D_j}$, with any $a_j \in \mathbb{R}$, $A_j \in \mathbb{R}^{1\times k}$, entails that this $\tilde\eta_j$ is the solution to the problem

$$\mathrm{E}\,\bigl|\eta_j - (A_j\Lambda - a_j)\bigr|^2 = \min! \qquad \eta_j \in L_2(P),\ b_j' \le \eta_j \le b_j'' \tag{43}$$

As a consequence, $\tilde\eta_j$ minimizes $\mathrm{E}\,\eta_j^2$ among all $\eta_j \in \Psi_2^{D_j}$ subject to (40). As another consequence, comparing $\tilde\eta_j$ with $\eta_j = 0$, we record that

$$2\,A_j D_j{}' \ge \mathrm{E}\,\tilde\eta_j^2 \ge 0 \tag{44}$$

Thus

$$2\,\mathrm{tr}\,AD' = 2\sum\nolimits_j A_j D_j{}' \ge \sum\nolimits_j \mathrm{E}\,\tilde\eta_j^2 = \mathrm{E}\,|\tilde\eta|^2 > 0 \tag{45}$$

in case all coordinates of $\tilde{\eta}$ are of form (41).

Form (42) of some $\tilde{\eta}_j \in \Psi_2^{D_j}$ is dictated by the side conditions rather than the objective function [Remark B.2.7 b]. If $\tilde{\eta}_j$ in addition may be chosen constant on $\{A_j\Lambda = a_j\}$, utilizing (30), such $\tilde{\eta}_j$ minimizes $\mathrm{E}\,\eta_j^2$ among all $\eta_j \in \Psi_2^{D_j}$ subject to bound (40). Since $\tilde{\eta}_j \in \Psi_2^{D_j}$ of form (42) is the solution to the problem

$$\mathrm{E}\,\eta_j\,(A_j\Lambda - a_j) = \max! \qquad \eta_j \in L_2(P),\ b_j' \le \eta_j \le b_j'' \tag{46}$$

we also get

$$A_j D_j{}' \ge 0 \tag{47}$$

by comparison of $\tilde{\eta}_j$ with $\eta_j = 0$. ////

Problem $O_v^{\mathrm{tr}}(b)$, with exact bias, can be solved for one dimension $p = 1$.

Theorem 5.5.5 [Problem $O_v^{\mathrm{tr}}(b)$; $p = 1$]

(a) *In case* $\omega_v^{\min} < b < \omega_v(\eta_h)$, *there exist* $c \in (-b, 0)$, $A \in \mathbb{R}^{1\times k} \setminus \{0\}$ *such that*

$$\tilde{\eta} = c \vee A\Lambda \wedge (c + b) \tag{48}$$

is the solution, and

$$\omega_v(\tilde{\eta}) = b \tag{49}$$

Conversely, if some $\tilde{\eta} \in \Psi_2^D$ *is of form* (48) *for any* $b \in (0, \infty)$, $c \in \mathbb{R}$ *and any* $A \in \mathbb{R}^{1\times k}$, *then* $\tilde{\eta}$ *is the solution.*

(b) *It holds that*

$$\omega_v^{\min} = \max\Big\{ \frac{AD'}{\mathrm{E}\,(A\Lambda)^+} \;\Big|\; A \in \mathbb{R}^{1\times k} \setminus \{0\} \Big\} \tag{50}$$

There exist $A \in \mathbb{R}^{1\times k} \setminus \{0\}$ *and* $\bar{\eta} \in \Psi_2^D$ *achieving* $\omega_v^{\min} = b$, *respectively. And then necessarily*

$$\bar{\eta}\,\mathbf{I}(A\Lambda \ne 0) = c\,\mathbf{I}(A\Lambda < 0) + (c + b)\,\mathbf{I}(A\Lambda > 0) \tag{51}$$

for some $c \in (-b, 0)$. *In the case* $k = 1$, *the solution is*

$$\tilde{\eta} = b\,\mathrm{sign}(D)\,\Big(\frac{P(\Lambda < 0)}{P(\Lambda \ne 0)}\,\mathbf{I}(\Lambda > 0) - \frac{P(\Lambda > 0)}{P(\Lambda \ne 0)}\,\mathbf{I}(\Lambda < 0)\Big) \tag{52}$$

PROOF Recall that $\omega_v(\eta) = \sup_P \eta - \inf_P \eta$.

(a) The solution $\tilde{\eta}$ for $\omega_v^{\min} < b < \omega_v(\eta_h)$ solves the convex, well-posed problem

$$\mathrm{E}\,\eta^2 = \min! \qquad \eta \in \mathbb{L},\ \mathrm{E}\,\eta\Lambda' = D,\ \omega_v(\eta) \le b \tag{53}$$

where the side condition $\mathrm{E}\,\eta = 0$ could be dropped. Then Theorem B.2.6 gives us multipliers $A \in \mathbb{R}^{1\times k}$ and $\beta \in [0, \infty)$ such that for all $\eta \in \mathbb{L}$,

$$L(\eta) = \mathrm{E}\,\eta^2 - 2\,\mathrm{E}\,\eta A\Lambda + \beta\omega_v^2(\eta) \ge \mathrm{E}\,\tilde{\eta}^2 - 2AD' + \beta b^2 = L(\tilde{\eta}) \tag{54}$$

Necessarily $\beta > 0$, hence (49) holds, since $\mathrm{E}\,\eta^2 \geq \mathrm{E}\,\tilde{\eta}^2$ for all $\eta \in \Psi_\infty^D$, hence all $\eta \in \Psi_2^D$ otherwise; whence $\tilde{\eta} = \eta_h$ would follow, contradicting the choice of b. Moreover, (54) for $\eta = 0$ shows that $2AD' \geq \mathrm{E}\,\tilde{\eta}^2 + \beta b^2 > 0$. Furthermore, $\tilde{\eta}$ must also solve the problem

$$L_1(\eta) = \mathrm{E}\,(\eta - A\Lambda)^2 = \min! \qquad \eta \in \mathbb{H}\,,\ \omega_v(\eta) \leq b \tag{55}$$

For such functions η the Lagrangian L_1 is only decreased by passing to $\eta_c = c \vee A\Lambda \wedge (c+b)$ with $c = \inf_P \eta$; and $L_1(\eta_c) < L_1(\eta)$ unless $\eta = \eta_c$. Therefore $\tilde{\eta}$ itself must be of this form. Since $\mathrm{E}\,\tilde{\eta} = 0$ but $\tilde{\eta} \neq 0$ the corresponding $\tilde{c} = \inf_P \tilde{\eta}$ falls into $(-b, 0)$. This proves (48).

Conversely, let $\tilde{\eta} \in \Psi_2^D$ be of form (48) for any $b \in (0, \infty)$, $\tilde{c} \in \mathbb{R}$ and any $A \in \mathbb{R}^{1\times k}$. Writing out $\mathrm{E}(\tilde{\eta} - A\Lambda) = 0$ we get $dL_2(\tilde{c}) = 0$ for

$$dL_2(c) = 2\,\mathrm{E}\,(c - A\Lambda)^+ - 2\,\mathrm{E}\,(A\Lambda - c - b)^+ \tag{56}$$

By Lemma C.2.3 this is the (increasing) derivative of the (convex) Lagrangian

$$L_2(c) = L_1(\eta_c) = \mathrm{E}\,\big|(c - A\Lambda)^+\big|^2 + \mathrm{E}\,\big|(A\Lambda - c - b)^+\big|^2 \tag{57}$$

Therefore, $\tilde{\eta}$ minimizes the corresponding Lagrangian L_1 among all $\eta \in \mathbb{H}$ subject to $\omega_v(\eta) \leq b$. For $\eta \in \Psi_2^D$ however $L_1(\eta) = \mathrm{E}\,\eta^2 + \text{const}$.

(b) Any $\bar{\eta} \in \Psi_2^D$ achieving $\omega_v^{\min} = b$ solves the convex, well-posed problem

$$\omega_v(\eta) = \min! \qquad \eta \in \mathbb{L}\,,\ \mathrm{E}\,\eta\Lambda' = D \tag{58}$$

where the centering condition could be ignored. Then Theorem B.2.3 gives us a multiplier $A \in \mathbb{R}^{1\times k}$ such that for all $\eta \in \mathbb{L}$,

$$L_3(\eta) = \omega_v(\eta) - \mathrm{E}\,\eta A\Lambda \geq b - AD' = L_3(\bar{\eta}) \tag{59}$$

We note that $A \neq 0$ since otherwise $L_3(0) = 0 \geq b$ contradicting $\bar{\eta} \neq 0$. Moreover, $\bar{\eta}$ solves the problem

$$L_4(\eta) = \mathrm{E}\,\eta A\Lambda = \max! \qquad \eta \in \mathbb{L}\,,\ \omega_v(\eta) \leq b \tag{60}$$

Denoting $c = \inf_P \bar{\eta}$ we conclude that $\bar{\eta}$ must be of form (51). Therefore,

$$AD' = \mathrm{E}\,\bar{\eta}A\Lambda = (c+b)\,\mathrm{E}\,(A\Lambda)^+ - c\,\mathrm{E}\,(A\Lambda)^- = b\,\mathrm{E}\,(A\Lambda)^+ \tag{61}$$

so that this A achieves equality in (50). But then (50) is proved since for arbitrary $A \in \mathbb{R}^{1\times k} \setminus \{0\}$ and $\eta \in \Psi_2^D$ we have

$$AD' = \mathrm{E}\,\eta A\Lambda \leq \mathrm{E}\,(A\Lambda)^+ \sup\nolimits_P \eta - \mathrm{E}\,(A\Lambda)^- \inf\nolimits_P \eta = \mathrm{E}\,(A\Lambda)^+ \omega_v(\eta) \tag{62}$$

Now pick any $A \in \mathbb{R}^{1\times k}$, $\eta \in \Psi_2^D$ achieving $\omega_v^{\min} = b$, respectively. Then equality holds in (62), so η must be of form (51), with $c = \inf_P \eta \in (-b, 0)$.

For any such $\eta_c \in \mathbb{L}$ with arbitrary $c \in \mathbb{R}$ we have

$$\mathrm{E}\,\eta_c^2 = \int_{\{A\Lambda=0\}} \eta^2\,dP + L_5(c) \tag{63}$$

where

$$L_5(c) = c^2 P(A\Lambda < 0) + (c+b)^2 P(A\Lambda > 0) \tag{64}$$

L_5 is convex with unique minimum at $\tilde{c} = -bP(A\Lambda > 0)/P(A\Lambda \neq 0)$. The corresponding $\eta_{\tilde{c}}$ that in addition vanishes on $\{A\Lambda = 0\}$ certainly minimizes $\mathrm{E}\,\eta_c^2$. However, this $\eta_{\tilde{c}}$ also verifies $\mathrm{E}\,\eta_{\tilde{c}} = 0$. And because $\mathrm{E}\,\eta_c A\Lambda = b\,\mathrm{E}\,(A\Lambda)^+ = AD'$ with $A \neq 0$, also $\mathrm{E}\,\eta_{\tilde{c}}\Lambda' = D$ holds, at least if the dimension is $k = 1$. ////

To deal with exact total variation bias ω_v, in case $p > 1$, one might try the convex map

$$G\colon \mathbb{L} \longrightarrow \boldsymbol{\ell}_\infty, \qquad G(\eta) = \bigl(\omega_v^2({e_n}'\eta)\bigr) \tag{65}$$

using a dense sequence of unit vectors $e_n \in \mathbb{R}^p$, $|e_n| = 1$. The range space is the nonseparable Banach space $Z = \boldsymbol{\ell}_\infty$ of all bounded real-valued sequences with sup norm, which indeed has a positive cone of nonempty interior. Its topological dual is the Banach space $Z^* = ba(\mathrm{N})$ of all bounded finitely additive functions $z^* = M$ on 2^{N} with total variation norm [Dunford and Schwartz (1957; Vol. I, Corollary IV 5.3)]. So the corresponding Lagrangian term z^*G would be the Radon integral

$$z^*G(\eta) = \int_{\mathrm{N}} \omega_v^2({e_n}'\eta)\,M(dn) \tag{66}$$

Anyway, the infinitely many terms $\omega_v({e_n}'\eta)$ are difficult to disentangle. We therefore resort to the bias variants $\omega_{v;s}$ that are based on the p canonical coordinates only. Problem $O_{v;\infty}^{\mathrm{tr}}(b)$ being solved coordinatewise by $O_v^{\mathrm{tr}}(b)$ for $p = 1$, problem $O_{v;2}^{\mathrm{tr}}(b)$ remains.

Theorem 5.5.6 [Problem $O_{v;2}^{\mathrm{tr}}(b)$]

(a) *In case $\omega_{v;2}^{\min} < b < \omega_{v;2}(\eta_h)$, there exist rows $A_j \in \mathbb{R}^{1\times k} \setminus \{0\}$ and numbers $c_j \in (-\infty, 0)$, $b_j, \beta \in (0, \infty)$, such that the solution has the coordinates*

$$\tilde{\eta}_j = c_j \vee A_j\Lambda \wedge (c_j + b_j) \tag{67}$$

where

$$\beta b_j = \mathrm{E}\,(c_j - A_j\Lambda)^+ \tag{68}$$

and

$$b_1^2 + \cdots + b_p^2 = b^2 \tag{69}$$

Conversely, if some $\tilde{\eta} \in \Psi_2^D$ is of form (67)–(69) for any $A_j \in \mathbb{R}^{1\times k}$, and numbers $\beta, b, b_j \in [0, \infty)$, $c_j \in \mathbb{R}$, then $\tilde{\eta}$ solves problem $O_{v;2}^{\mathrm{tr}}(b)$.

(b) *It holds that*

$$\left(\omega_{v;2}^{\min}\right)^2 = \sum_{j=1}^{p} \max\Bigl\{ \Bigl|\frac{A_j D_j{}'}{\mathrm{E}\,(A_j\Lambda)^+}\Bigr|^2 \Bigm| A_j \in \mathbb{R}^{1\times k} \setminus \{0\} \Bigr\} \tag{70}$$

There exist $A_j \in \mathbb{R}^{1\times k} \setminus \{0\}$ *and* $\bar{\eta} \in \Psi_2^D$ *attaining* $\omega_{v;2}^{\min}$, *respectively. And then, with* b_j^2 *denoting the* j^{th} max *in* (70), *necessarily*

$$\bar{\eta}_j\, \mathbf{I}(A_j\Lambda \neq 0) = c_j\, \mathbf{I}(A_j\Lambda < 0) + (c_j + b_j)\, \mathbf{I}(A_j\Lambda > 0) \tag{71}$$

for some numbers $c_j \in (-b_j, 0)$.

PROOF In view of (12) and Theorem 5.5.5 b, only (a) needs to be proved. Thus let $\omega_{v;2}^{\min} < b < \omega_{v;2}(\eta_h)$. Then problem $O_{v;2}^{\mathrm{tr}}(b)$,

$$\mathrm{E}\,|\eta|^2 = \min! \qquad \eta \in \mathbb{L},\ \mathrm{E}\,\eta\Lambda' = D,\ \omega_{v;2}(\eta) \le b \tag{72}$$

where the centering condition could be removed, is convex and well-posed. Thus Theorem B.2.6 supplies multipliers $A \in \mathbb{R}^{p\times k}$ and $\beta \in [0,\infty)$ such that for all $\eta \in \mathbb{L}$, and the solution $\tilde{\eta}$,

$$L(\eta) = \mathrm{E}\,|\eta|^2 - 2\,\mathrm{E}\,\eta' A\Lambda + \beta\omega_{v;2}^2(\eta) \ge \mathrm{E}\,|\tilde{\eta}|^2 - 2\,\mathrm{tr}\,AD' + \beta b^2 = L(\tilde{\eta}) \tag{73}$$

Since $\tilde{\eta} \neq \eta_h$ as $b < \omega_{v;2}(\eta_h)$ necessarily $\beta > 0$; hence $\omega_{v;2}(\tilde{\eta}) = b$. Moreover, $\tilde{\eta}$ solves the problem

$$L_1(\eta) = \mathrm{E}\,|\eta - A\Lambda|^2 = \min! \qquad \eta \in \mathbb{H},\ \omega_{v;2}(\eta) \le b \tag{74}$$

Denote by A_j the jth row of A, and set $c_j = \inf_P \eta_j$, $b_j = \omega_v(\eta_j)$. Then any such function η is improved by equating its coordinates with form (67); so $\tilde{\eta}$ itself has this form. Its c_j and b_j values minimize the convex Lagrangian

$$L_2(\boldsymbol{c},\boldsymbol{b}) = \sum_{j=1}^{p} \mathrm{E}\,\bigl|(c_j - A_j\Lambda)^+\bigr|^2 + \mathrm{E}\,\bigl|(A_j\Lambda - c_j - b_j)^+\bigr|^2 + \beta b_j^2 \tag{75}$$

Differentiation w.r.t. c_j and b_j [Lemma C.2.3] yields

$$\mathrm{E}\,(A_j\Lambda - c_j - b_j)^+ = \mathrm{E}\,(c_j - A_j\Lambda)^+ \tag{76}$$

that is, $\mathrm{E}\,\tilde{\eta}_j = 0$, and

$$\mathrm{E}\,(A_j\Lambda - c_j - b_j)^+ = \beta b_j \tag{77}$$

Conversely, any $\tilde{\eta} \in \Psi_2^D$ of this form minimizes the Lagrangians L_2 and L; hence $\mathrm{E}\,|\eta|^2$ among all $\eta \in \Psi_2^D$ subject to $\omega_{v;2}(\eta) \le b$. ////

5.5.2 Mean Square Error

Both criteria can be treated more symmetrically by minimizing some convex combination, which leads us to the as. mean square error problems $O^{\rm ms}_{*;s}(\beta)$,

$$\mathrm{E}\,|\eta|^2 + \beta\omega^2_{*;s} = \min! \qquad \eta \in \mathbb{H},\ \mathrm{E}\eta = 0,\ \mathrm{E}\eta\Lambda' = D \tag{78}$$

where $\beta \in (0,\infty)$ is any fixed weight. The name is estimator terminology: If $S = (S_n)$ is as. linear at P_θ with influence curve $\psi_\theta \in \Psi_2(\theta)$, then the objective function of problem $O^{\rm ms}_{*;s}(\beta)$ arises as the following limiting risk,

$$\sup_{q\in\mathcal{G}_*(\theta)} \lim_{b\to\infty} \lim_{n\to\infty} \int b \wedge n\big|\tau\circ S_n - \tau(\theta)\big|^2\, dQ^n_n(q,r) \tag{79}$$

where $\tau(\theta)$ remains the estimand, the loss function is $\ell(z) = |z|^2$, and the simple perturbations 5.3(2) of radius $r = \beta$ are employed. Depending on suitable constructions, this as. risk will not increase if the $\sup_q$ is taken inside (that is, interchanged with $\lim_b$ and $\lim_n$), and the laws Q_n at sample size n may range over the full balls $U_*(\theta, r/\sqrt{n}\,)$ of the corresponding neighborhood system $\mathcal{U}_*(\theta)$. [We now suppress θ again.]

The classical partial scores $\eta_h = D\psi_h = D\mathcal{I}^{-1}\Lambda$, by (6) and (14), is the unique solution of the Hellinger mean square problems $O^{\rm ms}_{h;s}(\beta)$.

In the case $p = 1$, the solutions to the problems $O^{\rm ms}_{*;s}(\beta)$ of course coincide for $s = 0, 2, \infty$. In particular, if $p = 1$, the solution to $O^{\rm ms}_{v;0}(\beta)$ is provided by parts (c) or (d) for the analogous statement in case $* = v$ to part (b) of the following theorem.

Theorem 5.5.7 [Problems $O^{\rm ms}_{*;s}(\beta)$, $\beta \in (0,\infty)$; $* = c, v$; $s = 0, 2, \infty$]

(a) *The problems $O^{\rm ms}_{*;s}(\beta)$ have unique solutions.*

(b) *The solution to problem $O^{\rm ms}_{c;0}(\beta)$ coincides with the solution* (15) *to problem $O^{\rm tr}_{c;0}(b)$, with $b \in (0,\infty)$ and β related by*

$$\beta b = \mathrm{E}\left(|A\Lambda - a| - b\right)^+ \tag{80}$$

(c) *The solutions to the problems $O^{\rm ms}_{*;2}(\beta)$ coincide with the solutions to the corresponding problems $O^{\rm tr}_{*;2}(b)$ of the forms* (32), (33), *and* (69), *respectively* (67), (68), *and* (69).

(d) *The solutions to the problems $O^{\rm ms}_{*;\infty}(\beta)$ are of forms* (32) *and* (67), *respectively, with constant $b_1 = \cdots = b_p = b \in (0,\infty)$ related to β via*

$$\beta b = \sum_{j=1}^{p} \mathrm{E}\left(|A_j\Lambda - a_j| - b\right)^+ \tag{c81}$$

respectively,

$$\beta b = \sum_{j=1}^{p} \mathrm{E}\,(c_j - A_j\Lambda)^+ \tag{v81}$$

(e) *In each case considered, the stated form of* $\tilde{\eta}$, *in terms of the given* β, *in combination with any bounds* $b_j, b \in [0,\infty]$, $c_j \in \mathbb{R}$, *and further multipliers* $a \in \mathbb{R}^p$, $A \in \mathbb{R}^{p\times k}$, *such that the cited conditions hold, is also sufficient for some* $\tilde{\eta} \in \Psi_2^D$ *to solve problem* $O^{\rm ms}_{*;s}(\beta)$.

PROOF As $\Psi_\infty^D \neq \emptyset$ the minimum m in (78) is finite. It is attained by the automatic (norm) boundedness, hence weak sequential compactness [Lemma B.1.2 c], of an approximating sequence, and by weak l.s.c. of both norm [Lemma B.1.2 b] and bias term, hence of the objective function. Uniqueness of the solution $\tilde{\eta}$ is a consequence of the convexity of Ψ_2^D and $\omega_{*;s}$ and the strict convexity of the norm (Cauchy–Schwarz equality).

Problem (78) being convex, well-posed, there exist multipliers $a \in \mathbb{R}^p$ and $A \in \mathbb{R}^{p\times k}$ such that for all $\eta \in \mathbb{L}$,

$$\begin{aligned} L(\eta) &= \mathrm{E}\,|\eta - (A\Lambda - a)|^2 + \beta\omega^2_{*;s}(\eta) \\ &\geq m - 2\,\mathrm{tr}\,AD' + \mathrm{E}\,|A\Lambda - a|^2 = L(\tilde{\eta}) \end{aligned} \tag{82}$$

Putting $b = \omega_{*;s}(\tilde{\eta})$, hence $\tilde{\eta}$ also solves the problem

$$L_1(\eta) = \mathrm{E}\,|\eta - (A\Lambda - a)|^2 = \min! \qquad \eta \in \mathbb{H},\ \omega_{*;s}(\eta) \leq b \tag{83}$$

which has already occurred as problems (22), (38), (55), and (74), respectively. Corresponding essentially complete classes of functions η are of the forms (15), (32), and (67), which are indexed by bounds $b \in (0,\infty)$, and $b_j \in (0,\infty)$ or $c_j \in \mathbb{R}$, respectively; $b_1 = \cdots = b_p = b$ in case $s = \infty$. Plugged back into the Lagrangian L, these special functions η define a convex function L_2 of the bounds. In the case $(c;0)$, L_2 reads

$$L_2(b) = \beta b^2 + \int \Big|\big(|A\Lambda - a| - b\big)^+\Big|^2\,dP \tag{84}$$

In the cases $(*;2)$, L_2 is given by (39) and (75). And in the cases $(*;\infty)$, we have

$$L_2(b) = \beta b^2 + \sum_{j=1}^{p} \int \Big|\big(|A_j\Lambda - a_j| - b\big)^+\Big|^2\,dP \tag{c85}$$

respectively,

$$L_2(\mathbf{c}, b) = \beta b^2 + \sum_{j=1}^{p} \int \big|(c_j - A_j\Lambda)^+\big|^2 + \big|(A_j\Lambda - c_j - b)^+\big|^2\,dP \tag{v85}$$

Appealing to Lemma C.2.3, differentiation of L_2 yields the necessary and sufficient conditions (80), and (33), respectively (76), (77), hence (68), and (c81), (v81), for a global minimum of the respective Lagrangian L_2.

Conversely, any function $\tilde{\eta}$ of the form thus derived achieves the minimum of the corresponding Lagrangian L_2, and then minimizes the corresponding Lagrangian L over $\mathbb{H}$. But, for $\eta \in \Psi_2^D$, the Lagrangian $L(\eta)$ equals $\mathrm{E}\,|\eta|^2 + \beta\omega^2_{*;s}(\eta)$, up to some constant. ////

Remark 5.5.8 The parameters $b_j, b \in [0,\infty]$ and $c_j \in \mathbb{R}$ in the solutions $\tilde{\eta}$ given by Theorems 5.5.1–5.5.7 are uniquely determined since these bounds have been derived in terms of the unique solution $\tilde{\eta}$ as

$$b = \omega_{*;s}(\tilde{\eta}), \qquad b_j = \omega_*(\tilde{\eta}_j), \qquad c_j = \inf_P \tilde{\eta}_j \tag{86}$$

The remaining parameters $a \in \mathbb{R}^p$ and $A \in \mathbb{R}^{p\times k}$ need not be unique as Lagrange multipliers [Remark B.2.10 a]. However, if

$$\operatorname{support} \Lambda(P) = \mathbb{R}^k \tag{87}$$

then the following identity a.e. P for a pair of matrices $A, A'' \in \mathbb{R}^{p\times k}$ and a pair of vectors $a = Az$, $a'' = A''z''$ with $z, z'' \in \mathbb{R}^k$,

$$(A''\Lambda - a'')\min\Bigl\{1, \frac{b}{|A''\Lambda - a''|}\Bigr\} = (A\Lambda - a)\min\Bigl\{1, \frac{b}{|A\Lambda - a|}\Bigr\} \tag{88}$$

forces $A'' = A$ and $a'' = a$. Likewise, if the following identity for some rows $A_j, A_j'' \in \mathbb{R}^{1\times k}\setminus\{0\}$ and numbers $a_j, a_j'' \in \mathbb{R}$ holds a.e. P,

$$c_j \vee (A_j''\Lambda - a_j'') \wedge (c_j + b_j) = c_j \vee (A_j\Lambda - a_j) \wedge (c_j + b_j) \tag{89}$$

then, under condition (87), necessarily $A_j'' = A_j$ and $a_j'' = a_j$.

Assumption (87) is stronger than $\mathcal{I} > 0$; that $P(\Lambda \in H) < 1$ for all $(k-1)$-dimensional hyperplanes H of $\mathbb{R}^k$, contrary to the case of the classical partial scores η_h, does not seem to suffice for uniqueness in the case of the robust solutions. Other conditions to this effect are (90) and (91) below. ////

5.5.3 Nonexistence of Strong Solution

The strong optimality (6) of $\eta_h = D\mathcal{I}^{-1}\Lambda$ in the parametric and Hellinger cases, minimizing not only the trace but the covariance matrix itself in the positive definite sense, can in general not be expected from the minimum trace solutions in robustness. We prove this for problem $O_c^{\mathrm{tr}}(b)$, under condition (91) on the law $\Lambda(P)$, to obtain unique Lagrange multipliers. This condition is certainly satisfied if

$$e \in \mathbb{R}^k,\ \alpha \in \mathbb{R},\ P(e'\Lambda = \alpha) > 0 \implies e = 0 \tag{90}$$

that is, $P(\Lambda \in H) = 0$ for all $(k-1)$-dimensional hyperplanes $H \subset \mathbb{R}^k$. In turn, (90) entails that $\mathcal{I} > 0$ [that is, $P(e'\Lambda = 0) = 1 \implies e = 0$].

Theorem 5.5.9 *For $\omega_c^{\min} < b < \omega_c(\eta_h)$, let $\tilde{\eta}$ be the solution to $O_c^{\mathrm{tr}}(b)$. Suppose that $p > 1$. Assume that for all $(k-1)$-dimensional hyperplanes H of $\mathbb{R}^k$,*

$$P(\Lambda \in H) < P(|\tilde{\eta}| < b) \tag{91}$$

Then $\tilde{\eta}$ does not *achieve: $\mathcal{C}(\tilde{\eta}) \le \mathcal{C}(\eta)$ for all $\eta \in \Psi_2^D$ subject to $\omega_c(\eta) \le b$.*

The proof needs a lemma, which has already been used implicitly with unit matrix $M = \mathbb{I}_p$ to solve problem (22).

Lemma 5.5.10 *Given $M \in \mathbb{R}^{p\times p}$, $M = M' > 0$, and $y \in \mathbb{R}^p$, $b \in (0,\infty)$. Then the unique solution x_0 of the problem*

$$(x-y)'M(x-y) = \min! \qquad x \in \mathbb{R}^p,\ |x| \le b \tag{92}$$

is

$$x_0 = \bigl(M + \gamma_M(y)\,\mathbb{I}_p\bigr)^{-1} M y \tag{93}$$

where

$$\begin{aligned} \gamma_M(y) &= 0 \quad \text{if } |y| \le b; \quad \text{and then } y = x_0 \\ \gamma_M(y) &> 0 \quad \text{if } |y| > b; \quad \text{and then } b = |x_0| \end{aligned} \tag{94}$$

PROOF Unconstrained minimization, by differentiation, of the convex Lagrangian

$$L(x) = (x-y)'M(x-y) + \gamma_M(y)|x|^2 \tag{95}$$

with nonnegative multiplier [Theorem B.2.1]. ////

PROOF [Theorem 5.5.9] Suppose that, on the contrary, $\mathcal{C}(\tilde\eta) \le \mathcal{C}(\eta)$ for all $\eta \in \Psi_2^D$ subject to $\omega_c(\eta) \le b$, and consider any symmetric, positive definite matrix $M \in \mathbb{R}^{p\times p}$. Writing $M = \sum_j \lambda_j z_j z_j'$ in terms of an ONB of p eigenvectors z_j and corresponding eigenvalues $\lambda_j \in (0,\infty)$, we see that

$$\mathrm{E}\,\eta' M\eta = \sum\nolimits_j \lambda_j z_j'\mathcal{C}(\eta) z_j \ge \sum\nolimits_j \lambda_j z_j'\mathcal{C}(\tilde\eta) z_j = \mathrm{E}\,\tilde\eta' M\tilde\eta \tag{96}$$

for all $\eta \in \Psi_2^D$ subject to $\omega_c(\eta) \le b$. Hence $\tilde\eta$ would solve

$$\mathrm{E}\,\eta' M\eta = \min! \qquad \eta \in \mathbb{L},\ \mathrm{E}\,\eta = 0,\ \mathrm{E}\,\eta\Lambda' = D,\ \omega_c(\eta) \le b \tag{97}$$

which is again a minimum norm problem, convex and well-posed. Thus Theorem B.2.6 provides some multipliers $Ma_M \in \mathbb{R}^p$, $MA_M \in \mathbb{R}^{p\times k}$, and $\beta_M \in [0,\infty)$ such that, for all $\eta \in \mathbb{L}$,

$$\begin{aligned} L_M(\eta) &= \mathrm{E}\,\eta' M\eta + 2a_M' M\,\mathrm{E}\,\eta - 2\,\mathrm{E}\,\eta' M A_M\Lambda + \beta_M\omega_c^2(\eta) \\ &\ge \mathrm{E}\,\tilde\eta' M\tilde\eta - 2\,\mathrm{tr}\,MA_M D' + \beta_M b^2 = L_M(\tilde\eta) \end{aligned} \tag{98}$$

Setting $Y_M = A_M\Lambda - a_M$ therefore $\tilde\eta$ solves the problem

$$\mathrm{E}\,(\eta - Y_M)' M(\eta - Y_M) = \min! \qquad \eta \in \mathbb{H},\ |\eta| \le b \tag{99}$$

Pointwise minimization by means of Lemma 5.5.10 ($y = Y_M$, $x = \eta$) yields

$$\tilde\eta = \bigl(M + \gamma_M(Y_M)\,\mathbb{I}_p\bigr)^{-1} M\,Y_M = Y \min\Bigl\{1, \frac{b}{|Y|}\Bigr\} \tag{100}$$

where $Y = A\Lambda - a$ corresponds to the former choice $M = \mathbb{I}_p$.

Note that $\mathrm{E}\,|\tilde{\eta}|^2 < b^2$ since $\omega_c^{\min} < b$, and hence $P(|\tilde{\eta}| < b) > 0$. But the events $\{|Y_M| < b\}$, $\{|Y| < b\}$ are each the same as $\{|\tilde{\eta}| < b\}$. Therefore

$$A_M\Lambda - a_M = A\Lambda - a \qquad \text{on } \{|\tilde{\eta}| < b\} \tag{101}$$

Condition (91) now ensures that $a_M = a$ and $A_M = A$.

Since $b < \omega_c(\eta_h)$ also $P(|Y| > b) > 0$. For any such value $|Y| > b$, however, (100) shows that $Y = Y_M$ is an eigenvector of M, as

$$b\frac{Y}{|Y|} = \bigl(M + \gamma_M(Y)\,\mathbb{I}_p\bigr)^{-1} M Y$$

hence

$$MY = b\frac{\gamma_M(Y)}{|Y| - b}\,Y \tag{102}$$

Now iterate the argument for another symmetric, positive definite matrix M'' that has no eigenvectors in common with M. Such two matrices exist in dimension $p > 1$: Starting with any ONB (u_j), complement the vector $v_1 = (u_1 + \cdots + u_p)/\sqrt{p}$ by Gram–Schmidt to another ONB (v_j), and take $M = \sum_j j\,u_j u_j{}'$, and $M'' = \sum_j j\,v_j v_j{}'$; for example. ////

Admissibility

The minimum trace solutions $\tilde{\eta}$ derived in Theorems 5.5.1 to 5.5.6 are at least *admissible*: No other partial influence curve satisfying the respective oscillation bound can improve the covariance in the positive definite sense,

$$\eta \in \Psi_2^D\,,\;\; \omega_{*;s}(\eta) \le b\,,\;\; \mathcal{C}(\eta) \le \mathcal{C}(\tilde{\eta}) \implies \eta = \tilde{\eta} \tag{103}$$

This is fairly trivial since $\mathcal{C}(\eta) \le \mathcal{C}(\tilde{\eta})$ entails that $\operatorname{tr}\mathcal{C}(\eta) \le \operatorname{tr}\mathcal{C}(\tilde{\eta})$, and the minimum trace solution $\tilde{\eta}$ is unique.

5.5.4 Equivariance Under Reparametrizations

Functionals and estimators are usually defined by recipes, like zeros of M equations, or the minimum distance principle. In our approach, as. linear functionals and estimators have to be (re)constructed from their influence curves, which are obtained as solutions to certain optimization problems. Although some nonparametric methods (mean, median) do not depend on the assumption of any central parametric model, the methods in general do refer to the central model $\mathcal{P} = \{\, P_\theta \mid \theta \in \Theta \,\}$, if not by construction (like general M equations based on influence curves, or minimum distance), then at least by the Fisher consistency requirement. But the central model may simply be relabelled, applying transformations of the parameter space.

Consider some group (under composition) of differentiable transformations $g\colon \Theta \to \Theta$ of the parameter space. The ideal model reparametrized by g is

$$\mathcal{P}^g = \{\, P_\theta^g \mid P_\theta^g = P_{g(\theta)}\,,\; \theta \in \Theta \,\} \tag{104}$$

Then a method to define functionals and estimators may be applied with $\mathcal{P}$ as central model on one hand, which yields a functional T and estimator S, or assuming $\mathcal{P}^g$ as central model on the other hand, and then yields a functional T^g and estimator S^g. The method is called *equivariant* (under the transformation group) if, for every group element g with inverse denoted by i, the functionals T^g, T and estimators S^g, S thus obtained are naturally connected by, respectively,

$$T^g = i \circ T, \qquad S^g = i \circ S \tag{105}$$

Remark 5.5.11 If $\bar{\imath}$ is a sample space transformation that achieves image measures $\bar{\imath}(P_\theta) = P_{i(\theta)}$, the usual definition of equivariance would require that $T^g = T \circ \bar{\imath}$ and, similarly, $S^g = S \circ \bar{\imath}$. Condition (105) does not imply this representation by means of T and S, which is questionable anyway.

For example, in the normal location model $P_\theta = N(\theta, \mathbb{I}_k)$ with $\theta \in \mathbb{R}^k$, which may be relabelled by translations $g(\theta) = \theta + t$, the James–Stein estimator $\tilde{S}_t(x) = \bigl(1-(k-2)/|x+t|^2\bigr)(x+t)-t$ with shrinkage point $-t$, in dimension $k \geq 3$, is just as good as the one with $t = 0$; their risks are merely shifted: $\mathrm{E}_\theta \bigl|\tilde{S}_t - \theta\bigr|^2 = \mathrm{E}_{\theta+t} \bigl|\tilde{S}_0 - (\theta+t)\bigr|^2$. Thus, employing w.r.t. model $\mathcal{P}^g$ the estimator $\tilde{S}_0^g = \tilde{S}_t \circ \bar{\imath}$, the James–Stein method (with shrinkage point up to one's choice) can be made to comply with the equivariance condition (105): $\tilde{S}_t(x-t) = \tilde{S}_0(x)-t$, although $\tilde{S}_0$ itself is certainly not translation equivariant: $\tilde{S}_0(x-t) \neq \tilde{S}_0(x) - t$ (nor is $\tilde{S}_t$). ////

Due to the assumed smoothness of g and i, the models $\mathcal{P}^g$ and $\mathcal{P}$ are simultaneously L_2 differentiable at $i(\theta)$, respectively θ, with the following relations, in obvious notation, between L_2 derivatives and Fisher informations,

$$\Lambda_\theta = di(\theta)'\Lambda^g_{i(\theta)}, \qquad \mathcal{I}_\theta = di(\theta)'\mathcal{I}^g_{i(\theta)}di(\theta) \tag{106}$$

The respective classes of influence curves $\Psi_2^g(i(\theta))$ and $\Psi_2(\theta)$ are mapped onto one another via the following correspondence within the same integration space $L_2^k(P^g_{i(\theta)}) = L_2^k(P_\theta)$,

$$\psi^g_{i(\theta)} = di(\theta)\psi_\theta \tag{107}$$

because

$$\mathrm{E}^g_{i(\theta)}\, \psi^g_{i(\theta)} = di(\theta)\, \mathrm{E}_\theta\, \psi_\theta$$

$$\mathrm{E}^g_{i(\theta)}\, \psi^g_{i(\theta)}(\Lambda^g_{i(\theta)})'di(\theta) = di(\theta)\, \mathrm{E}_\theta\, \psi_\theta\Lambda_\theta'$$

where $\det di(\theta) \neq 0$. In particular, given any $\psi_\theta \in \Psi_2(\theta)$, (107) does define an influence curve $\psi^g_{i(\theta)} \in \Psi_2^g(i(\theta))$, and vice versa. Expectation, hence also covariance below, agree under $P^g_{i(\theta)} = P_\theta$: $\mathrm{E}^g_{i(\theta)} = \mathrm{E}_\theta$, $\mathcal{C}^g_{i(\theta)} = \mathcal{C}_\theta$.

In view of Definitions 4.2.12 and 4.2.16, and applying the chain rule, we realize: The functionals T^g, T and estimators S^g, S that are connected by (105), respectively, are simultaneously as. linear at $P^g_{i(\theta)} = P_\theta$, in the respective models, and their influence curves $\psi^g_{i(\theta)}$, ψ_θ are related by (107). Thus the relation (107) between the corresponding influence curves, holding for all $\theta \in \Theta$ and every group element $g: \Theta \to \Theta$ with inverse denoted by i, is the *local* version of *equivariance*.

The classical scores, for example, verify condition (107) since

$$di(\theta)\psi_{h,\theta} \overset{(106)}{=} di(\theta)\mathcal{I}_\theta^{-1} di(\theta)'\Lambda^g_{i(\theta)} \overset{(106)}{=} \bigl(\mathcal{I}^g_{i(\theta)}\bigr)^{-1}\Lambda^g_{i(\theta)} = \psi^g_{h,i(\theta)} \tag{108}$$

By use of exact M equations (possibly not one-step constructions), local equivariance (107) in principle ensures global equivariance (105), since

$$0 = di(t)\int \psi_t \, dQ = \int \psi^g_{i(t)} \, dQ, \qquad t = T(Q) \tag{109}$$

That the minimum distance method is equivariant, in principle, is a consequence of the global definitions since, setting $t = T(Q)$,

$$\min_{\theta\in\Theta} d(Q, P^g_\theta) = \min_{\theta\in\Theta} d(Q, P_\theta) = d(Q, P_t) = d(Q, P^g_{i(t)}) \tag{110}$$

In (109) and (110), the empirical measure $Q = \hat{P}_n$ may also be inserted.

On the local level, equivariance of the Hellinger and CvM MD functionals, in view of their expansions 4.2(47), respectively 4.2(49), is expressed by (108), respectively (112), and implied by (111) below.

Like L_2 differentiability, also CvM differentiability w.r.t. some arbitrary weight $\mu \in \mathcal{M}(\mathbb{B}^m)$ of $\mathcal{P}^g$ at $i(\theta)$, and of $\mathcal{P}$ at θ, holds simultaneously, and the CvM derivatives and CvM informations, in obvious notation, are related via

$$\Delta_\theta = di(\theta)'\Delta^g_{i(\theta)}, \qquad \mathcal{J}_\theta = di(\theta)'\mathcal{J}^g_{i(\theta)} di(\theta) \tag{111}$$

The respective classes of CvM influence curves $\Phi^g_\mu(i(\theta))$ and $\Phi_\mu(\theta)$ can be mapped onto one another by the following correspondence within $L^k_2(\mu)$,

$$\varphi^g_{i(\theta)} = di(\theta)\varphi_\theta \tag{112}$$

Then functionals T^g and T that are connected via (105) are simultaneously CvM differentiable at $P^g_{i(\theta)} = P_\theta$, in the respective models, and their CvM influence curves $\varphi^g_{i(\theta)}$ and φ_θ are related by (112). Under the assumptions $\mu \in \mathcal{M}_\sigma(\mathbb{B}^m)$ and 4.2(20), (112) implies the equivariance relation (107) between the corresponding ψ functions of form 4.2(51).

The robust influence curves defined as solutions to certain optimality problems do in general not verify the local equivariance relation (107). We assume the full parameter case $\tau = \mathrm{id}_{\mathbb{R}^k}$. Then for

$$\varrho_\theta = Y_\theta \min\Bigl\{1, \frac{b}{|Y_\theta|}\Bigr\}, \qquad Y_\theta = A_\theta\Lambda_\theta - a_\theta \tag{113}$$

which is the optimal influence curve given in Theorem 5.5.1 a, we obtain that

$$di(\theta)\varrho_\theta = Y^g_{i(\theta)} \min\Bigl\{1, \frac{b}{\bigl|di(\theta)^{-1}Y^g_{i(\theta)}\bigr|}\Bigr\} \tag{114}$$

where

$$\begin{aligned} Y^g_{i(\theta)} &= A^g_{i(\theta)}\Lambda^g_{i(\theta)} - a^g_{i(\theta)} = di(\theta)Y_\theta \\ A^g_{i(\theta)} &= di(\theta)A_\theta di(\theta)', \qquad a^g_{i(\theta)} = di(\theta)a_\theta \end{aligned} \tag{115}$$

For clipping, however, the norm is now standardized by $di(\theta)di(\theta)'$, which in general cannot even be compensated for by a different bound $b^g \neq b$. Thus, the influence curve $di(\theta)\varrho_\theta$ given by (114) and (115) does not attain the necessary forms (15) or (19) of a solution $\varrho^g_{i(\theta)}$ in model $\mathcal{P}^g$—unless we restrict the transformation group; for example, so that $di(\theta)$ be orthogonal, or that $di(\theta)$ be a scalar multiple of the identity $\mathbb{I}_k$. In general, therefore, the solution to problem $O^{\rm tr}_c(b)$ is not equivariant.

M Standardization

Equivariance of optimal influence curves can be enforced by employing objective functions and bias versions that themselves stay invariant under the correspondence (107). The appropriate *M standardizations* and corresponding norms $|\,.\,|_{M_\theta^{-1}}$ use model dependent symmetric, positive definite matrices $M_\theta \in \mathbb{R}^{k\times k}$ which, like Fisher information $\mathcal{I}_\theta$ in view of (106), are equivariant under reparametrizations:

$$di(\theta)' M^g_{i(\theta)} di(\theta) = M_\theta \tag{116}$$

As a special case, one could choose $M_\theta = \mathbb{I}_k$ and, depending on the transformation, define $M^g_{i(\theta)}$ to be the inverse of $di(\theta)di(\theta)'$. In general, if Y_θ and $Y^g_{i(\theta)} = di(\theta)Y_\theta$ are given by (113) and (115), relation (116) ensures that

$$(Y^g_{i(\theta)})' M^g_{i(\theta)} Y^g_{i(\theta)} = Y_\theta{}' M_\theta Y_\theta \tag{117}$$

Assuming (116), let $\psi^g_{i(\theta)} \in \Psi^g_2(i(\theta))$ and $\psi_\theta \in \Psi_2(\theta)$ be any two influence curves related by (107). Then, similarly to (117), we get

$$(\psi^g_{i(\theta)})' M^g_{i(\theta)} \psi^g_{i(\theta)} = \psi_\theta{}' M_\theta \psi_\theta \tag{118}$$

First, this implies invariance of the M standardized covariance,

$$\mathrm{E}^g_{i(\theta)}\, (\psi^g_{i(\theta)})' M^g_{i(\theta)} \psi^g_{i(\theta)} = \mathrm{E}_\theta\, \psi_\theta{}' M_\theta \psi_\theta \tag{119}$$

hence of the trace,

$$\mathrm{E}^g_{i(\theta)}\, \bigl|(M^g_{i(\theta)})^{1/2} \psi^g_{i(\theta)}\bigr|^2 = \mathrm{E}_\theta\, \bigl|M_\theta^{1/2}\psi_\theta\bigr|^2 \tag{120}$$

As for oscillation, in view of definition 5.3(1), we observe that the tangent classes $\mathcal{G}_*^g(i(\theta))$ and $\mathcal{G}_*(\theta)$ are actually the same, and for every their member q we have

$$\begin{aligned} \left|\mathrm{E}_{i(\theta)}^g\,(M_{i(\theta)}^g)^{1/2}\,\psi_{i(\theta)}^g\,q\right|^2 &= \mathrm{E}_{i(\theta)}^g\,q\,(\psi_{i(\theta)}^g)'\,M_{i(\theta)}^g\,\mathrm{E}_{i(\theta)}^g\,\psi_{i(\theta)}^g\,q \\ &\underset{(116)}{\overset{(107)}{=}} \mathrm{E}_\theta\,q\psi_\theta{}'\,M_\theta\,\mathrm{E}_\theta\,\psi_\theta\,q = \left|\mathrm{E}_\theta\,M_\theta^{1/2}\,\psi_\theta\,q\right|^2 \end{aligned} \tag{121}$$

Secondly, this implies invariance of the M standardized oscillation $\omega_{*,\theta}^M$ [the exact $(s=0)$, not the coordinatewise $(s=2,\infty)$, variants!],

$$\begin{aligned} \omega_{*,i(\theta)}^{M,g}(\psi_{i(\theta)}^g) &= \omega_{*,i(\theta)}^g\big((M_{i(\theta)}^g)^{1/2}\,\psi_{i(\theta)}^g\big) \\ &= \omega_{*,\theta}\big(M_\theta^{1/2}\,\psi_\theta\big) = \omega_{*,\theta}^M(\psi_\theta) \end{aligned} \tag{122}$$

Consequentially, the optimization problem $O_{*,\theta}^{M,\mathrm{tr}}(b)$, which is defined by

$$\operatorname{tr} M_\theta\,\mathcal{C}_\theta(\psi_\theta) = \mathrm{E}_\theta\,\psi_\theta{}'M_\theta\,\psi_\theta = \min! \qquad \psi_\theta \in \Psi_2(\theta),\ \omega_{*,\theta}^M(\psi_\theta) \le b \tag{123}$$

and the corresponding transformed minimum norm problem $O_{*,i(\theta)}^{M,g,\mathrm{tr}}(b)$,

$$\operatorname{tr} M_{i(\theta)}^g\,\mathcal{C}_{i(\theta)}^g(\psi_{i(\theta)}^g) = \min! \qquad \psi_{i(\theta)}^g \in \Psi_2^g(i(\theta)),\ \omega_{*,i(\theta)}^{M,g}(\psi_{i(\theta)}^g) \le b \tag{124}$$

simultaneously have (unique) solutions ϱ_θ, respectively $\varrho_{i(\theta)}^g$, and these are necessarily related by (107). Therefore, problem $O_{*,\theta}^{M,\mathrm{tr}}(b)$ defines an equivariant influence curve.

An equivariant influence curve, for the same reason, is defined by the corresponding invariant mean square error problem $O_{*,\theta}^{M,\mathrm{ms}}(\beta)$,

$$\mathrm{E}_\theta\,\psi_\theta{}'M_\theta\psi_\theta + \beta\left|\omega_{*,\theta}^M(\psi_\theta)\right|^2 = \min! \qquad \psi_\theta \in \Psi_2(\theta) \tag{125}$$

with weight $\beta \in (0,\infty)$. The objective function—invariant MSE—arises as limiting risk if in (79) the norm $|\,.\,|_{M_\theta^{-1}}$ is employed (then β means a neighborhood radius).

For explicit solutions, $\theta \in \Theta$ is now fixed and omitted from notation. Minimal standardized bias is denoted by

$$\omega_*^{M,\min} = \inf\{\,\omega_*^M(\psi) \mid \psi \in \Psi_2\,\}, \qquad \omega_*^M(\psi) = \omega_*(M^{1/2}\psi) \tag{126}$$

Then the solutions can be derived from corresponding unstandardized results (for $p=k$) by first substituting formally

$$\eta \rightsquigarrow M^{1/2}\eta, \qquad D \rightsquigarrow M^{1/2}D \tag{127}$$

and then equating $D = \mathbb{I}_k$. In case $* = c$, the (equivariant) solutions to the invariant minimum trace and mean square error problems may thus be obtained from Theorems 5.5.1 and 5.5.7 b, respectively.

Theorem 5.5.12 [Problems $O_c^{M,\mathrm{tr}}(b)$, $O_c^{M,\mathrm{ms}}(\beta)$; $D = \mathbb{I}_k$]

(a) *In case* $\omega_c^{M,\min} < b \le \omega_c^M(\psi_h)$, *there exist some* $a \in \mathbb{R}^k$, $A \in \mathbb{R}^{k\times k}$ *such that the solution to problem* $O_c^{M,\mathrm{tr}}(b)$ *is of the form*

$$\tilde{\varrho} = (A\Lambda - a)w\,, \qquad w = \min\Bigl\{1, \frac{b}{|M^{1/2}(A\Lambda - a)|}\Bigr\} \tag{128}$$

and

$$\omega_c^M(\tilde{\varrho}) = \omega_c(M^{1/2}\tilde{\varrho}) = b \tag{129}$$

Conversely, if some $\tilde{\varrho} \in \Psi_2$ *is of form (128) for any* $b \in (0,\infty)$, $a \in \mathbb{R}^k$, *and* $A \in \mathbb{R}^{k\times k}$, *then* $\tilde{\varrho}$ *solves problem* $O_c^{M,\mathrm{tr}}(b)$, *and the following representations hold,*

$$a = Az, \quad z = \frac{\mathrm{E}\,\Lambda w}{\mathrm{E}\,w}, \qquad A^{-1} = \mathrm{E}\,(\Lambda - z)(\Lambda - z)'w \tag{130}$$

where $A = A' > 0$.

(b) *The solution to problem* $O_{c;0}^{M,\mathrm{ms}}(\beta)$ *is unique, and agrees with the solution (128) to problem* $O_{c;0}^{M,\mathrm{tr}}(b)$, *with bias bound* $b \in (0,\infty)$ *and bias weight* $\beta \in (0,\infty)$ *related by*

$$\beta b = \mathrm{E}\,\bigl(|M^{1/2}(A\Lambda - a)| - b\bigr)^+ \tag{131}$$

(c) *We have*

$$\omega_c^{M,\min} = \max\Bigl\{\frac{\mathrm{tr}\,AM}{\mathrm{E}\,|M^{1/2}(A\Lambda - a)|}\;\Bigm|\; a \in \mathbb{R}^k,\ A \in \mathbb{R}^{k\times k}\setminus\{0\}\Bigr\} \tag{132}$$

There exist $a \in \mathbb{R}^k$, $A \in \mathbb{R}^{k\times k}\setminus\{0\}$ *and* $\bar{\varrho} \in \Psi_2$ *achieving* $\omega_c^{M,\min} = b$, *respectively. And then necessarily*

$$\bar{\varrho} = b\frac{A\Lambda - a}{|M^{1/2}(A\Lambda - a)|} \qquad \text{on } \{A\Lambda \ne a\} \tag{133}$$

Moreover, $a = Az$ *for some* $z \in \mathbb{R}^k$, *and* $A = A' \ge 0$.

If $\bar{\varrho}$ *in addition is constant on* $\{A\Lambda = a\}$, *then it is the solution.*

Equivariance (107) of the solutions given through (a)–(c) and, in particular, invariance of equation (131), may be verified directly, using (117).

Admissibility

As in the unstandardized case, the solutions $\tilde{\varrho}$ to the problems $O_*^{M,\mathrm{tr}}(b)$ are admissible, in the sense that

$$\psi \in \Psi_2,\ \omega_*^M(\psi) \le b,\ \mathcal{C}(\psi) \le \mathcal{C}(\tilde{\varrho}) \implies \psi = \tilde{\varrho} \tag{134}$$

For $\mathcal{C}(\psi) \le \mathcal{C}(\tilde{\varrho})$ implies that

$$\mathrm{E}\,|\psi|_{M^{-1}}^2 = \mathrm{tr}\,M^{1/2}\mathcal{C}(\psi)M^{1/2} \le \mathrm{tr}\,M^{1/2}\mathcal{C}(\tilde{\varrho})M^{1/2} = \mathrm{E}\,|\tilde{\varrho}|_{M^{-1}}^2$$

and the minimum norm solution $\tilde{\varrho}$ is unique.

Self-Standardization

Another invariant oscillation term $\omega_{*,\theta}^{\mathcal{C}}$ is obtained by *self-standardization*; that is, standardization of the influence curves $\psi_\theta \in \Psi_2(\theta)$ through their own covariances,

$$\omega_{*,\theta}^{\mathcal{C}}(\psi_\theta) = \omega_{*,\theta}\bigl(\mathcal{C}_\theta(\psi_\theta)^{-1/2}\psi_\theta\bigr) \tag{135}$$

Indeed, if $\psi_{i(\theta)}^g \in \Psi_2^g(i(\theta))$, $\psi_\theta \in \Psi_2(\theta)$ are any two influence curves related by (107), then

$$di(\theta)'\mathcal{C}_\theta^g(\psi_{i(\theta)}^g)^{-1}di(\theta) = \mathcal{C}_\theta(\psi_\theta)^{-1} \tag{136}$$

Thus (116) is fulfilled by the inverse covariance $M_\theta = \mathcal{C}_\theta(\psi_\theta)^{-1}$. Therefore, (121) applies and, similarly to (122), we obtain that

$$\omega_{*,i(\theta)}^{\mathcal{C},g}(\psi_{i(\theta)}^g) = \omega_{*,\theta}^{\mathcal{C}}(\psi_\theta) \tag{137}$$

In case $* = c$, self-standardized (contamination) bias is given by

$$\bigl|\omega_{c,\theta}^{\mathcal{C}}(\psi_\theta)\bigr|^2 = \sup\nolimits_{P_\theta} \psi_\theta'\mathcal{C}_\theta(\psi_\theta)^{-1}\psi_\theta \tag{138}$$

Always,

$$\bigl|\omega_{c,\theta}^{\mathcal{C}}(\psi_\theta)\bigr|^2 \ge \mathrm{E}_\theta\, \psi_\theta'\mathcal{C}_\theta(\psi_\theta)^{-1}\psi_\theta = \operatorname{tr}\mathbb{I}_k = k \tag{139}$$

If $b > \sqrt{k}$, and the law $\Lambda_\theta(P_\theta)$ fulfills condition (90), a result on robust covariances says that there exist some $A_\theta \in \mathbb{R}^{k\times k}$ and $a_\theta \in \mathbb{R}^k$ such that

$$\hat{\varrho}_\theta = Y_\theta \min\Bigl\{1,\ \frac{b}{|Y_\theta|_{\mathcal{C}_\theta(\hat{\varrho}_\theta)}}\Bigr\}, \qquad Y_\theta = A_\theta\Lambda_\theta - a_\theta \tag{140}$$

is in fact an influence curve at P_θ [Proposition C.2.8]. Then, using (136), we verify that, with $Y_{i(\theta)}^g$ given as in (115),

$$di(\theta)\hat{\varrho}_\theta = Y_{i(\theta)}^g \min\Bigl\{1,\ \frac{b}{|Y_{i(\theta)}^g|_{\mathcal{C}_\theta(\hat{\varrho}_{i(\theta)}^g)}}\Bigr\} = \hat{\varrho}_{i(\theta)}^g \tag{141}$$

This form of self-standardized influence curve, therefore, is equivariant.

Optimality and Admissibility

Fix $\theta \in \Theta$ and drop it from notation again. Then, setting $M = \mathcal{C}(\hat{\varrho})^{-1}$, it follows from Theorem 5.5.12 a, converse part, that $\hat{\varrho}$ is the unique solution to the following minimum norm problem

$$\operatorname{tr}\mathcal{C}(\hat{\varrho})^{-1}\mathcal{C}(\psi) = \min! \qquad \psi \in \Psi_2,\ \sup\nolimits_P |\psi|_{\mathcal{C}(\hat{\varrho})} \le b \tag{142}$$

which is standardized by the covariance of its own solution. But then $\hat{\varrho}$ is also admissible for the corresponding minimum trace problem with self-standardized bias bound,

$$\operatorname{tr}\mathcal{C}(\hat{\varrho})^{-1}\mathcal{C}(\psi) = \min! \qquad \psi \in \Psi_2,\ \omega_c^{\mathcal{C}}(\psi) \le b \tag{143}$$

Indeed, if any $\psi \in \Psi_2$ achieves

$$\omega_c^{\mathcal{C}}(\psi) \le b\,, \qquad \mathcal{C}(\psi) \le \mathcal{C}(\hat{\varrho}) \tag{144}$$

then $\mathcal{C}(\hat{\varrho})^{-1} \le \mathcal{C}(\psi)^{-1}$ and hence

$$\psi'\mathcal{C}(\hat{\varrho})^{-1}\psi \le \psi'\mathcal{C}(\psi)^{-1}\psi \le b^2 \tag{145}$$

Now $\psi = \hat{\varrho}$ follows from the admissibility (134) of $\hat{\varrho}$ being the solution to the minimum trace problem $O_c^{M,\mathrm{tr}}(b)$, with norming matrix $M = \mathcal{C}(\hat{\varrho})^{-1}$.

Chapter 6

Stable Constructions

6.1 The Construction Problem

Having solved for certain (optimal) influence curves $\psi_\theta \in \Psi_2(\theta)$, one for each $\theta \in \Theta$, the problem remains to construct a functional T and an as. estimator S, without knowing the parameter θ, such that both T and S are as. linear at P_θ with influence curve ψ_θ. In regard to the nonparametric optimality results [Chapter 4] moreover, S should be regular for T on certain full neighborhood systems $\mathcal{U}(\theta)$ about P_θ, so that for all $r \in (0,\infty)$ and all sequences $Q_n \in U(\theta, r/\sqrt{n})$,

$$\sqrt{n}\,\bigl(S_n - T_n(Q_n)\bigr)(Q_n^n) \xrightarrow{\;\mathrm{w}\;} \mathcal{N}(0, \mathcal{C}_\theta(\psi_\theta)) \tag{1}$$

By the finite-dimensional delta method, if $\tau\colon \mathbb{R}^k \to \mathbb{R}^p$ is differentiable at θ, this entails regularity of $\tau\circ S$ for $\tau\circ T$ on $\mathcal{U}(\theta)$ with limit law $\mathcal{N}(0, \Gamma_\theta(\psi_\theta))$; that is,

$$\sqrt{n}\,\bigl(\tau\circ S_n - \tau\circ T_n(Q_n)\bigr)(Q_n^n) \xrightarrow{\;\mathrm{w}\;} \mathcal{N}(0, \Gamma_\theta(\psi_\theta)) \tag{2}$$

for all such sequences (Q_n), provided that

$$\limsup_{n\to\infty} \sqrt{n}\,\bigl|T_n(Q_n) - \theta\bigr| < \infty \tag{3}$$

The covariance $\Gamma_\theta(\psi_\theta) = \mathcal{C}_\theta(\eta_\theta)$ of $\eta_\theta = d\tau(\theta)\psi_\theta$ has been introduced by 4.3(3). Covariance and expectation under P_θ are and will subsequently be denoted by $\mathcal{C}_\theta$ and E_θ, respectively.

We are assuming a parametric model

$$\mathcal{P} = \{\, P_\theta \mid \theta \in \Theta \,\} \subset \mathcal{M}_1(\mathcal{A}) \tag{4}$$

on a general sample space $(\Omega, \mathcal{A})$, whose parameter space Θ is an open subset of some finite-dimensional $\mathbb{R}^k$. In principle, $\mathcal{P}$ is supposed to be identifiable, and L_2 differentiable at $\theta \in \Theta$ with derivative $\Lambda_\theta \in L_2^k(P_\theta)$. However, the constructions by maximum likelihood type equations go through if $\mathcal{P}$ is only L_1 differentiable at θ with derivative $\Lambda_\theta \in L_1^k(P_\theta)$, and the bounded influence curves $\Psi_\infty(\theta)$ are employed. For the minimum distance constructions, we require $\mathcal{P}$ identifiable and differentiable in other norms.

Example 6.1.1 For certain functionals and estimators, the regularity (1) may not be achievable under arbitrarily small perturbations of P_θ.

As total variation distance of product measures is bounded by

$$d_v\Big(\bigotimes_{i=1}^n Q_i\,, \bigotimes_{i=1}^n P_i\Big) \le \sum_{i=1}^n d_v(Q_i, P_i) \tag{5}$$

not any as. estimator S will asymptotically be affected in law under sequences $Q_n \in B_v(P_\theta, r_n)$ with radius $nr_n \to 0$ since then

$$d_v\big(S_n(Q_n^n), S_n(P_\theta^n)\big) \le d_v(Q_n^n, P_\theta^n) \le nr_n \longrightarrow 0 \tag{6}$$

The rate $1/n$ seems critical. Beyond that rate, if $r_n \to 0$ but $nr_n \to \infty$, the product measures may already become as. orthogonal. For example, consider contaminations

$$Q_n = (1 - r_n)P_\theta + r_n \mathbf{I}_{z_n} \tag{7}$$

Assume $z_n \in \Omega \setminus A_n \in \mathcal{A}$ exist such that $P_\theta(A_n) = 1$. Then indeed,

$$P_\theta^n(A_n^n) = 1\,, \qquad Q_n^n(A_n^n) = (1 - r_n)^n \longrightarrow 0 \tag{8}$$

Let us investigate the behavior of mean $\bar{X}$ and expectation E under perturbations of this size $r_n \to 0$, $nr_n \to \infty$. Fix some $P_0 \in \mathcal{M}_1(\mathbb{B})$ of expectation $\theta = 0$ and variance 1. Given any $z \in \mathbb{R}$ set $z_n = z/\sqrt{r_n}$. Then, for the contaminations (7), we get

$$\mathrm{E}(Q_n) = r_n z_n = \sqrt{r_n}\, z \to 0\,, \quad \mathrm{Var}(Q_n) = (1 - r_n)(1 + r_n z_n^2) \to 1 + z^2$$

Using $z_n^2/n = z^2/(nr_n) \to 0$ the Lindeberg condition can be verified. So

$$\sqrt{n}\,\big(\bar{X}_n - \mathrm{E}(Q_n)\big)(Q_n^n) \xrightarrow{\mathrm{w}} \mathcal{N}(0, 1 + z^2) \tag{9}$$

Thus (1) cannot hold.

What can happen to $\bar{X}$ and E under perturbations of P_0 at the critical rate $r_n = r/n$, for some $r \in (0,\infty)$? In this case, given any $z \in \mathbb{R}$ we employ $z_n = \sqrt{n}\, z$. Then, for the contaminations (7),

$$\mathrm{E}(Q_n) = rz/\sqrt{n}\,, \qquad \mathrm{Var}(Q_n) \to 1 + rz^2$$

But as $z_n^2/n \not\to 0$ the Lindeberg condition is not fulfilled for $z \ne 0$. Calculating the Fourier transform g_n of $\sqrt{n}\,\bar{X}_n(Q_n^n)$, we obtain for all $t \in \mathbb{R}$ that

$$\begin{aligned} g_n(t) &= \Big(\big(1 - \tfrac{r}{n}\big)\big(1 - \tfrac{1}{2n}(t^2 + \mathrm{o}(n^0))\big) + \tfrac{r}{n}\, e^{\,izt}\Big)^n \\ &\longrightarrow \exp\big(-\tfrac{1}{2}t^2 + r(e^{\,izt} - 1)\big) \end{aligned} \tag{10}$$

Invoking two stochastically independent random variables $Z \sim \mathcal{N}(0,1)$ standard normal and $Y \sim P(r)$ Poisson with expectation/variance r we have thus shown that

$$\sqrt{n}\,\bar{X}_n(Q_n^n) \xrightarrow{\;\mathrm{w}\;} \mathcal{L}(Z + zY) \tag{11}$$

hence

$$\sqrt{n}\,\bigl(\bar{X}_n - \mathrm{E}(Q_n)\bigr)(Q_n^n) \xrightarrow{\;\mathrm{w}\;} \mathcal{L}\bigl(Z + z(Y-r)\bigr) \tag{12}$$

So (1) is violated again.

It should be noted that the contaminations (7), with $z_n \in \Omega \setminus A_n \in \mathcal{A}$ and $P_\theta(A_n) = 1$, satisfy $d_v(Q_n, P_\theta) = r_n$ but, as $r_n \to 0$,

$$d_h(Q_n, P_\theta) = \Bigl|1 - \sqrt{1 - r_n}\Bigr|^{1/2} \approx \sqrt{\tfrac{1}{2} r_n} \tag{13}$$

Thus, the critical $d_v = \mathrm{O}(1/n)$ rate translates into a $d_h = \mathrm{O}(1/\sqrt{n}\,)$ rate.

To conclude, observe that $\bar{X}$ and E cannot even satisfy (1) under perturbations $Q_n \in B_v(P_\theta, r_n)$ with radius $nr_n \to 0$. In view of (6), $\bar{X}$ won't be affected. But given such $r_n > 0$ it is possible to choose z_n in the contaminations (7) so that $\sqrt{n}\, r_n z_n \not\to 0$. Then $\sqrt{n}\,\bigl(\bar{X}_n - \mathrm{E}(Q_n)\bigr)(Q_n^n)$ is not as. normal $\mathcal{N}(0,1)$. And a superposition of such $\mathrm{o}(1/n)$ contaminations with the previous ones will also destroy convergence in (9) and (12). ////

Actually, we shall work under triangular arrays $Q_{n,1}, \ldots, Q_{n,n} \in \mathcal{M}_1(\mathcal{A})$. Then, for the product measure at sample size n, which is denoted by

$$Q_n^{(n)} = Q_{n,1} \otimes \cdots \otimes Q_{n,n} \tag{14}$$

the functional value is defined as the average,

$$T_n(Q_n^{(n)}) = \frac{1}{n} \sum_{i=1}^{n} T_n(Q_{n,i}) \tag{15}$$

The subsequent constructions will ensure that for all radii $r \in (0, \infty)$ and all triangular arrays $Q_{n,i} \in U(\theta, r/\sqrt{n}\,)$,

$$\sqrt{n}\,\bigl(S_n - T_n(Q_n^{(n)})\bigr)(Q_n^{(n)}) \xrightarrow{\;\mathrm{w}\;} \mathcal{N}(0, \mathcal{C}_\theta(\psi_\theta)) \tag{16}$$

The constructions come in pairs: the first is the functional, the second is the estimator. The estimator construction usually requires similar though somewhat stronger conditions than the construction of the corresponding functional. Linked together, they yield the nonparametric regularity (16).

6.2 M Equations

Huber's (1964, 1967) general proofs of the as. behavior of M estimates may be modified to achieve, by the method of M equations, the nonparametric regularity 6.1(16) for the total variation system $\mathcal{U}_v(\theta)$. In technical respects, we have to deal with sums of summands $\psi_n(x_i)$ under triangular arrays $Q_{n,i} \in \mathcal{M}_1(\mathcal{A})$. The following proposition summarizes the suitable uniform weak law of large numbers and central limit theorem.

Expectations and covariances under $Q_{n,i}$ and P_θ are denoted by $\mathrm{E}_{n,i}$, E_θ, $\mathcal{C}_{n,i}$, and $\mathcal{C}_\theta$, respectively. In the subsequent proofs, summation and maximization over $i = 1, \ldots, n$ may be abbreviated by $\sum_i$ and $\max_i$, respectively.

Proposition 6.2.1 *Let $\psi_n \colon (\Omega, \mathcal{A}) \to (\mathbb{R}^k, \mathbb{B}^k)$ and $c_n \in (0, \infty)$ be two sequences of functions, respectively constants, such that*

$$\sup |\psi_n| = \mathrm{O}(c_n), \qquad c_n = \mathrm{o}(\sqrt{n}) \tag{1}$$

(a) *Then, for all arrays $Q_{n,i} \in \mathcal{M}_1(\mathcal{A})$ and all $\varepsilon \in (0,1)$,*

$$\lim_{n\to\infty} Q_n^{(n)} \left(\left| \frac{1}{n} \sum_{i=1}^n \bigl(\psi_n(x_i) - \mathrm{E}_{n,i}\, \psi_n\bigr) \right| \le \varepsilon \right) = 1 \tag{2}$$

(b) *Suppose radii $r_n \in (0,1)$ such that*

$$\lim_{n\to\infty} r_n c_n = 0 \tag{3}$$

Then

$$\lim_{n\to\infty} Q_n^{(n)} \left(\left| \frac{1}{n} \sum_{i=1}^n \psi_n(x_i) - \mathrm{E}_\theta\, \psi_n \right| \le 2\varepsilon \right) = 1 \tag{4}$$

holds for all arrays $Q_{n,i} \in B_v(P_\theta, r_n)$ and all $\varepsilon \in (0,1)$.

(c) *Assume an array $Q_{n,i} \in \mathcal{M}_1(\mathcal{A})$ and let $0 < C' = C \in \mathbb{R}^{k \times k}$ be some covariance of full rank such that*

$$\frac{1}{n} \sum_{i=1}^n \mathcal{C}_{n,i}(\psi_n) = C + \mathrm{o}(n^0) \tag{5}$$

Then

$$\left(\frac{1}{\sqrt{n}} \sum_{i=1}^n \bigl(\psi_n(x_i) - \mathrm{E}_{n,i}\, \psi_n\bigr) \right) (Q_n^{(n)}) \xrightarrow{\ \mathrm{w}\ } \mathcal{N}(0, C) \tag{6}$$

(d) *Suppose radii $r_n \in (0,1)$ such that*

$$\lim_{n\to\infty} r_n c_n^2 = 0 \tag{7}$$

Then

$$\frac{1}{n} \sum_{i=1}^n \mathcal{C}_{n,i}(\psi_n) = \mathcal{C}_\theta(\psi_n) + \mathrm{o}(n^0) \tag{8}$$

holds for all arrays $Q_{n,i} \in B_v(P_\theta, r_n)$.

PROOF

(a) By Chebyshev's inequality,

$$Q_n^{(n)}\Big(\Big|\frac{1}{n}\sum\nolimits_i\big(\psi_n(x_i)-\mathrm{E}_{n,i}\,\psi_n\big)\Big|>\varepsilon\Big) \qquad (9)$$
$$\leq \frac{1}{\varepsilon^2 n^2}\sum\nolimits_i \operatorname{tr}\mathcal{C}_{n,i}(\psi_n) \leq \frac{1}{\varepsilon^2 n^2}\sum\nolimits_i \mathrm{E}_{n,i}\,|\psi_n|^2 \leq \frac{\mathrm{O}(c_n^2)}{\varepsilon^2 n}$$

and the upper bound goes to 0 due to (1).

(b) Part (a) is in force. Moreover, if $d_v(Q_{n,i},P_\theta)\leq r_n$ then

$$\Big|\frac{1}{n}\sum\nolimits_i\int\psi_n\,d(Q_{n,i}-P_\theta)\Big|\leq 2r_n\,\mathrm{O}(c_n)\xrightarrow{(3)}0 \qquad (10)$$

(c) Consider any unit vector $e\in\mathbb{R}^k$, $|e|=1$. Then

$$\sigma_n^2=\frac{1}{n}\sum\nolimits_i e'\mathcal{C}_{n,i}(\psi_n)e\xrightarrow{(5)}e'\mathcal{C}e>0 \qquad (11)$$

For any $\delta\in(0,1)$, the Lindeberg expression of $e'\sum_i\psi_n(x_i)$ is

$$\frac{1}{n\sigma_n^2}\sum\nolimits_i\int_{\{(e'\psi_n-e'\,\mathrm{E}_{n,i}\,\psi_n)^2>n\delta^2\sigma_n^2\}}\big(e'\psi_n-e'\,\mathrm{E}_{n,i}\,\psi_n\big)^2\,dQ_{n,i} \qquad (12)$$

As $|e'\psi_n-e'\,\mathrm{E}_{n,i}\,\psi_n|\leq 2c_n$, the integration domain, due to (1), eventually become empty. So, the Lindeberg condition is fulfilled. Then, by the Lindeberg–Feller theorem,

$$\Big(\frac{e'}{\sqrt{n}}\sum\nolimits_i\big(\psi_n(x_i)-\mathrm{E}_{n,i}\,\psi_n\big)\Big)(Q_n^{(n)})\xrightarrow{\ \mathrm{w}\ }\mathcal{N}(0,e'\mathcal{C}e) \qquad (13)$$

Letting e vary over all unit vectors, this proves (6).

(d) If $d_v(Q_{n,i},P_\theta)\leq r_n$, then, for all coordinates $j,l=1,\dots,k$,

$$\Big|\frac{1}{n}\sum\nolimits_i\int\psi_{n,j}\psi_{n,l}\,d(Q_{n,i}-P_\theta)\Big|\leq 2r_n\,\mathrm{O}(c_n^2)\xrightarrow{(7)}0 \qquad (14)$$

Moreover, (7) implying (3), convergence (10) of expectations is available.////

6.2.1 Location Parameter

This subsection supplies the location M functionals and corresponding location M estimates relevant to Section 5.2.

Lemma 6.2.2 *Let $\psi\colon\mathbb{R}\to\mathbb{R}$ be increasing and $F\in\mathcal{M}_1(\mathbb{B})$ such that*

$$F\{x\in\mathbb{R}\mid\psi(x-t)<\psi(x)\}\wedge F\{x\in\mathbb{R}\mid\psi(x)<\psi(x+t)\}>0 \qquad (15)$$

for all $t\in(0,\infty)$. Then, for all $t\in\mathbb{R}$,

$$\int\psi(x-t)\,F(dx)=\int\psi(x)\,F(dx)\implies t=0 \qquad (16)$$

The lemma, whose proof is obvious, ensures uniqueness of the zero of M equations. For example, condition (15) is fulfilled by all F in the contamination ball $B_c(G,\varepsilon)$ of radius $\varepsilon \in [0,1)$, if ψ is strictly increasing on some interval to which the center $G \in \mathcal{M}_1(\mathbb{B})$ assigns positive probability.

Recall the L_2 differentiable location setup 5.2(1)–5.2(4), 5.2(7) of Section 5.2; especially, the influence curves $\psi_F = \psi / \mathrm{E}_F\, \psi\Lambda_f \in \Psi_2(F)$ for distribution functions $F \in \mathcal{F}_2(\psi)$, and their variances $\mathcal{C}_F(\psi_F) = \mathrm{E}_F\, |\psi_F|^2$; expectation and (co)variance under F are denoted by E_F, respectively by $\mathcal{C}_F$. The following theorem supplies functionals T_ψ that are as. linear at every $F \in \mathcal{F}_2(\psi)$ with influence curve ψ_F.

Actually, we rather work under L_1 differentiability conditions. Thus let $\mathcal{F}_1 \subset \mathcal{M}_1(\mathbb{B})$ denote the set of all absolutely continuous distribution functions F on the real line whose Lebesgue densities f are absolutely continuous on every bounded interval and have integrable derivatives f',

$$dF = f\, d\lambda\,, \qquad \mathrm{E}_F\, |\Lambda_f| < \infty\,, \qquad \Lambda_f = -f'/f \tag{17}$$

Note that the M equation (21) below can be solved to an exact 0 since assumptions (18) and (19) enforce the intermediate value theorem. Assumptions (18) and (19) also imply uniqueness (16) of the zero.

Theorem 6.2.3 *Assume $F \in \mathcal{F}_1$ and $\psi\colon \mathbb{R} \to \mathbb{R}$ such that*

$$\psi \ \text{ bounded, increasing} \tag{18}$$

$$\mathrm{E}_F\, \psi = 0, \qquad \mathrm{E}_F\, \psi\Lambda_f > 0 \tag{19}$$

$$\psi \ \text{ uniformly continuous} \tag{20}$$

Let $T_\psi = (T_n)$ be a functional such that

$$\int \psi(x - T_n(G_n))\, G_n(dx) = \mathrm{o}\Big(\frac{1}{\sqrt{n}}\Big) \tag{21}$$

for all $r \in (0,\infty)$, $G_n \in B_v(F, r/\sqrt{n})$. Then, for all such sequences (G_n),

$$T_n(G_n) = \int \psi_F\, dG_n + \mathrm{o}\Big(\frac{1}{\sqrt{n}}\Big) \tag{22}$$

PROOF The function $L\colon \mathbb{R} \to \mathbb{R}$ defined by

$$L(t) = \int \psi(x - t)\, F(dx) \tag{23}$$

which is bounded decreasing by assumption (18), is differentiable at $t = 0$,

$$L'(0) = -\, \mathrm{E}_F\, \psi\Lambda_f \tag{24}$$

Indeed,

$$\frac{L(t)}{t} = \int \psi(x)\, \frac{f(x+t) - f(x)}{t}\, \lambda(dx)$$

As ψ is bounded and $\Lambda_f = -f'/f$, it suffices to show that

$$\lim_{t\to 0}\int\Bigl|\frac{f(x+t)-f(x)}{t} - f'(x)\Bigr|\,\lambda(dx) = 0 \tag{25}$$

But this is a consequence of Vitali's theorem [Proposition A.2.2] since

$$\lim_{t\to 0}\frac{f(x+t)-f(x)}{t} = f'(x) \qquad \text{a.e. } \lambda(dx)$$

and

$$\begin{aligned}\int\Bigl|\frac{f(x+t)-f(x)}{t}\Bigr|\,\lambda(dx) &\le \frac{1}{|t|}\iint_{-t^-}^{t^+}|f'(x+s)|\,\lambda(ds)\,\lambda(dx)\\ &\le \int |f'(x)|\,\lambda(dx) = \mathrm{E}_F\,|\Lambda_f| < \infty\end{aligned}$$

Now let $\sqrt{n}\,d_v(G_n,F) \le r < \infty$. We first show that

$$\limsup_{n\to\infty}\sqrt{n}\,\bigl|T_n(G_n)\bigr| < \infty \tag{26}$$

In fact, writing $t_n = T_n(G_n)$, boundedness of ψ and (21) imply that

$$\bigl|L(t_n)\bigr| \le \Bigl|\int \psi(x-t_n)\,G_n(dx)\Bigr| + 2d_v(G_n,F)\sup|\psi| = \mathrm{O}\Bigl(\frac{1}{\sqrt{n}}\Bigr) \tag{27}$$

Suppose $|t_n| \ge \delta > 0$ infinitely often. By monotony of L it would follow that $|L(t_n)| \ge |L(\delta)| \wedge L(-\delta) > 0$ infinitely often, the lower bound being strictly positive by (24) and (19). Thus $t_n \to 0$. But then (26) follows from (27) since eventually $2|L(t_n)| \ge |L'(0)|\,|t_n|$. Furthermore,

$$\begin{aligned}L(t_n) &= L'(0)\,t_n + \mathrm{o}(t_n)\\ &= \int \psi(x-t_n)\,G_n(dx) - \int \psi(x-t_n)\,(G_n-F)(dx)\\ &\underset{(19)}{\overset{(21)}{=}} \mathrm{o}\Bigl(\frac{1}{\sqrt{n}}\Bigr) - \int \bigl(\psi(x-t_n)-\psi(x)\bigr)\,(G_n-F)(dx) - \int \psi\,dG_n\\ &\underset{(20)}{=} -\int \psi\,dG_n + \mathrm{o}\Bigl(\frac{1}{\sqrt{n}}\Bigr)\end{aligned} \tag{28}$$

Since $\sqrt{n}\,\mathrm{o}(t_n) \to 0$ due to (26), division through $L'(0)$ gives (22). ////

Theorem 6.2.4 *Let $F \in \mathcal{F}_1$ and $\psi\colon \mathbb{R} \to \mathbb{R}$ satisfy (18) and (19), as well as*

$$\psi \ \ \textit{Lipschitz bounded} \tag{29}$$

Let $S_\psi = (S_n)$ be an as. estimator such that

$$\frac{1}{\sqrt{n}}\sum_{i=1}^{n}\psi(x_i - S_n) \xrightarrow{G_n^{(n)}} 0 \tag{30}$$

for all arrays $G_{n,i} \in B_v(F, r/\sqrt{n})$, *all* $r \in (0,\infty)$. *Then, for all such* $(G_{n,i})$,

$$\sqrt{n}\,\Big(S_n - \frac{1}{n}\sum_{i=1}^{n}\int \psi_F\,dG_{n,i}\Big)(G_n^{(n)}) \xrightarrow{\;\mathrm{w}\;} \mathcal{N}(0, \mathcal{C}_F(\psi_F)) \tag{31}$$

PROOF Fix $r \in (0,\infty)$, an array $G_{n,i} \in B_v(F, r/\sqrt{n})$, and any bounded sequence $s_n \in \mathbb{R}$. Under $G_n^{(n)}$ consider the array of random variables

$$\psi_n(x_i) = \psi\Big(x_i - \frac{s_n}{\sqrt{n}}\Big) \tag{32}$$

Note that

$$\Big|\int \psi_n\,d(G_{n,i} - F)\Big| \le 2\,d_v(G_{n,i}, F)\sup|\psi|$$

$$\Big|\int \psi_n^2\,d(G_{n,i} - F)\Big| \le 2\,d_v(G_{n,i}, F)\sup|\psi|^2$$

while

$$\Big|\int (\psi_n - \psi)\,dF\Big| \le K\,\frac{|s_n|}{\sqrt{n}}$$

$$\Big|\int (\psi_n^2 - \psi^2)\,dF\Big| \le 2K\,\frac{|s_n|}{\sqrt{n}}\sup|\psi|$$

where K is the Lipschitz constant of ψ. By (18), (19), and (29), it thus follows that

$$\sigma_n^2 = \frac{1}{n}\sum\nolimits_i \mathcal{C}_{n,i}(\psi_n) \longrightarrow \mathrm{E}_F\,\psi^2 \in (0,\infty) \tag{33}$$

where $\mathcal{C}_{n,i}$ denotes variance, and likewise $\mathrm{E}_{n,i}$ will denote expectation, under $G_{n,i}$. Therefore, Proposition 6.2.1 applies with $r_n = r/\sqrt{n}$. By the Lindeberg–Feller theorem, then

$$\Big(\frac{1}{\sqrt{n}}\sum\nolimits_i \big(\psi_n(x_i) - \mathrm{E}_{n,i}\,\psi_n\big)\Big)\,(G_n^{(n)}) \xrightarrow{\;\mathrm{w}\;} \mathcal{N}(0, \mathcal{C}_F(\psi)) \tag{34}$$

With L given by (23), the centering terms behave as follows,

$$\begin{aligned}\frac{1}{\sqrt{n}}\sum\nolimits_i \int \psi_n\,dG_{n,i} &= \frac{1}{\sqrt{n}}\sum\nolimits_i \int \psi\,dG_{n,i} + \sqrt{n}\,L\Big(\frac{s_n}{\sqrt{n}}\Big) \\ &\quad + \frac{1}{\sqrt{n}}\sum\nolimits_i \int (\psi_n - \psi)\,d(G_{n,i} - F) \\ &= \frac{1}{\sqrt{n}}\sum\nolimits_i \int \psi\,dG_{n,i} - s_n\,\mathrm{E}_F\,\psi\Lambda_f + \mathrm{o}(n^0)\end{aligned}$$

using the differentiability (24) and the bound

$$n\Big|\int (\psi_n - \psi)\,d(G_{n,i} - F)\Big| \overset{(29)}{\le} 2\,K|s_n|\,r$$

Thus we have shown that

$$\left(\frac{1}{\sqrt{n}}\sum\nolimits_i \big(\psi_n(x_i) - \mathrm{E}_{n,i}\,\psi\big) + s_n\,\mathrm{E}_F\,\psi\Lambda_f\right)(G_n^{(n)}) \xrightarrow{\;\mathrm{w}\;} \mathcal{N}(0, \mathcal{C}_F(\psi)) \tag{35}$$

Given $s \in \mathbb{R}$ now set

$$s_n = s + \frac{1}{\sqrt{n}}\sum\nolimits_i \mathrm{E}_{n,i}\,\psi_F$$

which defines a bounded sequence, as $|s_n| \le |s| + 2r\sup|\psi_F|$. Then

$$G_n^{(n)}\left(\sqrt{n}\left(S_n - \frac{1}{n}\sum\nolimits_i \mathrm{E}_{n,i}\,\psi_F\right) \le s\right) = G_n^{(n)}\big(\sqrt{n}\,S_n \le s_n\big) \tag{36}$$

Using the monotony of ψ and denoting by U_n the LHS in (30), we obtain that the distribution functions (36) are bounded from above and below by

$$G_n^{(n)}\left(\frac{1}{\sqrt{n}}\sum\nolimits_i \psi_n(x_i) \le U_n\right) \xrightarrow{(35)} \mathcal{N}(0, \mathcal{C}_F(\psi_F))(-\infty, s] \tag{37}$$

and

$$G_n^{(n)}\left(\frac{1}{\sqrt{n}}\sum\nolimits_i \psi_n(x_i) < U_n\right) \xrightarrow{(35)} \mathcal{N}(0, \mathcal{C}_F(\psi_F))(-\infty, s] \tag{38}$$

respectively. Thus (31) has been proved. ////

6.2.2 General Parameter

We are given a family of influence curves $\psi_\theta \in \Psi_\infty(\theta)$ in the general parametric model $\mathcal{P}$ [cf. 6.1(4)]. Since the true parameter is unknown, the following assumptions will effectively have to be made for all θ in the open parameter set $\Theta \subset \mathbb{R}^k$. Now fix any $\theta \in \Theta$. The following conditions (40)–(43) are formulated to hold at the fixed θ:

$$\sup_{\zeta\in\Theta}\sup|\psi_\zeta| < \infty \tag{39}$$

$$\zeta \in \Theta,\ \int \psi_\zeta\,dP_\theta = 0 \implies \zeta = \theta \tag{40}$$

There exist some compact subset $D = D_\theta \subset \Theta$ such that

$$\inf_{\zeta\in\Theta\setminus D}\left|\int \psi_\zeta\,dP_\theta\right| > 0 \tag{41}$$

For all $\zeta \in \Theta$,

$$\lim_{t\to 0}\psi_{\zeta+t}(x) = \psi_\zeta(x) \qquad \text{a.e. } P_\theta(dx) \tag{42}$$

Moreover,

$$\lim_{t\to 0}\sup|\psi_{\theta+t} - \psi_\theta| = 0 \tag{43}$$

In connection with bounded influence curves, it actually suffices to assume L_1 differentiability of $\mathcal{P}$ at θ with derivative $\Lambda_\theta \in L_1^k(P_\theta)$,

$$\int \bigl|dP_{\theta+t} - dP_\theta(1+t'\Lambda_\theta)\bigr| = \mathrm{o}(|t|) \tag{44}$$

Under these conditions, if the M equation (45) can also be solved, M functionals can be constructed with the desired expansion.

Theorem 6.2.5 *Assume (39)–(44), and let $T_\psi = (T_n)$ be a functional such that, for all $r \in (0,\infty)$ and all sequences $Q_n \in B_v(P_\theta, r/\sqrt{n})$,*

$$\int \psi_{T_n(Q_n)}\, dQ_n = \mathrm{o}\Bigl(\frac{1}{\sqrt{n}}\Bigr) \tag{45}$$

Then, for all such sequences (Q_n),

$$T_n(Q_n) = \theta + \int \psi_\theta\, dQ_n + \mathrm{o}\Bigl(\frac{1}{\sqrt{n}}\Bigr) \tag{46}$$

PROOF Introduce the function $L\colon \Theta \to \mathbb{R}^k$,

$$L(\zeta) = \int \psi_\zeta\, dP_\theta \tag{47}$$

Given $\sqrt{n}\, d_v(Q_n, P_\theta) \le r < \infty$ put $\zeta_n = T_n(Q_n)$. Then eventually

$$|L(\zeta_n)| \le \Bigl|\int \psi_{\zeta_n}\, dQ_n\Bigr| + 2\sup|\psi_{\zeta_n}|\, d_v(Q_n,P_\theta) \underset{(45)}{\overset{(39)}{\le}} \frac{3r}{\sqrt{n}} \sup_{\zeta\in\Theta}\sup|\psi_\zeta| \tag{48}$$

In particular, it follows that $L(\zeta_n) \to 0$, and then $\zeta_n \in D$ eventually, by assumption (41). Now let U be any open neighborhood of θ. Due to (42) and (39), the function L is continuous on Θ. By (40), it has no zero in the compact set $D \setminus U$, hence must be bounded away from 0 there. As $L(\zeta_n) \to 0$, it follows that $\zeta_n \in U$ eventually. Thus, $t_n = \zeta_n - \theta \to 0$. Moreover, using $L(\theta) = 0$ we have

$$\begin{aligned} L(\theta + t_n) &= \int \psi_{\zeta_n}\, dQ_n - \int (\psi_{\theta+t_n} - \psi_\theta)\, d(Q_n - P_\theta) - \int \psi_\theta\, dQ_n \\ &\underset{(43)}{\overset{(45)}{=}} - \int \psi_\theta\, dQ_n + \mathrm{o}\Bigl(\frac{1}{\sqrt{n}}\Bigr) \end{aligned} \tag{49}$$

But L_1 differentiability of $\mathcal{P}$ at θ and the properties of $\Psi_\infty(\theta)$ imply that L has bounded derivative $dL(\theta) = -\mathbb{I}_k$ at θ since, as $t \to 0$,

$$\begin{aligned} L(\theta + t) &= -\int \psi_{\theta+t}\, d(P_{\theta+t} - P_\theta) \\ &= -\int \psi_\theta\, d(P_{\theta+t} - P_\theta) - \int (\psi_{\theta+t} - \psi_\theta)\, d(P_{\theta+t} - P_\theta) \\ &\underset{(43)}{\overset{(39)}{=}} - \int \psi_\theta \Lambda_\theta'\, dP_\theta\, t + \mathrm{o}(|t|) + \mathrm{o}(|t|^0) \int |dP_{\theta+t} - dP_\theta| \\ &= -t + \mathrm{o}(|t|) \end{aligned} \tag{50}$$

Plugging in $t = t_n$, (50) and (49) give (46), because $\sqrt{n}\, t_n$ is bounded by (48) so that indeed $\mathrm{o}(|t_n|) = \mathrm{o}(1/\sqrt{n}\,)$. ////

Locally Uniform Consistency

The corresponding M estimate construction first needs to ensure locally uniform consistency. For measurability reasons, all x sections of ψ are required continuous,

$$\psi(x,.\,) \in \mathcal{C}^k(\Theta) \tag{51}$$

which strengthens condition (42). Condition (41) is strengthened by the requirement that there exist some compact subset $D = D_\theta \subset \Theta$ such that

$$\inf_{\zeta\in\Theta\setminus D} |L(\zeta)| > \int \sup_{\zeta\in\Theta\setminus D} \bigl|\psi_\zeta(x) - L(\zeta)\bigr|\, P_\theta(dx) \tag{52}$$

where L is the function (47). The following consistency result corresponds to Huber (1967, Lemma 2, Theorem 3; 1981, Lemma 2.3, Theorem 2.4).

Proposition 6.2.6 *Assume* (39), (40), (51), (52). *Let* $S_\psi = (S_n)$ *be an as. estimator such that, for all* $r \in (0,\infty)$ *and arrays* $Q_{n,i} \in B_v(P_\theta, r/\sqrt{n}\,)$,

$$\frac{1}{n}\sum_{i=1}^{n} \psi_{S_n}(x_i) \xrightarrow{Q_n^{(n)}} 0 \tag{53}$$

Then, for all such arrays $(Q_{n,i})$,

$$S_n \xrightarrow{Q_n^{(n)}} \theta \tag{54}$$

PROOF Choose $D \subset \Theta$ and numbers c', c'' according to (52) such that

$$\inf_{\zeta\in\Theta\setminus D} |L(\zeta)| > c'' > c' > \mathrm{E}_\theta V \tag{55}$$

where

$$V(x) = \sup_{\zeta\in\Theta\setminus D} \bigl|\psi_\zeta(x) - L(\zeta)\bigr| \tag{56}$$

Due to assumptions (39) and (51), which also imply that L is continuous, V is indeed a random variable and bounded.

Given an array $(Q_{n,i})$ with $\sqrt{n}\, d_v(Q_{n,i}, P_\theta) \le r < \infty$, Proposition 6.2.1 applies with $r_n = r/\sqrt{n}$ and $\psi_n = V$ so that, by Chebyshev's inequality,

$$\lim_{n\to\infty} Q_n^{(n)}\Bigl(\Bigl|\frac{1}{n}\sum\nolimits_i V(x_i) - \mathrm{E}_\theta V\Bigr| > 2\varepsilon\Bigr) = 0 \tag{57}$$

Since for all $\zeta \in \Theta \setminus D$,

$$\Bigl|\frac{1}{n}\sum\nolimits_i \psi_\zeta(x_i)\Bigr| = \Bigl|\frac{1}{n}\sum\nolimits_i \psi_\zeta(x_i) - L(\zeta) + L(\zeta)\Bigr| \ge |L(\zeta)| - \frac{1}{n}\sum\nolimits_i V(x_i) \tag{58}$$

it follows from (55) and (57) that

$$\lim_{n\to\infty} Q_n^{(n)}\left(\inf_{\zeta\in\Theta\setminus D}\Bigl|\frac{1}{n}\sum\nolimits_i \psi_\zeta(x_i)\Bigr| > c'' - c'\right) = 1 \tag{59}$$

Hence, in view of (53),

$$\lim_{n\to\infty} Q_n^{(n)}(S_n \in D) = 1 \tag{60}$$

To proceed, observe that by assumption (51), for all $\zeta\in\Theta$ and $x\in\Omega$,

$$\lim_{\delta\to 0}\sup_{|t|<\delta}\bigl|\psi_{\zeta+t}(x) - \psi_\zeta(x)\bigr| = 0$$

By (39) and dominated convergence, it follows that for all $\zeta\in\Theta$,

$$\lim_{\delta\to 0}\int \sup_{|t|<\delta}\bigl|\psi_{\zeta+t}(x) - \psi_\zeta(x)\bigr|\, P_\theta(dx) = 0 \tag{61}$$

Now let U be any open neighborhood of θ. As L is continuous and by (40) has no zero in the compact set $D\setminus U$, there exists some $\varepsilon\in(0,1)$ such that $|L(\zeta)|\ge 7\varepsilon$ for all $\zeta\in D\setminus U$. According to (61), all $\zeta\in\Theta$ have open neighborhoods $U(\zeta)$ such that

$$\int \sup_{\xi\in U(\zeta)}\bigl|\psi_\xi(x) - \psi_\zeta(x)\bigr|\, P_\theta(dx) \le \varepsilon \tag{62}$$

Consequentially, $|L(\xi) - L(\zeta)|\le\varepsilon$ for $\xi\in U(\zeta)$. Select a finite subcover $U(\zeta_1),\dots,U(\zeta_m)$ of the compact set $D\setminus U$. Then

$$\begin{aligned}\sup_{\xi\in D\setminus U}\Bigl|\frac{1}{n}\sum\nolimits_i \psi_\xi(x_i) - L(\xi)\Bigr| \le{}& \sup_{j=1,\dots,m}\frac{1}{n}\sum\nolimits_i \sup_{\xi\in U(\zeta_j)}\bigl|\psi_\xi(x_i) - \psi_{\zeta_j}(x_i)\bigr| \\ &+ \sup_{j=1,\dots,m}\Bigl|\frac{1}{n}\sum\nolimits_i \psi_{\zeta_j}(x_i) - L(\zeta_j)\Bigr| + \varepsilon\end{aligned} \tag{63}$$

To each of the $2m$ averages on the RHS, Proposition 6.2.1 applies with ψ_n chosen as $\sup_{\xi\in U(\zeta_j)}|\psi_\xi - \psi_{\zeta_j}|$, for which (62) is available, or $\psi_n = \psi_{\zeta_j}$, respectively. Thus the uniform boundedness (39) and Chebyshev's inequality give

$$\lim_{n\to\infty} Q_n^{(n)}\left(\sup_{\xi\in D\setminus U}\Bigl|\frac{1}{n}\sum\nolimits_i \psi_\xi(x_i) - L(\xi)\Bigr| \le 6\varepsilon\right) = 1 \tag{64}$$

Using $|L(\xi)|\ge 7\varepsilon$ for $\xi\in D\setminus U$ and the triangle inequality as in (58), this entails

$$\lim_{n\to\infty} Q_n^{(n)}\left(\inf_{\xi\in D\setminus U}\Bigl|\frac{1}{n}\sum\nolimits_i \psi_\xi(x_i)\Bigr| \ge \varepsilon\right) = 1 \tag{65}$$

hence

$$\lim_{n\to\infty} Q_n^{(n)}(S_n \in U) = 1 \tag{66}$$

in view of (53). ////

Locally Uniform Asymptotic Normality

Based on this consistency result, locally uniform as. normality of M estimates will be proved. We need that ψ_θ be bounded,

$$\sup |\psi_\theta| < \infty \tag{67}$$

and a Lipschitz condition about θ in sup norm: There exist $\delta_0, K \in (0,\infty)$ such that for all $|\eta - \zeta| + |\zeta - \theta| \le \delta_0$,

$$\sup |\psi_\eta - \psi_\zeta| \le K|\eta - \zeta| \tag{68}$$

hence

$$|L(\eta) - L(\zeta)| \le K|\eta - \zeta|$$

where L is the function (47). Under assumptions (44), (67), and (68) implying (43), $dL(\theta) = -\mathbb{I}_k$ by (50). In particular, δ_0 may be assumed so small that there is some $a \in (0,\infty)$ such that for all $|\eta - \theta| \le \delta_0$,

$$|L(\eta)| \ge a|\eta - \theta| \tag{69}$$

Introduce the variable

$$Z_n(\eta,\zeta) = \left| \sum_{i=1}^n \frac{\psi_\eta(x_i) - \psi_\zeta(x_i) - L(\eta) + L(\zeta)}{n|L(\eta)| + \sqrt{n}} \right| \tag{70}$$

The following lemma corresponds to Huber's (1967) crucial Lemma 3.

Lemma 6.2.7 *Assume* (68) *and* (69). *Then*

$$\sup_{|\eta-\theta|\le\delta_0} Z_n(\eta,\theta) \xrightarrow{Q_n^{(n)}} 0 \tag{71}$$

for all $r \in (0,\infty)$ *and all arrays* $Q_{n,i} \in B_v(P_\theta, r/\sqrt{n})$.

PROOF Following Huber (1967), the cube $|\eta - \theta| \le \delta_0$ is subdivided into a slowly increasing number of smaller cubes, and on each of those smaller cubes $Z_n(\eta,\theta)$ is bounded in $Q_n^{(n)}$ probability. The norm $|.|$ on $\mathbb{R}^k$ can in this proof be taken the sup norm.

For some $0 < q < 1/2$ such that $2/q \in \mathbb{N}$ and some $M \in \mathbb{N}$ to be further specified below, consider the concentric cubes

$$C_m = \{ \eta \in \mathbb{R}^k \mid |\eta - \theta| \le \delta_0(1-q)^m \}, \qquad m = 0,\dots,M \tag{72}$$

Subdivide the difference $C_{m-1} \setminus C_m$ into smaller cubes with edges of length

$$2d = \delta_0(1-q)^{m-1} - \delta_0(1-q)^m = \delta_0(1-q)^{m-1}q \tag{73}$$

and centers ξ so that the coordinates of $\xi - \theta$ are odd multiples of d, and

$$|\xi - \theta| = \frac{\delta_0}{2}\left((1-q)^m + (1-q)^{m-1}\right) = \delta_0(1-q)^{m-1}\left(1 - \frac{q}{2}\right) \tag{74}$$

For each value of m there are less than $\delta_0(1-q)^{m-1}/d = 2/q$ to the k such small cubes. So the number N of small cubes $D_1, \ldots, D_N$ in $C_0 \backslash C_M$ is bounded by

$$q^k N \leq 2^k M \tag{75}$$

Given $\sqrt{n}\, d_v(Q_{n,i}, P_\theta) \leq r < \infty$ and any $\varepsilon \in (0,1)$ we have

$$\begin{aligned} Q_n^{(n)}\Big(\sup_{\eta \in C_0} Z_n(\eta,\theta) \geq 2\varepsilon\Big) \leq Q_n^{(n)}\Big(\sup_{\eta \in C_M} Z_n(\eta,\theta) \geq 2\varepsilon\Big) \\ + \sum_{j=1}^{N} Q_n^{(n)}\Big(\sup_{\eta \in D_j} Z_n(\eta,\theta) \geq 2\varepsilon\Big) \end{aligned} \tag{76}$$

The proper specification of q and $M = M_n$ so that the RHS for $n \to \infty$ tends to 0 will be

$$q \leq \frac{\varepsilon a}{3K} \tag{77}$$

and, with any fixed $1/2 < \gamma < 1$,

$$(1-q)^M \leq \frac{1}{n^\gamma} < (1-q)^{M-1} \tag{78}$$

thus

$$M_n - 1 < \frac{\gamma \log n}{|\log(1-q)|} \leq M_n \tag{79}$$

Consequentially,

$$N = \mathrm{O}(\log n) \tag{80}$$

Now take any of the cubes $D_j \subset C_{m-1} \setminus C_m$ with center ξ and edges of length $2d$ according to (73) and (74). By the triangle inequality, we have

$$Z_n(\eta,\theta) \leq Z_n(\eta,\xi) + \left|\sum\nolimits_i \frac{\psi_\xi(x_i) - \psi_\theta(x_i) - L(\xi)}{n|L(\eta)| + \sqrt{n}}\right| \tag{81}$$

for all $\eta \in D_j$, for which (69) and (68) ensure that

$$|L(\eta)| \geq a|\eta - \theta| \geq a\delta_0(1-q)^m \tag{82}$$

and

$$2 \sup|\psi_\eta - \psi_\xi| \leq 2Kd = K\delta_0(1-q)^{m-1}q \tag{83}$$

Therefore,

$$Z_n(\eta,\xi) \leq \frac{2nKd}{na\delta_0(1-q)^m} = \frac{Kq}{a(1-q)} \underset{(77)}{\leq} \frac{2\varepsilon}{3} \tag{84}$$

Thus,

$$\sup_{\eta \in D_j} Z_n(\eta,\theta) \geq 2\varepsilon \implies \left|\sum\nolimits_i \frac{\psi_\xi(x_i) - \psi_\theta(x_i) - L(\xi)}{na\delta_0(1-q)^m}\right| \geq \varepsilon \tag{85}$$

But by (68) and (74),

$$\left|\sum\nolimits_i \int \frac{\psi_\xi - \psi_\theta}{n(1-q)^m}\, d(Q_{n,i} - P_\theta)\right| \le \frac{K\delta_0(1-q)^{m-1}}{(1-q)^m}\frac{2r}{\sqrt{n}} \le \frac{4rK\delta_0}{\sqrt{n}} \tag{86}$$

By Chebyshev's inequality, eventually

$$\begin{aligned}
Q_n^{(n)}\Bigl(\sup_{\eta\in D_j} Z_n(\eta,\theta) \ge 2\varepsilon\Bigr) &\overset{(85)}{\le} Q_n^{(n)}\left(\left|\sum\nolimits_i \frac{\psi_\xi(x_i) - \psi_\theta(x_i) - L(\xi)}{na\delta_0(1-q)^m}\right| \ge \varepsilon\right) \\
&\underset{(86)}{\le} Q_n^{(n)}\left(\left|\frac{3}{n}\sum\nolimits_i \bigl(\psi_\xi(x_i) - \psi_\theta(x_i) - \mathrm{E}_{n,i}(\psi_\xi - \psi_\theta)\bigr)\right| > a\delta_0(1-q)^m\varepsilon\right) \\
&\le \sum\nolimits_i \frac{9\,\mathrm{tr}\,\mathcal{C}_{n,i}(\psi_\xi - \psi_\theta)}{n^2a^2\delta_0^2(1-q)^{2m}\varepsilon^2} \overset{(68)}{\le} \frac{9K^2|\xi-\theta|^2}{na^2\delta_0^2(1-q)^{2m}\varepsilon^2} \\
&\underset{(74)}{\le} \frac{9K^2(1-q)^{2m-2}}{na^2(1-q)^{2m}\varepsilon^2} = \frac{9K^2}{na^2(1-q)^2\varepsilon^2} \overset{(77)}{\le} \frac{1}{nq^2(1-q)^2}
\end{aligned} \tag{87}$$

Using (80) it follows that

$$\sum_{j=1}^{N} Q_n^{(n)}\Bigl(\sup_{\eta\in D_j} Z_n(\eta,\theta) \ge 2\varepsilon\Bigr) = \mathrm{O}\Bigl(\frac{\log n}{n}\Bigr) \longrightarrow 0 \tag{88}$$

Furthermore, using (68) for $\eta \in C_M$, we have

$$\begin{aligned}
\sup_{\eta\in C_M} Z_n(\eta,\theta) &\le \frac{2}{\sqrt{n}}\sum_{i=1}^{n} \sup_{\eta\in C_M} \sup|\psi_\eta - \psi_\theta| \\
&\le 2\sqrt{n}\,K\delta_0(1-q)^M \underset{(78)}{\le} 2K\delta_0\frac{\sqrt{n}}{n^\gamma} \longrightarrow 0
\end{aligned} \tag{89}$$

In view of (76), (88) and (89) imply (71). ////

The following result extends Huber's (1967) Theorem 3 and Corollary to the locally uniform setup.

Theorem 6.2.8 *Under assumptions* (44), (67), *and* (68), *let* $S_\psi = (S_n)$ *be an as. estimator such that, for all* $r \in (0,\infty)$ *and all* $Q_{n,i} \in B_v(P_\theta, r/\sqrt{n})$,

$$\frac{1}{\sqrt{n}}\sum_{i=1}^{n} \psi_{S_n}(x_i) \xrightarrow{Q_n^{(n)}} 0 \tag{90}$$

and

$$\lim_{n\to\infty} Q_n^{(n)}\bigl(|S_n - \theta| \le \delta_0\bigr) = 1 \tag{91}$$

Then, for all such arrays $(Q_{n,i})$,

$$\left(\sqrt{n}\,(S_n - \theta) - \frac{1}{\sqrt{n}}\sum_{i=1}^{n}\int \psi_\theta\, dQ_{n,i}\right)(Q_n^{(n)}) \xrightarrow{\;\mathrm{w}\;} \mathcal{N}(0, \mathcal{C}_\theta(\psi_\theta)) \tag{92}$$

PROOF Writing $\psi_{S_n}(x_i) = \big(\psi_{S_n}(x_i) - \psi_\theta(x_i) - L(S_n)\big) + \big(\psi_\theta(x_i) + L(S_n)\big)$ and adding up, (91) ensures that with $Q_n^{(n)}$ probability tending to 1,

$$\Big|\sum\nolimits_i \frac{\psi_\theta(x_i) + L(S_n)}{n|L(S_n)| + \sqrt{n}}\Big| \le \sup_{|\eta-\theta|\le\delta_0} Z_n(\eta,\theta) + \Big|\frac{1}{\sqrt{n}}\sum\nolimits_i \psi_{S_n}(x_i)\Big| \tag{93}$$

By Lemma 6.2.7 and (90) respectively, the terms on the RHS go to 0. Therefore, given any $\varepsilon \in (0,1)$, the bound

$$\Big|\frac{1}{\sqrt{n}}\sum\nolimits_i \big(\psi_\theta(x_i) + L(S_n)\big)\Big| \le \varepsilon\big(\sqrt{n}\,|L(S_n)| + 1\big) \tag{94}$$

is violated with $Q_n^{(n)}$ probability eventually smaller than ε. Moreover, for any $M \in (0,\infty)$ and all $n \ge 1$, Chebyshev's inequality tells us that

$$Q_n^{(n)}\Big(\Big|\frac{1}{\sqrt{n}}\sum\nolimits_i \big(\psi_\theta(x_i) - \mathrm{E}_{n,i}\,\psi_\theta\big)\Big| > M\Big) \le \frac{1}{M^2}\sup|\psi_\theta|^2 \tag{95}$$

where

$$\Big|\frac{1}{\sqrt{n}}\sum\nolimits_i \int \psi_\theta\, d(Q_{n,i} - P_\theta)\Big| \le 2r\,\sup|\psi_\theta| \tag{96}$$

Using (67) choose

$$M_\varepsilon = \Big(2r \vee \frac{1}{\sqrt{\varepsilon}}\Big)\sup|\psi_\theta| \tag{97}$$

Then, for all $n \ge 1$,

$$\Big|\frac{1}{\sqrt{n}}\sum\nolimits_i \psi_\theta(x_i)\Big| \le 2M_\varepsilon \tag{98}$$

is violated with $Q_n^{(n)}$ probability smaller than ε. So both (94) and (98) hold simultaneously with probability eventually larger than $1 - 2\varepsilon$. But, by the triangle inequality, (94) implies that

$$\sqrt{n}\,|L(S_n)|(1-\varepsilon) \le \varepsilon + \Big|\frac{1}{\sqrt{n}}\sum\nolimits_i \psi_\theta(x_i)\Big| \tag{99}$$

hence, by (98),

$$\sqrt{n}\,|L(S_n)| \le \frac{2M_\varepsilon + \varepsilon}{1-\varepsilon} \tag{100}$$

Turning back again to (94), this gives

$$\Big|\frac{1}{\sqrt{n}}\sum\nolimits_i \psi_\theta(x_i) + \sqrt{n}\,L(S_n)\Big| \le \frac{(2M_\varepsilon + 1)\varepsilon}{1-\varepsilon} \tag{101}$$

which holds with $Q_n^{(n)}$ probability eventually larger than $1 - 2\varepsilon$. Since the upper bound goes to 0 for $\varepsilon \to 0$ it follows that

$$\frac{1}{\sqrt{n}}\sum\nolimits_i \psi_\theta(x_i) + \sqrt{n}\,L(S_n) \xrightarrow{Q_n^{(n)}} 0 \tag{102}$$

Furthermore, in view of (50), assumptions (44), (67), and (68) implying (43), ensure that $dL(\theta) = -\mathbb{I}_k$. In particular, we conclude from (69), (91), and (100) that the bound

$$\sqrt{n}\,|S_n - \theta| \le \frac{2M_\varepsilon + \varepsilon}{a(1-\varepsilon)} < \infty \tag{103}$$

comes true with $Q_n^{(n)}$ probability eventually larger than $1-2\varepsilon$. Using this tightness and $dL(\theta) = -\mathbb{I}_k$, the finite-dimensional delta method yields that

$$\sqrt{n}\,L(S_n) = -\sqrt{n}\,(S_n - \theta) + \mathrm{o}_{Q_n^{(n)}}(n^0) \tag{104}$$

which, connected to (102), implies that

$$\sqrt{n}\,(S_n - \theta) = \frac{1}{\sqrt{n}} \sum\nolimits_i \psi_\theta(x_i) + \mathrm{o}_{Q_n^{(n)}}(n^0) \tag{105}$$

Now appeal to Proposition 6.2.1 c, d, with $r_n = r/\sqrt{n}$, and $\psi_n = \psi_\theta$ bounded. Then, especially (6) gives (92). ////

Remark 6.2.9 Using these results, it should be possible to extend the work on regression M estimates by Huber (1973, 1981) and, more generally, Maronna and Yohai (1981) to the locally uniform regression setup, with its variety of (errors-in-variables, error-free-variables) infinitesimal neighborhood systems and possibly unbounded optimally robust influence curves. Then, such an extension would also be due for the regression L and R estimates of Bickel (1973), respectively, Jaeckel (1972) and Jurečková (1971). ////

6.3 Minimum Distance

Assuming the general parametric model $\mathcal{P}$ introduced by 6.1(4), the minimum distance (MD) idea is to determine the value θ so that P_θ fits best a given probability, respectively the empirical measure.

6.3.1 MD Functionals

To make this precise, the set of probabilites $\mathcal{M}_1(\mathcal{A})$ has to be mapped, and the parametric model $\mathcal{P}$ embedded, into some metric space (Ξ, d). The following conditions are imposed on the parametrization $\theta \mapsto P_\theta$:

$$\zeta \ne \theta \implies d(P_\zeta, P_\theta) > 0 \tag{1}$$

$$\zeta \to \theta \implies d(P_\zeta, P_\theta) \to 0 \tag{2}$$

For every $\theta \in \Theta$ there exist numbers $\eta_\theta, K_\theta \in (0, \infty)$ such that

$$|\zeta - \theta| \le \eta_\theta \implies d(P_\zeta, P_\theta) \ge K_\theta |\zeta - \theta| \tag{3}$$

Analytic Construction

The open parameter space $\Theta \subset \mathbb{R}^k$ being locally compact separable, it has a representation

$$\Theta = \bigcup_{\nu=1}^{\infty} \Theta_\nu \tag{4}$$

with Θ_ν open, the closure $\overline{\Theta}_\nu$ compact, and $\overline{\Theta}_\nu \subset \Theta_{\nu+1}$ for all $\nu \geq 1$.

Due to (2), the parametrization $\theta \mapsto P_\theta$ is uniformly continuous on each compact $\overline{\Theta}_\nu$: For every $\delta \in (0,\infty)$ there exists some $\varepsilon_\nu(\delta) \in (0,\infty)$ such that for all $\zeta, \theta \in \overline{\Theta}_\nu$,

$$|\zeta - \theta| < \varepsilon_\nu(\delta) \implies d(P_\zeta, P_\theta) < \delta \tag{5}$$

But then the same is true for the inverse $P_\theta \mapsto \theta$ restricted to the image set $\{\, P_\theta \mid \theta \in \overline{\Theta}_\nu \,\}$: For every $\varepsilon \in (0,\infty)$ there is some $\delta_\nu(\varepsilon) \in (0,\infty)$ such that for all $\zeta, \theta \in \overline{\Theta}_\nu$,

$$d(P_\zeta, P_\theta) < \delta_\nu(\varepsilon) \implies |\zeta - \theta| < \varepsilon \tag{6}$$

Choose three sequences $r_\nu, z_\nu, \rho_n \in (0,\infty)$ such that

$$\lim_{\nu\to\infty} r_\nu = \infty\,, \qquad \lim_{\nu\to\infty} z_\nu = 0\,, \qquad \lim_{n\to\infty} \sqrt{n}\,\rho_n = 0 \tag{7}$$

For $\delta = \rho_n$ in (5), by the compactness of $\overline{\Theta}_\nu$, there exist finite subsets

$$\Theta_{\nu,n} = \{\theta_{\nu,n;1}, \ldots, \theta_{\nu,n;q_{\nu,n}}\} \subset \overline{\Theta}_\nu \subset \Theta$$

such that for all $n, \nu \geq 1$,

$$\sup_{\zeta\in\overline{\Theta}_\nu} \inf\{\, d(P_\zeta, P_\theta) \mid \theta \in \Theta_{\nu,n} \,\} \leq \rho_n \tag{8}$$

Define $T_{\nu,n}\colon \mathcal{M}_1(\mathcal{A}) \to \Theta_{\nu,n}$ so that $T_{\nu,n}(Q)$ denotes the first element $\theta_{\nu,n;j}$ of $\Theta_{\nu,n}$ to achieve $\inf\{\, d(Q, P_\theta) \mid \theta \in \Theta_{\nu,n} \,\}$. This means that

$$T_{\nu,n}(Q) = \theta_{\nu,n;j} \iff \begin{cases} d(Q, P_{\theta_{\nu,n;i}}) > d(Q, P_{\theta_{\nu,n;j}})\,, & i < j \\ d(Q, P_{\theta_{\nu,n;i}}) \geq d(Q, P_{\theta_{\nu,n;j}})\,, & i \geq j \end{cases} \tag{9}$$

Given any $Q \in \mathcal{M}_1(\mathcal{A})$, by (2) and $\overline{\Theta}_\nu$ compact, there is some $\zeta \in \overline{\Theta}_\nu$ which minimizes $d(Q, P_\theta)$ for $\theta \in \overline{\Theta}_\nu$. For this ζ choose $\theta_{\nu,n;i} \in \Theta_{\nu,n}$ according to (8). Then, by (8) and (9),

$$d(Q, P_{\theta_{\nu,n;j}}) \leq d(Q, P_{\theta_{\nu,n;i}}) \leq d(Q, P_\zeta) + \rho_n \tag{10}$$

Thus, for all $Q \in \mathcal{M}_1(\mathcal{A})$ and all $n, \nu \geq 1$,

$$d(Q, P_{T_{\nu,n}(Q)}) \leq \inf_{\theta\in\overline{\Theta}_\nu} d(Q, P_\theta) + \rho_n \tag{11}$$

Let $B_d(P_\theta, r) = \{\, Q \in \mathcal{M}_1(\mathcal{A}) \mid d(Q, P_\theta) \leq r \,\}$.

Lemma 6.3.1 *Assume* (1) *and* (2), *and let* $T_{\nu,n}$ *be defined by* (9). *Then for every* $\nu \in \mathbb{N}$ *there exists some* $m_\nu \in \mathbb{N}$ *such that for all* $n \geq m_\nu$,

$$\sup_{\theta \in \overline{\Theta}_\nu} \sup\{ |T_{\nu,n}(Q) - \theta| \mid Q \in B_d(P_\theta, r_\nu/\sqrt{n}) \} \leq z_\nu \tag{12}$$

PROOF Fix any $\nu \geq 1$, $\theta \in \overline{\Theta}_\nu$, and any sequence $Q_n \in B_d(P_\theta, r_\nu/\sqrt{n})$. Put $\zeta_n = T_{\nu,n}(Q_n)$. Then by (11),

$$d(P_{\zeta_n}, P_\theta) \leq d(Q_n, P_{\zeta_n}) + d(Q_n, P_\theta) \leq 2d(Q_n, P_\theta) + \rho_n \leq \frac{2r_\nu}{\sqrt{n}} + \rho_n \tag{13}$$

Choose m_ν so that this bound becomes less than $\delta_\nu(z_\nu)$ for all $n \geq m_\nu$. Then $|\zeta_n - \theta| \leq z_\nu$ by (6). ////

We can arrange that $m_\nu < m_{\nu+1}$ for all $\nu \geq 1$. Then define

$$\nu(n) = \max\{ \nu \geq 1 \mid m_\nu \leq n \} \tag{14}$$

Thus $m_{\nu(n)} \leq n < m_{\nu(n)+1}$ for all $n \geq 1$. The MD functional $T_d = (T_{d,n})$ is now obtained from $(T_{\nu,n})$ by a certain diagonalization:

$$T_{d,n} = T_{\nu(n),n} \colon \mathcal{M}_1(\mathcal{A}) \longrightarrow \Theta_{\nu(n),n} \subset \Theta \tag{15}$$

$\sqrt{n}$ Boundedness

This construction, among other things, achieves bounded infinitesimal oscillation of T_d on the neighborhood system $\mathcal{U}_d$.

Theorem 6.3.2 *Assume* (1), (2) *and* (3), *and let* $T_d = (T_{d,n})$ *be defined by* (15). *Then, for all* $\theta \in \Theta$ *and* $r \in (0, \infty)$,

$$\limsup_{n \to \infty} \sup\{ \sqrt{n}\, |T_{d,n}(Q) - \theta| \mid Q \in B_d(P_\theta, r/\sqrt{n}) \} < \infty \tag{16}$$

PROOF Fix any $\theta \in \Theta$, $r \in (0, \infty)$, and any sequence $Q_n \in B_d(P_\theta, r/\sqrt{n})$. Choose ν such that $\theta \in \overline{\Theta}_\nu$ and then n_0 so that $\nu(n) \geq \nu$ and $r_{\nu(n)} \geq r$ for all $n \geq n_0$. Put $\zeta_n = T_{d,n}(Q_n)$. Then

$$\begin{aligned} d(P_{\zeta_n}, P_\theta) &\leq d(Q_n, P_{\zeta_n}) + d(Q_n, P_\theta) \\ &\overset{(11)}{\leq} \inf_{\zeta \in \overline{\Theta}_{\nu(n)}} d(Q_n, P_\zeta) + \rho_n + d(Q_n, P_\theta) \\ &\leq \inf_{\zeta \in \overline{\Theta}_\nu} d(Q_n, P_\zeta) + \rho_n + d(Q_n, P_\theta) \leq \frac{2r}{\sqrt{n}} + \rho_n \end{aligned} \tag{17}$$

At the same time, $|\zeta_n - \theta| \leq z_{\nu(n)}$ by Lemma 6.3.1. Increase n_0 so that also $z_{\nu(n)} \leq \eta_\theta$ for all $n \geq n_0$. Then it follows from (3) and (17) that

$$K_\theta \sqrt{n}\, |\zeta_n - \theta| \leq 2r + \sqrt{n}\, \rho_n$$

which proves (16). ////

Special Metrics

The metric $d = d_*$ is now specialized to the total variation ($* = v$), Kolmogorov ($* = \kappa$), Hellinger ($* = h$), and Cramér–von Mises ($* = \mu$) distances introduced in 4.2(7)–4.2(10). The corresponding metric spaces are: $\Xi_h = \mathcal{L}_2(\mathcal{A})$, the Hilbert space of square roots defined in 2.3(5); Ξ_v, the Banach space of real measures on $\mathcal{A}$; Ξ_κ, the Banach space of bounded measurable real-valued functions on $\mathbb{R}^m$; and $\Xi_\mu = L_2(\mu)$. We can allow a possibly infinite weight $\mu \in \mathcal{M}(\mathbb{B}^m)$ (as the arguments go through, and the results are not affected, if d_μ is modified to $1 \wedge d_\mu$).

For $* = v, \kappa, h$, identifiability (1) of $\mathcal{P}$ in the metric d_* is the same as identifiability of $\mathcal{P}$,

$$\zeta \neq \theta \implies P_\zeta \neq P_\theta \tag{18}$$

and this suffices for identifiability of $\mathcal{P}$ in $L_2(\mu)$ if μ has full support [see Example 2.3.12 a, however]. The continuity and local Lipschitz properties (2) and (3) will for these metrics be ensured by the corresponding global differentiability of $\mathcal{P}$. Recall Definitions 2.3.6, 2.3.10, and 2.3.11.

Lemma 6.3.3 *Assume that $\mathcal{P}$ is L_1, L_2,* CvM *differentiable at all $\theta \in \Theta$, respectively. Then conditions* (2) *and* (3) *are fulfilled for the metrics d_v and d_κ, for d_h, and for d_μ, respectively.*

PROOF In the cases $* = h, \mu$, differentiability implies continuity (2). Moreover,

$$\begin{aligned} 8d_h^2(P_\zeta, P_\theta) &= (\zeta - \theta)'\mathcal{I}_\theta(\zeta - \theta) + \mathrm{o}(|\zeta - \theta|^2) \\ d_\mu^2(P_\zeta, P_\theta) &= (\zeta - \theta)'\mathcal{J}_\theta(\zeta - \theta) + \mathrm{o}(|\zeta - \theta|^2) \end{aligned} \tag{19}$$

as $\zeta \to \theta$. Since $\mathcal{I}_\theta > 0$, $\mathcal{J}_\theta > 0$ are part of the differentiability definitions, condition (3) is verified. L_1 differentiability in the cases $* = v, \kappa$ ensures that

$$2d_v(P_\zeta, P_\theta) + \mathrm{o}(|\zeta - \theta|) = \mathrm{E}_\theta\, |(\zeta - \theta)'\Lambda_\theta| \le |\zeta - \theta|\, \mathrm{E}_\theta\, |\Lambda_\theta| \tag{20}$$

Hence (2) is fulfilled, also for $d_\kappa \le d_v$. Furthermore,

$$2d_*(P_\zeta, P_\theta) \ge K_{*,\theta}|\zeta - \theta| + \mathrm{o}(|\zeta - \theta|) \tag{21}$$

with the constants

$$K_{v,\theta} = \inf_{|t|=1} \int |t'\Lambda_\theta|\, dP_\theta\,, \quad K_{\kappa,\theta} = \inf_{|t|=1} \sup_{y \in \mathbb{R}^m} \Bigl| \int_{\{x \le y\}} t'\Lambda_\theta(x)\, P_\theta(dx) \Bigr| \tag{22}$$

The nondegeneracy condition 2.3(64) being part of L_1 differentiability entails that for each $t \neq 0$ the two terms

$$\int |t'\Lambda_\theta|\, dP_\theta\,, \qquad \sup_{y \in \mathbb{R}^m} \Bigl| \int_{\{x \le y\}} t'\Lambda_\theta(x)\, P_\theta(dx) \Bigr|$$

are stricly positive, the latter by the uniqueness theorem for the distribution functions of the two finite measures $(t'\Lambda_\theta)^+ \, dP_\theta$, $(t'\Lambda_\theta)^- \, dP_\theta$. Since

$$\bigl| \mathrm{E}_\theta \, |t'\Lambda_\theta| - \mathrm{E}_\theta \, |s'\Lambda_\theta| \bigr| \le |t-s| \, \mathrm{E}_\theta \, |\Lambda_\theta|$$

$$\left| \int_{\{x \le y\}} t'\Lambda_\theta(x) \, P_\theta(dx) - \int_{\{x \le y\}} s'\Lambda_\theta(x) \, P_\theta(dx) \right| \le |t-s| \int |\Lambda_\theta| \, dP_\theta$$

both terms are continuous. It follows that their minimal values $K_{v,\theta}$ and $K_{\kappa,\theta}$ on the compact set $|t| = 1$ are strictly positive. ////

Hellinger and Cramér–von Mises Expansions

The Hellinger and Cramér–von Mises MD functionals T_h and T_μ thus constructed, in addition to being $\sqrt{n}$ bounded, have the announced expansions 4.2(47) and 4.2(49), respectively.

Theorem 6.3.4

(a) *Assume $\mathcal{P}$ is identifiable, and L_2 differentiable at every $\theta \in \Theta$ with L_2 derivative Λ_θ and Fisher information $\mathcal{I}_\theta$. Then T_h is at every $\theta \in \Theta$* Hellinger *differentiable with influence curve $\psi_{h,\theta} = \mathcal{I}_\theta^{-1}\Lambda_\theta$.*

(b) *Assume $\mathcal{P}$,* w.r.t. *some weight $\mu \in \mathcal{M}(\mathbb{B}^m)$, is identifiable in $L_2(\mu)$ and* CvM *differentiable at every $\theta \in \Theta$ with* CvM *derivative Δ_θ and* CvM *information $\mathcal{J}_\theta > 0$. Then T_μ is at every $\theta \in \Theta$* CvM *differentiable with* CvM *influence curve $\varphi_{\mu,\theta} = \mathcal{J}_\theta^{-1}\Delta_\theta$.*

PROOF The proof being similar in both cases, it is carried out for (b): Fix any $\theta \in \Theta$, $r \in (0,\infty)$, and any sequence $Q_n \in B_\mu(P_\theta, r/\sqrt{n})$. Write $\zeta_n = T_{\mu,n}(Q_n)$. We are comparing $\zeta_n - \theta$ with, essentially, the vector u_n of Fourier coefficients of $Q_n - P_\theta$ relative to the ONS $\mathcal{J}_\theta^{-1/2}\Delta_\theta$,

$$u_n = \mathcal{J}_\theta^{-1} \int \Delta_\theta (Q_n - P_\theta) \, d\mu \tag{23}$$

so that

$$\int (Q_n - P_\theta - u_n{}'\Delta_\theta)\Delta_\theta{}' \, d\mu = 0 \tag{24}$$

By Cauchy–Schwarz, $|u_n|$ is of the order $1/\sqrt{n}$. Now plug into the proof of Theorem 6.3.2: Denoting by β_ν the positive distance of $\overline{\Theta}_\nu$ to the complement of $\Theta_{\nu+1}$, increase n_0 there such that $\nu(n) > \nu$, and $|u_n| < \beta_\nu$ hence $\theta + u_n \in \Theta_{\nu+1}$, for all $n \ge n_0$. By Pythagoras and CvM differentiability, it follows that, for all $n \ge n_0$,

$$\Bigl| \bigl\| Q_n - P_\theta - u_n{}'\Delta_\theta \bigr\|^2 + \bigl\| (\zeta_n - \theta - u_n)'\Delta_\theta \bigr\|^2 \Bigr|^{1/2} \tag{25}$$

$$\overset{(24)}{=} \bigl\| Q_n - P_\theta - (\zeta_n - \theta)'\Delta_\theta \bigr\| = \| Q_n - P_{\zeta_n} \| + \mathrm{o}(|\zeta_n - \theta|)$$

$$\underset{(17)}{\overset{(11)}{\le}} \| Q_n - P_{\theta + u_n} \| + \rho_n + \mathrm{o}(|\zeta_n - \theta|) = \bigl\| Q_n - P_\theta - u_n{}'\Delta_\theta \bigr\| + \mathrm{o}\Bigl(\frac{1}{\sqrt{n}}\Bigr)$$

Therefore, $\|(\zeta_n - \theta - u_n)'\Delta_\theta\| = \mathrm{o}(1/\sqrt{n})$ and, since $\mathcal{J}_\theta > 0$, this implies that $\zeta_n = \theta + u_n + \mathrm{o}(1/\sqrt{n})$; which is the asserted CvM expansion. ////

Non-Euclidean Sample Space

Employing some sufficient transformation, the Kolmogorov and Cramér–von Mises MD functionals T_κ and T_μ, which so far are bound to a finite-dimensional Euclidean $(\mathbb{R}^m, \mathbb{B}^m)$, can in principle be extended to an arbitrary sample space $(\Omega, \mathcal{A})$. Conditions (26) and (26′) further below are certainly fulfilled if $d_v(Q_n, P_\theta)$, resp. $\max_i d_v(Q_{n,i}, P_\theta)$, are some $\mathrm{O}(1/\sqrt{n})$.

Proposition 6.3.5 *Assume that $\mathcal{P}$ is identifiable, and L_1 differentiable at every $\theta \in \Theta$. Let $\mu \in \mathcal{M}_b(\mathbb{B})$ be of support $[0,1]$. Then there exists a statistic $\chi\colon (\Omega, \mathcal{A}) \to ([0,1], \mathbb{B})$ such that for $* = \kappa, \mu$ and the* MD *functional $T_* = (T_{*,n})$ the following holds true at every $\theta \in \Theta$: If $Q_n \in \mathcal{M}_1(\mathcal{A})$ is a sequence of probabilities whose image measures $Q_n^\chi = \chi(Q_n)$ under χ fulfill*

$$\limsup_{n\to\infty} \sqrt{n}\, d_*(Q_n^\chi, P_\theta^\chi) < \infty \tag{26}$$

then

$$\limsup_{n\to\infty} \sqrt{n}\, \big|T_{*,n}(Q_n^\chi) - \theta\big| < \infty \tag{27}$$

Moreover,

$$\sqrt{n}\,\big(T_{\mu,n}(Q_n^\chi) - \theta\big) = \sqrt{n} \int \big(Q_n^\chi(y) - P_\theta^\chi(y)\big)\varphi_{\mu,\theta}^\chi(y)\, \mu(dy) \; + \mathrm{o}(n^0) \tag{28}$$

with $\varphi_{\mu,\theta}^\chi = \mathcal{J}_\theta^{-1}\Delta_\theta$ derived from the L_1 derivative Λ_θ of $\mathcal{P}$ at θ via

$$\Delta_\theta(y) = \int \mathbf{I}(\chi \le y)\, \mathrm{E}_\theta(\Lambda_\theta|\chi)\, dP_\theta\,, \qquad \mathcal{J}_\theta = \int \Delta_\theta \Delta_\theta{}' d\mu \tag{29}$$

PROOF Pick (θ_n) dense in Θ and coefficients $\gamma_n \in (0,1)$, $\sum_n \gamma_n = 1$. L_1 differentiability implying d_v continuity (2), (P_{θ_n}) is d_v dense in $\mathcal{P}$, which therefore is dominated by $\nu = \sum_n \gamma_n P_{\theta_n}$. Choosing any versions p_θ of the likelihoods $dP_\theta/d\nu$, define the map Π,

$$\Pi = (p_{\theta_n})\colon (\Omega, \mathcal{A}) \longrightarrow (\mathbb{R}^{\mathrm{N}}, \mathbb{B}^{\mathrm{N}}) \tag{30}$$

The image measures under Π, for all $B \in \mathbb{B}^{\mathrm{N}}$, have the representation

$$\begin{aligned} P_\theta^\Pi(B) &= \int_{\{\Pi\in B\}} p_\theta\, d\nu = \int_{\{\Pi\in B\}} \mathrm{E}_\nu(p_\theta|\Pi)\, d\nu \\ &= \int_B \mathrm{E}_\nu(p_\theta|\Pi = y)\, \nu^\Pi(dy) \end{aligned}$$

hence

$$P_\theta^\Pi(dy) = \mathrm{E}_\nu(p_\theta|\Pi = y)\, \nu^\Pi(dy) \tag{31}$$

Jensen's inequality for conditional expectations tells us that

$$\bigl|\mathrm{E}_\nu(p_{\theta+t}|\Pi)-\mathrm{E}_\nu(p_\theta|\Pi)-\mathrm{E}_\nu(t'\Lambda_\theta p_\theta|\Pi)\bigr| \le \mathrm{E}_\nu\bigl(|p_{\theta+t}-p_\theta-t'\Lambda_\theta p_\theta|\big|\Pi\bigr) \tag{32}$$

But for all $B\in\mathbb{B}^{\mathrm{N}}$,

$$\begin{aligned}\int_{\{\Pi\in B\}} \mathrm{E}_\nu(\Lambda_\theta p_\theta|\Pi)\,d\nu &= \int_{\{\Pi\in B\}} \Lambda_\theta\,dP_\theta = \int_{\{\Pi\in B\}} \mathrm{E}_\theta(\Lambda_\theta|\Pi)\,dP_\theta \\ &= \int_{\{\Pi\in B\}} \mathrm{E}_\theta(\Lambda_\theta|\Pi)\,\mathrm{E}_\nu(p_\theta|\Pi)\,d\nu\end{aligned}$$

hence

$$\mathrm{E}_\nu(\Lambda_\theta p_\theta|\Pi) = \mathrm{E}_\theta(\Lambda_\theta|\Pi)\,\mathrm{E}_\nu(p_\theta|\Pi) \qquad \text{a.e. } \nu \tag{33}$$

Inserting this into (32) an integration w.r.t. ν shows that the family $\Pi(\mathcal{P})$ of image measures (31) fulfills the L_1 differentiability condition 2.3(63) with derivative $\mathrm{E}_\theta(\Lambda_\theta|\Pi=.\,)$, at every $\theta\in\Theta$.

The nondegeneracy 2.3(64) of this derivative is a consequence of the sufficiency of Π. In fact, each p_{θ_n} is $\sigma(\Pi)$ measurable by the definition of Π. And if $\theta_m\to\theta\in\Theta$ along some subsequence, then $p_{\theta_m}\to p_\theta$ in $L_1(\nu)$, hence in ν probability, and then a.e. ν along some further subsequence. This shows that every density p_θ can be modified on a set of ν measure 0 to become $\sigma(\Pi)$ measurable, which proves Π sufficient for $\mathcal{P}$ [Neyman's criterion]. If this measurability is inserted into (32) and (33), uniqueness of the L_1 derivative of $\mathcal{P}$ at θ [to be shown like 2.3(52), p 58] implies that

$$\mathrm{E}_\theta(\Lambda_\theta|\Pi) = \Lambda_\theta \qquad \text{a.e. } P_\theta \tag{34}$$

Especially, the nondegeneracy of Λ_θ carries over. Identifiability of $\mathcal{P}$ is inherited to $\Pi(\mathcal{P})$ as well since 'a reduction by sufficiency does not reduce the parameter space'.

Now the same considerations as for Π apply to any Borel isomorphism

$$\pi\colon(\mathbb{R}^{\mathrm{N}},\mathbb{B}^{\mathrm{N}}) \longrightarrow ([0,1],\mathbb{B})$$

[Parthasaraty (1967; Chapter I.2)], and then to the composition $\chi=\pi\circ\Pi$.

Thus Lemma 6.3.3 and Theorem 6.3.2 are in force for the family $\chi(\mathcal{P})$ of image measures $\chi(P_\theta)=P_\theta^\chi$ on the standard space, and yield (26) for T_κ. But our assumptions on μ ensure that L_1 differentiability and identifiability of $\chi(\mathcal{P})$ also imply CvM differentiability and identifiability of $\chi(\mathcal{P})$ in $L_2(\mu)$. Thus Lemma 6.3.3 and Theorems 6.3.2, 6.3.4 b are in force to yield the assertions concerning T_μ. ////

Remark 6.3.6 The construction draws on LeCam's (1969) ideas concerning the Kolmogorov MD functional T_κ. In the cases $*=h,\mu$, Beran (1981, 1982) and Millar (1981, 1983), also refer to that source and skip the construction of the MD functional T_*, or are generous on the existence of a minimum or, for the minimization of $d_*(Q, P_\zeta)$, presume the substitutions

$$\sqrt{dP_\zeta} \rightsquigarrow \sqrt{dP_\theta}\bigl(1 + \tfrac{1}{2}(\zeta - \theta)'\Lambda_\theta\bigr), \qquad P_\zeta \rightsquigarrow P_\theta + (\zeta - \theta)'\Delta_\theta$$

of roots and distribution functions right away. ////

6.3.2 MD Estimates

Minimum distance estimates $S_* = (S_{*,n})$ are obtained by evaluating minimum distance functionals $T_* = (T_{*,n})$ at the empirical measure,

$$S_{*,n}(x_1, \ldots, x_n) = T_{*,n}(\hat{P}_n), \qquad \hat{P}_n(x_1, \ldots, x_n) = \frac{1}{n}\sum_{i=1}^{n} \mathbf{I}_{x_i} \tag{35}$$

To translate the properties derived in Theorems 6.3.2 and 6.3.4 to MD estimates, the empirical $\hat{P}_n$ must in the underlying metric d_* approximate a theoretical measure $\bar{Q}_n$ at (stochastic) rate $1/\sqrt{n}$. For the Hellinger distance d_h, this requires some (kernel) smoothing of the empirical as in Beran (1977), and likewise for total variation distance d_v, since in these two metrics $d_*(\hat{P}_n, \bar{Q}_n) = 1$, if $\bar{Q}_n$ has no atoms. Let us rather concentrate on the Kolmogorov and Cramér–von Mises MD estimates S_κ and S_μ. By the results of Sections A.3 and A.4, the empirical is sufficiently close to the theoretical distribution function in d_κ and d_μ distance. The sample space is assumed to be some finite-dimensional Euclidean $(\mathbb{R}^m, \mathbb{B}^m)$.

Kolmogorov MD Estimate

Theorem 6.3.7 *Let $\mathcal{P}$ be identifiable, and L_1 differentiable at all $\theta \in \Theta$. Then $S_\kappa = (S_{\kappa,n})$ is at all sample sizes $n \geq 1$ a random variable,*

$$S_{\kappa,n}\colon (\mathbb{R}^{mn}, \mathbb{B}^{mn}) \longrightarrow (\Theta, \mathbb{B}^k) \tag{36}$$

Moreover, for all $\theta \in \Theta$, all $r \in (0, \infty)$ and all arrays $Q_{n,i} \in B_\kappa(P_\theta, r/\sqrt{n})$, the sequence of laws $\sqrt{n}\,(S_{\kappa,n} - \theta)(Q_n^{(n)})$ is tight in $\mathbb{R}^k$.

PROOF The Kolomogorov distance $d_\kappa(\hat{P}_n, P)$ between the empirical and any other distribution function $P \in \mathcal{M}_1(\mathbb{B}^m)$ is Borel measurable since it is the pointwise supremum of the functions

$$(x_1, \ldots, x_n)' \longmapsto \Bigl|\frac{1}{n}\sum\nolimits_i \mathbf{I}(x_i \leq y) - P(y)\Bigr|$$

where y ranges over any countable dense subset of $\mathbb{R}^m$. In view of definitions (9) and (15), each $T_{\nu,n}(\hat{P}_n)$, hence also $S_{\kappa,n} = T_{\nu(n),n}(\hat{P}_n)$, is Borel measurable. $S_{\kappa,n}$ does not take values outside the parameter space Θ since each functional $T_{\nu,n}$ takes its values in $\Theta_{\nu,n} \subset \overline{\Theta}_\nu \subset \Theta$.

Let $\sqrt{n}\, d_\kappa(Q_{n,i}, P_\theta) \le r < \infty$ and $\varepsilon \in (0,1)$. By Proposition A.3.6, there exists an $m \in (0,\infty)$ such that

$$\liminf_{n\to\infty} Q_n^{(n)}\bigl(\sqrt{n}\, d_\kappa(\hat{P}_n, \bar{Q}_n) \le m\bigr) > 1 - \varepsilon \tag{37}$$

where $\bar{Q}_n = 1/n \sum_i Q_{n,i}$ like $Q_{n,i}$ satisfies $d_\kappa(\bar{Q}_n, P_\theta) \le r/\sqrt{n}$. It follows that

$$\liminf_{n\to\infty} Q_n^{(n)}\bigl(\sqrt{n}\, d_\kappa(\hat{P}_n, P_\theta) \le r + m\bigr) > 1 - \varepsilon \tag{38}$$

Theorem 6.3.2 for the enlarged radius $r+m$ gives us a bound $M \in (0,\infty)$ such that

$$\liminf_{n\to\infty} Q_n^{(n)}\bigl(\sqrt{n}\, |S_{\kappa,n} - \theta| \le M\bigr) > 1 - \varepsilon \tag{39}$$

and this proves the tightness assertion. ////

Cramér–von Mises MD Estimate

The Cramér–von Mises MD estimate S_μ will not only be $\sqrt{n}$ consistent but also as. normal and as. minimax for T_μ. The weight μ is assumed to be σ finite. Moreover, it is required that for all $\theta \in \Theta$, as by 4.2(20) for $P = P_\theta$,

$$\int P_\theta(1 - P_\theta)\, d\mu < \infty \tag{40}$$

Assuming CvM differentiability of $\mathcal{P}$ at θ, let the function $\psi_{\mu,\theta}$ be related to $\varphi_{\mu,\theta} = \mathcal{J}_\theta^{-1}\Delta_\theta$ via 4.2(51); its covariance $\mathcal{C}_\theta(\psi_{\mu,\theta})$ under P_θ is of form 4.2(53). In addition, confer Lemma A.4.6.

Theorem 6.3.8 *Let $\mu \in \mathcal{M}_\sigma(\mathbb{B}^m)$. Assume (40), and that $\mathcal{P}$ is identifiable in $L_2(\mu)$, and CvM differentiable at every $\theta \in \Theta$ with derivative Δ_θ and CvM information $\mathcal{J}_\theta > 0$. Then $S_\mu = (S_{\mu,n})$ is at all sample sizes $n \ge 1$ a random variable,*

$$S_{\mu,n} \colon (\mathbb{R}^{mn}, \mathbb{B}^{mn}) \longrightarrow (\Theta, \mathbb{B}^k) \tag{41}$$

Moreover, for all $\theta \in \Theta$, all $r \in (0,\infty)$, and all arrays $Q_{n,i} \in U(\theta, r/\sqrt{n})$, where

$$U(\theta, r/\sqrt{n}) = B_\mu(P_\theta, r/\sqrt{n}) \cap B_{\mu,1}(P_\theta, r) \tag{42}$$

the sequence of laws $\sqrt{n}\,(S_{\mu,n} - \theta)(Q_n^{(n)})$ is tight in $\mathbb{R}^k$. Furthermore,

$$\sqrt{n}\,\bigl(S_{\mu,n} - T_{\mu,n}(Q_n^{(n)})\bigr)(Q_n^{(n)}) \xrightarrow{\;\mathrm{w}\;} \mathcal{N}(0, \mathcal{C}_\theta(\psi_{\mu,\theta})) \tag{43}$$

provided that, in case $m > 1$, in addition A.4(60) or A.4(61) are assumed for μ and $P = P_\theta$, or

$$U(\theta, r/\sqrt{n}) = B_\mu(P_\theta, r/\sqrt{n}) \cap B_{\mu,1}(P_\theta, r) \cap B_\kappa(P_\theta, r_n) \tag{44}$$

for any sequence $r_n \in (0,\infty)$ tending to 0.

PROOF As in the proof to Theorem 6.3.7, the construction of MD functionals in (9) and (15) ensures that $S_{\mu,n}$ does not take values outside the parameter space Θ, and that $S_{\mu,n}$ is Borel measurable provided the CvM distance between the empirical and any other distribution function $P \in \mathcal{M}_1(\mathbb{B}^m)$ is Borel measurable. This measurability, however, holds by Fubini's theorem because of the nonnegative and product measurable integrand in

$$d_\mu^2(\hat{P}_n, P) = \int \Bigl|\frac{1}{n}\sum\nolimits_i \mathbf{I}(x_i \le y) - P(y)\Bigr|^2 \mu(dy)$$

Given $\theta \in \Theta$, $r \in (0,\infty)$ and an array $Q_{n,i} \in U(\theta, r/\sqrt{n})$ with $U(\theta, r/\sqrt{n})$ defined by (42). Then $\int |Q_{n,i} - P_\theta|\,d\mu \le r$ where, by assumption (40), $\int P_\theta(1-P_\theta)\,d\mu < \infty$. Thus Lemma A.4.6 a and Theorem A.4.4 a imply norm boundedness of the empirical process $Y_n = \sqrt{n}\,(\hat{P}_n - \bar{Q}_n)$: For every $\varepsilon \in (0,\infty)$ there exists some $m \in (0,\infty)$ such that

$$\liminf_{n\to\infty} Q_n^{(n)}\bigl(\sqrt{n}\,d_\mu(\hat{P}_n, \bar{Q}_n) \le m\bigr) > 1 - \varepsilon \tag{45}$$

where $\bar{Q}_n = 1/n \sum_i Q_{n,i}$ like $Q_{n,i}$ satisfies $d_\mu(\bar{Q}_n, P_\theta) \le r/\sqrt{n}$. It follows that

$$\liminf_{n\to\infty} Q_n^{(n)}\bigl(\sqrt{n}\,d_\mu(\hat{P}_n, P_\theta) \le r + m\bigr) > 1 - \varepsilon \tag{46}$$

Theorem 6.3.2 for the enlarged radius $r+m$ gives us a bound $M \in (0,\infty)$ such that

$$\liminf_{n\to\infty} Q_n^{(n)}\bigl(\sqrt{n}\,|S_{\mu,n} - \theta| \le M\bigr) > 1 - \varepsilon \tag{47}$$

proving tightness. Furthermore, Theorem 6.3.4 b for the radius $r+m$ supplies a sequence $\delta_n \downarrow 0$ such that, because $Q_{n,i} \in B_\mu(P_\theta, r/\sqrt{n})$,

$$\sqrt{n}\,\Bigl|T_{\mu,n}(Q_{n,i}) - \theta - \int (Q_{n,i} - P_\theta)\varphi_{\mu,\theta}\,d\mu\Bigr| \le \delta_n \tag{48}$$

and in view of (46), with $Q_n^{(n)}$ probability eventually larger than $1-\varepsilon$,

$$\sqrt{n}\,\Bigl|S_{\mu,n} - \theta - \int (\hat{P}_n - P_\theta)\varphi_{\mu,\theta}\,d\mu\Bigr| \le \delta_n \tag{49}$$

Therefore, with $Q_n^{(n)}$ probability eventually larger than $1-\varepsilon$,

$$\Bigl|\sqrt{n}\,(S_{\mu,n} - T_{\mu,n}(Q_n^{(n)})) - \int Y_n \varphi_{\mu,\theta}\,d\mu\Bigr| \le 2\delta_n \tag{50}$$

Thus we have proved that

$$\sqrt{n}\,(S_{\mu,n} - T_{\mu,n}(Q_n^{(n)})) = \int Y_n \varphi_{\mu,\theta}\,d\mu + \mathrm{o}_{Q_n^{(n)}}(n^0) \tag{51}$$

Invoking Lemma A.4.6 b and Theorem A.4.4 b, we obtain that

$$\langle Y_n | e\rangle \, (Q_n^{(n)}) \xrightarrow{\;\mathrm{w}\;} \mathcal{N}(0, V(P_\theta, e)), \qquad e \in L_2(\mu) \tag{52}$$

where $V(P_\theta, e)$ is defined by A.4(21) and, in view of A.4(22), satisfies

$$V(P_\theta, u'\varphi_{\mu,\theta}) = u'\mathcal{C}_\theta(\psi_{\mu,\theta})u \tag{53}$$

for linear combinations $e = u'\varphi_{\mu,\theta}$ with any $u \in \mathbb{R}^k$. By the Cramèr–Wold device, (43) has been proved. ////

Remark 6.3.9

(a) If $\mathcal{P}$ is L_1 or L_2 differentiable at θ and μ is finite, then S_μ is as. linear at P_θ with bounded influence curve $\psi_{\mu,\theta}$.

(b) If μ is finite, the intersection of $B_\mu(P_\theta, r/\sqrt{n})$ with $B_{\mu,1}(P_\theta, r)$ is not needed in (42) or (44).

(c) Instead of weak convergence of the empirical process Y_n in $L_2(\mu)$, we have used only its norm boundedness and weak convergence of the continuous linear functionals $\langle Y_n | e\rangle$. In Theorem 6.3.8, this distinction allows the balls $B_{\mu,1}(P_\theta, r)$ in (42) and (44), with fixed radius $r \in (0, \infty)$ instead of radii $r_n \to 0$ or even $r_n = r/\sqrt{n}$ [Millar (1981)]. ////

Remark 6.3.10 Under the assumptions of Theorem 6.3.8, and if in addition

$$\mathcal{C}_\theta(\psi_{\mu,\theta}) > 0 \tag{54}$$

the CvM MD estimate S_μ is optimum for estimating and testing the CvM MD functional T_μ on the neighborhood systems (42), (44), in the sense of Theorems 4.3.2, 4.3.4, and 4.3.8.

Indeed, by Lemma 4.2.8, the neighborhood systems (42) and (44) are full. The function $\psi_{\mu,\theta}$, for σ finite weight and under assumption (40), is well defined and, in view of 4.2(54), has the properties 4.3(104). The as. expansion 4.2(49) of the functional T_μ, in view of 4.2(50), implies the as. expansion 4.2(44).

If $\psi_{\mu,\theta}$ is bounded, simple perturbations along the least favorable tangents $A\psi_{\mu,\theta}$, with $A \in \mathbb{R}^{k\times k}$ nonsingular, can be considered directly in the proofs of the above-mentioned theorems. An unbounded $\psi_{\mu,\theta}$ may be approximated according to Remark 4.2.3. In the present setup, another approximation may be based on Lemma C.2.6: The functions ψ corresponding via 4.2(51) to simple functions $\varphi = e$ of type C.2(23), in view of 4.2(52), satisfy

$$\int |\psi - \psi_{\mu,\theta}|^2 \, dP_\theta \le \int P_\theta(1 - P_\theta)\, d\mu \int |\varphi - \varphi_{\mu,\theta}|^2 \, d\mu \tag{55}$$

and are bounded by

$$|\psi| \le \mu(\varphi \ne 0) \sup |\varphi| \tag{56}$$

Thus the desired approximation is obtained if $\varphi \to \varphi_{\mu,\theta}$ in $L_2^k(\mu)$.

Theorem 4.3.8 holds in the fixed radius version [Remark 4.3.9 a]. This is clear from the proof to Lemma 4.2.8 in the cases $* = \mu$ and $* = v, \kappa$, and since the balls $B_{\mu,1}(P_\theta, r)$ in (42) and (44) are not scaled down. ////

Transforming the observations by the statistic $\chi\colon (\Omega, \mathcal{A}) \to ([0,1], \mathbb{B})$ given in Proposition 6.3.5, the definition of Kolmogorov and Cramèr–von Mises MD estimates can in principle be extended to a general sample space,

$$S_{*,n}(x_1, \dots, x_n) = T_{*,n}(\hat{P}_n^\chi), \qquad \hat{P}_n^\chi(x_1, \dots, x_n) = \frac{1}{n} \sum_{i=1}^n \mathbf{I}_{\chi(x_i)} \tag{57}$$

Based on Proposition 6.3.5, we thus obtain the following corollary to Theorems 6.3.7 and 6.3.8.

Corollary 6.3.11 *Assume $\mathcal{P}$ is identifiable, and L_1 differentiable at every $\theta \in \Theta$. Let $\mu \in \mathcal{M}_b(\mathbb{B})$ be of support $[0,1]$. Then the* MD *estimate $S_* = (S_{*,n})$ of type $* = \kappa, \mu$ is at all sample sizes $n \ge 1$ a random variable,*

$$S_{*,n}\colon (\Omega^n, \mathcal{A}^n) \longrightarrow (\Theta, \mathbb{B}^k) \tag{58}$$

For every $\theta \in \Theta$, and all arrays $Q_{n,i} \in \mathcal{M}_1(\mathcal{A})$ satisfying

$$\limsup_{n\to\infty} \sqrt{n} \max_{i=1,\dots,n} d_*(Q_{n,i}^\chi, P_\theta^\chi) < \infty \tag{26'}$$

the sequence of laws $\sqrt{n}\,(S_{*,n} - \theta)(Q_n^{(n)})$ is tight in $\mathbb{R}^k$. Moreover,

$$\sqrt{n}\,\bigl(S_{\mu,n} - T_{\mu,n}(Q_n^{\chi(n)})\bigr)(Q_n^{(n)}) \xrightarrow{\mathrm{w}} \mathcal{N}\bigl(0, \mathcal{C}_\theta(\psi_{\mu,\theta}^\chi)\bigr) \tag{59}$$

where

$$\psi_{\mu,\theta}^\chi(x) = \int_0^1 \bigl(\mathbf{I}(\chi(x) \le y) - P_\theta^\chi(y)\bigr) \varphi_{\mu,\theta}^\chi(y)\, \mu(dy) \tag{60}$$

and $\varphi_{\mu,\theta}^\chi = \mathcal{J}_\theta^{-1} \Delta_\theta$ is defined by (29).

Remark 6.3.12 In the non-i.i.d. regression setup, MD estimates of CvM type have been constructed by Millar (1982, 1984) and Koul (1985), while the related paper by Beran (1982) omits the construction issue. ////

6.4 One-Steps

M and MD functionals and estimators may serve as starting points for one-step constructions of more general functionals and estimators. We are given a family of influence curves $\psi_\theta \in \Psi_2(\theta)$ in the general parametric model $\mathcal{P}$ introduced by 6.1(4). The influence curves ψ_θ will need an approximation by a sequence of families of suitably smooth and bounded functions

$$\psi_{n,\theta} \in L_2^k(P_\theta), \qquad \theta \in \Theta,\ n \ge 1 \tag{1}$$

Since the true θ is unknown, the following conditions will in the end have to be imposed at all $\theta \in \Theta$. Fix any $\theta \in \Theta$. Model $\mathcal{P}$ is assumed L_2 differentiable at θ with derivative $\Lambda_\theta \in L_2^k(P_\theta)$ and Fisher information $\mathcal{I}_\theta > 0$. Moreover, the following conditions are formulated about θ and understood to hold for all bounded sequences $\sqrt{n}\,(\theta_n - \theta)$ in $\mathbb{R}^k$, as $n \to \infty$:

$$\lim_{n\to\infty} \int \left|\psi_{n,\theta_n}\sqrt{dP_{\theta_n}} - \psi_\theta\sqrt{dP_\theta}\,\right|^2 = 0 \tag{2}$$

$$\lim_{n\to\infty} \int |\psi_{n,\theta_n} - \psi_\theta|^2\, dP_\theta = 0 \tag{3}$$

$$\sup|\psi_{n,\theta_n}| = \mathrm{O}(c_{n,\theta}), \qquad c_{n,\theta} = \mathrm{o}(\sqrt{n}\,) \tag{4}$$

$$\int \psi_{n,\theta_n}\, dP_{\theta_n} = \mathrm{o}\Big(\frac{1}{\sqrt{n}}\Big) \tag{5}$$

$$\lim_{n\to\infty} \sup|\psi_{n,\theta_n} - \psi_{n,\theta}| = 0 \tag{6}$$

and

$$\sup|\psi_{n,\theta_n} - \psi_{n,\theta}| \le K_{n,\theta}|\theta_n - \theta| \tag{7}$$

eventually, for some $K_{n,\theta} \in (0,\infty)$ that are subject to further assumptions.

Lemma 6.4.1

(a) *Under conditions* (3)–(5), *if* $\sqrt{n}\,(\theta_n-\theta) \in \mathbb{R}^k$ *is a bounded sequence,*

$$\lim_{n\to\infty} \sqrt{n}\,(\theta_n - \theta) + \sqrt{n}\int \psi_{n,\theta_n}\, dP_\theta = 0 \tag{8}$$

(b) *Under assumption* (4), *conditions* (2) *and* (3) *are equivalent.*

(c) *Under* (2), *there exist versions* $(\psi_{n,\theta}^\flat)$ *satisfying conditions* (3)–(5).

Remark 6.4.2 If the map $\zeta \mapsto \psi_\zeta\sqrt{dP_\zeta} \in \mathcal{L}_2^k(\mathcal{A})$ is continuous at θ, condition (2) holds with $\psi_{n,\theta_n} = \psi_{\theta_n}$. Then construction (15), (16) in part (c) below will show that approximations $\psi_{n,\zeta}^\flat$ with the further properties (3)–(5) can be obtained by downweighting large values of ψ_ζ. ////

PROOF

(a) For any sequence $\sqrt{n}\,(\theta_n - \theta)$ tending to some h in $\mathbb{R}^k$ we have

$$-\int \psi_{n,\theta_n}\, dP_\theta + \mathrm{o}\Big(\frac{1}{\sqrt{n}}\Big) \overset{(5)}{=} \int \psi_{n,\theta_n}\,(dP_{\theta_n} - dP_\theta) \tag{9}$$

$$= \int \psi_{n,\theta_n}\big(\sqrt{dP_{\theta_n}} + \sqrt{dP_\theta}\,\big)\big(\sqrt{dP_{\theta_n}} - \sqrt{dP_\theta}\,\big)$$

$$= \int \psi_{n,\theta_n}\,\big(\sqrt{dP_{\theta_n}} - \sqrt{dP_\theta}\,\big)^2 + 2\int \psi_{n,\theta_n}\sqrt{dP_\theta}\,\big(\sqrt{dP_{\theta_n}} - \sqrt{dP_\theta}\,\big)$$

$$= \int \psi_{n,\theta_n}\,\big(\sqrt{dP_{\theta_n}} - \sqrt{dP_\theta}\,\big)^2 + 2\int \psi_\theta\sqrt{dP_\theta}\,\big(\sqrt{dP_{\theta_n}} - \sqrt{dP_\theta}\,\big)$$

$$+\, 2\int (\psi_{n,\theta_n} - \psi_\theta)\sqrt{dP_\theta}\,\big(\sqrt{dP_{\theta_n}} - \sqrt{dP_\theta}\,\big)$$

where

$$\Bigl|\int \psi_{n,\theta_n}\bigl(\sqrt{dP_{\theta_n}}-\sqrt{dP_\theta}\bigr)^2\Bigr| \tag{10}$$
$$\le \sup|\psi_{n,\theta_n}|\int\bigl(\sqrt{dP_{\theta_n}}-\sqrt{dP_\theta}\bigr)^2 \underset{(4)}{=} \mathrm{o}(\sqrt{n}\,)\,\mathrm{O}\Bigl(\frac{1}{n}\Bigr)=\mathrm{o}\Bigl(\frac{1}{\sqrt{n}}\Bigr)$$

respectively,

$$\Bigl|\int(\psi_{n,\theta_n}-\psi_\theta)\sqrt{dP_\theta}\bigl(\sqrt{dP_{\theta_n}}-\sqrt{dP_\theta}\,\bigr)\Bigr|^2 \tag{11}$$
$$\le\int|\psi_{n,\theta_n}-\psi_\theta|^2\,dP_\theta\int\bigl(\sqrt{dP_{\theta_n}}-\sqrt{dP_\theta}\bigr)^2 \underset{(3)}{=}\mathrm{o}\Bigl(\frac{1}{n}\Bigr)$$

and, by L_2 differentiability,

$$2\sqrt{n}\int\psi_\theta\sqrt{dP_\theta}\bigl(\sqrt{dP_{\theta_n}}-\sqrt{dP_\theta}\,\bigr)\longrightarrow\langle\psi_\theta\sqrt{dP_\theta}\,|h'\Lambda_\theta\sqrt{dP_\theta}\,\rangle=h \tag{12}$$

(b) By the triangle inequality in $\mathcal{L}_2^k(\mathcal{A})$, we have

$$\Bigl|\bigl\|\psi_{n,\theta_n}\sqrt{dP_{\theta_n}}-\psi_\theta\sqrt{dP_\theta}\,\bigr\|\Bigr|-\Bigl|\bigl\|\psi_{n,\theta_n}\sqrt{dP_\theta}-\psi_\theta\sqrt{dP_\theta}\,\bigr\|\Bigr| \tag{13}$$
$$\le\bigl\|\psi_{n,\theta_n}\bigl(\sqrt{dP_{\theta_n}}-\sqrt{dP_\theta}\,\bigr)\bigr\|\le\sup|\psi_{n,\theta_n}|\bigl\|\sqrt{dP_{\theta_n}}-\sqrt{dP_\theta}\,\bigr\|$$

which, by (4) and L_2 differentiability, is some $\mathrm{o}(\sqrt{n}\,)\,\mathrm{O}(1/\sqrt{n}\,)=\mathrm{o}(n^0)$.

(c) Take any function $m\colon[0,\infty)\to\mathbb{R}$, absolutely continuous with derivative dm, and satisfying

$$0\le m\le m(0)=1,\qquad \sup m\,\mathrm{id}_{(0,\infty)}<\infty,\qquad \sup|dm|<\infty \tag{14}$$

Choose any sequence $c_n\in(0,\infty)$ tending to ∞ at rate $\mathrm{o}(\sqrt{n}\,)$. Using the kernel m and the clipping constants c_n, define weighted and recentered versions $\psi_{n,\zeta}^\flat$ of $\psi_{n,\zeta}$ by

$$\psi_{n,\zeta}^\flat=w_{n,\zeta}\psi_{n,\zeta}-\int w_{n,\zeta}\psi_{n,\zeta}\,dP_\zeta \tag{15}$$

with random weights

$$w_{n,\zeta}=m\Bigl(\frac{|\psi_{n,\zeta}|}{c_n}\Bigr) \tag{16}$$

Then

$$|w_{n,\zeta}\psi_{n,\zeta}|=\Bigl|m\Bigl(\frac{|\psi_{n,\zeta}|}{c_n}\Bigr)\frac{\psi_{n,\zeta}}{c_n}c_n\Bigr|\le c_n\sup m\,\mathrm{id}_{(0,\infty)}=\mathrm{O}(c_n)=\mathrm{o}(\sqrt{n}\,)$$

hence also

$$\sup_{\zeta\in\Theta}\sup|\psi_{n,\zeta}^\flat|=\mathrm{o}(\sqrt{n}\,) \tag{17}$$

Thus (4) and (5) hold for $(\psi^\flat_{n,\varsigma})$. It remains to verify (3) or, equivalently, (2) using the approximations $(\psi^\flat_{n,\varsigma})$. By triangle inequality, as the family $(\psi_{n,\varsigma})$ satisfies condition (2) by assumption, we must show that

$$\lim_{n\to\infty} \int \Bigl|\psi^\flat_{n,\theta_n}\sqrt{dP_{\theta_n}} - \psi_{n,\theta_n}\sqrt{dP_{\theta_n}}\,\Bigr|^2 = 0 \tag{18}$$

for every bounded sequence $\sqrt{n}\,(\theta_n - \theta)$ in $\mathbb{R}^k$. To prove this, write

$$\begin{aligned}
&\bigl\|\psi^\flat_{n,\theta_n}\sqrt{dP_{\theta_n}} - \psi_{n,\theta_n}\sqrt{dP_{\theta_n}}\,\bigr\| \qquad\qquad \text{in the space } \mathcal{L}_2^k(\mathcal{A})\\
&\quad= \Bigl\|w_{n,\theta_n}\psi_{n,\theta_n}\sqrt{dP_{\theta_n}} - \int w_{n,\theta_n}\psi_{n,\theta_n}\,dP_{\theta_n} - \psi_{n,\theta_n}\sqrt{dP_{\theta_n}}\,\Bigr\|\\
&\quad\le \bigl\|(w_{n,\theta_n}-1)\psi_{n,\theta_n}\sqrt{dP_{\theta_n}}\,\bigr\| + \Bigl|\int w_{n,\theta_n}\psi_{n,\theta_n}\,dP_{\theta_n}\Bigr|\\
&\quad\le 2\,\bigl\|(w_{n,\theta_n}-1)\psi_{n,\theta_n}\sqrt{dP_{\theta_n}}\,\bigr\| + \mathrm{o}(n^0)
\end{aligned} \tag{19}$$

The latter bound is indeed true because

$$\begin{aligned}
&\Bigl|\int w_{n,\theta_n}\psi_{n,\theta_n}\,dP_{\theta_n}\Bigr|\\
&\qquad\le \int |w_{n,\theta_n}-1||\psi_{n,\theta_n}|\,dP_{\theta_n} + \int \bigl|\psi_{n,\theta_n}\,dP_{\theta_n} - \psi_\theta\,dP_\theta\bigr|
\end{aligned} \tag{20}$$

Then, by Cauchy–Schwarz,

$$\int |w_{n,\theta_n}-1||\psi_{n,\theta_n}|\,dP_{\theta_n} \le \bigl\|(w_{n,\theta_n}-1)\,\psi_{n,\theta_n}\sqrt{dP_{\theta_n}}\,\bigr\| \tag{21}$$

while

$$\begin{aligned}
&\int \bigl|\psi_{n,\theta_n}\,dP_{\theta_n} - \psi_\theta\,dP_\theta\bigr|\\
&\le \int \bigl|\psi_{n,\theta_n}\sqrt{dP_{\theta_n}} - \psi_\theta\sqrt{dP_\theta}\,\bigr|\sqrt{dP_{\theta_n}} + \int \bigl|\psi_\theta\sqrt{dP_\theta}\,\bigr|\bigl|\sqrt{dP_{\theta_n}} - \sqrt{dP_\theta}\,\bigr|\\
&\le \bigl\|\psi_{n,\theta_n}\sqrt{dP_{\theta_n}} - \psi_\theta\sqrt{dP_\theta}\,\bigr\| + \bigl\|\psi_\theta\sqrt{dP_\theta}\,\bigr\|\bigl\|\sqrt{dP_{\theta_n}} - \sqrt{dP_\theta}\,\bigr\|
\end{aligned} \tag{22}$$

which goes to 0 due to (2), $\psi_\theta \in \Psi_2(\theta)$, and by L_2 differentiability of $\mathcal{P}$. Thus (19) is proved. Furthermore, using $0 \le w_{n,\theta_n} \le 1$, we have

$$\begin{aligned}
&\bigl\|(w_{n,\theta_n}-1)\bigl(\psi_{n,\theta_n}\sqrt{dP_{\theta_n}} - \psi_\theta\sqrt{dP_\theta}\,\bigr)\bigr\|^2\\
&\qquad= \int (w_{n,\theta_n}-1)^2\bigl|\psi_{n,\theta_n}\sqrt{dP_{\theta_n}} - \psi_\theta\sqrt{dP_\theta}\,\bigr|^2\\
&\qquad\le \bigl\|\psi_{n,\theta_n}\sqrt{dP_{\theta_n}} - \psi_\theta\sqrt{dP_\theta}\,\bigr\|^2 \underset{(2)}{\longrightarrow} 0
\end{aligned} \tag{23}$$

Therefore, in view of (19), to prove (18) it remains to show that

$$\lim_{n\to\infty}\bigl\|(w_{n,\theta_n}-1)\,\psi_\theta\sqrt{dP_\theta}\,\bigr\| = 0 \tag{24}$$

However, by the properties (14) of the function m, we have

$$\begin{aligned}\int (w_{n,\theta_n} - 1)^2 \, dP_{\theta_n} &= \int \Bigl| m\Bigl(\frac{|\psi_{n,\theta_n}|}{c_n}\Bigr) - m(0)\Bigr|^2 \, dP_{\theta_n} \\ &\le \frac{1}{c_n^2} \sup |dm|^2 \, \bigl\| \psi_{n,\theta_n} \sqrt{dP_{\theta_n}} \, \bigr\|^2 \end{aligned} \tag{25}$$

where $c_n \to \infty$, $\sup |dm| < \infty$, and $\bigl\|\psi_{n,\theta_n}\sqrt{dP_{\theta_n}}\,\bigr\| \to \bigl\|\psi_\theta\sqrt{dP_\theta}\,\bigr\|$ by (2). Hence w_{n,θ_n} tends to 1 in P_{θ_n} probability, and then also in P_θ probability, since $d_v(P_{\theta_n}, P_\theta) \to 0$. Now assertion (24) follows by dominated convergence because $|w_{n,\theta_n} - 1||\psi_\theta| \le |\psi_\theta| \in L_2(P_\theta)$. ////

Example 6.4.3 To robustify influence curves as in (15) and (16) by downweighting, the following kernel functions have been suggested by Tukey, Welsh, Andrews, Huber and Hampel, and Hampel, respectively:

$$m_{\mathrm{T}}(u) = \bigl|(1-u^2)^+\bigr|^2 \tag{26}$$

$$m_{\mathrm{W}}(u) = \exp(-u^2) \tag{27}$$

$$m_{\mathrm{A}}(u) = \frac{\sin u}{u} \, \mathbf{I}_{[0,\pi)}(u) \tag{28}$$

$$m_{\mathrm{HH}}(u) = 1 \wedge \frac{1}{u} \tag{29}$$

and

$$m_{\mathrm{H}}(u) = \mathbf{I}_{[0,a)}(u) + \frac{a}{u} \, \mathbf{I}_{[a,b)}(u) + \frac{a(c-u)}{u(c-b)} \, \mathbf{I}_{[b,c)}(u) \tag{30}$$

for some constants $0 < a < b < c < \infty$. (These authors, however, have certainly not intended the passage $c_n \to \infty$!) ////

Regularization of Log Likelihoods

Approximation (2) can be achieved for the classical scores $\psi_{h,\theta} = \mathcal{I}_\theta^{-1}\Lambda_\theta$.

Lemma 6.4.4 *If $\mathcal{P}$ is L_2 differentiable at $\theta \in \Theta$, there exist a sequence of families of functions $\psi_{h,n,\theta} \in L_2^k(P_\theta)$ that achieve the approximation (2) of $\psi_{h,\theta}$ at θ.*

PROOF The parameter space Θ being separable, and $\theta \mapsto P_\theta$ being d_v continuous by L_2 differentiability, $\mathcal{P}$ is dominated by some finite measure μ, such that $dP_\theta = p_\theta \, d\mu$ $(\theta \in \Theta)$. Employing these densities p_θ and the canonical basis $\{e_1, \dots, e_k\}$ of $\mathbb{R}^k$, define the sequence of families of functions $\psi_{h,n,\theta} \in L_2^k(P_\theta)$ by

$$\psi_{h,n,\theta} = \mathcal{I}_{n,\theta}^{-1}\Lambda_{n,\theta}\,\mathbf{I}(\mathcal{I}_{n,\theta} > 0), \qquad \mathcal{I}_{n,\theta} = \mathrm{E}_\theta \, \Lambda_{n,\theta}\Lambda_{n,\theta}{}' \tag{31}$$

where

$$e_j{}'\Lambda_{n,\theta} = \sqrt{n}\left(\sqrt{\frac{p_{\theta+e_{n,j}}}{p_\theta}} - 1\right)\mathbf{I}(p_\theta > 0), \qquad e_{n,j} = \frac{e_j}{\sqrt{n}} \tag{32}$$

Then, if $\sqrt{n}\,(\theta_n - \theta) = h_n$ is any bounded sequence in $\mathbb{R}^k$, it holds that

$$\lim_{n\to\infty} \big\| \Lambda_\theta \, \mathbf{I}_{\{p_{\theta_n}=0\}} \sqrt{dP_\theta} \, \big\| = 0 \tag{33}$$

by dominated convergence, as $P_\theta(p_{\theta_n} = 0) \le d_v(P_{\theta_n}, P_\theta) \to 0$. Moreover,

$$\begin{aligned} &\big\| \Lambda_{n,\theta_n} \sqrt{dP_{\theta_n}} - \Lambda_\theta \, \mathbf{I}_{\{p_{\theta_n}>0\}} \sqrt{dP_\theta} \, \big\|^2 \\ &\quad = \sum_{j=1}^k \int_{\{p_{\theta_n}>0\}} \Big| \sqrt{n}\, \big(\sqrt{p_{\theta_n+e_{n,j}}} - \sqrt{p_{\theta_n}} \big) - e_j{}'\Lambda_\theta \sqrt{p_\theta} \Big|^2 \, d\mu \\ &\quad \le \sum_{j=1}^k \big\| \sqrt{n}\, \big(\sqrt{dP_{\theta_n+e_{n,j}}} - \sqrt{dP_{\theta_n}} \big) - e_j{}'\Lambda_\theta \sqrt{dP_\theta} \, \big\|^2 \longrightarrow 0 \end{aligned} \tag{34}$$

since each summand, by L_2 differentiability, tends to 0 according to

$$\begin{aligned} &\big\| \sqrt{n}\, \big(\sqrt{dP_{\theta_n+e_{n,j}}} - \sqrt{dP_{\theta_n}} \big) - e_j{}'\Lambda_\theta \sqrt{dP_\theta} \, \big\| \\ &\quad \le \big\| \sqrt{n}\, \big(\sqrt{dP_{\theta_n+e_{n,j}}} - \sqrt{dP_\theta} \big) - (h_n + e_j)'\Lambda_\theta \sqrt{dP_\theta} \, \big\| \\ &\qquad + \big\| \sqrt{n}\, \big(\sqrt{dP_{\theta_n}} - \sqrt{dP_\theta} \big) - h_n{}'\Lambda_\theta \sqrt{dP_\theta} \, \big\| \longrightarrow 0 \end{aligned} \tag{35}$$

Putting (33) and (34) together, we have shown that

$$\lim_{n\to\infty} \big\| \Lambda_{n,\theta_n} \sqrt{dP_{\theta_n}} - \Lambda_\theta \sqrt{dP_\theta} \, \big\| = 0 \tag{36}$$

In particular, $\mathcal{I}_{n,\theta_n} \to \mathcal{I}_\theta$ in operator norm. It follows that

$$\begin{aligned} &\big\| \psi_{h,n,\theta_n} \sqrt{dP_{\theta_n}} - \psi_{h,\theta} \sqrt{dP_\theta} \, \big\| \\ &\quad \le \| \mathcal{I}_{n,\theta_n}^{-1} \| \big\| \Lambda_{n,\theta_n} \sqrt{dP_{\theta_n}} - \Lambda_\theta \sqrt{dP_\theta} \, \big\| + \sqrt{\mathcal{I}_\theta} \, \| \mathcal{I}_{n,\theta_n}^{-1} - \mathcal{I}_\theta^{-1} \| \end{aligned} \tag{37}$$

As $n \to \infty$, the upper bound tends to 0. ////

6.4.1 Functionals

Given influence curves $\psi_\theta \in \Psi_2(\theta)$ and the family (1) of approximations $(\psi_{n,\theta})$, two one-step constructions T and $\widetilde{T}$, based on some preliminary functional τ, are as follows:

$$T_n(Q) = \tau_n(Q) + \int \psi_{n,\tau_n(Q)} \, dQ \tag{38}$$

$$\widetilde{T}_n(Q) = \tau_n(Q) + 2 \int \psi_{n,\tau_n(Q)} \sqrt{dP_{\tau_n(Q)}} \sqrt{dQ} \tag{39}$$

The initial functional τ must be *strict*; that is, each $\tau_n(\mathcal{M}_1(\mathcal{A})) \subset \Theta$. Moreover, given a neighborhood system $\mathcal{U}(\theta)$ about P_θ, τ is assumed to be $\sqrt{n}$ *bounded on* $\mathcal{U}(\theta)$, in the sense that for all $r \in (0, \infty)$,

$$\limsup_{n\to\infty} \sqrt{n} \, \sup\big\{ |\tau_n(Q) - \theta| \;\big|\; Q \in U(\theta, r/\sqrt{n}) \big\} < \infty \tag{40}$$

For example, the MD functionals $\tau = T_d$ constructed in Subsection 6.3.1 are strict and $\sqrt{n}$ bounded on the systems $\mathcal{U}_d(\theta)$ of d balls about P_θ, for all $\theta \in \Theta$ [see 6.3(15) and Theorem 6.3.2].

The Main Expansion

The following result provides as. expansions of T and $\widetilde{T}$ for a hierarchy of neighborhood systems, under increasingly stronger boundedness and smoothness conditions on influence curves (or their approximations).

Theorem 6.4.5 *Let $\mathcal{P}$ be L_2 differentiable at $\theta \in \Theta$. Assume τ is strict and $\sqrt{n}$ bounded on $\mathcal{U}(\theta)$.*

(a) $U(\theta, r/\sqrt{n}) = B_h(P_\theta, r/\sqrt{n})$: *Under conditions (3)–(5) for T, and under conditions (2) and (5) for $\widetilde{T}$, respectively, both functionals T and $\widetilde{T}$ are Hellinger differentiable, hence as. linear, at P_θ with influence curve ψ_θ.*

(b) $U(\theta, r/\sqrt{n}) = B_v(P_\theta, r/\sqrt{n})$: *Under conditions (3)–(6), the functional T has the following as. expansion:*

$$\sqrt{n}\,\big(T_n(Q_n) - \theta\big) = \sqrt{n} \int \psi_{n,\theta}\, dQ_n + \mathrm{o}(n^0) \tag{41}$$

holding for all sequences $Q_n \in B_v(P_\theta, r/\sqrt{n})$, all $r \in (0,\infty)$.

(c) $U(\theta, r/\sqrt{n}) = B_\mu(P_\theta, r/\sqrt{n}) \cap B_v(P_\theta, r_n)$: *Let $(\Omega, \mathcal{A}) = (\mathbb{R}^m, \mathbb{B}^m)$ of finite dimension, and let $\mu \in \mathcal{M}(\mathbb{B}^m)$, $r_n \in (0,\infty)$. Assume (3)–(5), and (7) such that*

$$\lim_{n\to\infty} r_n K_{n,\theta} = 0 \tag{42}$$

Then the expansion (41) of the functional T extends to all $r \in (0,\infty)$ and all sequences $Q_n \in U(\theta, r/\sqrt{n})$.

PROOF Fix $r \in (0,\infty)$ and $Q_n \in U(\theta, r/\sqrt{n})$. In any case, by $\sqrt{n}$ boundedness of τ and after passing to subsequences, setting $t_n = \tau_n(Q_n)$ we may assume $\sqrt{n}\,(t_n - \theta) \to h$ in $\mathbb{R}^k$. Then for the functional $\widetilde{T}$ we write

$$\begin{aligned}
\sqrt{n}\,\big(\widetilde{T}_n(Q_n) - \theta\big) &= \sqrt{n}\,(t_n - \theta) + 2\sqrt{n} \int \psi_{n,t_n} \sqrt{dP_{t_n}}\sqrt{dQ_n} \\
&\overset{(5)}{=} \sqrt{n}\,(t_n - \theta) - 2\sqrt{n} \int \psi_{n,t_n} \sqrt{dP_{t_n}}\big(\sqrt{dP_{t_n}} - \sqrt{dP_\theta}\,\big) \\
&\quad + \mathrm{o}(n^0) + 2\sqrt{n} \int \psi_{n,t_n} \sqrt{dP_{t_n}}\big(\sqrt{dQ_n} - \sqrt{dP_\theta}\,\big)
\end{aligned} \tag{43}$$

But

$$\begin{aligned}
&\Big| \int \psi_{n,t_n} \sqrt{dP_{t_n}}\big(\sqrt{dP_{t_n}} - \sqrt{dP_\theta}\,\big) - \int \psi_\theta \sqrt{dP_\theta}\,\big(\sqrt{dP_{t_n}} - \sqrt{dP_\theta}\,\big)\Big| \\
&\quad \le \big\| \psi_{n,t_n}\sqrt{dP_{t_n}} - \psi_\theta\sqrt{dP_\theta}\,\big\|\,\big\|\sqrt{dP_{t_n}} - \sqrt{dP_\theta}\,\big\| \underset{(2)}{=} \mathrm{o}\Big(\frac{1}{\sqrt{n}}\Big)
\end{aligned} \tag{44}$$

where

$$\sqrt{n}\,(t_n-\theta)-2\sqrt{n}\int\psi_\theta\sqrt{dP_\theta}\,\bigl(\sqrt{dP_{t_n}}-\sqrt{dP_\theta}\,\bigr)\longrightarrow h-h=0 \tag{45}$$

by L_2 differentiability. Therefore $\widetilde{T}$ has the following general expansion

$$\sqrt{n}\,\bigl(\widetilde{T}_n(Q_n)-\theta\bigr)=2\sqrt{n}\int\psi_{n,t_n}\sqrt{dP_{t_n}}\bigl(\sqrt{dQ_n}-\sqrt{dP_\theta}\,\bigr)+\mathrm{o}(n^0) \tag{46}$$

In case (a), when $d_h(Q_n,P_\theta)=\mathrm{O}(1/\sqrt{n}\,)$, this yields the expansion 4.2(47) for $\widetilde{T}$, since

$$\begin{aligned}&\Bigl|\int\bigl(\psi_{n,t_n}\sqrt{dP_{t_n}}-\psi_\theta\sqrt{dP_\theta}\,\bigr)\bigl(\sqrt{dQ_n}-\sqrt{dP_\theta}\,\bigr)\Bigr|\\ &\qquad\le\bigl\|\psi_{n,t_n}\sqrt{dP_{t_n}}-\psi_\theta\sqrt{dP_\theta}\,\bigr\|\bigl\|\sqrt{dQ_n}-\sqrt{dP_\theta}\,\bigr\|\underset{(2)}{=}\mathrm{o}\Bigl(\frac{1}{\sqrt{n}}\Bigr)\end{aligned} \tag{47}$$

The functional T, by Lemma 6.4.1 a, has the following general expansion,

$$\begin{aligned}\sqrt{n}\,\bigl(T_n(Q_n)-\theta\bigr)&=\sqrt{n}\,(t_n-\theta)+\sqrt{n}\int\psi_{n,t_n}\,dQ_n\\ &=\sqrt{n}\,(t_n-\theta)+\sqrt{n}\int\psi_{n,t_n}\,dP_\theta+\sqrt{n}\int\psi_{n,t_n}\,d(Q_n-P_\theta)\\ &=\sqrt{n}\int\psi_{n,t_n}\,d(Q_n-P_\theta)+\mathrm{o}(n^0)\end{aligned} \tag{48}$$

In case (a), when $\sqrt{n}\,d_h(Q_n,P_\theta)\le r<\infty$, we write as in (9),

$$\begin{aligned}\int\psi_{n,t_n}\,d(Q_n-P_\theta)&=\int\psi_{n,t_n}\bigl(\sqrt{dQ_n}+\sqrt{dP_\theta}\,\bigr)\bigl(\sqrt{dQ_n}-\sqrt{dP_\theta}\,\bigr)\\ &=\int\psi_{n,t_n}\bigl(\sqrt{dQ_n}-\sqrt{dP_\theta}\,\bigr)^2+2\int\psi_\theta\sqrt{dP_\theta}\bigl(\sqrt{dQ_n}-\sqrt{dP_\theta}\,\bigr)\\ &\quad+2\int(\psi_{n,t_n}-\psi_\theta)\sqrt{dP_\theta}\,\bigl(\sqrt{dQ_n}-\sqrt{dP_\theta}\,\bigr)\end{aligned} \tag{49}$$

By (4) and (3), respectively, both

$$\Bigl|\int\psi_{n,t_n}\bigl(\sqrt{dQ_n}-\sqrt{dP_\theta}\,\bigr)^2\Bigr|\le\frac{2r^2}{n}\sup|\psi_{n,t_n}| \tag{50}$$

and

$$\Bigl|\int(\psi_{n,t_n}-\psi_\theta)\sqrt{dP_\theta}\,\bigl(\sqrt{dQ_n}-\sqrt{dP_\theta}\,\bigr)\Bigr|\le\frac{\sqrt{2}\,r}{\sqrt{n}}\bigl\|(\psi_{n,t_n}-\psi_\theta)\sqrt{dP_\theta}\,\bigr\| \tag{51}$$

are of the order $\mathrm{o}(1/\sqrt{n}\,)$. Thus expansion 4.2(47) follows for T.

In case (b), when $\sqrt{n}\,d_v(Q_n,P_\theta)\le r<\infty$, the general expansion (48) simplifies to (41) according to

$$\Bigl|\int(\psi_{n,t_n}-\psi_{n,\theta})\,d(Q_n-P_\theta)\Bigr|\le\frac{2r}{\sqrt{n}}\sup|\psi_{n,t_n}-\psi_{n,\theta}|\overset{(6)}{=}\mathrm{o}\Bigl(\frac{1}{\sqrt{n}}\Bigr) \tag{52}$$

In case (c), because

$$\Bigl|\int (\psi_{n,t_n} - \psi_{n,\theta})\, d(Q_n - P_\theta)\Bigr| \le 2 r_n K_{n,\theta} |t_n - \theta| = \mathrm{o}\Bigl(\frac{1}{\sqrt{n}}\Bigr) \tag{53}$$

by (7) and (42), again (41) follows from (48). ////

Remark 6.4.6 Let τ be a strict and $\sqrt{n}$ bounded functional on $\mathcal{U}_d(\theta)$, the system of d balls about P_θ, where d is some metric on $\mathcal{M}_1(\mathcal{A})$ such that

$$\limsup_{n\to\infty} \sqrt{n}\, |\theta_n - \theta| < \infty \implies \limsup_{n\to\infty} \sqrt{n}\, d(P_{\theta_n}, P_\theta) < \infty \tag{54}$$

Then the clipping constants c_n employed in the construction (15), (16), and eventually in the constructions (38) and (39) of functionals, may be adapted via

$$\bar{c}_n(Q) = \frac{c_n}{c \vee \sqrt{n}\, d(Q_n, P_{\tau_n(Q)})} \tag{55}$$

with any fixed $c \in (0,\infty)$, to the discrepancy between the underlying measure and the ideal model. Indeed, if $d(Q_n, P_\theta)$ is of the order $\mathrm{O}(1/\sqrt{n}\,)$ then also $\tau_n(Q_n) - \theta = \mathrm{O}(1/\sqrt{n}\,)$, hence $d(P_{t_n}, P_\theta) = \mathrm{O}(1/\sqrt{n}\,)$ by (54), where $t_n = \tau_n(Q_n)$. Hence $d(Q_n, P_{t_n}) \le d(Q_n, P_\theta) + d(P_{t_n}, P_\theta)$ is of the order $\mathrm{O}(1/\sqrt{n}\,)$. From $c_n \to \infty$ it follows that $\bar{c}_n(Q_n) \to \infty$. Moreover, $\bar{c}_n(Q_n) \le c_n/c = \mathrm{o}(\sqrt{n}\,)$. Thus Theorem 6.4.5 obtains with this adaptive truncation. ////

CvM Expansion

The construction of CvM differentiable functionals with prescribed CvM influence curves is also feasible. For a finite-dimensional Euclidean sample space $(\Omega,\mathcal{A}) = (\mathbb{R}^m, \mathbb{B}^m)$ and any weight $\mu \in \mathcal{M}(\mathbb{B}^m)$, assume that $\mathcal{P}$ is CvM differentiable at $\theta \in \Theta$. Let $\varphi_\theta \in \Phi_\mu(\theta)$ be given. φ_θ needs to be approximated by a sequence of families of smoothly parametrized functions $\varphi_{n,\zeta} \in L_2^k(\mu)$ such that for all bounded sequences $\sqrt{n}\,(\theta_n - \theta)$ in $\mathbb{R}^k$,

$$\lim_{n\to\infty} \int |\varphi_{n,\theta_n} - \varphi_\theta|^2\, d\mu = 0 \tag{56}$$

Using for example the CvM MD functional $\tau = T_\mu$ as a starting point, the one-step construction is

$$T_n(Q) = \tau_n(Q) + \int \varphi_{n,\tau_n(Q)} \bigl(Q - P_{\tau_n(Q)}\bigr)\, d\mu \tag{57}$$

where, as usual, we identify probabilities and distribution functions.

Proposition 6.4.7 *Let $\mathcal{P}$ be* CvM *differentiable at $\theta \in \Theta$ relative to some weight $\mu \in \mathcal{M}(\mathbb{B}^m)$. Assume $\varphi_\theta \in \Phi_\mu(\theta)$ such that condition* (56) *can be fulfilled. Suppose τ strict and $\sqrt{n}$ bounded on $\mathcal{U}_\mu(\theta)$. Then T defined by* (57) *is* CvM *differentiable at P_θ with* CvM *influence curve φ_θ.*

PROOF For any sequence $Q_n \in B_\mu(P_\theta, r/\sqrt{n})$ with $r \in (0,\infty)$, we have

$$\sqrt{n}\,\bigl(T_n(Q_n) - \theta\bigr) = \sqrt{n}\,(t_n - \theta) + \sqrt{n}\int \varphi_{n,t_n}(Q_n - P_{t_n})\,d\mu \tag{58}$$

where $t_n = \tau(Q_n)$, and $\sqrt{n}\,(t_n - \theta) \to h \in \mathbb{R}^k$ can be assumed. Then write

$$\begin{aligned}\int \varphi_{n,t_n}(Q_n - P_{t_n})\,d\mu &= \int \varphi_\theta(Q_n - P_\theta)\,d\mu - \int \varphi_\theta(P_{t_n} - P_\theta)\,d\mu \\ &\quad + \int (\varphi_{n,t_n} - \varphi_\theta)(Q_n - P_\theta)\,d\mu - \int (\varphi_{n,t_n} - \varphi_\theta)(P_{t_n} - P_\theta)\,d\mu\end{aligned} \tag{59}$$

The first term on the RHS makes the leading term in the asserted expansion. Denoting by Δ_θ the CvM derivative of $\mathcal{P}$ at θ the second term satisfies

$$\lim_{n\to\infty} \sqrt{n}\int \varphi_\theta(P_{t_n} - P_\theta)\,d\mu = \int \varphi_\theta\, h'\Delta_\theta\,d\mu \overset{4.2(48)}{=} h \tag{60}$$

The remaining two terms can be bounded by Cauchy–Schwarz. From (56) and the orders $\mathrm{O}(1/\sqrt{n})$ of $d_\mu(Q_n, P_\theta)$ and $d_\mu(P_{t_n}, P_\theta)$ it follows that the last two integrals in (59) are each of the order $\mathrm{o}(1/\sqrt{n})$. Thus the expansion 4.2(49) is proved. ////

6.4.2 Estimators

The following one-step estimator construction requires initial estimators that are strict, $\sqrt{n}$ consistent, and suitably discretized. An estimator σ is called *strict* if each σ_n takes values in Θ. Given a neighborhood system $\mathcal{U}(\theta)$ about P_θ, σ is called *$\sqrt{n}$ consistent on $\mathcal{U}(\theta)$* if for all $r \in (0,\infty)$,

$$\lim_{M\to\infty}\limsup_{n\to\infty}\sup\bigl\{\, Q_n^{(n)}\bigl(\sqrt{n}\,|\sigma_n - \theta| > M\bigr) \bigm| Q_{n,i} \in U(\theta, r/\sqrt{n})\,\bigr\} = 0 \tag{61}$$

For example, the MD estimates $\sigma = S_\kappa, S_\mu$ are strict [by 6.3(35) and strictness of MD functionals] and $\sqrt{n}$ consistent on $\mathcal{U}_\kappa(\theta)$ [Theorem 6.3.7] and $\mathcal{U}_\mu(\theta)$ [Theorem 6.3.8], respectively.

Discretization

In view of 6.3(9), MD functionals and estimates are discrete already by construction, as are all estimators in practice. The following discretization of a general estimator σ dispenses with certain measurability and uniformity assumptions in theory.

Fix some $b \in (0,\infty)$. For each $n \ge 1$ pave $\mathbb{R}^k$ by the cubes

$$Q_n(a) = \prod_{j=1}^k \Bigl(a_j - \frac{b}{\sqrt{n}}\,,\, a_j + \frac{b}{\sqrt{n}}\Bigr]\,, \qquad a \in \mathcal{G}_n \tag{62}$$

using the grids $\mathcal{G}_n = \bigl(2b\,\mathbb{Z}/\sqrt{n}\bigr)^k$. Select any points $a_n^* \in Q_n(a)$ such that $a_n^* \in \Theta$ in case $Q_n(a) \cap \Theta \neq \emptyset$. Then the discretized version σ^* of σ is

$$\sigma_n^* = \sum_{a \in \mathcal{G}_n} a_n^* \, \mathbf{I}\bigl(\sigma_n \in Q_n(a)\bigr) \tag{63}$$

Strictness is inherited from σ to σ^*, and the same goes for $\sqrt{n}$ consistency since $\sqrt{n}\,\sup|\sigma_n^* - \sigma_n| \le 2\sqrt{k}\,b$.

The discretization achieves the following: If K is a compact in $\mathbb{R}^k$ choose any $m \in \mathbb{N}$ such that $K \subset (-bm, bm]^k$. Then, for every $n \ge 1$, there exist at most $q = m^k$ different values $h_{n,s} \in K$ (this number q depending on K but not on n) such that

$$\sqrt{n}\,(\sigma_n^* - \theta) \in K \implies \sqrt{n}\,(\sigma_n^* - \theta) = h_{n,1}, \ldots, h_{n,q} \tag{64}$$

One-Step M Estimate

Now let influence curves $\psi_\theta \in \Psi_2(\theta)$ and the family (1) of approximations $\psi_{n,\theta}$ be given. Using a strict estimator σ and its discretized version σ^* define the estimator $S = (S_n)$ by

$$S_n = \sigma_n^* + \frac{1}{n}\sum_{i=1}^{n} \psi_{n,\sigma_n^*(x_1,\ldots,x_n)}(x_i) \tag{65}$$

With the slight strengthening of condition (4) to

$$\sup|\psi_{n,\theta_n}| = \mathrm{O}(c_{n,\theta}), \qquad c_{n,\theta} = \mathrm{o}(\sqrt[4]{n}) \tag{4'}$$

the conditions (2)–(7) at $\theta \in \Theta$ will be taken over. They guarantee uniform as. normality of the estimator S given by (65).

Theorem 6.4.8 *Let $\mathcal{P}$ be L_2 differentiable at $\theta \in \Theta$. Assume σ is strict and $\sqrt{n}$ consistent on $\mathcal{U}(\theta)$.*

(a) *$U(\theta, r/\sqrt{n}) = B_h(P_\theta, r/\sqrt{n})$: Assume (3), (4′), and (5). Then, for all $r \in (0,\infty)$ and all arrays $Q_{n,i} \in B_h(P_\theta, r/\sqrt{n})$,*

$$\sqrt{n}\left(S_n - \theta - \frac{2}{n}\sum_{i=1}^{n}\int \psi_\theta \sqrt{dP_\theta}\sqrt{dQ_{n,i}}\right)(Q_n^{(n)}) \xrightarrow{\mathrm{w}} \mathcal{N}(0, C_\theta(\psi_\theta)) \tag{66}$$

In particular, S is as. linear at P_θ with influence curve ψ_θ.

(b) *$U(\theta, r/\sqrt{n}) = B_v(P_\theta, r/\sqrt{n})$: Assume (3), (4′), (5), and (6). Then, for all $r \in (0,\infty)$ and all arrays $Q_{n,i} \in B_v(P_\theta, r/\sqrt{n})$,*

$$\sqrt{n}\left(S_n - \theta - \frac{1}{n}\sum_{i=1}^{n}\int \psi_{n,\theta}\, dQ_{n,i}\right)(Q_n^{(n)}) \xrightarrow{\mathrm{w}} \mathcal{N}(0, C_\theta(\psi_\theta)) \tag{67}$$

(c) $U(\theta, r/\sqrt{n}\,) = B_\mu(P_\theta, r/\sqrt{n}\,) \cap B_v(P_\theta, r_n)$: *Let* $(\Omega, \mathcal{A}) = (\mathbb{R}^m, \mathbb{B}^m)$ *of finite dimension, and let* $\mu \in \mathcal{M}_\sigma(\mathbb{B}^m)$, $r_n \in (0,\infty)$. *Assume* (3), (4′), (5), *and* (7) *such that*

$$\lim_{n\to\infty} r_n\,(K_{n,\theta} + c_{n,\theta}^2) = 0 \tag{68}$$

Then the as. normality (67) *of* S *extends to all arrays* $Q_{n,i} \in U(\theta, r/\sqrt{n}\,)$ *and all* $r \in (0,\infty)$.

PROOF Fix $r \in (0,\infty)$, an array $Q_{n,i} \in U(\theta, r/\sqrt{n}\,)$, and $\varepsilon \in (0,1)$. By assumption, the sequence of laws $\sqrt{n}\,(\sigma_n^* - \theta)(Q_n^{(n)})$ is tight in $\mathbb{R}^k$. Hence we can choose some compact $K \subset \mathbb{R}^k$ such that for all $n \ge 1$,

$$Q_n^{(n)}\bigl(\sqrt{n}\,(\sigma_n^* - \theta) \in K\bigr) > 1 - \varepsilon \tag{69}$$

Due to discretization, there is some finite q and for every $n \ge 1$ there are at most q elements $h_{n,s} \in K$ such that

$$Q_n^{(n)}\bigl(\sqrt{n}\,(\sigma_n^* - \theta) = h_{n,1}, \dots, h_{n,q}\bigr) > 1 - \varepsilon \tag{70}$$

Thus, writing $\theta_{n,s} = \theta + h_{n,s}/\sqrt{n}$, we have

$$\begin{aligned} &Q_n^{(n)}\bigl(\bigl|\sqrt{n}\,(\sigma_n^* - \theta) + \sqrt{n}\,\mathrm{E}_\theta\,\psi_{n,\sigma_n^*}\bigr| > \varepsilon\bigr) \\ &\qquad \le \varepsilon + \sum_{s=1}^q Q_n^{(n)}\bigl(\bigl|h_{n,s} + \sqrt{n}\,\mathrm{E}_\theta\,\psi_{n,\theta_{n,s}}\bigr| > \varepsilon\bigr) \end{aligned} \tag{71}$$

By Lemma 6.4.1 a, each of the q sequences $h_{n,s} + \sqrt{n}\,\mathrm{E}_\theta\,\psi_{n,\theta_{n,s}}$ tends to 0. As $\varepsilon \in (0,1)$ is arbitrary, we have proved that

$$\sqrt{n}\,(\sigma_n^* - \theta) = -\sqrt{n}\,\mathrm{E}_\theta\,\psi_{n,\sigma_n^*} + \mathrm{o}_{Q_n^{(n)}}(n^0) \tag{72}$$

Moreover, let $\varrho_n = \max_i d_v(Q_{n,i}, P_\theta)$. Then, in any case, there exist a sequence of numbers $\gamma_n \in (0,\infty)$ so that

$$\gamma_n = \mathrm{o}(\sqrt[4]{n}\,), \qquad \gamma_n \longrightarrow \infty, \qquad \varrho_n \gamma_n^2 \longrightarrow 0 \tag{73}$$

Introduce the auxiliary functions

$$\psi_n^\natural = \psi_\theta\,\mathbf{I}(|\psi_\theta| \le \gamma_n) - \int_{\{|\psi_\theta| \le \gamma_n\}} \psi_\theta\,dP_\theta \tag{74}$$

Then

$$\lim_{n\to\infty} \mathrm{E}_\theta\,|\psi_n^\natural - \psi_\theta|^2 = 0 \tag{75}$$

since $\gamma_n \to \infty$. Furthermore, by discretization and Chebyshev,

$$\begin{aligned} &Q_n^{(n)}\biggl(\biggl|\frac{1}{\sqrt{n}}\sum_{i=1}^n (\psi_{n,\sigma_n^*} - \psi_n^\natural)(x_i) - \int (\psi_{n,\sigma_n^*} - \psi_n^\natural)\,dQ_{n,i}\biggr| > \varepsilon\biggr) \\ &\qquad \le \varepsilon + \sum_{s=1}^q \frac{1}{n\varepsilon^2}\sum_{i=1}^n \int |\psi_{n,\theta_{n,s}} - \psi_n^\natural|^2\,dQ_{n,i} \end{aligned} \tag{76}$$

where for each $s = 1, \ldots, q$ and $i = 1, \ldots, n$,

$$\int \left|\psi_{n,\theta_{n,s}} - \psi_n^\natural\right|^2 dQ_{n,i} \tag{77}$$

$$= \int \left|\psi_{n,\theta_{n,s}} - \psi_n^\natural\right|^2 dP_\theta + \int \left|\psi_{n,\theta_{n,s}} - \psi_n^\natural\right|^2 d(Q_{n,i} - P_\theta)$$

$$\leq 2 \int \left|\psi_{n,\theta_{n,s}} - \psi_\theta\right|^2 dP_\theta + 2 \int \left|\psi_n^\natural - \psi_\theta\right|^2 dP_\theta + \varrho_n \big(\mathrm{O}(c_{n,\theta}^2) + 8\gamma_n^2\big)$$

By (3) and (4′), this bound goes to 0. Putting (65) and (72) and (76) and (77) together, the estimator S has the following general as. expansion,

$$\begin{aligned} &\sqrt{n}\,\Big(S_n - \theta - \frac{1}{n}\sum_{i=1}^n \int \psi_{n,\sigma_n^*}\, d(Q_{n,i} - P_\theta)\Big) \\ &\quad = \frac{1}{\sqrt{n}}\sum_{i=1}^n \Big(\psi_n^\natural(x_i) - \int \psi_n^\natural\, dQ_{n,i}\Big) + \mathrm{o}_{Q_n^{(n)}}(n^0) \end{aligned} \tag{78}$$

Especially under $Q_{n,i} = P_\theta$, the approximating sum, by Chebyshev's inequality and (75), equals $\sum_i \psi_\theta(x_i)$ up to some $\mathrm{o}_{P_\theta^n}(n^0)$; hence S is as. linear at P_θ with influence curve ψ_θ . In general, in view of (73)–(75), Proposition 6.2.1 applies. Thus

$$\Big(\frac{1}{\sqrt{n}}\sum_{i=1}^n \Big(\psi_n^\natural(x_i) - \int \psi_n^\natural\, dQ_{n,i}\Big)\Big)(Q_n^{(n)}) \xrightarrow{\;\mathrm{w}\;} \mathcal{N}(0, C_\theta(\psi_\theta)) \tag{79}$$

It remains to prove that the difference between the random centering term in (78) and the asserted deterministic expressions is in $Q_n^{(n)}$ probability as. negligible. But by (70), we can with $Q_n^{(n)}$ probability larger than $1 - \varepsilon$ replace σ_n^* by one of the q sequences $\theta_{n,s}$. And upon this replacement, the centering term coincides with one of those evaluated in (49)–(53). ////

Remark 6.4.9 Let the estimator σ be strict and $\sqrt{n}$ consistent on $\mathcal{U}_d(\theta)$, the system of d balls about P_θ , where d is some metric on $\mathcal{M}_1(\mathcal{A})$ such that (54) holds and, moreover, the empirical process is tight on $\mathcal{U}_d(\theta)$; that is, for every $r, \varepsilon \in (0, \infty)$ there exists some $M \in (0, \infty)$ such that

$$\limsup_{n\to\infty} \sup\big\{\, Q_n^{(n)}\big(\sqrt{n}\, d(\hat{P}_n, \bar{Q}_n) > M\big) \;\big|\; Q_{n,i} \in B_d(\theta, r/\sqrt{n}\,)\,\big\} < \varepsilon \tag{80}$$

Then the clipping constants $c_n = \mathrm{o}(\sqrt[4]{n}\,)$ employed in (15) and (16), and eventually in the estimator construction (65), may be adapted via

$$\hat{c}_n = \frac{c_n}{c \vee \sqrt{n}\, d(\hat{P}_n, P_{\sigma_n^*})} \tag{81}$$

with any fixed $c \in (0, \infty)$, to the discrepancy between the empirical and the ideal model. Indeed, if $\max_i d(Q_{n,i}, P_\theta) = \mathrm{O}(1/\sqrt{n}\,)$, the two sequences of laws of $\sqrt{n}\, d(P_{\sigma_n^*}, P_\theta)$, $\sqrt{n}\, d(\hat{P}_n, \bar{Q}_n)$ under $Q_n^{(n)}$ are tight, by discretization and (54), (80), respectively. Hence, also the sequence of laws of

$$\sqrt{n}\, d(\hat{P}_n, P_{\sigma_n^*}) \le \sqrt{n}\, d(\hat{P}_n, \bar{Q}_n) + \sqrt{n}\, d(\bar{Q}_n, P_\theta) + \sqrt{n}\, d(P_{\sigma_n^*}, P_\theta)$$

is tight under $Q_n^{(n)}$, provided that $d(\bar{Q}_n, P_\theta) = \mathrm{O}(1/\sqrt{n})$ [which is the case if d comes from some norm]. Hence it follows from $c_n \to \infty$ that, on the one hand, $\hat{c}_n \to \infty$ in $Q_n^{(n)}$ probability. On the other hand, we have $\hat{c}_n \le c_n/c = \mathrm{o}(\sqrt[4]{n})$. Thus Theorem 6.4.8 obtains with this adaptive truncation.

The assumptions on the metric are satisfied by the Kolmogorov distance d_κ [L_1 differentiability of $\mathcal{P}$ at θ, and Proposition A.3.6]. Under CvM differentiability of $\mathcal{P}$ at θ, also the CvM distance d_μ fulfills these assumptions [Theorem A.4.4 a; with certain modifications as by 6.3(42) if the weight μ is not finite but only σ finite]. ////

Remark 6.4.10 Under more special assumptions, Bickel (1975) has constructed one-step Huber regression M estimates, at one fixed distribution. Under similar conditions, Müller (1993) constructs more general one-step regression M estimates for infinitesimal mean conditional contamination, locally uniformly. This neighborhood model ($* = c$, $t = \alpha = 1$) is very close to the basic errors-in-variables contamination model ($* = c$, $t = 0$), however, particularly leading to bounded influence curves [Section 7.2]. Both papers ignore the stability issue concerning the initial estimator and, at some extra conditions, again dispense with the discretization (which, however, was invented just to make the theory easier with no cost in practice).

Our general construction techniques in principle settle the construction problem also for the regression model and, in addition, for the possibly unbounded influence curves that will be obtained as optimally robust for other infinitesimal error-free-variables regression neighborhood systems. ////

CvM Construction

Unbiased estimators S for the CvM differentiable one-step functionals T constructed by (57) can be obtained by direct evaluation at the empirical measure,

$$S_n = T_n(\hat{P}_n) = \tau_n(\hat{P}_n) + \int \varphi_{n,\tau_n(\hat{P}_n)} \big(\hat{P}_n - P_{\tau_n(\hat{P}_n)}\big)\, d\mu \tag{82}$$

There is no measurability problem in this definition if $\tau_n(\hat{P}_n)$ is discrete, measurable. This and the following assumptions on the initial functional are met by the CvM functional $\tau = T_\mu$, for whose construction $\mathcal{P}$ is required identifiable in $L_2(\mu)$ and CvM differentiable at all $\theta \in \Theta$. To the given CvM influence curve $\varphi_\theta \in \Phi_\mu(\theta)$, the function ψ_θ is related via 4.2(51).

Proposition 6.4.11 *Assume that, relative to some weight* $\mu \in \mathcal{M}_\sigma(\mathbb{B}^m)$, $\mathcal{P}$ *is* CvM *differentiable and satisfies condition* 6.3(40) *at* $\theta \in \Theta$. *Suppose condition* (56) *can be fulfilled for* $\varphi_\theta \in \Phi_\mu(\theta)$. *Assume* τ *is strict, and* $\sqrt{n}$ *bounded w.r.t. the neighborhoods* $U(\theta, r/\sqrt{n})$ *given by* 6.3(42), *and each* $\tau_n(\hat{P}_n)$ *is a disrete random variable. Then, for all* $r \in (0, \infty)$ *and all arrays* $Q_{n,i} \in U(\theta, r/\sqrt{n})$,

$$\sqrt{n}\,\bigl(S_n - T_n(Q_n^{(n)})\bigr)(Q_n^{(n)}) \xrightarrow{\mathrm{w}} \mathcal{N}(0, C_\theta(\psi_\theta)) \tag{83}$$

provided that, in case $m > 1$, *in addition* A.4(60) *or* A.4(61) *are assumed for* μ *and* $P = P_\theta$, *or the neighborhoods* $U(\theta, r/\sqrt{n})$ *of form* 6.3(44).

PROOF Based on the CvM differentiability of T [Proposition 6.4.7], the proof to Theorem 6.3.8 carries over: Starting out with 6.3(46), simply replace T_μ, S_μ, $\varphi_{\mu,\theta}$, and $\psi_{\mu,\theta}$, in each one of the steps 6.3(48)–6.3(53), by T, S, φ_θ, and ψ_θ, respectively. ////

Chapter 7

Robust Regression

7.1 The Ideal Model

The linear regression model, with carriers treated random, has been defined through 2.4(36)–2.4(40):

$$P_\theta(dx,dy) = f(y - x'\theta)\,\lambda(dy)\,K(dx)\,, \qquad \theta \in \mathbb{R}^k \tag{1}$$

By Theorem 2.4.6, the model is, at every $\theta \in \mathbb{R}^k$, L_2 differentiable with L_2 derivative and Fisher information of full rank given by

$$\Lambda_\theta(x,y) = \Lambda_f(y - x'\theta)\,x\,, \qquad \mathcal{I}_\theta = \mathcal{I}_f \mathcal{K}\,, \qquad \mathcal{K} = \int xx'\,K(dx) \tag{2}$$

Thus the robustness results for general parameter apply with this ideal center model. Due to the regression structure, however, additional variants of neighborhoods, bias terms, and corresponding optimization (and, in principle, construction) problems arise.

7.2 Regression Neighborhoods

Every probability $Q = Q(dx,dy)$ on $\mathbb{B}^{k+1}$ may be factorized into the regressor marginal $Q(dx)$, which is a probability on $\mathbb{B}^k$, and into the conditional distribution $Q(dy|x)$ of the measurement y given the regressor x, which can be chosen as a Markov kernel from $\mathbb{R}^k$ to $\mathbb{B}$. Thus,

$$Q(dx,dy) = Q(dy|x)\,Q(dx) \tag{1}$$

in the sense that for all $g\colon \mathbb{R}^{k+1} \to [0,\infty]$ Borel measurable,

$$\int g(x,y)\,Q(dx,dy) = \iint g(x,y)\,Q(dy|x)\,Q(dx) \tag{2}$$

especially, for all $A \in \mathbb{B}^k$ and $B \in \mathbb{B}$,

$$Q(A \times B) = \int_A Q(B|x)\, Q(dx) \tag{3}$$

In particular, for the model distributions themselves this says that

$$P_\theta(dx, dy) = F(dy - x'\theta)\, K(dx) \tag{4}$$

where

$$F(dy - x'\theta) = f(y - x'\theta)\, \lambda(dy) = P_\theta(dy|x) \tag{5}$$

To some extent, the following neighborhoods also cover deviations from the ideal model that may be caused by an actual nonlinearity of the regression.

7.2.1 Errors-in-Variables

Under elements $Q(dx, dy) = Q(dy|x)\, Q(dx)$ of the usual neighborhoods $U_*(\theta, r) = B_*(P_\theta, r)$ introduced in Subsection 4.2.1, both the ideal regressor marginal $P_\theta(dx) = K(dx)$ and $P_\theta(dy|x) = F(dy - x'\theta)$, the ideal conditional distribution of y given x, may be distorted, to another probability $Q(dx)$ on $\mathbb{B}^k$, respectively, Markov kernel $Q(dy|x)$ from $\mathbb{R}^k$ to $\mathbb{B}$.

In the regression setup, therefore, these neighborhoods are called *errors-in-variables*, or *unconditional*, neighborhoods. Adding a subscript $t = 0$ artificially, they will be denoted by $U_*(\theta, r) = U_{*,0}(\theta, r)$.

Remark 7.2.1 Although he thinks it too late to correct the development, Huber (1991) argues that, in regression,

(a) x outliers mostly correspond to very precious observations; and,

(b) treating regressors x and errors u in the same combined way, moves robust regression conceptually towards principal components analysis.

At least in the infinitesimal setup, this opinion is somewhat relativated by Lemma 7.3.2 below, which shows that similar effects as by errors-in-variables may be created by error-free-variables. ////

7.2.2 Error-Free-Variables

Let $\mathbb{M}_\bullet$ denote the set of all $\bar{\mathbb{R}}$-valued Borel measurable functions on $\mathbb{R}^k$, and put

$$\mathbb{M}_\bullet^+ = \{\, m \in \mathbb{M}_\bullet \mid m \ge 0 \,\}, \qquad \mathbb{M}_\bullet^- = -\mathbb{M}_\bullet^+ \tag{6}$$

Error-free-variables, or *conditional*, neighborhoods $U_{*,\varepsilon}(\theta, r)$ about P_θ, of type $*$, radius $r \in [0, \infty]$, and with any *contamination curve* $\varepsilon \in \mathbb{M}_\bullet^+$, by definition consist of all probabilities $Q(dx, dy) = Q(dy|x)\, Q(dx)$ on $\mathbb{B}^{k+1}$ such that

$$Q(dx) = K(dx) \tag{7}$$

and

$$Q(dy|x) \in B_*\bigl(F(dy - x'\theta), r\varepsilon(x)\bigr) \qquad \text{a.e. } K(dx) \tag{8}$$

Moreover, for any $\alpha \in [1, \infty]$, *average* conditional neighborhoods $U_{*,\alpha}(\theta, r)$ of exponent α are defined as the unions of all $U_{*,\varepsilon}(\theta, r)$ whose contamination curves $\varepsilon \in \mathbb{M}_{\bullet}^{+}$ satisfy the $L_\alpha(K)$ norm constraint

$$1 \geq \|\varepsilon\|_\alpha = \begin{cases} \left|\mathrm{E}\,\varepsilon^\alpha\right|^{1/\alpha} & (\alpha < \infty) \\ \sup_K \varepsilon & (\alpha = \infty) \end{cases} \tag{9}$$

Inclusion into Unconditional Balls

The relations between conditional and unconditional neighborhoods are illuminated by the following equivalences, which in the cases $* = v, h, c$ lead to the average/supremum contamination constraints for $\alpha = 1, 2, \infty$.

Proposition 7.2.2 *Given* $\varepsilon \in \mathbb{M}_{\bullet}^{+}$, $r \in (0, 1)$, *put* $\varepsilon_r = \varepsilon \wedge (1/r)$. *Then*

$$U_{v,0}(\theta, r) \supset U_{v,\varepsilon}(\theta, r) \iff 1 \geq \|\varepsilon_r\|_1 \tag{10}$$

$$U_{h,0}(\theta, r) \supset U_{h,\varepsilon}(\theta, r) \iff 1 \geq \|\varepsilon_r\|_2 \tag{11}$$

$$U_{c,0}(\theta, r) \supset U_{c,\varepsilon}(\theta, r) \iff 1 \geq \|\varepsilon_r\|_\infty \tag{12}$$

PROOF In case $* = c$, if $Q \in U_{c,\varepsilon}(\theta, r)$ then

$$Q(dy|x) = \big(1 - r\,\varepsilon_r(x)\big)\, P_\theta(dy|x) + r\,\varepsilon_r(x)\, M(dy|x)$$

for some Markov kernel $M(dy|x)$ from $\mathbb{R}^k$ to $\mathbb{B}$. Thus

$$\begin{aligned} Q(dx, dy) &= (1 - r)\, P_\theta(dx, dy) + r\, H(dy|x)\, K(dx) \\ H(dy|x) &= \big(1 - \varepsilon_r(x)\big) P_\theta(dy|x) + \varepsilon_r(x)\, M(dy|x) \end{aligned}$$

If $\sup_K \varepsilon_r \leq 1$ then $H(dy|x)$ is a Markov kernel, hence $H(dy|x)\, K(dx)$ a probability, and $Q \in U_{c,0}(\theta, r)$ follows. Conversely, if $\sup_K \varepsilon_r > 1$, choose any discrete $G \in \mathcal{M}_1(\mathbb{B})$ such that $G(dy) \perp P_\theta(dy|x) = F(dy - x'\theta)$ for all $x \in \mathbb{R}^k$, and employ the kernel

$$M(dy|x) = P_\theta(dy|x)\, \mathbf{I}\big(\varepsilon_r(x) \leq 1\big) + G(dy)\, \mathbf{I}\big(\varepsilon_r(x) > 1\big)$$

Then $Q \in U_{c,\varepsilon}(\theta, r)$, however $Q \notin U_{c,0}(\theta, r)$ since $H(dy|x)\, K(dx)$ is not a probability.

In case $* = h$, if $\mathrm{E}\,\varepsilon_r^2 \leq 1$ and $Q \in U_{h,\varepsilon}(\theta, r)$, then $Q \in U_{h,0}(\theta, r)$ as

$$\begin{aligned} 2\, d_h^2(Q, P_\theta) &= \iint \left|\sqrt{Q(dy|x)} - \sqrt{P_\theta(dy|x)}\,\right|^2 K(dx) \\ &\leq 2 \int 1 \wedge r^2 \varepsilon^2(x)\, K(dx) = 2r^2\, \mathrm{E}\,\varepsilon_r^2 \leq 2r^2 \end{aligned}$$

Conversely, if $\mathrm{E}\,\varepsilon_r^2 > 1$, choose any discrete probability G on $\mathbb{B}$ such that $G(dy) \perp P_\theta(dy|x) = F(dy - x'\theta)$ for all $x \in \mathbb{R}^k$, and set

$$Q(dy|x) = \gamma^2(x)\, P_\theta(dy|x) + \big(1 - \gamma^2(x)\big)\, G(dy), \qquad \gamma(x) = 1 - r^2 \varepsilon_r^2(x)$$

Then $Q \in U_{h,\varepsilon}(\theta, r)$ but $Q \notin U_{h,0}(\theta, r)$ because

$$\int \left|\sqrt{Q(dy|x)} - \sqrt{P_\theta(dy|x)}\,\right|^2$$
$$= \int \left|(1-\gamma(x))\sqrt{P_\theta(dy|x)} - \sqrt{1-\gamma^2(x)}\,\sqrt{G(dy)}\,\right|^2$$
$$= r^4\varepsilon_r^4(x) + 1 - \bigl(1 - r^2\varepsilon_r^2(x)\bigr)^2 = 2r^2\varepsilon_r^2(x)$$

The proof in case $* = v$ runs similarly. ////

Linear Contamination Rate

In the infinitesimal setup, where $r = \mathrm{O}(1/\sqrt{n}\,)$, the linear contamination rate $\varepsilon(x) \propto |x|$ is necessary to include the parametric local alternatives into the conditional shrinking neighborhoods. Given $r, c \in (0, \infty)$ the condition is that

$$P_{\theta+t/\sqrt{n}} \in U_{*,\varepsilon}(\theta, r/\sqrt{n}\,), \qquad |t| \le c \tag{13}$$

eventually. Note also the quadratic Fisher information $\mathcal{I}_f\, xx'$ of the conditional (location) model (5) given $x \in \mathbb{R}^k$. This is relevant for Section 7.5.

Proposition 7.2.3 *Given* $r, c \in (0, \infty)$, *the eventual inclusion* (13) *in the cases* $* = v, h$ *with* $\varepsilon = \varepsilon_*$ *is equivalent to, respectively,*

$$r\varepsilon_v(x) \ge c|x|\,\mathrm{E}_F|\Lambda_f| \qquad \text{a.e. } K(dx) \tag{14}$$
$$r\varepsilon_h(x) \ge c|x|\sqrt{\mathcal{I}_f/8} \qquad \text{a.e. } K(dx) \tag{15}$$

PROOF L_1 differentiability follows from Proposition 2.4.1 and 2.3(67),

$$\int \bigl|f(u-s) - f(u) - s\Lambda_f(u)f(u)\bigr|\,\lambda(du) = \mathrm{o}(|s|)$$

Moreover, by absolute continuity of f and C.2(4),

$$\int \bigl|f(u-s) - f(u)\bigr|\,\lambda(du) = \int\Bigl|\int_u^{u-s} f'(v)\,\lambda(dv)\Bigr|\,\lambda(du)$$
$$\le \iint_{-s^+}^{s^-} \bigl|f'(u+v)\bigr|\,\lambda(dv)\,\lambda(du) = |s|\,\mathrm{E}_F|\Lambda_f|$$

Inserting $s = x't/\sqrt{n}$ it follows that for all $x \in \mathbb{R}^k$, $n \ge 1$,

$$\sqrt{n}\,\sup_{|t|\le c} \int \bigl|P_{\theta+t/\sqrt{n}}(dy|x) - P_\theta(dy|x)\bigr| \le c|x|\,\mathrm{E}_F|\Lambda_f|$$

and the upper bound is approximated as $n \to \infty$ (set $t = cx/|x|$ if $x \ne 0$).

Thus, on the one hand, (14) implies (13) for $* = v$. On the other hand, (13) for $* = v$ entails that $c|x|\,\mathrm{E}_F|\Lambda_f| - \mathrm{o}(n^0) \le r\varepsilon_v(x)$ a.e. $K(dx)$, which yields (14) if we let $n \to \infty$.

Based on 2.4(4) and 2.4(16), the proof in case $* = h$ runs similarly. ////

Subsequently, the neighborhoods are restricted to types $* = c, v, h$ (contamination, total variation, Hellinger).

7.2.3 Translation Invariance

According to 2.4(49), the linear regression model 2.4(36)–2.4(40) is invariant under translations, the sample transformations $g_\theta(x, y) = (x, x'\theta + y)$, in the sense that $P_\theta = g_\theta(P_0)$, and hence $\Lambda_\theta = \Lambda_0 \circ g_{-\theta}$. This invariance entails the following correspondence

$$\Psi_\alpha(\theta) = \Psi_\alpha(0) \circ g_{-\theta}, \qquad \psi_\theta(x, y) = \psi_0(x, y - x'\theta) \tag{16}$$

between the influence curves $\psi_\theta \in \Psi_\alpha(\theta)$ and $\psi_0 \in \Psi_\alpha(0)$ at P_θ and P_0, respectively, for every $\theta \in \mathbb{R}^k$.

Remark 7.2.4 The correspondence is merely set-theoretic. It does not imply that a functional or as. estimator that is as. linear at P_θ must also be as. linear at P_0, and even if they are, their influence curves at P_θ and P_0, respectively, need not be related by (16).

However, if the same optimization problems $O^{\mathrm{tr/ms}}_{*,t;s}(b)$ are considered at P_θ as we do in Section 7.4 at P_0, the solutions $\tilde{\eta}_\theta$ and $\tilde{\eta}_0$, respectively, will necessarily be related via (16): $\tilde{\eta}_\theta(x, y) = \tilde{\eta}_0(x, y - x'\theta)$, and the zero $S_n\big((x_i, y_i)\big)$ at sample size n of $\sum_i \tilde{\eta}_\theta(x_i, y_i)$ (the corresponding M equation and M estimate, respectively) will, in principle, be *translation equivariant*:

$$S_n\big((x_i, y_i)\big) = S_n\big((x_i, y_i - x_i'\theta)\big) + \theta \tag{17}$$

for all $(x_1, y_1), \dots, (x_n, y_n) \in \mathbb{R}^{k+1}$ and $\theta \in \mathbb{R}^k$. ////

The neighborhoods $U_{*,t}(\theta, r)$, for $* = c, v, h$ and $t = 0, \varepsilon, \alpha$, transform under the regression translations $g_\theta(x, y) = (x, x'\theta + y)$ in the same way as the center measures do: $P_\theta = g_\theta(P_0)$. For the image measure under g_θ of any probability $Q(dx, dy) = Q(dy|x)\, Q(dx)$ on $\mathbb{B}^{k+1}$ is given by

$$g_\theta(Q)(dx, dy) = Q(dy - x'\theta|x)\, Q(dx) \tag{18}$$

as, in view of (3), $Q(x \in A, x'\theta + y \in B) = \int_A Q(B - x'\theta|x)\, Q(dx)$ holds for all $A \in \mathbb{B}^k$ and $B \in \mathbb{B}$. Thus the regression neighborhoods correspond via

$$U_{*,t}(\theta, r) = g_\theta \circ U_{*,t}(0, r), \qquad Q_\theta = g_\theta(Q_0) \tag{19}$$

Therefore, θ may be restricted to $\theta = 0$, and omitted from notation. Then: y (measurements) $= u$ (errors), and $P(dx, du) = F(du)\, K(dx)$ is the product probability of the error and regressor distributions.

7.2.4 Neighborhood Submodels

As submodel of shrinking neighborhoods $U_{*,t}(0, r/\sqrt{n}\,)$ we employ the simple perturbations

$$dQ_n(q,r) = \Bigl(1 + \frac{r}{\sqrt{n}}\,q\Bigr)\,dP \tag{20}$$

with suitable one-dimensional bounded tangents $q \in Z_\infty$. Denoting by E expectation under P, the classes $\mathcal{G}_{*,0} \subset Z_\infty$ that according to Lemma 5.3.1 define simple perturbations (20) which are representative for the unconditional neighborhoods $U_{*,0}(0, r/\sqrt{n}\,)$ are, in view of 5.3(1), given by

$$\mathcal{G}_{*,0} = \bigl\{\, q \in Z_\infty \bigm| (21\,*) \,\bigr\} \tag{21}$$

where

$$(21\,c)\ \inf\nolimits_P q \ge -1\,, \qquad (21\,v)\ \mathrm{E}\,|q| \le 2\,, \qquad (21\,h)\ \mathrm{E}\,q^2 \le 8$$

For the formulation of the classes $\mathcal{G}_{*,\varepsilon} \subset Z_\infty$ that define simple perturbations (20) in the conditional neighborhoods $U_{*,\varepsilon}(0, r/\sqrt{n}\,)$ with contamination curve $\varepsilon \in \mathbb{M}_{\bullet}^{+}$, we introduce conditional expectation $\mathrm{E}_{\bullet}$ given x:

$$\mathrm{E}_{\bullet}\, g = \int g(x,u)\, F(du) \qquad \text{a.e. } K(dx) \tag{22}$$

for $g \in L_1^p(P)$. Likewise, for $g\colon \mathbb{R}^{k+1} \to \bar{\mathbb{R}}$ Borel measurable, the conditional essential extrema $\inf_{\bullet}$ and $\sup_{\bullet}$ given x, are defined by

$$\inf\nolimits_{\bullet} g = \inf\nolimits_{F(du)} g(x,u)\,, \quad \sup\nolimits_{\bullet} g = \sup\nolimits_{F(du)} g(x,u) \qquad \text{a.e. } K(dx) \tag{23}$$

Since $P(g > z) = \int F\{\, u \mid g(x,u) > z \,\}\, K(dx)$ [Fubini's theorem], the overall P essential extrema, like expectation, can be evaluated iteratively,

$$\mathrm{E}\,\mathrm{E}_{\bullet} = \mathrm{E}\,, \qquad \inf\nolimits_P \inf\nolimits_{\bullet} = \inf\nolimits_P\,, \qquad \sup\nolimits_P \sup\nolimits_{\bullet} = \sup\nolimits_P \tag{24}$$

where the outer operators may equally be evaluated under K as under P.
For $Q = Q_n(q,r)$ given by (20) and $g = rq/\sqrt{n}$, we have

$$Q(dx) = (1 + \mathrm{E}_{\bullet}\, g)\, K(dx)\,, \qquad Q(du|x) = \frac{1 + g(x,u)}{1 + \mathrm{E}_{\bullet}\, g}\, F(du) \tag{25}$$

Therefore $Q(dx) = K(dx)$ iff $\mathrm{E}_{\bullet}\, g = 0$ (i.e., g conditionally centered).

Definition 7.2.5 *For $p \in \mathbb{N}$ and $\alpha \in [1, \infty]$ let*

$$Z_{\alpha\bullet}^p = \bigl\{\, \zeta \in L_\alpha^p(P) \bigm| \mathrm{E}_{\bullet}\, \zeta = 0 \,\bigr\} \tag{26}$$

For exponents $\alpha = \infty, 2, 1$, by definition $Z_{\alpha\bullet}^p \subset Z_\alpha^p$ are the subspaces of conditionally centered tangents in Z_α^p, respectively.

Remark 7.2.6

(a) The parametric tangent verifies $\mathrm{E}_{\bullet}\Lambda = x\,\mathrm{E}_F\Lambda_f = 0$.

(b) $Z^p_{\alpha\bullet}$ are closed linear subspaces of Z^p_α in $L^p_\alpha(P)$; $\alpha \in [1,\infty]$. ////

Given any contamination curve $\varepsilon \in \mathbb{M}^+_\bullet$, we thus introduce the following classes of one-dimensional, conditionally centered, and bounded tangents,

$$\mathcal{G}_{*,\varepsilon} = \{\, q \in Z_{\infty\bullet} \mid (27*) \,\} \tag{27}$$

with

$$(27\,c)\ \inf_\bullet q \ge -\varepsilon, \qquad (27\,v)\ \mathrm{E}_\bullet |q| \le 2\varepsilon, \qquad (27\,h)\ \mathrm{E}_\bullet q^2 \le 8\varepsilon^2$$

For exponent $\alpha \in [1,\infty]$, we set

$$\mathcal{G}_{*,\alpha} = \bigcup \{\, \mathcal{G}_{*,\varepsilon} \mid \varepsilon \in \mathbb{M}^+_\bullet,\ \|\varepsilon\|_\alpha \le 1 \,\} \tag{28}$$

Then Lemma 5.3.1 extends to conditional regression balls.

Lemma 7.2.7 *Given $q \in Z_\infty$, $\varepsilon \in \mathbb{M}^+_\bullet$, $\alpha \in [1,\infty]$, let $r \in (0,\infty)$ and $n \in \mathbb{N}$ be such that $\sqrt{n} \ge -r \inf_P q$. Then, for $t = \varepsilon, \alpha$ and $* = c, v$,*

$$Q_n(q,r) \in U_{*,t}(0, r/\sqrt{n}) \iff q \in \mathcal{G}_{*,t} \tag{29}$$

while

$$Q_n(q,r) \in U_{h,t}(0, r_n/\sqrt{n}),\ \limsup_{n\to\infty} r_n \le r \iff q \in \mathcal{G}_{h,t} \tag{30}$$

PROOF Similar to, and actually based on, the proof to Lemma 5.3.1. ////

7.2.5 Tangent Subspaces

In regard to Section 4.4, we determine the tangent subspaces $V_{2;*,t}$ that belong to the conditional neighborhoods $\mathcal{U}_{*,t}$, as well as the (closed) linear span $W_{2;*,t}$ in $Z_{2\bullet} \subset L_2(P)$ of $V_{2;*,t}$ and the coordinates of the L_2 derivative Λ (the parametric tangent, built into the definition of influence curves and as. linearity). The corresponding orthogonal projections from $Z_{2\bullet}$ onto $V_{2;*,t}$ and $W_{2;*,t}$, respectively, are denoted by Π_V, Π_W.

Proposition 7.2.8 *Let $\varepsilon \in \mathbb{M}^+_\bullet$, $\alpha \in [1,\infty]$, and $* = c, v, h$.*

(a) *Then, in case $t = \varepsilon$, the two tangent subspaces are*

$$V_{2;*,\varepsilon} = \{\, q \in Z_{2\bullet} \mid P(\varepsilon = 0, q \ne 0) = 0 \,\} \tag{31}$$

$$W_{2;*,\varepsilon} = \{\, q \in Z_{2\bullet} \mid \exists\, a \in \mathbb{R}^k\colon P(\varepsilon = 0, q \ne a'\Lambda) = 0 \,\} \tag{32}$$

Hence

$$\Lambda \in V^k_{2;*,\varepsilon} \iff K\big(\varepsilon(x) = 0, x \ne 0\big) = 0 \tag{33}$$

The orthogonal projections $\Pi_V\colon Z_{2\bullet} \to V_{2;*,\varepsilon}$ and $\Pi_W\colon Z_{2\bullet} \to W_{2;*,\varepsilon}$ are given by

$$\Pi_V(\zeta) = \zeta\, \mathbf{I}(\varepsilon \ne 0) \tag{34}$$

respectively

$$\Pi_W(\zeta) = \zeta\,\mathbf{I}(\varepsilon \neq 0) + a'\Lambda\,\mathbf{I}(\varepsilon = 0) \tag{35}$$

with any $a \in \mathbb{R}^k$ *chosen such that* $\mathrm{E}\,\zeta\Lambda'\,\mathbf{I}(\varepsilon = 0) = a'\,\mathrm{E}\,\Lambda\Lambda'\,\mathbf{I}(\varepsilon = 0)$.

(b) *In case* $t = \alpha$,

$$V_{2;*,\alpha} = W_{2;*,\alpha} = Z_{2.} \tag{36}$$

Hence $\Lambda \in V^k_{2;*,\alpha}$ *always, and* $\Pi_V = \Pi_W$ *both are the identity on* $Z_{2.}$.

PROOF In any case, $V_{2;*,t}$ is the closed linear span of $\mathcal{G}_{*,t} \subset Z_{2.}$ in $L_2(P)$.

(a) By (27) all $q \in \mathcal{G}_{*,\varepsilon}$ satisfy (†): $\mathrm{E}.\,q = 0$, $P(\varepsilon = 0, q \neq 0) = 0$. (†) carries over to linear combinations and $L_2(P)$ limits, hence to $V_{2;*,\varepsilon}$.

Conversely, if $q \in L_2(P)$ satisfies (†) consider $q_m = q\,\mathbf{I}(m\varepsilon \geq 1)$ which, for $m \in \mathbb{N}$ sufficiently large, is arbitrarily close to q in $L_2(P)$. Furthermore, for $n \in \mathbb{N}$ set $q_{m,n} = q_m\,\mathbf{I}(|q_m| \leq n) - \mathrm{E}.\,q_m\,\mathbf{I}(|q_m| \leq n)$. Then $\mathrm{E}.\,q_{m,n} = 0$ and $q_{m,n} \to q_m$ in $L_2(P)$ as $n \to \infty$. Moreover, $|q_{m,n}| \leq 2n$ and $q_{m,n} \neq 0$ only if $m\varepsilon \geq 1$; hence $|q_{m,n}| \leq 2mn\varepsilon$. It follows that $q_{m,n}/(2mn) \in \mathcal{G}_{*,\varepsilon}$ and therefore $q \in V_{2;*,\varepsilon}$.

Apparently, the linear span of $V_{2;*,\varepsilon}$ and the coordinates of Λ is

$$W_{2;*,\varepsilon} = \left\{ \bar{q} = q + a'\Lambda \mid q \in V_{2;*,\varepsilon},\ a \in \mathbb{R}^k \right\} \tag{37}$$

which is closed [Dieudonné (1960; Theorem 5.9.2)]. The functions q, $\bar{q}$ in this representation, since $\Lambda \in Z^k_{2.}$, belong to $Z_{2.}$ simultaneously. Furthermore, $q = 0$ a.e. P on $\{\varepsilon = 0\}$ iff $\bar{q} = a'\Lambda$ a.e. P on $\{\varepsilon = 0\}$.

Given any $\zeta \in Z_{2.}$, obviously $\Pi_V(\zeta)$ as defined by (34) is itself in $V_{2;*,\varepsilon}$, and achieves the best approximation of ζ by elements of $V_{2;*,\varepsilon}$. Likewise, $\Pi_W(\zeta)$ as defined by (35) is itself a member of $W_{2;*,\varepsilon}$, and the condition on the vector $a \in \mathbb{R}^k$ (which exists since Π_W exists, and is determined only up to uniqueness of $a'\Lambda$ on $\{\varepsilon = 0\}$ a.e. P) is just $\zeta - \Pi_W(\zeta) \perp W_{2;*,\varepsilon}$.

(b) Property $\mathrm{E}. = 0$ is clearly inherited from $\mathcal{G}_{*,\alpha}$ to $V_{2;*,\alpha}$. Conversely, approximate $q \in Z_{2.}$ by $q_m = q\,\mathbf{I}(|q| \leq m) - \mathrm{E}.\,q\,\mathbf{I}(|q| \leq m)$ in $L_2(P)$. We have $|q_m| \leq 2m$ and so $q_m/(2m) \in \mathcal{G}_{*,\varepsilon_1} \subset \mathcal{G}_{*,\alpha}$ for ε_1 constant 1. Thus $q \in V_{2;*,\alpha}$. ////

The unconditional neighborhood systems $\mathcal{U}_* = \mathcal{U}_{*,0}$, of course, have full tangent spaces $V_{2;*,0} = W_{2;*,0} = Z_2$ (with projection the identity on Z_2).

7.3 Conditional Bias

The tangent classes $\mathcal{G}_{*,t}$, for $* = c, v, h$ and $t = 0, \varepsilon, \alpha$, generate corresponding regression oscillation terms and coordinatewise variants, which can be defined for all $\eta = (\eta_1, \ldots, \eta_p)' \in L_1^p(P)$,

$$\omega_{*,t}(\eta) = \omega_{*,t;0}(\eta) = \sup\{\, |\mathrm{E}\,\eta q| \mid q \in \mathcal{G}_{*,t} \,\} \tag{1}$$

and

$$\omega_{*,t;2}^2(\eta) = \sum_{j=1}^{p} \omega_{*,t}^2(\eta_j)\,, \qquad \omega_{*,t;\infty}(\eta) = \max_{j=1,\dots,p} \omega_{*,t}(\eta_j) \tag{2}$$

Then, for $\alpha \in [1,\infty]$ and all combinations of $* = c, v, h$ and $s = 0, 2, \infty$,

$$\omega_{*,\alpha;s}(\eta) = \sup\{\, \omega_{*,\varepsilon;s}(\eta) \mid \|\varepsilon\|_\alpha \le 1 \,\} \tag{3}$$

except that $\omega_{*,\alpha;2}$, for $\alpha < \infty$, would be larger than the right-hand sup. In case $s = 2$ and $t = \alpha < \infty$, therefore, we define $\omega_{*,\alpha;2}$ by (3). (3) will be realized also for $s = e$, in view of (22), (23) below.

7.3.1 General Properties

Some bias relations can be derived without explicit computation already from the structure of neighborhoods. The following lemma thus extends Lemma 5.3.2 to conditional regression neighborhoods.

Lemma 7.3.1 *Given $\varepsilon \in \mathbb{M}_{\bullet}^{+}$ and $\alpha \in [1,\infty]$. Then the terms $\omega_{*,t;s}$, for $* = c, v, h$, $t = \varepsilon, \alpha$, $s = 0, 2, \infty$, are positively homogeneous, subadditive, hence convex on $L_1^p(P)$, and weakly lower semicontinuous on $L_2^p(P)$. Moreover, for every $\eta \in L_1^p(P)$,*

$$\omega_{*,t;s}(\eta) = \omega_{*,t;s}(\eta - \mathrm{E}_{\bullet}\,\eta) \tag{4}$$

$$\omega_{*,t}(\eta) = \sup\{\, \omega_{*,t}(e'\eta) \mid e \in \mathbb{R}^p,\ |e| = 1 \,\} \tag{5}$$

$$\omega_{*,t;\infty}(\eta) \le \omega_{*,t;0}(\eta) \le \omega_{*,t;2}(\eta) \le \sqrt{p}\,\omega_{*,t;\infty}(\eta) \tag{6}$$

$$\omega_{c,t;s}(\eta) \le \omega_{v,t;s}(\eta) \le 2\,\omega_{c,t;s}(\eta) \tag{7}$$

PROOF If $q \in \mathcal{G}_{*,\varepsilon}$ then $\mathrm{E}_{\bullet}\, q = 0$ hence $\mathrm{E}(q\,\mathrm{E}_{\bullet}\,\eta) = \mathrm{E}\,\mathrm{E}_{\bullet}\, q\,\mathrm{E}_{\bullet}\,\eta = 0$.

If $q \in \mathcal{G}_{c,\varepsilon}$ then $q^- \le \varepsilon$ hence $\mathrm{E}_{\bullet}\,|q| = 2\,\mathrm{E}_{\bullet}\, q^- \le 2\varepsilon$, thus $q \in \mathcal{G}_{v,\varepsilon}$. Conversely, if $q \in \mathcal{G}_{v,\varepsilon}$ then $q = (q^+ - \mathrm{E}_{\bullet}\, q^+) - (q^- - \mathrm{E}_{\bullet}\, q^-)$ with both summands in $\mathcal{G}_{c,\varepsilon}$, since $\mathrm{E}_{\bullet}\, q = 0$ and $\mathrm{E}_{\bullet}\,|q| \le 2\varepsilon$.

This proves (4) and (7). The arguments to prove the remaining assertions can be taken literally from the proof of Lemma 5.3.2. ////

As the next lemma shows, Nature is able to mimic errors-in-variables by error-free-variables, using most of her allocated $\varepsilon(x)$ of contamination to create very skew conditional given x distributions of u, for the 'largest' x.

Lemma 7.3.2 *For every $\eta \in L_1^p(P)$ it holds that*

$$\omega_{c,1}(\eta) = \omega_{c,0}(\eta) \qquad \text{if } \mathrm{E}_{\bullet}\,\eta = 0 \tag{8}$$

$$\omega_{v,1}(\eta) \le \omega_{v,0}(\eta)\,, \qquad \text{and } \omega_{v,0}(\eta) \le 2\,\omega_{v,1}(\eta) \text{ if } \mathrm{E}_{\bullet}\,\eta = 0 \tag{9}$$

$$\omega_{h,2}(\eta) \le \omega_{h,0}(\eta)\,, \qquad \text{with equality if } \mathrm{E}_{\bullet}\,\eta = 0 \tag{10}$$

PROOF As $\mathcal{G}_{*,\varepsilon} \subset L_\infty(P)$ we have $\mathcal{G}_{*,\varepsilon} = \bigcup_m \mathcal{G}_{*,\varepsilon_m}$ with $\varepsilon_m = \varepsilon \wedge m$. For $q \in \mathcal{G}_{c,\varepsilon}$ use this fact to choose m so large that $q \in \mathcal{G}_{c,\varepsilon_m}$ and consider $\tilde{q} = q+\varepsilon_m - \mathrm{E}\,\varepsilon_m$. If $\mathrm{E}\,\varepsilon \le 1$ then $\tilde{q} \in \mathcal{G}_{c,0}$ since $\mathrm{E}\,\varepsilon_m \le 1$ and $q+\varepsilon_m \ge 0$. Moreover, $\mathrm{E}\,(\tilde{q}-q)\eta = \mathrm{E}\,(\varepsilon_m - \mathrm{E}\,\varepsilon_m)\,\mathrm{E}_{\bullet}\,\eta = 0$ if $\mathrm{E}_{\bullet}\,\eta = 0$.

The corresponding first inequalities in (9) and (10), even without the restriction $\mathrm{E}_{\bullet} = 0$, are obtained from the following equivalences, which correspond to Proposition 7.2.2 and can be shown likewise,

$$\mathcal{G}_{v,\varepsilon} \subset \mathcal{G}_{v,0} \iff \mathrm{E}\,\varepsilon \le 1, \qquad \mathcal{G}_{h,\varepsilon} \subset \mathcal{G}_{h,0} \iff \mathrm{E}\,\varepsilon^2 \le 1 \tag{11}$$

To prove the reverse inequalities suppose that $\mathrm{E}_{\bullet}\,\eta = 0$. The idea is to assign to each errors-in-variables $Q(du|x)\,Q(dx)$ the error-free-variables measure with the same kernel $Q(du|x)$. For the simple perturbations 7.2(20): $Q_n = Q_n(q,r)$ with $q \in \mathcal{G}_{*,0}$, we have $g_n = rq/\sqrt{n}$ in 7.2(25) so that $(1+g_n)/(1+\mathrm{E}_{\bullet}\,g_n) \approx 1 + g_n - \mathrm{E}_{\bullet}\,g_n$. Thus, we pass from each $q \in \mathcal{G}_{*,0}$ to $q_{\bullet} = q - \mathrm{E}_{\bullet}\,q$. Then actually $\mathrm{E}\,(q_{\bullet} - q)\eta = \mathrm{E}\,\mathrm{E}_{\bullet}\,q\,\mathrm{E}_{\bullet}\,\eta = 0$.

In case $* = c$, the minimum choice of ε to force $q_{\bullet}$ into $\mathcal{G}_{c,\varepsilon}$ (so as to achieve $q_{\bullet} \ge -\varepsilon$) is obviously $\varepsilon = -\inf_{\bullet} q + \mathrm{E}_{\bullet}\,q$. As $q \ge -1$ and $\mathrm{E}\,q = 0$, then $\mathrm{E}\,\varepsilon \le 1$ holds. Hence $q_{\bullet} \in \mathcal{G}_{c,1}$.

In case $* = v$, the minimum choice of ε to force $q_{\bullet}$ into $\mathcal{G}_{v,\varepsilon}$ is of course by $2\varepsilon = \mathrm{E}_{\bullet}\,|q_{\bullet}|$. Since only $\mathrm{E}\,|q_{\bullet}| \le 2\,\mathrm{E}\,|q| \le 4$, the factor 2 arises.

In case $* = h$, the minimum choice of ε to force $q_{\bullet}$ into $\mathcal{G}_{h,\varepsilon}$ is naturally by $8\varepsilon^2 = \mathrm{E}_{\bullet}\,q_{\bullet}^2$. Then $\mathrm{E}\,\varepsilon^2 \le 1$ since $\mathrm{E}\,q_{\bullet}^2 = \mathrm{E}\,q^2 - |\mathrm{E}_{\bullet}\,q|^2 \le \mathrm{E}\,q^2$. ////

7.3.2 Explicit Terms

Explicit expressions for the conditional bias $\omega_{*,t}$ of type $* = c, v, h$ are now derived that complement the unconditional bias terms $\omega_* = \omega_{*,0}$ of Proposition 5.3.3. Conjugate exponents $\alpha, \tilde{\alpha} \in [1,\infty]$ by definition satisfy $1/\alpha + 1/\tilde{\alpha} = 1$, and $\|\,.\,\|_{\tilde{\alpha}}$ denotes the integration norm of $L_{\tilde{\alpha}}(K)$. The following suprema $\sup_{|e|=1}$ are taken over all unit vectors in $\mathbb{R}^p$.

Proposition 7.3.3 *Consider any $\varepsilon \in \mathrm{M}_{\bullet}^{+}$, and let $\alpha, \tilde{\alpha} \in [1,\infty]$ be conjugate exponents. Then, for $\eta \in Z_{1\bullet}^p$,*

$$\omega_{c,\varepsilon}(\eta) = \sup_{|e|=1} \mathrm{E}\,\varepsilon \sup_{\bullet} e'\eta \tag{12}$$

$$\omega_{v,\varepsilon}(\eta) = \sup_{|e|=1} \mathrm{E}\,\varepsilon (\sup_{\bullet} e'\eta - \inf_{\bullet} e'\eta) \tag{13}$$

$$\omega_{h,\varepsilon}(\eta) = \sqrt{8}\,\sup_{|e|=1} \mathrm{E}\,\varepsilon \big|\mathrm{E}_{\bullet}(e'\eta)^2\big|^{1/2} \tag{14}$$

and

$$\omega_{c,\alpha}(\eta) = \sup_{|e|=1} \|\sup_{\bullet} e'\eta\,\|_{\tilde{\alpha}} \tag{15}$$

$$\omega_{v,\alpha}(\eta) = \sup_{|e|=1} \|\sup_{\bullet} e'\eta - \inf_{\bullet} e'\eta\,\|_{\tilde{\alpha}} \tag{16}$$

$$\omega_{h,\alpha}(\eta) = \sqrt{8}\sup_{|e|=1}\left\| \left(\mathrm{E}_{\bullet}(e'\eta)^2\right)^{1/2}\right\|_{\tilde{\alpha}} \tag{17}$$

Especially,

$$\omega_{c,1}(\eta) = \sup_P|\eta| = \omega_{c,0}(\eta) \tag{18}$$

$$\omega_{h,2}^2(\eta) = 8\,\mathrm{maxev}\,\mathcal{C}(\eta) = \omega_{h,0}^2(\eta) \tag{19}$$

PROOF We can assume $p=1$ in view of (5). To prove (12) pick $q \in \mathcal{G}_{c,\varepsilon}$. Then $\varepsilon + q \geq 0$ and $\mathrm{E}_{\bullet}\,\eta = 0 = \mathrm{E}_{\bullet}\,q$, hence

$$\mathrm{E}\,\eta q = \mathrm{E}\,\mathrm{E}_{\bullet}\,\eta q = \mathrm{E}\,\mathrm{E}_{\bullet}\,(\varepsilon+q)\eta \leq \mathrm{E}\,\mathrm{E}_{\bullet}\,(\varepsilon+q)\sup_{\bullet}\eta = \mathrm{E}\,\varepsilon\sup_{\bullet}\eta$$

Moreover, for $\delta \in (0,\infty)$ consider the event $A = \{\eta > \sup_{\bullet}\eta - \delta\} \in \mathbb{B}^{k+1}$. By definition and Fubini, its x sections have measure $F(A_{\bullet}) > 0$, which is Borel measurable. Then, setting $\varepsilon_m = \varepsilon \wedge m$ for $m \in (0,\infty)$, the function

$$q(x,u) = \frac{\varepsilon_m(x)}{F(A_{\bullet})}\mathbf{I}_A(x,u) - \varepsilon_m(x)$$

is in $\mathcal{G}_{c,\varepsilon}$, and

$$\mathrm{E}\,\eta q = \mathrm{E}\,\mathrm{E}_{\bullet}\,(\varepsilon_m + q)\eta \geq \mathrm{E}\,\varepsilon_m(\sup_{\bullet}\eta - \delta)$$

If $\delta \to 0$ and then $m \to \infty$, this lower bound tends to $\mathrm{E}\,\varepsilon\sup_{\bullet}\eta$. The same argument applied to $-\eta$ completes the proof of (12).

The proof of (13) is omitted, being entirely similar to the proof of (12).

To prove (14), pick any $q \in \mathcal{G}_{h,\varepsilon}$. Then the Jensen and Cauchy–Schwarz inequalities for conditional expectations show that

$$|\mathrm{E}\,\eta q| \leq \mathrm{E}\,|\mathrm{E}_{\bullet}\,\eta q| \leq \mathrm{E}\,\big|\mathrm{E}_{\bullet}\,\eta^2\big|^{1/2}\big|\mathrm{E}_{\bullet}\,q^2\big|^{1/2} \leq \sqrt{8}\,\mathrm{E}\,\varepsilon\big|\mathrm{E}_{\bullet}\,\eta^2\big|^{1/2}$$

To approximate the bound, let $m,n,\delta \in (0,\infty)$, and introduce the functions

$$\eta_n = \eta\,\mathbf{I}(|\eta| \leq n), \qquad \bar{\eta}_n = \eta_n - \mathrm{E}_{\bullet}\,\eta_n$$

$$\gamma_n^2 = 8\big|\mathrm{E}_{\bullet}\,\bar{\eta}_n^2\big|^{-1}\,\mathbf{I}(\mathrm{E}_{\bullet}\,\bar{\eta}_n^2 > \delta), \qquad q = \varepsilon_m\gamma_n\bar{\eta}_n$$

where $\varepsilon_m = \varepsilon \wedge m$. Then $q \in \mathcal{G}_{h,\varepsilon}$ and $\mathrm{E}\,\eta q = \mathrm{E}\,\varepsilon_m\gamma_n\,\mathrm{E}_{\bullet}\,\eta\eta_n$.

By dominated convergence and Minkowski's inequality for conditional expectations, we obtain that, as $n \to \infty$, a.e. $K(dx)$,

$$\mathrm{E}_{\bullet}\,\eta\eta_n = \mathrm{E}_{\bullet}\,\eta^2\,\mathbf{I}(|\eta| \leq n) \longrightarrow \mathrm{E}_{\bullet}\,\eta^2$$

$$\left(\big|\mathrm{E}_{\bullet}\,\bar{\eta}_n^2\big|^{1/2} - \big|\mathrm{E}_{\bullet}\,\eta^2\big|^{1/2}\right)^2 \leq \mathrm{E}_{\bullet}\,(\bar{\eta}_n - \eta)^2 \leq \mathrm{E}_{\bullet}\,\eta^2\,\mathbf{I}(|\eta| > n) \longrightarrow 0$$

Hence

$$\liminf_{n\to\infty}\gamma_n^2 \geq 8\big|\mathrm{E}_{\bullet}\,\eta^2\big|^{-1}\,\mathbf{I}(\mathrm{E}_{\bullet}\,\eta^2 > \delta)$$

And

$$\liminf_{n\to\infty}\mathrm{E}\,\eta q \geq \sqrt{8}\,\mathrm{E}\,\varepsilon_m\big|\mathrm{E}_{\bullet}\,\eta^2\big|^{1/2}\,\mathbf{I}(\mathrm{E}_{\bullet}\,\eta^2 > \delta)$$

by Fatou. Now let $\delta \to 0$, $m \to \infty$, and use monotone convergence.

Obviously (12)–(14) imply (15)–(17) by Hölder's (in)equality. ////

By interchange of $\sup_{|e|=1}$ and E, the following simplified expressions, which are indexed by $s = e$, suggest themselves as approximations:

$$\omega_{c,\varepsilon;e}(\eta) = \mathrm{E}\,\varepsilon \sup_{.} |\eta| = \frac{1}{2}\,\omega_{v,\varepsilon;e}(\eta) \tag{20}$$

$$\omega_{h,\varepsilon;e}(\eta) = \sqrt{8}\ \mathrm{E}\,\varepsilon\bigl(\mathrm{E}_{.}\,|\eta|^2\bigr)^{1/2} \tag{21}$$

Accordingly, to achieve (3), we define

$$\omega_{c,\alpha;e}(\eta) = \bigl\|\sup_{.} |\eta|\,\bigr\|_{\tilde\alpha} = \frac{1}{2}\,\omega_{v,\alpha;e}(\eta) \tag{22}$$

$$\omega_{h,\alpha;e}(\eta) = \sqrt{8}\,\bigl\|\bigl(\mathrm{E}_{.}\,|\eta|^2\bigr)^{1/2}\bigr\|_{\tilde\alpha} \tag{23}$$

Note that, in case $p = 1$, for $\eta \in Z_{1.}$,

$$\omega_{c,\varepsilon;0}(\eta) = \mathrm{E}\,\varepsilon \sup_{.} \eta \vee \mathrm{E}\,\varepsilon \sup_{.}(-\eta) \le \mathrm{E}\,\varepsilon \sup_{.} |\eta| = \omega_{c,\varepsilon;e}(\eta) \tag{24}$$

while

$$\omega_{h,\varepsilon;0}(\eta) = \sqrt{8}\ \mathrm{E}\,\varepsilon\bigl|\mathrm{E}_{.}\,\eta^2\bigr|^{1/2} = \omega_{h,\varepsilon;e}(\eta) \tag{25}$$

which shows the kind of increase $(* = c)$, respectively, coincidence $(* = h)$.

Lemma 7.3.4 *Given $\varepsilon \in \mathrm{M}_{.}^{+}$ and $\alpha \in [1,\infty]$. Then the terms $\omega_{*,t;e}$, for $* = c, v, h$ and $t = \varepsilon, \alpha$, are positively homogeneous, subadditive, hence convex on $L_1^p(P)$, and weakly lower semicontinuous on $L_2^p(P)$. Moreover, for every $\eta = (\eta_1, \dots, \eta_p)' \in Z_{1.}^p$,*

$$\omega_{*,t;0}(\eta) \le \omega_{*,t;e}(\eta) \le \omega_{*,t;e}(\eta_1) + \cdots + \omega_{*,t;e}(\eta_p) \tag{26}$$

Especially,

$$\omega_{h,t;e}(\eta) \le p\,\omega_{h,t;0}(\eta) \tag{27}$$

and

$$\omega_{h,2;e}^2(\eta) = 8\,\mathrm{tr}\,\mathcal{C}(\eta) = 8\,\mathrm{E}\,|\eta|^2 \tag{28}$$

$$\omega_{c,1;e}(\eta) = \sup_P |\eta| = \omega_{c,1;0}(\eta) = \omega_{c,0;0}(\eta) \tag{29}$$

PROOF Obviously $\omega_{*,t;e}(\eta) \ge \omega_{*,t;0}(\eta)$. Even if $\mathrm{E}_{.}\,\eta \ne 0$, it follows from $|\eta| \le \sum_j |\eta_j|$ that $\sup_{.} |\eta| \le \sum_j \sup_{.} |\eta_j|$ and from $\mathrm{E}_{.}\,|\eta|^2 = \sum_j \mathrm{E}_{.}\,\eta_j^2$ that $\bigl(\mathrm{E}_{.}\,|\eta|^2\bigr)^{1/2} \le \sum_j \bigl|\mathrm{E}_{.}\,\eta_j^2\bigr|^{1/2}$. Hence the second bound in (26).

For $* = h$, in view of (25) and (6), (26) implies (27); (28) is the special case $\alpha = \tilde\alpha = 2$ of (23). For $* = c$, (29) is the special case $\alpha = 1$, $\tilde\alpha = \infty$ of (22), and iterates (18).

The following representations are obvious,

$$\omega_{c,\varepsilon;e}(\eta) = \sup\bigl\{\,\mathrm{E}\,|\eta|q \bigm| q \in L_\infty(P),\ \mathrm{E}_{.}\,|q| \le \varepsilon\,\bigr\} \tag{30}$$

$$\omega_{h,\varepsilon;e}(\eta) = \sup\bigl\{\,\mathrm{E}\,|\eta|q \bigm| q \in L_\infty(P),\ \mathrm{E}_{.}\,q^2 \le 8\varepsilon^2\,\bigr\} \tag{31}$$

Thus the terms $\omega_{*,\varepsilon;e}$ are positive homogeneous, subadditive, hence convex. To show l.s.c., let $\eta_n \to \eta$ in $L_2^p(P)$, hence $|\eta_n| \to |\eta|$ in $L_2(P)$. Then $\lim_n \mathrm{E}\,|\eta_n|q = \mathrm{E}\,|\eta|q$ for each q, hence $\liminf_n \omega_{*,\varepsilon;e}(\eta_n) \ge \omega_{*,\varepsilon;e}(\eta)$. But for convex functions, strong and weak l.s.c. are the same [Lemma B.1.2 b]. The terms $\omega_{*,\alpha;e}$, as sup of functions $\omega_{*,\varepsilon;e}$, inherit these properties. ////

In regard to Section 4.4, we observe that the maximal oscillation over average conditional neighborhoods can already be achieved by a restriction to contamination curves whose tangent spaces $V_{2;*,\varepsilon}$ contain the given partial influence curve.

Proposition 7.3.5 *Let* $\eta \in Z_{2\bullet}^p$ *and* $\alpha \in [1,\infty]$, $* = c, v, h$. *Then*

$$\omega_{*,\alpha;e}(\eta) = \sup\{\, \omega_{*,\varepsilon;e}(\eta) \mid \|\varepsilon\|_\alpha \le 1,\ \eta \in V_{2;*,\varepsilon}^p \,\} \tag{32}$$

$$\omega_{*,\alpha;0}(\eta) = \sup_{|e|=1} \sup\{\, \omega_{*,\varepsilon;0}(e'\eta) \mid \|\varepsilon\|_\alpha \le 1,\ e'\eta \in V_{2;*,\varepsilon} \,\} \tag{33}$$

$$\omega_{*,\alpha;\infty}(\eta) = \sup_{j=1,\dots,p} \sup\{\, \omega_{*,\varepsilon;0}(\eta_j) \mid \|\varepsilon\|_\alpha \le 1,\ \eta_j \in V_{2;*,\varepsilon} \,\} \tag{34}$$

PROOF Consider any contamination curve $\varepsilon \in \mathbb{M}_\bullet^+$ such that $\|\varepsilon\|_\alpha \le 1$.

To prove (32), first let $\alpha \in (1,\infty)$. Then $\omega_{*,\varepsilon;e}(\eta) = \omega_{*,\alpha;e}(\eta)$ can be achieved [Hölder] iff $\|\varepsilon\|_\alpha = 1$ and, in $L_1(K)$,

$$\varepsilon^\alpha \propto \begin{cases} \sup_\bullet |\eta|^{\tilde\alpha} & (* = c, v) \\ (\mathrm{E}_\bullet |\eta|^2)^{\tilde\alpha} & (* = h) \end{cases} \tag{35}$$

For such ε, however, $P(\varepsilon = 0, \eta \ne 0) = 0$ hence $\eta \in V_{2;*,\varepsilon}^p$.

If $\alpha = \infty$ then $\omega_{*,\alpha;e}(\eta) = \omega_{*,\varepsilon_1;e}(\eta)$ and $\eta \in V_{2;*,\varepsilon_1}^p = Z_{2\bullet}^p$ for $\varepsilon_1 = 1$.

If $\alpha = 1$ consider $\bar\varepsilon = (\varepsilon + \kappa)/(1+\kappa)$ with $\kappa \in (0,1)$. Then $\|\bar\varepsilon\|_\alpha \le 1$, and we have $\eta \in V_{2;*,\bar\varepsilon}^p = Z_{2\bullet}^p$ since $\bar\varepsilon > 0$. Moreover,

$$\omega_{*,\bar\varepsilon;e}(\eta) = \frac{1}{1+\kappa}\,\omega_{*,\varepsilon;e}(\eta) + \frac{\kappa}{1+\kappa}\,\omega_{*,\varepsilon_1;e}(\eta) \tag{36}$$

where $\omega_{*,\varepsilon;e}(\eta)$ may tend to $\omega_{*,1;e}(\eta)$ and $\omega_{*,\varepsilon_1;e}(\eta) \le \omega_{*,1;e}(\eta)$. Thus the assertion follows if $\kappa \downarrow 0$.

Arguing for any fixed unit vector $e \in \mathbb{R}^p$, $|e| = 1$, (33) can likewise be proved. On passing to a coordinate achieving the $\max_j$, the proof of (34) may be based on (33). ////

7.4 Optimal Influence Curves

In this section, some of the optimality problems on as. variance (as. minimax risk with quadratic loss) and as. bias (infinitesimal oscillation) will be solved that are peculiar to robust regression.

7.4.1 Optimization Problems

Taking over the $p \times k$ matrix $D = d\tau(0)$ from 5.5(2), and imposing a bound on the conditional bias term $\omega_{*,t;s}$, the minimum trace problem $O_{*,t;s}^{\mathrm{tr}}(b)$ reads

$$\mathrm{E}\,|\eta|^2 = \min! \qquad \eta \in \Psi_2^D,\ \omega_{*,t;s}(\eta) \le b \tag{6°}$$

where $* = c, v, h$, $s = 0, 2, \infty, e$, and $t = \varepsilon, \alpha$; the case $t = 0$ being covered already by Section 5.5.

Formulation (6°) is preliminary: For conditional bias, actually the subclass of conditionally centered (partial) influence curves will be employed.

Definition 7.4.1 *The set* $\Psi^D_{2\bullet}$ *of all* square integrable, *and the subset* $\Psi^D_{\infty\bullet}$ *of all* bounded, conditionally centered (partial) influence curves at $P = P_0$ *are, for* $\alpha = 2, \infty$, *respectively,*

$$\Psi^D_{\alpha\bullet} = \{\, \eta \in L^p_\alpha(P) \mid \mathrm{E}_\bullet\, \eta = 0,\ \mathrm{E}\, \eta \Lambda' = D \,\} \tag{1}$$

Remark 7.4.2

(a) The attribute 'square integrable' is usually omitted.

(b) In case $D = \mathbb{I}_k$ we write $\Psi^{\mathbb{I}_k}_{\alpha\bullet} = \Psi_{\alpha\bullet}$.

(c) Since $\mathrm{E}_\bullet \Lambda = x\, \mathrm{E}_F \Lambda_f = 0$ we have

$$\psi_h = \mathcal{I}^{-1} \Lambda \in \Psi_{2\bullet}\,, \qquad \eta_h = D\psi_h \in \Psi^D_{2\bullet} \tag{2}$$

(d) Using any conditionally centered $\varrho \in \Psi_{\alpha\bullet}$ in 4.2(43), we obtain that

$$\Psi^D_{\alpha\bullet} = \{\, D\psi \mid \psi \in \Psi_{\alpha\bullet} \,\} \tag{3}$$

(e) $\Psi^D_{\alpha\bullet} \subset \Psi^D_\alpha$ are convex and (weakly) closed in $L^p_\alpha(P)$; $\alpha = 2, \infty$. ////

Moreover, we set

$$\mathbb{H}_\bullet = L^p_2(K) \tag{4}$$

If K has finite support then $\dim \mathbb{H}_\bullet < \infty$, but in general $\dim \mathbb{H}_\bullet = \infty$.

Aspects of Problem $O^{\mathrm{tr}}_{*,t;s}(b)$; $t = \varepsilon, \alpha$

We outline some general optimization features in the conditional case.

Conditional Centering: Comparing $\eta \in \Psi^D_2$ with $\eta_\bullet = \eta - \mathrm{E}_\bullet \eta$, which is in $\Psi^D_{2\bullet}$ since $\mathrm{E}_\bullet \Lambda = 0$, covariance decreases in the positive definite sense,

$$\mathcal{C}(\eta) - \mathcal{C}(\eta_\bullet) = \mathcal{C}(\mathrm{E}_\bullet \eta) \geq 0 \tag{5}$$

while conditional bias $\omega_{*,t;s}$ for $* = c, v, h$, $t = \varepsilon, \alpha$, $s = 0, 2, \infty$, in view of 7.3(4), stays the same. Thus Ψ^D_2 may in (6°) be replaced by the conditionally centered subclass $\Psi^D_{2\bullet}$, and the corresponding problems $O^{\mathrm{tr}}_{*,t;s}(b)$ read

$$\mathrm{E}\, |\eta|^2 = \min! \qquad \eta \in \Psi^D_{2\bullet}\,,\ \omega_{*,t;s}(\eta) \leq b \tag{6}$$

where $* = c, v, h$, $t = \varepsilon, \alpha$, $s = 0, 2, \infty$. And we take (6) as the definite formulation also for $s = e$.

Minimum Norm: The conditional bias terms $\omega_{*,t;s}$ being (weakly) l.s.c. [Lemmas 7.3.1 and 7.3.4], problem $O^{\mathrm{tr}}_{*,t;s}(b)$ is still a minimum norm problem over a convex closed subset of the Hilbert space $\mathbb{H} = L^p_2(P)$ with unique solution, since $\Psi^D_{2\bullet}$ like Ψ^D_2 is obviously convex and also weakly closed:

If $\eta_m \to \eta$ weakly in $\mathbb{H}$ then $\mathrm{E}_\bullet \eta_m \to \mathrm{E}_\bullet \eta$ weakly in $\mathbb{H}_\bullet$.

One At a Time Optimality: Also the classes $\Psi^D_{\alpha\bullet}$ are defined coordinatewise,

$$\Psi^D_{\alpha\bullet} = \Psi^{D_1}_{\alpha\bullet} \times \cdots \times \Psi^{D_p}_{\alpha\bullet}\,, \qquad \alpha = 2, \infty \tag{7}$$

Therefore, problem $O^{\rm tr}_{*,t;\infty}(b)$ falls apart into p separate such problems, one for each row vector D_j of D (with $p=1$). This entails that the solution to $O^{\rm tr}_{*,t;\infty}(b)$ minimizes not only the trace but each diagonal element of the covariance. Due to different side conditions [see bound 7.3(6)], the solutions to $O^{\rm tr}_{*,t;s}(b)$ for $s = 0, 2$ are however not rendered inadmissible.

Also the reduction formulas 5.5(12) and 5.5(13) for minimal bias, and its coordinatewise attainability, carry over to the conditional bias versions $\omega_{*,t;s}$ for $s = 2, \infty$ (with Ψ^D_α, $\Psi^{D_j}_\alpha$ replaced by $\Psi^D_{\alpha\bullet}$, $\Psi^{D_j}_{\alpha\bullet}$, respectively).

In particular, in view of 7.3(25), the solutions to the Hellinger problems $O^{\rm tr}_{h,t;\infty}(b)$, for $t = \varepsilon, \alpha$, may be built up coordinatewise by the solutions to the corresponding problems $O^{\rm tr}_{h,t;e}(b)$ $[\,p=1\,]$.

Convex Optimization: Because the explicit calculations of $\omega_{*,t;s}$ use the assumption $\mathrm{E}_\bullet = 0$, the functions f, G, and H change from 5.5(7) to

$$f(\eta) = \mathrm{E}\,|\eta|^2\,, \quad G(\eta) = \omega_{*,t;s}(\eta)\,, \quad H\eta = (\mathrm{E}_\bullet\,\eta, \mathrm{E}\,\eta\Lambda') \in \mathbb{H}_\bullet \times \mathbb{R}^{p\times k} \tag{8}$$

Then the oscillation terms may be extended to all of $\mathbb{H}$ using the explicit expressions derived under the condition $\mathrm{E}_\bullet = 0$, which will be taken care of separately by the linear operator H; thus (4) is given up.

Now the range of H becomes infinite-dimensional iff $\dim \mathbb{H}_\bullet = \infty$. Thus well-posedness, in addition to argument (10) below, requires the explicit verification of condition B.2(19), which can no longer be ignored.

Finite Values: The oscillation variants $\omega_{*,t;s}$ are finite on $\mathbb{L}$, provided in case $t = \varepsilon$ that

$$\|\varepsilon\|_1 < \infty \tag{9}$$

Still $\Psi^D_{\infty\bullet}$ is dense in $\Psi^D_{2\bullet} \subset \mathbb{H}$, which can be shown like 5.5(9) but only defining $\psi^\natural_m = \psi_m - \mathrm{E}_\bullet\,\psi_m$ in 5.5(8).

Neglect of Conditional Centering: For $t = \varepsilon, \alpha$, the constraint $\mathrm{E}_\bullet\,\eta = 0$ can be removed from the total variation problems $O^{\rm tr}_{v,t;s}(b)$ with $s = 0, 2, \infty$ and from the Hellinger problems $O^{\rm tr}_{h,t;s}(b)$ with $s = 0, 2, \infty, e$: If $\eta \in \mathbb{H}$ is recentered at $\mathrm{E}_\bullet\,\eta$, then $\mathrm{E}\,\eta\eta'$ decreases à la (5), the term $\mathrm{E}\,\eta\Lambda'$ stays the same, and the explicit expressions of $\omega_{*,t;s}(\eta)$ derived under the assumption that $\mathrm{E}_\bullet\,\eta = 0$ stay the same $(* = v)$, respectively decrease $(* = h)$.

From the contamination problems $O^{\rm tr}_{c,t;s}(b)$ $[\,t = \varepsilon, \alpha,\ s = 0, 2, \infty, e\,]$ the constraint $\mathrm{E}_\bullet = 0$ may be removed if the error distribution F satisfies the symmetry type condition 7.5(27) below.

Well-Posedness: If $\mathrm{E}_\bullet = 0$ can be removed, the earlier argument 5.5(10) with $a = 0$ suffices for the well-posedness conditions B.2(18) and B.2(22) on the linear constraints; only in case $t = \varepsilon$, integrability (9): $\mathrm{E}\,\varepsilon < \infty$, is assumed to keep oscillations, especially of $\eta = A\psi$ in 5.5(10), finite on $\mathbb{L}$.

If $\mathrm{E}_{\bullet} = 0$ may not be removed, pick any $\psi \in \Psi_{\infty\bullet}$. Then, for all $a \in \mathbb{H}_{\bullet}$ and $A \in \mathbb{R}^{p\times k}$, one achieves

$$\eta = a + A\psi \in \mathbb{H}, \qquad a = \mathrm{E}_{\bullet}\,\eta, \quad A = \mathrm{E}\,\eta\Lambda' \tag{10}$$

To keep the oscillations of $\eta = a + A\psi$ finite, in case $t = \varepsilon$, square integrability is assumed,

$$\|\varepsilon\|_2 < \infty \tag{11}$$

Thus the well-posedness conditions B.2(18) and B.2(22) on the linear constraints can be verified, which completely suffices in case $\dim \mathbb{H}_{\bullet} < \infty$.

In case $\dim \mathbb{H}_{\bullet} = \infty$, we restrict a and A to some bounded neighborhoods of 0 in $\mathbb{H}_{\bullet}$ and of D in $\mathbb{R}^{p\times k}$, respectively. Then conditions B.2(18) and B.2(22) are still fulfilled and, in addition, the norms stay bounded,

$$\mathrm{E}\,|\eta|^2 \le 2\,\mathrm{E}\,|a|^2 + 2\,\|A\|^2\,\mathrm{E}\,|\psi|^2 \tag{12}$$

At the same time, the oscillations $\omega_{*,\varepsilon;s}(\eta)$ for $t = \varepsilon$ and $s = 0, 2, \infty, e$ (especially in the only remaining case $* = c$ that $\mathrm{E}_{\bullet} = 0$ cannot be ignored) satisfy

$$\omega_{*,\varepsilon;s}(\eta) \le \sqrt{p}\,\bigl(\mathrm{E}\,\varepsilon|a| + \|A\|\,\omega_{*,\varepsilon;e}(\psi)\bigr) \tag{13}$$

hence stay bounded if $\mathrm{E}\,\varepsilon^2 < \infty$. Under assumption (11), therefore, the well-posedness condition B.2(19) can be verified for $* = c$ and $t = \varepsilon$.

For $t = \alpha \in [1, \infty]$ and $s = 0, 2, \infty, e$, the oscillations are bounded by

$$\omega_{*,\alpha;s}(\eta) \le \sqrt{p}\,\bigl(\|a\|_{\tilde\alpha} + \|A\|\,\omega_{*,\alpha;e}(\psi)\bigr) \tag{14}$$

The RHS, with a and A ranging over bounded neighborhoods of 0 in $\mathbb{H}_{\bullet}$ and of D in $\mathbb{R}^{p\times k}$, respectively, remains bounded iff the conjugate exponent $\tilde\alpha$ is ≤ 2; that is, $\alpha \ge 2$. Thus, for $* = c$ and $t = \alpha \in [2, \infty]$, condition B.2(19) is again fulfilled.

For $* = c$ and $t = \alpha \in [1, 2)$, however, condition B.2(19) hangs [unboundedness of $\|a\|_\infty$ for $\|a\|_2 < .01$ being typical]. In general, we thus have no Lagrange multiplier $a \in \mathbb{H}_{\bullet}$ for the $\mathrm{E}_{\bullet}$ operator. The difficulty does not arise if $\dim \mathbb{H}_{\bullet} < \infty$ [as by K of finite support], or if the constraint $\mathrm{E}_{\bullet} = 0$ may be removed [as by F verifying 7.5(27) below].

By postponing the constraint $\mathrm{E}_{\bullet} = 0$ after conditioning on x, the general problem $O^{\mathrm{tr}}_{*,t;s}(b)$ can in fact be reduced to a family of optimization problems in $L^p_2(F)$, each one with the $\mathbb{R}^p$-valued side condition $\mathrm{E}_F = 0$; which supplies a family of multipliers $a(x) \in \mathbb{R}^p$. Measurability of the map a, however, must then be ensured by extra arguments and conditions [cf. Theorem 7.4.13 a and its proof, in the case $* = c$, $t = \alpha = 1$, $s = 0, e$]. Measurability of a is indispensable, not only because a is part of the solution $\tilde\eta$ (product measurable) but, already before $\tilde\eta$ has been patched together from its x sections, in order to justify the derivation of $\tilde\eta(x, .)$ by pointwise minimization of integrands.

Classical Scores: The universal optimality of the classical partial scores $\eta_h = D\mathcal{I}^{-1}\Lambda$ for the unconditional Hellinger problems $O^{\mathrm{tr}}_{h,0;s}(b)$ will not carry over to the conditional problems $O^{\mathrm{tr}}_{h,t;s}(b)$ with $t = \varepsilon, \alpha$. In particular, η_h does not solve the mean conditional Hellinger problems $O^{\mathrm{tr}}_{h,1;s}(b)$ for $s = 0, e$ [Theorem 7.4.19]. However, η_h is still the solution to the square conditional Hellinger problem $O^{\mathrm{tr}}_{h,2;s}(b)$, all variants $s = 0, 2, \infty, e$. Since $\mathrm{E}_{\bullet}\, \eta_h = 0$, this follows from the restriction onto $\Psi^D_{2\bullet}$, from 7.3(10), 7.3(19), and 7.3(28), respectively, and from the Cramèr–Rao bound 5.5(6).

Mean Conditional Contamination: If the solution $\tilde\eta$ to the unconditional problem $O^{\mathrm{tr}}_{c,0;0}(b)$ turns out conditionally centered [e.g., under symmetry 7.5(27) of F], then $\tilde\eta$ also solves the conditional problems $O^{\mathrm{tr}}_{c,1;s}(b)$ for $s = 0, e$. Likewise, for $s = 2, \infty$, the solution $\tilde\eta$ to the unconditional problem $O^{\mathrm{tr}}_{c,0;s}(b)$ also solves the corresponding conditional problem $O^{\mathrm{tr}}_{c,1;s}(b)$, provided that $\mathrm{E}_{\bullet}\, \tilde\eta = 0$. This follows from the restriction onto $\Psi^D_{2\bullet}$, and from 7.3(8), 7.3(18), and 7.3(29).

Sup Conditional Contamination: The boundary case $t = \alpha = \infty$ of average conditional contamination may also be represented by some fixed (least favorable) contamination curve $t = \varepsilon_1$; namely, $\varepsilon_1(x)$ constant 1. By definition, the tangent classes $\mathcal{G}_{*,\infty}$ and $\mathcal{G}_{*,\varepsilon_1}$ coincide, hence also the oscillations $\omega_{*,\infty;s} = \omega_{*,\varepsilon_1;s}$, and the problems $O^{\mathrm{tr}}_{*,\infty;s}(b) = O^{\mathrm{tr}}_{*,\varepsilon_1;s}(b)$.

In this model, as Nature is required to spread her contamination evenly, it will pay to take chances and use high clipping points $c(x)$, respectively, weights $h(x)$ close to 1, at the large x, which are informative if they are not contaminated. It will not pay to take any chances at the small and uninformative values of x, but to simply ignore observations there.

Minimal Bias: Finiteness of $\omega^{\min}_{*,t;s}$, by considering $\omega_{*,t;s}$ on $\mathbb{L}$, is always realized for $t = \alpha$, and for $t = \varepsilon$ is ensured by assumption (9): $\mathrm{E}\,\varepsilon < \infty$.

Attainability of $\omega^{\min}_{*,t;s}$, contrary to the unconditional case $t = 0$, is not automatic when $t = \varepsilon, \alpha$; except for $t = \alpha \in [1, 2]$. For such exponents we have the bound

$$\mathrm{E}\,|\eta|^2 \le p^2\, \omega^2_{*,\alpha;s}(\eta) \tag{15}$$

Therefore, any sequence in $\Psi^D_{2\bullet}$ approximating $\omega^{\min}_{*,\alpha;s}$ (finite) is necessarily (norm) bounded in $\mathbb{H}$. Attainability of $\omega^{\min}_{*,\alpha;s}$ for $\alpha \in [1, 2]$ thus follows by the previous argument [p 198].

Tangent Subspaces: For $\varepsilon \in \mathbb{M}^+_{\bullet}$ such that $K(\varepsilon = 0) > 0$, the tangent space $V_{2;*,\varepsilon}$ of the infinitesimal neighborhoods $\mathcal{U}_{*,\varepsilon}$ is strictly smaller than $Z_{2\bullet}$, if $K\bigl(\varepsilon(x) = 0, x \ne 0\bigr) > 0$ not even containing the coordinates of Λ [Proposition 7.2.8 a]. As the parametric tangent has been built into the definition of influence curves and as. linearity, the smallest tangent space to consider is the (closed) linear span $W_{2;*,\varepsilon}$ of $V_{2;*,\varepsilon}$ and the k coordinates of Λ in $Z_{2\bullet} \subset L_2(P)$, which is given by 7.2(32), so that

$$W^p_{2;*,\varepsilon} = \bigl\{\, \eta \in Z^p_{2\bullet} \;\big|\; \exists\, A \in \mathbb{R}^{p\times k} \colon P(\varepsilon = 0, \eta \ne A\Lambda) = 0 \,\bigr\} \tag{16}$$

The orthogonal projection $\Pi^p_W\colon \mathbb{H} \to W^p_{2;*,\varepsilon} \subset Z^p_{2.}$ of $\eta \in \Psi^D_{2.}$ onto $W^p_{2;*,\varepsilon}$ decreases the norm $\mathrm{E}\,|\eta|^2$ while it leaves both $\mathrm{E}_{\bullet}\,\eta = 0$ and $\mathrm{E}\,\eta\Lambda' = D$ unchanged. Nor does Π^p_W affect the oscillations of $\eta \in \Psi^D_{2.}$,

$$\omega_{*,\varepsilon;s}\big(\Pi^p_W(\eta)\big) = \omega_{*,\varepsilon;s}(\eta) \tag{17}$$

This is apparent from definition 5.3(5) of the exact versions $s = 0$, and is inherited to the coordinatewise versions $s = 2, \infty$. That (17) also holds for version $s = e$, is not obvious from the representations 7.3(30), 7.3(31). But the computation 7.2(35) of Π_W in Proposition 7.2.8 a yields that $\Pi^p_W(\eta) \neq \eta$ at most on $\{\varepsilon = 0\}$. As $\Pi^p_W(\eta) = \eta$ on $\{\varepsilon \neq 0\}$, the explicit terms $\omega_{*,\varepsilon;e}$ given by 7.3(20), 7.3(21) thus agree for η and $\Pi^p_W(\eta)$.

Therefore, the solutions $\tilde\eta$ to the problems $O^{\mathrm{tr}}_{*,\varepsilon;s}(b)$ will automatically satisfy

$$\tilde\eta \in W^p_{2;*,\varepsilon} \tag{18}$$

The proportionality $\tilde\eta = A\Lambda$ implied on $\{\varepsilon = 0\}$ is mathematical proof to the intuition that the classical partial scores may essentially be taken over on that part of the design space which is free of contamination.

7.4.2 Auxiliary Results

Basic Optimality Lemmas

The first three lemmas are the basis for conditional neighborhood optimality in the cases $* = c, v, h$ (contamination, total variation, Hellinger).

Lemma 7.4.3 *Let $Y \in \mathbb{H}$ and $c \in \mathbb{M}^+_{\bullet}$ be given.*

(a) *Then the unique solution to the problem*

$$\mathrm{E}\,|\chi - Y|^2 = \min! \qquad \chi \in \mathbb{H},\ \sup_{\bullet} |\chi| \le c \tag{19}$$

is

$$\tilde\chi = Y \min\Big\{1, \frac{c}{|Y|}\Big\} \tag{20}$$

(b) *A function $\bar\chi \in \mathbb{H}$ satisfying $\sup_{\bullet} |\bar\chi| \le c$ solves the problem*

$$\mathrm{E}\,\chi' Y = \max! \qquad \chi \in \mathbb{H},\ \sup_{\bullet} |\chi| \le c \tag{21}$$

iff

$$\bar\chi = \frac{cY}{|Y|} \qquad \text{on } \{Y \neq 0\} \tag{22}$$

PROOF We argue modulo null sets in this proof.

(a) Pointwise minimization of the integrand in

$$\mathrm{E}\,|\chi - Y|^2 = \mathrm{E}\,\mathrm{E}_{\bullet}\,|\chi - Y|^2 \tag{23}$$

subject to $\sup_{\bullet} |\chi| \le c$ [Lemma 5.5.10; $M = \mathbb{I}_p$]. The resulting $\tilde\chi$ is obviously measurable, and also square integrable since $|\tilde\chi| \le |Y|$.

(b) For $\chi \in \mathbb{H}$ such that $\sup_{\bullet} |\chi| \le c$ we have

$$\mathrm{E}\,\chi' Y \le \mathrm{E}\,|\chi||Y| = \mathrm{E}\,\mathrm{E}_{\bullet}\,|\chi||Y| \le \mathrm{E}\,c\,\mathrm{E}_{\bullet}\,|Y| \tag{24}$$

where the first bound [Cauchy–Schwarz] is attained iff the values of χ and Y in $\mathbb{R}^p$ have the same direction. The second bound is attained if and only if $|\chi| = \sup_{\bullet} |\chi| = c$ whenever $Y \ne 0$.

If $\bar{\chi}$ of form (22) is just measurable and satisfies $|\bar{\chi}| \le c$, also on the event $\{Y = 0\}$, then $\mathrm{E}\,\bar{\chi}^2 \le \mathrm{E}\,c^2$, hence $\bar{\chi} \in \mathbb{H}$ if $\mathrm{E}\,c^2 < \infty$. ////

Lemma 7.4.4 *Suppose that $p = 1$, and let $Y \in \mathbb{H}$, $c' \in \mathrm{M}_{\bullet}^-$, $c'' \in \mathrm{M}_{\bullet}^+$.*

(a) *Then the unique solution to the problem*

$$\mathrm{E}\,(\chi - Y)^2 = \min! \qquad \chi \in \mathbb{H},\ c' \le \chi \le c'' \tag{25}$$

is

$$\tilde{\chi} = c' \vee Y \wedge c'' \tag{26}$$

(b) *A function $\bar{\chi} \in \mathbb{H}$ satisfying $c' \le \bar{\chi} \le c''$ solves the problem*

$$\mathrm{E}\,\chi' Y = \max! \qquad \chi \in \mathbb{H},\ c' \le \chi \le c'' \tag{27}$$

iff

$$\bar{\chi}\,\mathbf{I}(Y \ne 0) = c'\,\mathbf{I}(Y < 0) + c''\,\mathbf{I}(Y > 0) \tag{28}$$

PROOF Similar to the preceeding proof. ////

Lemma 7.4.5 *Let $Y \in \mathbb{H}$ and $h \in \mathrm{M}_{\bullet}^+$ be given.*

(a) *The unique solution to the problem*

$$\mathrm{E}\,|\chi - Y|^2 = \min! \qquad \chi \in \mathbb{H},\ \mathrm{E}_{\bullet}\,|\chi|^2 \le h^2 \tag{29}$$

is

$$\tilde{\chi} = Y \min\Big\{1, \frac{h}{\sqrt{\mathrm{E}_{\bullet}\,|Y|^2}}\Big\} \tag{30}$$

(b) *A function $\bar{\chi} \in \mathbb{H}$ satisfying $\mathrm{E}_{\bullet}\,|\bar{\chi}|^2 \le h^2$ solves the problem*

$$\mathrm{E}\,\chi' Y = \max! \qquad \chi \in \mathbb{H},\ \mathrm{E}_{\bullet}\,|\chi|^2 \le h^2 \tag{31}$$

iff

$$\bar{\chi} = \frac{hY}{\sqrt{\mathrm{E}_{\bullet}\,|Y|^2}} \qquad \text{on } \{\sup_{\bullet} |Y| \ne 0\} \tag{32}$$

PROOF We argue modulo null sets.

(a) Pointwise minimization of the integrand in

$$\mathrm{E}\,|\chi - Y|^2 = \mathrm{E}\,\mathrm{E}_{\bullet}\,|\chi - Y|^2 \tag{33}$$

subject to $\mathrm{E}_{\bullet}\,|\chi|^2 \le h^2$ [which are convex and, if $h(x) > 0$, well-posed minimum norm problems in $L_2^p(F)$]. The resulting $\tilde{\chi}$ is obviously product measurable, and square integrable since $\mathrm{E}_{\bullet}\,|\chi|^2 \le \mathrm{E}_{\bullet}\,|Y|^2$.

(b) For any $\chi \in \mathbb{H}$ such that $\mathrm{E}_{\bullet} |\chi|^2 \le h^2$ we have

$$\mathrm{E}\,\chi' Y \le \mathrm{E}\,|\chi||Y| = \mathrm{E}\,\mathrm{E}_{\bullet}\,|\chi||Y| \le \mathrm{E}\,h\big(\mathrm{E}_{\bullet}\,|Y|^2\big)^{1/2} \tag{34}$$

where the first bound [Cauchy–Schwarz] is attained iff the values of χ and Y in $\mathbb{R}^p$ have the same direction. The second bound is attained iff $|\chi| \propto |Y|$, with proportionality factor in $\mathbb{M}_{\bullet}^{+}$, and also $\mathrm{E}_{\bullet}\,|\chi|^2 = h^2$ whenever $\sup_{\bullet} |Y| \neq 0$.

If $\bar{\chi}$ of form (32) is just measurable and satisfies $\mathrm{E}_{\bullet}\,|\bar{\chi}|^2 \le h^2$, also where $\sup_{\bullet} Y = 0$, then $\mathrm{E}\,\bar{\chi}^2 \le \mathrm{E}\,h^2$, hence $\bar{\chi} \in \mathbb{H}$ if $\mathrm{E}\,h^2 < \infty$. ////

Conditional Averaging

The next lemma provides an improvement by conditional averaging, which will be applied in solving the problems $O^{\mathrm{tr}}_{*,\varepsilon;s}(b)$ with $b = \omega^{\min}_{*,\varepsilon;s}$; $* = c, v$.

Lemma 7.4.6 *Imagine a partial influence curve $\bar{\eta} \in \Psi^D_{2\bullet}$ of the form*

$$\bar{\eta} = c\frac{Ax}{|Ax|}\,\mathbf{I}(Ax \neq 0)\,\mathrm{sign}(\Lambda_f - m) \qquad \text{on } \{\Lambda_f \neq m\} \tag{35}$$

with $A \in \mathbb{R}^{p\times k}$, $c \in \mathbb{M}_{\bullet}$, and any $m = \mathrm{med}\,\Lambda_f(F)$. Then

$$\hat{\eta} = c\frac{Ax}{|Ax|}\,\mathbf{I}(Ax \neq 0)\big(\mathrm{sign}(\Lambda_f - m) + \gamma\,\mathbf{I}(\Lambda_f = m)\big) \tag{36}$$

with $\gamma \in [-1, 1]$ determined from $\mathrm{E}\,\mathrm{sign}(\Lambda_f - m) + \gamma P(\Lambda_f = m) = 0$, defines a partial influence curve $\hat{\eta} \in \Psi^D_{2\bullet}$ that achieves

$$\mathcal{C}(\hat{\eta}) \le \mathcal{C}(\bar{\eta}) \tag{37}$$

In subsequent proofs, the unit vector in direction of $y \in \mathbb{R}^p \setminus \{0\}$ is abbreviated by

$$e_y = \frac{y}{|y|}\,\mathbf{I}(y \neq 0) \tag{38}$$

Then always $y'e_y = |y|$ and $y = |y|\,e_y$.

PROOF Averaging $\bar{\eta}$ over $\{\Lambda_f = m\}$ conditionally on x, we obtain the function

$$\begin{aligned} \hat{\eta} &= \bar{\eta}\,\mathbf{I}(\Lambda_f \neq m) + \bar{a}\,\mathbf{I}(\Lambda_f = m) \\ \bar{a}\,\mathrm{E}_{\bullet}\,\mathbf{I}(\Lambda_f = m) &= \mathrm{E}_{\bullet}\,\bar{\eta}\,\mathbf{I}(\Lambda_f = m) \end{aligned} \tag{39}$$

where $\mathrm{E}_{\bullet}\,\mathbf{I}(\Lambda_f = m) = P(\Lambda_f = m)$ and

$$\mathrm{E}_{\bullet}\,\bar{\eta}\,\mathbf{I}(\Lambda_f = m) = -\,\mathrm{E}_{\bullet}\,\bar{\eta}\,\mathbf{I}(\Lambda_f \neq m) = -c\,e_{Ax}\,\mathrm{E}\,\mathrm{sign}(\Lambda_f - m) \tag{40}$$

Hence,

$$\bar{a} = \gamma c\, e_{Ax}\,, \qquad \gamma P(\Lambda_f = m) = -\,\mathrm{E}\,\mathrm{sign}(\Lambda_f - m) \tag{41}$$

From (39), $\mathrm{E}_{\bullet}\hat{\eta} = \mathrm{E}_{\bullet}\bar{\eta} = 0$ is immediate. $\mathrm{E}\,\hat{\eta}\Lambda' = \mathrm{E}\,\bar{\eta}\Lambda' = D$ is verified by

$$\begin{aligned} \mathrm{E}\,\hat{\eta}\Lambda' &= \mathrm{E}\,\hat{\eta}\,\mathbf{I}(\Lambda_f \neq m)\Lambda' + \mathrm{E}\,\bar{a}\,\mathrm{E}_{\bullet}\,\mathbf{I}(\Lambda_f = m)\Lambda_f x' \\ &\stackrel{(39)}{=} \mathrm{E}\,\bar{\eta}\,\mathbf{I}(\Lambda_f \neq m)\Lambda' + \mathrm{E}\,\mathrm{E}_{\bullet}\,\bar{\eta}\,\mathbf{I}(\Lambda_f = m) m x' = \mathrm{E}\,\bar{\eta}\Lambda' \end{aligned} \tag{42}$$

Moreover,

$$\begin{aligned} 0 &\le \mathrm{E}_{\bullet}(\bar{\eta} - \hat{\eta})(\bar{\eta} - \hat{\eta})'\,\mathbf{I}(\Lambda_f = m) \\ &= \mathrm{E}_{\bullet}\,\bar{\eta}\bar{\eta}'\,\mathbf{I}(\Lambda_f = m) - \mathrm{E}_{\bullet}\,\hat{\eta}\hat{\eta}'\,\mathbf{I}(\Lambda_f = m) \end{aligned} \tag{43}$$

as

$$\begin{aligned} \mathrm{E}_{\bullet}\,\hat{\eta}\bar{\eta}'\,\mathbf{I}(\Lambda_f = m) &= \bar{a}\,\mathrm{E}_{\bullet}\,\bar{\eta}'\,\mathbf{I}(\Lambda_f = m) \\ &= \bar{a}\bar{a}'\,\mathrm{E}_{\bullet}\,\mathbf{I}(\Lambda_f = m) = \mathrm{E}_{\bullet}\,\hat{\eta}\hat{\eta}'\,\mathbf{I}(\Lambda_f = m) \end{aligned} \tag{44}$$

while $\mathrm{E}\,(\bar{\eta} - \hat{\eta})(\bar{\eta} - \hat{\eta})'\,\mathbf{I}(\Lambda_f \neq m) = 0$ by (39). This implies (37). ////

Remark 7.4.7 Whenever we get a partial influence curve $\bar{\eta} \in \Psi_{2\bullet}^D$ of form

$$\bar{\eta} = c\,\frac{Ax}{|Ax|}\,\operatorname{sign}(\Lambda_f - m) \qquad \text{on } \{Ax(\Lambda_f - m) \neq 0\} \tag{45}$$

with some $A \in \mathbb{R}^{p \times k} \setminus \{0\}$, we need that $\bar{\eta}$ vanishes for $Ax = 0$ to obtain the form (35). This condition is trivially fulfilled under the subsequent assumption (46), which ensures that $K(Ax = 0) = 0$ for $A \neq 0$. ////

Continuous Design

A design assumption somewhat stronger than just $\mathcal{K} = \mathrm{E}\,xx' > 0$ turns out useful in the lower bias cases,

$$e \in \mathbb{R}^k\,, \quad K(e'x = 0) > 0 \implies e = 0 \tag{46}$$

that is, $K(V) = 0$ for all $(k-1)$-dimensional linear subspaces $V \subset \mathbb{R}^k$. In dimension $k = 1$ this simply means that $K(x = 0) = 0$. Condition (46) is implied by condition 5.5(90) in the regression case. Conversely, if the distribution function of $\Lambda_f(F)$ is continuous, (46) implies 5.5(90). Condition (46) has the disadvantage that, for dimension $k > 1$, it cannot be fulfilled by any molecular (empirical) design measure K.

Minimal Bias Zero

The solution in case $\omega_{*,\varepsilon;s}^{\min} = 0$ is summarized by the following lemma.

Lemma 7.4.8 *Let $\varepsilon \in \mathbb{M}_{\bullet}^{+}$ and $* = c, v, h$. Then the following holds:*

(a) *If $\omega_{*,\varepsilon;s}^{\min} = 0$ is achieved, then $K(\varepsilon = 0) > 0$.*

(b) *Conversely, if $K(\varepsilon = 0) > 0$, and condition (46) is satisfied, then*

$$\tilde{\eta} = \mathcal{I}_f^{-1} D \mathcal{K}_0^{-1} \Lambda\,\mathbf{I}(\varepsilon = 0)\,, \qquad \mathcal{K}_0 = \mathrm{E}\,xx'\,\mathbf{I}(\varepsilon = 0) \tag{47}$$

is the solution to problem $O_{*,\varepsilon;s}^{\mathrm{tr}}(b)$, with $b = \omega_{*,\varepsilon;s}^{\min} = 0$.

PROOF Obviously, $\bar\eta \in \Psi_{2.}^{D}$ achieves $\omega_{*,\varepsilon;s}(\bar\eta) = 0$ iff $P(\varepsilon\bar\eta = 0) = 1$. And then $K(\varepsilon = 0) > 0$ since $\bar\eta \neq 0$. This proves (a).

Conversely, if $K(\varepsilon = 0) > 0$, some $\tilde\eta \in \Psi_{2.}^{D}$ is defined by (47) under condition (46) that achieves $\omega_{*,\varepsilon;s}(\tilde\eta) = 0$. Moreover, for all other $\bar\eta \in \Psi_{2.}^{D}$ satisfying $\omega_{*,\varepsilon;s}(\bar\eta) = 0$; that is, $P(\varepsilon\bar\eta = 0) = 1$, we have

$$\mathrm{E}\,\bar\eta\tilde\eta' = \mathrm{E}\,\bar\eta\,\mathbf{I}(\varepsilon = 0)\Lambda'\mathcal{K}_0^{-1}D'\mathcal{I}_f^{-1} = \mathcal{I}_f^{-1}D\mathcal{K}_0^{-1}D' = \mathrm{E}\,\tilde\eta\tilde\eta' \tag{48}$$

hence

$$0 \le \mathrm{E}\,(\bar\eta - \tilde\eta)(\bar\eta - \tilde\eta)' = \mathrm{E}\,\bar\eta\bar\eta' - \mathrm{E}\,\tilde\eta\tilde\eta' \tag{49}$$

Therefore $\tilde\eta$ is optimum in the positive definite sense. ////

Properties of Lagrange Multipliers

Positive (semi)definiteness of $AD' = DA'$ and some representations of Lagrange multipliers are the concern of the following remark.

Remark 7.4.9 [$AD' = DA' \ge 0$, and multiplier representations]
(a) Writing out $AD' = \mathrm{E}\,Ax(\Lambda_f - m)\bar\eta'$ for $\bar\eta \in \Psi_{2.}^{D}$ of the lower case forms (45) or (75) below, respectively, we obtain that

$$AD' = A\,\mathrm{E}\,\frac{cxx'}{|Ax|}\,\mathbf{I}(Ax \neq 0)\,\mathrm{E}\,|\Lambda_f - m|\,A' \ge 0 \tag{50}$$

Likewise,

$$AD' = \mathrm{E}\left(-c'(A\Lambda - a)^{-} + c''(A\Lambda - a)^{+}\right) \ge 0 \tag{51}$$

and

$$AD' = \mathrm{E}\,|Ax|\,(c'' - c')\,\mathrm{E}\,\Lambda_f^{+} \ge 0 \tag{52}$$

hold for $\bar\eta \in \Psi_{2.}^{D}$ of lower case forms (96) and (169) below, respectively.

Spelling out $D = \mathrm{E}\,\bar\eta\Lambda'$ for $\bar\eta \in \Psi_{2.}^{D}$ of lower case forms (35) and (76) below, respectively, yields even more that

$$D = A\,\mathrm{E}\,\frac{cxx'}{|Ax|}\,\mathbf{I}(Ax \neq 0)\,\mathrm{E}\,|\Lambda_f - m| \tag{53}$$

Such a representation $D = A\widetilde{\mathcal{I}}$ with any matrix $\widetilde{\mathcal{I}} = \widetilde{\mathcal{I}}' \ge 0$, as implied by (53) [or (59) or (195) below], entails that $A\widetilde{\mathcal{I}}\widetilde{\mathcal{I}}A' = DD' > 0$, because $\operatorname{rk} D = p$, hence also $AD' = DA' = A\widetilde{\mathcal{I}}A' > 0$; in particular, $\operatorname{rk} A = p$.

This concerns the lower case solutions for neighborhood type $* = c, v$.

(b) In view of the subsequent relations (55), this argument also applies to the main case solutions $\tilde\eta \in \Psi_{2.}^{D}$ for neighborhood type $* = c, v$, which will attain the form

$$\tilde\eta = (A\Lambda - a)\,w\,, \qquad w = \min\Bigl\{1,\ \frac{c}{|A\Lambda - a|}\Bigr\} \tag{54}$$

with some conditional centering term $a \in \mathbb{H}_{\bullet}$, some matrix $A \in \mathbb{R}^{p\times k}$, and some clipping function $c \in \mathbb{M}_{\bullet}^{+}$. Then the following representations hold,

$$
\begin{aligned}
a &= \vartheta A x \quad \text{on } \{c>0\} \\
D &= A \operatorname{E} x x' \operatorname{E}_{\bullet} (\Lambda_f - \vartheta)^2 w
\end{aligned}
\qquad \vartheta = \frac{\operatorname{E}_{\bullet} \Lambda_f w}{\operatorname{E}_{\bullet} w} \mathbf{I}(c>0) \tag{55}
$$

In particular, we can rewrite form (54) of $\tilde{\eta}$ using the *residuals conditional centering* function $\vartheta \in \mathbb{M}_{\bullet}$ and a *residuals clipping* function $r \in \mathbb{M}_{\bullet}^{+}$,

$$
\tilde{\eta} = Ax(\Lambda_f - \vartheta)w\,, \quad w = \min\Bigl\{1, \frac{r}{|\Lambda_f - \vartheta|}\Bigr\}, \quad r = \frac{c}{|Ax|}\mathbf{I}(Ax \neq 0) \tag{56}
$$

For a proof of (55), observe that $\operatorname{E}_{\bullet} w > 0$ iff $c > 0$, as well as

$$
\begin{aligned}
0 = \operatorname{E}\tilde{\eta} &= \operatorname{E}_{\bullet}(A\Lambda - a)w = Ax \operatorname{E}_{\bullet} \Lambda_f w - a(x) \operatorname{E}_{\bullet} w \\
D = \operatorname{E}\tilde{\eta}\Lambda' &= \operatorname{E} Axx'(\Lambda_f - \vartheta)\Lambda_f w = A \operatorname{E} xx' \operatorname{E}_{\bullet} (\Lambda_f - \vartheta)^2 w
\end{aligned}
$$

where $\operatorname{E}_{\bullet}(\Lambda_f - \vartheta)w = \operatorname{E}_{\bullet} \Lambda_f w - \vartheta \operatorname{E}_{\bullet} w = 0$ by the definition of ϑ.

In view of (214) and (220) and (232) and (238) below, the main and lower case solutions $\tilde{\eta} \in \Psi_{2\bullet}^{D}$ for neighborhood type $* = h$ (Hellinger) will attain the form

$$
\tilde{\eta} = A\Lambda h \tag{57}
$$

with some $A \in \mathbb{R}^{p\times k}$, and some weight $h \in \mathbb{M}_{\bullet}^{+}$ depending only on x. Thus

$$
a = 0 \in \mathbb{H}_{\bullet}\,, \qquad \vartheta = 0 \in \mathbb{M}_{\bullet}\,, \qquad D = \mathcal{I}_f A \operatorname{E} xx'h(x) \tag{58}
$$

as (54) and (55) apply with w replaced by h. ////

Existence of a Lower Case Matrix

The following lemma provides us with a suitable matrix for the lower case solutions of the problems $O_{*,1;0}^{\mathrm{tr}}(b)$, for $b = \omega_{*,1;0}^{\min}$ and the types $* = v, h$.

Lemma 7.4.10 *Under condition* (46), *there exists a matrix* $A \in \mathbb{R}^{p\times k}$ *such that*

$$
D = \frac{\operatorname{tr} AD'}{\operatorname{E}|Ax|} \operatorname{E} \frac{Axx'}{|Ax|} \mathbf{I}(Ax \neq 0) \tag{59}
$$

PROOF In view of Theorem 5.5.1 b, equation (59) resembles the relation $D = \operatorname{E}\tilde{\eta}\Lambda'$ for $\tilde{\eta} \in \Psi_2^D$ of form 5.5(19) achieving the minimal bias $\omega_{c,0;0}^{\min}$ given by 5.5(18), if the u component part there is simply stripped off.

Thus we prove this fixed point type result by a Lagrangian argument, solving in $\mathbb{H}_{\bullet} = L_2^p(K)$ the following problem,

$$
\sup\nolimits_K |\chi| = \min! \qquad \chi \in \mathbb{H}_{\bullet}\,, \ \operatorname{E}\chi x' = D \tag{60}
$$

which is convex and, by familiar arguments, well-posed with a bounded solution $\bar{\chi} \neq 0$ (as $D \neq 0$). Writing $m = \sup_K |\bar{\chi}| \in (0, \infty)$, there exists some matrix $A \in \mathbb{R}^{p \times k}$ ($A \neq 0$ as $m > 0$) such that

$$\begin{aligned} m - \operatorname{tr} AD' &= \min\{ \sup_K |\chi| - \mathrm{E}\, \chi' Ax \mid \chi \in \mathbb{H}_{\bullet} \} \\ &\leq \inf\{ \kappa(1 - \mathrm{E}\,|Ax|) \mid \kappa \in [0, \infty) \} \end{aligned} \tag{61}$$

The bound (actually, an equality) is obtained by inserting the special functions $\chi_\kappa = \kappa\, e_{Ax}$. Since $m - \operatorname{tr} AD' > -\infty$, we conclude that necessarily

$$\mathrm{E}\,|Ax| \leq 1 \tag{62}$$

Hence the inf in (61) equals 0, and so

$$m \leq \operatorname{tr} AD' \leq \frac{\operatorname{tr} AD'}{\mathrm{E}\,|Ax|} \tag{63}$$

But for all $\chi \in \mathbb{H}_{\bullet}$ satisfying $\mathrm{E}\, \chi x' = D$, we have the bound

$$\operatorname{tr} AD' = \mathrm{E}\, \chi' Ax \leq \sup_K |\chi|\, \mathrm{E}\,|Ax| \tag{64}$$

Consequentially,

$$m \geq \frac{\operatorname{tr} AD'}{\mathrm{E}\,|Ax|} \tag{65}$$

Thus, equalities must hold in (62) and (63),

$$\mathrm{E}\,|Ax| = 1, \qquad m = \operatorname{tr} AD' \tag{66}$$

Moreover, equality is achieved in (64) iff the values of χ and Ax have the same direction, and $|\chi| = \sup_K |\chi| = m$ whenever $Ax \neq 0$. Thus

$$\bar{\chi} = m\, e_{Ax} \qquad \text{on } \{Ax \neq 0\} \tag{67}$$

Now $K(Ax \neq 0) = 1$ by condition (46) because $A \neq 0$. Therefore,

$$D = \mathrm{E}\, \bar{\chi} x' = m\, \mathrm{E}\, e_{Ax} x' \tag{68}$$

In view of the identities (66), this is just (59). ////

We shall now derive the solutions to the problems $O^{\mathrm{tr}}_{*,t;s}(b)$ successively in the cases $* = c, v, h$. The reader should bear in mind that $t = 0, \varepsilon, \alpha$ stands for errors-in-variables, respectively error-free-variables balls with fixed contamination curve $\varepsilon \in \mathbb{M}_{\bullet}^{+}$, respectively any $\varepsilon \in \mathbb{M}_{\bullet}^{+}$ subject to $\mathrm{E}\, \varepsilon^{\alpha} \leq 1$ [Section 7.2], and $s = 0, 2, \infty, e$ indicates the exact, coordinatewise, and approximate bias versions, respectively [Section 7.3].

By inspection of the Lagrangians, without explicit proof, we also record the corresponding MSE type problems $O^{\mathrm{ms}}_{*,t;s}(\beta)$ under (b) for which the minimum trace solutions derived under (a) are optimal; a power other than 2 of $\omega_{*,t;s}$ in these MSE type objective functions could be compensated for by a different value of the bias weight β.

7.4.3 Contamination Optimality

$* = c,\ t = \varepsilon,\ s = e$

Theorem 7.4.11 [Problem $O^{\mathrm{tr}}_{c,\varepsilon;e}(b)$] *Assume* (11): $\mathrm{E}\,\varepsilon^2 < \infty$.

(a) *Then, in case* $\omega^{\min}_{c,\varepsilon;e} < b < \omega_{c,\varepsilon;e}(\eta_h)$, *there exist some multipliers* $a \in \mathbb{H}_{\bullet}$, $A \in \mathbb{R}^{p\times k}$, $c \in \mathbb{M}^{+}_{\bullet}$, *and* $\beta \in (0,\infty)$, *such that the solution is given by*

$$\tilde\eta = (A\Lambda - a)w, \qquad w = \min\Bigl\{1,\ \frac{c}{|A\Lambda - a|}\Bigr\} \tag{69}$$

where

$$\begin{aligned} c &= 0 && \text{if } \mathrm{E}_{\bullet}\,|A\Lambda - a| \le \beta\varepsilon \\ \mathrm{E}_{\bullet}\,\bigl(|A\Lambda - a| - c\bigr)^{+} &= \beta\varepsilon && \text{if } \mathrm{E}_{\bullet}\,|A\Lambda - a| > \beta\varepsilon \end{aligned} \tag{70}$$

and

$$\mathrm{E}\,\varepsilon c = b \tag{71}$$

Conversely, for any $a \in \mathbb{H}_{\bullet}$, $A \in \mathbb{R}^{p\times k}$, *and* $\beta \in [0,\infty)$, *some* $c \in \mathbb{M}^{+}_{\bullet}$ *is defined by* (70), *and if some* $\tilde\eta \in \Psi^{D}_{2\bullet}$ *is of form* (69), *then it is the solution to problem* $O^{\mathrm{tr}}_{c,\varepsilon;e}(b)$ *with* b *defined by* (71).

(b) *Given* $\beta \in [0,\infty)$, *the unique solution* $\tilde\eta \in \Psi^{D}_{2\bullet}$ *to the problem*

$$\mathrm{E}\,|\eta|^2 + 2\beta\,\omega_{c,\varepsilon;e}(\eta) = \min! \qquad \eta \in \Psi^{D}_{2\bullet} \tag{72}$$

is of necessary and sufficient form (69) *and* (70).

(c) *With any* $m = \operatorname{med}\Lambda_f(F)$, *and*

$$\sigma_A = \sup\bigl\{\,\tau \in [0,\infty) \bigm| K(\varepsilon \ge \tau|Ax|) = 1\,\bigr\} \tag{73}$$

we have

$$\omega^{\min}_{c,\varepsilon;e} = \max\Bigl\{\,\frac{\sigma_A \operatorname{tr} AD'}{\mathrm{E}\,|\Lambda_f - m|} \Bigm| A \in \mathbb{R}^{p\times k}\,\Bigr\} \tag{74}$$

Suppose that $\omega^{\min}_{c,\varepsilon;e} > 0$, *and that* $\omega^{\min}_{c,\varepsilon;e}$ *is achieved by some* $\bar\eta \in \Psi^{D}_{2\bullet}$. *Choose any matrix* $A \in \mathbb{R}^{p\times k}$ *attaining the max in* (74). *Then necessarily*

$$\bar\eta = \begin{cases} c\,\dfrac{Ax}{|Ax|}\,\operatorname{sign}(\Lambda_f - m) & \text{if } Ax(\Lambda_f - m) \ne 0 \\[2ex] 0 & \text{if } \varepsilon > \sigma_A|Ax| \end{cases} \tag{75}$$

where $c = \sup_{\bullet}|\bar\eta| \in \mathbb{M}^{+}_{\bullet}$. *Under condition* (46), *the solution is of form*

$$\tilde\eta = \begin{cases} c\,\dfrac{Ax}{|Ax|}\bigl(\operatorname{sign}(\Lambda_f - m) + \gamma\,\mathbf{I}(\Lambda_f = m)\bigr) & \text{if } Ax \ne 0 \\[2ex] 0 & \text{if } \varepsilon > \sigma_A|Ax| \end{cases} \tag{76}$$

with $\gamma \in [-1,1]$ *determined from* $\mathrm{E}\operatorname{sign}(\Lambda_f - m) + \gamma P(\Lambda_f = m) = 0$.

PROOF Recall that $\omega_{c,\varepsilon;e}(\eta) = \mathrm{E}\,\varepsilon \sup_{.} |\eta|$.

(a) As $\mathrm{E}\,\varepsilon^2 < \infty$, the convex problem

$$\mathrm{E}\,|\eta|^2 = \min! \qquad \eta \in \mathbb{H},\ \mathrm{E}_{.}\,\eta = 0,\ \mathrm{E}\,\eta\Lambda' = D,\ \omega_{c,\varepsilon;e}(\eta) \le b \tag{77}$$

is well-posed [p 276]. Thus there exist some multipliers $a \in \mathbb{H}_{.}$, $A \in \mathbb{R}^{p\times k}$ and $\beta \in [0,\infty)$ such that the solution $\tilde\eta$, when compared with all $\eta \in \mathbb{H}$ of finite $\omega_{c,\varepsilon;e}(\eta)$, satisfies

$$\begin{aligned} L(\eta) &= \mathrm{E}\,|\eta|^2 + 2\,\mathrm{E}\,a'\,\mathrm{E}_{.}\,\eta - 2\,\mathrm{E}\,\eta' A\Lambda + 2\beta\omega_{c,\varepsilon;e}(\eta) \\ &\ge \mathrm{E}\,|\tilde\eta|^2 - 2\,\mathrm{tr}\,AD' + 2\beta b = L(\tilde\eta) \end{aligned} \tag{78}$$

Necessarily $\beta > 0$, hence $\omega_{c,\varepsilon;e}(\tilde\eta) = b$. Otherwise, as η_h is in the closure of $\Psi_\infty^D \subset \mathbb{H}$, and $\omega_{c,\varepsilon;e}$ is finite on $\mathbb{L}$ as $\mathrm{E}\,\varepsilon < \infty$, it would follow from (78) that $\mathrm{E}\,|\tilde\eta|^2 \le \mathrm{E}\,|\eta_h|^2$, hence $\tilde\eta = \eta_h$ contradicting $b < \omega_{c,\varepsilon;e}(\eta_h)$.

For each $\eta \in \mathbb{H}$ apply Lemma 7.4.3 a with $Y = A\Lambda - a$, $c = \sup_{.} |\eta|$. The solution $\tilde\chi$, since $\mathrm{E}\,|\tilde\chi - Y|^2 \le \mathrm{E}\,|\eta - Y|^2$ and $\omega_{c,\varepsilon;e}(\tilde\chi) \le \omega_{c,\varepsilon;e}(\eta)$, improves on η, hence $\tilde\eta$ itself is of this form. To determine the optimal clipping function $c \in \mathbb{M}_{.}^{+}$ in $\chi_c = Y \min\{1, c/|Y|\}$, write out the Lagrangian

$$L_1(c) = L(\chi_c) + \mathrm{E}\,|Y|^2 = \mathrm{E}\,\mathrm{E}_{.}\,\bigl|(|Y| - c)^+\bigr|^2 + 2\beta\,\mathrm{E}\,\varepsilon c \tag{79}$$

Pointwise minimization by differentiation w.r.t. $c(x)$ [Lemma C.2.3] yields that $\mathrm{E}_{.}\,(|Y| - c)^+ = \beta\varepsilon$, or $\mathrm{E}_{.}\,|Y| \le \beta\varepsilon$; hence (70).

Conversely, form (69), (70) of some $\tilde\eta \in \Psi_{2.}^D$, for any contamination curve ε, any $a \in \mathbb{H}_{.}$, $A \in \mathbb{R}^{p\times k}$, $b, \beta \in [0,\infty)$, and $c \in \mathbb{M}_{.}^{+}$, implies that $\tilde\eta$ minimizes the corresponding Lagrangian L_1 on $\mathbb{M}_{.}^{+}$, hence L on $\mathbb{H}$, and therefore solves problem $O_{c,\varepsilon;e}^{\mathrm{tr}}(b)$ with its own bias b.

Concerning c, observe that

$$\begin{aligned} c &= \inf\bigl\{\, \tau \in [0,\infty] \bigm| \mathrm{E}_{.}\,\bigl(|A\Lambda - a| - \tau\bigr)^+ \le \beta\varepsilon \,\bigr\} \\ c > \tau &\iff \mathrm{E}_{.}\,\bigl(|A\Lambda - a| - \tau\bigr)^+ > \beta\varepsilon \end{aligned} \tag{80}$$

where the functions $\mathrm{E}_{.}\,\bigl(|A\Lambda - a| - \tau\bigr)^+$, given the fact that $a \in \mathbb{H}_{.}$, are all measurable functions of x. This implies that $c \in \mathbb{M}_{.}^{+}$.

(c) Given any $A \in \mathbb{R}^{p\times k}$, setting $Y = Ax(\Lambda_f - m)$, and by the definition of σ_A, the following bound is true for all $\eta \in \Psi_{2.}^D$,

$$\begin{aligned} \mathrm{E}\,\varepsilon \sup_{.} |\eta| &\ge \sigma_A\,\mathrm{E}\,|Ax| \sup_{.} |\eta| = \frac{\sigma_A\,\mathrm{E}\,|Y| \sup_{.} |\eta|}{\mathrm{E}\,|\Lambda_f - m|} \\ &\ge \frac{\sigma_A\,\mathrm{E}\,|\eta||Y|}{\mathrm{E}\,|\Lambda_f - m|} \ge \frac{\sigma_A\,\mathrm{E}\,\eta' Y}{\mathrm{E}\,|\Lambda_f - m|} = \frac{\sigma_A\,\mathrm{tr}\,AD'}{\mathrm{E}\,|\Lambda_f - m|} \end{aligned} \tag{81}$$

This proves $\ge$ in (74). To verify attainability, consider the convex problem

$$\mathrm{E}\,\varepsilon \sup_{.} |\eta| = \min! \qquad \eta \in \mathbb{H},\ \mathrm{E}_{.}\,\eta = 0,\ \mathrm{E}\,\eta\Lambda' = D \tag{82}$$

which has a finite value and is well-posed, as $\mathrm{E}\,\varepsilon^2 < \infty$. Thus there exist multipliers $a \in \mathbb{H}_{\textstyle .}$, $A \in \mathbb{R}^{p\times k}$ such that, whether $\omega_{c,\varepsilon;e}^{\min}$ is achieved or not,

$$\begin{aligned}\omega_{c,\varepsilon;e}^{\min} - \operatorname{tr} AD' &= \inf\bigl\{\,\mathrm{E}\,\varepsilon \sup_{\textstyle .} |\eta| - \mathrm{E}\,\eta' Y \bigm| \eta \in \mathbb{H}\,\bigr\} \\ &= \inf\bigl\{\,\mathrm{E}\,\varepsilon \sup_{\textstyle .} |\eta| - \mathrm{E}\,\eta' Y \bigm| \eta \in \mathbb{L}\,\bigr\} \\ &= \inf\bigl\{\,\mathrm{E}\,c(\varepsilon - \mathrm{E}_{\textstyle .}\,|Y|) \bigm| c \in \mathbb{M}_{\textstyle .}^{+},\ \mathrm{E}\,c^2 < \infty\,\bigr\}\end{aligned} \tag{83}$$

where $Y = A\Lambda - a$, and the last identity holds by Lemma 7.4.3 b. We conclude that

$$\varepsilon \ge \mathrm{E}_{\textstyle .}\,|Y| \tag{84}$$

and then

$$\omega_{c,\varepsilon;e}^{\min} = \operatorname{tr} AD' \tag{85}$$

Decomposing $a(x) = \bar\vartheta(x)\,Ax + \hat a(x)$ orthogonally, we further obtain

$$\mathrm{E}_{\textstyle .}\,|Y| \ge |Ax|\,\mathrm{E}_{\textstyle .}\,|\Lambda_f - \bar\vartheta| \ge |Ax|\,\mathrm{E}_{\textstyle .}\,|\Lambda_f - m| \tag{86}$$

hence

$$\sigma_A \ge \mathrm{E}\,|\Lambda_f - m| \tag{87}$$

and therefore,

$$\omega_{c,\varepsilon;e}^{\min} = \operatorname{tr} AD' \le \frac{\sigma_A \operatorname{tr} AD'}{\mathrm{E}\,|\Lambda_f - m|} \tag{88}$$

Thus A achieves the max in (74). And if $\omega_{c,\varepsilon;e}^{\min} > 0$, the following equalities must hold for our multipliers $a \in \mathbb{H}_{\textstyle .}$ and $A \in \mathbb{R}^{p\times k}$,

$$\sigma_A = \mathrm{E}\,|\Lambda_f - m|\,, \qquad \hat a = 0, \qquad \bar\vartheta = m, \qquad Y = Ax(\Lambda_f - m) \tag{89}$$

Now suppose that $\omega_{c,\varepsilon;e}^{\min} > 0$, and let $A \in \mathbb{R}^{p\times k}$ be any matrix attaining the max in (74). Then, if $\bar\eta \in \Psi_{2\textstyle .}^{D}$ achieves $\omega_{c,\varepsilon;e}^{\min}$, equalities must hold in bound (81), which implies form (75); in particular, $K\bigl(\varepsilon = \sigma_A|Ax|\bigr) > 0$.

Under condition (46), every such $\bar\eta$ may be improved according to Lemma 7.4.6 and Remark 7.4.7. Hence the minimum norm solution $\tilde\eta$ itself must be of form (76). ////

$* = c,\ t = \varepsilon,\ s = 0;\ p = 1$

For dimension $p = 1$ we can deal with contamination oscillation exactly.

Theorem 7.4.12 [Problem $O_{c,\varepsilon;0}^{\mathrm{tr}}(b)$; $p = 1$] *Assume* (11): $\mathrm{E}\,\varepsilon^2 < \infty$.

(a) *Then, in case* $\omega_{c,\varepsilon;0}^{\min} < b < \omega_{c,\varepsilon;0}(\eta_h)$, *there exist some multipliers* $a \in \mathbb{H}_{\textstyle .}$, $A \in \mathbb{R}^{1\times k}$, $c' \in \mathbb{M}_{\textstyle .}^{-}$, $c'' \in \mathbb{M}_{\textstyle .}^{+}$, *and* $\beta', \beta'' \in [0,\infty)$, $\beta' + \beta'' > 0$, *such that the solution is given by*

$$\begin{aligned}\tilde\eta = c' \vee (A\Lambda - a) \wedge c'' &= (A\Lambda - a)\,w \\ w &= 1 \wedge \max\Bigl\{\frac{c'}{A\Lambda - a}, \frac{c''}{A\Lambda - a}\Bigr\}\end{aligned} \tag{90}$$

where

$$\begin{aligned} \mathrm{E}_{\bullet}\big((A\Lambda - a) - c''\big)^{+} &= \beta''\varepsilon \qquad \textit{if}\ \ \mathrm{E}_{\bullet}(A\Lambda - a)^{+} > \beta''\varepsilon \\ \mathrm{E}_{\bullet}\big(c' - (A\Lambda - a)\big)^{+} &= \beta'\varepsilon \qquad \textit{if}\ \ \mathrm{E}_{\bullet}(A\Lambda - a)^{-} > \beta'\varepsilon \\ c' = c'' &= 0 \qquad\qquad \textit{otherwise} \end{aligned} \tag{91}$$

and

$$\begin{aligned} \beta''\,\mathrm{E}\,\varepsilon c'' - \beta'\,\mathrm{E}\,\varepsilon c' &= (\beta' + \beta'')\,b \\ &= (\beta' + \beta'')\max\{\mathrm{E}\,\varepsilon c'', -\mathrm{E}\,\varepsilon c'\} \end{aligned} \tag{92}$$

Conversely, if some $\tilde\eta \in \Psi_{2\bullet}^D$ *is of this form, for any* $a \in \mathbb{H}_{\bullet}$, $A \in \mathbb{R}^{1\times k}$, *any* $c', c'' \in \mathbb{M}_{\bullet}$, *and* $\beta', \beta'' \in [0,\infty)$, *then* $\tilde\eta$ *solves problem* $O^{\mathrm{tr}}_{c,\varepsilon;0}(b)$.

(b) *Given* $\beta \in [0,\infty)$, *the unique solution* $\tilde\eta \in \Psi_{2\bullet}^D$ *to the problem*

$$\mathrm{E}\,\eta^2 + 2\beta\,\omega_{c,\varepsilon;0}(\eta) = \min! \qquad \eta \in \Psi_{2\bullet}^D \tag{93}$$

is of necessary and sufficient form (90) *and* (91), *with* $\beta = \beta' + \beta''$.

(c) *Defining*

$$\begin{aligned} \sigma'_{a,A} &= \sup\{\tau \in [0,\infty) \mid K\big(\varepsilon \ge \tau\,\mathrm{E}_{\bullet}(A\Lambda - a)^{-}\big) = 1\} \\ \sigma''_{a,A} &= \sup\{\tau \in [0,\infty) \mid K\big(\varepsilon \ge \tau\,\mathrm{E}_{\bullet}(A\Lambda - a)^{+}\big) = 1\} \end{aligned} \tag{94}$$

and formally interpreting $0/0 = 0$, *we have*

$$\omega^{\min}_{c,\varepsilon;0} = \max\Big\{\frac{AD'}{(1/\sigma'_{a,A}) + (1/\sigma''_{a,A})} \Bigm| a \in \mathbb{H}_{\bullet},\ A \in \mathbb{R}^{1\times k}\Big\} \tag{95}$$

Suppose that $\omega^{\min}_{c,\varepsilon;0} > 0$, *and* $\omega^{\min}_{c,\varepsilon;0}$ *is achieved by some* $\bar\eta \in \Psi_{2\bullet}^D$. *Choose any* $a \in \mathbb{H}_{\bullet}$ *and* $A \in \mathbb{R}^{1\times k}$ *attaining the max in* (95). *Then necessarily*

$$\bar\eta = \begin{cases} c'\,\mathbf{I}(A\Lambda < a) + c''\,\mathbf{I}(A\Lambda > a) & \textit{if}\ \ A\Lambda \ne a \\ 0 \quad \textit{if}\ \ \varepsilon > \sigma'_{a,A}\,\mathrm{E}_{\bullet}(A\Lambda - a)^{-}\ \ \textit{or/and}\ \ \varepsilon > \sigma''_{a,A}\,\mathrm{E}_{\bullet}(A\Lambda - a)^{+} \end{cases} \tag{96}$$

where $c' = \inf_{\bullet}\bar\eta \in \mathbb{M}^{-}_{\bullet}$, $c'' = \sup_{\bullet}\bar\eta \in \mathbb{M}^{+}_{\bullet}$, *and* $\mathrm{E}\,\varepsilon c'' = -\mathrm{E}\,\varepsilon c'$ *holds. Under condition* (46), *the solution* $\bar\eta$ *in addition satisfies* $\bar\eta = \bar a$ *on the event* $\{A\Lambda = a\}$, *for some function* $\bar a \in \mathbb{M}_{\bullet}$.

PROOF Recall that $\omega_{c,\varepsilon;0}(\eta) = \omega_{c,\varepsilon}(\eta) = \mathrm{E}\,\varepsilon\sup_{\bullet}\eta \vee \big(-\mathrm{E}\,\varepsilon\inf_{\bullet}\eta\big)$.

(a) Since $\mathrm{E}\,\varepsilon^2 < \infty$, the problem with $\mathbb{R}^2$-valued convex constraint,

$$\mathrm{E}\,\eta^2 = \min! \qquad \eta \in \mathbb{H},\ \mathrm{E}_{\bullet}\eta = 0,\ \mathrm{E}\,\eta\Lambda' = D,\quad \begin{aligned} \mathrm{E}\,\varepsilon\sup_{\bullet}\eta &\le b \\ -\mathrm{E}\,\varepsilon\inf_{\bullet}\eta &\le b \end{aligned} \tag{97}$$

is convex, well-posed. Thus there exist multipliers $a \in \mathbb{H}_{\bullet}$, $A \in \mathbb{R}^{1\times k}$, and $\beta', \beta'' \in [0,\infty)$ such that, setting $Y = A\Lambda - a$, the unique minimum norm solution $\tilde\eta$, in comparison with all other $\eta \in \mathbb{H}$ of finite $\omega_{c,\varepsilon}(\eta)$, satisfies

$$\begin{aligned} L(\eta) &= \mathrm{E}\,(\eta - Y)^2 + 2\beta''\,\mathrm{E}\,\varepsilon\sup_{\bullet}\eta - 2\beta'\,\mathrm{E}\,\varepsilon\inf_{\bullet}\eta \\ &\ge \mathrm{E}\,\tilde\eta^2 - 2AD' + \mathrm{E}\,Y^2 + 2(\beta' + \beta'')\,b = L(\tilde\eta) \end{aligned} \tag{98}$$

An approximation of η_h by $\Psi_{\infty.}^D$, where $\omega_{c,\varepsilon}$ is finite as $\mathrm{E}\,\varepsilon$ is finite, shows that necessarily $\beta' + \beta'' > 0$, and thus, bound (98) extends to all of $\mathbb{H}$. In view of (98), $\tilde{\eta}$ is also the solution to

$$\mathrm{E}\,(\eta - Y)^2 = \min! \qquad \eta \in \mathbb{H}, \ \tilde{c}' \le \eta \le \tilde{c}'' \tag{99}$$

with its own $\tilde{c}' = \inf_{\bullet} \tilde{\eta}$ and $\tilde{c}'' = \sup_{\bullet} \tilde{\eta}$. Therefore, we may restrict attention to functions $\eta_c = c' \vee Y \wedge c''$ with any $c' \in \mathbb{M}_{\bullet}^-$, $c'' \in \mathbb{M}_{\bullet}^+$. Minimization of the Lagrangian $L_1(c) = L(\eta_c)$,

$$L_1(c) = \mathrm{E}\Big(\mathrm{E}_{\bullet}\,|(Y - c'')^+|^2 + \mathrm{E}_{\bullet}\,|(c' - Y)^+|^2 + 2\varepsilon(\beta'' c'' - \beta' c')\Big) \tag{100}$$

by differentiation of the convex integrand w.r.t. $c'(x) \le 0$ and $c''(x) \ge 0$ [Lemma C.2.3] yields (91). Note that $\mathrm{E}_{\bullet} Y^+$ and $\mathrm{E}_{\bullet} Y^-$ are simultaneously larger, or less, than $\beta''\varepsilon$ and $\beta'\varepsilon$, respectively, since $\mathrm{E}_{\bullet}\tilde{\eta} = 0$.

Conversely, any such $\tilde{\eta} \in \Psi_{2.}^D$ minimizes the Lagrangian L_1, hence L over $\mathbb{H}$, and hence minimizes $\mathrm{E}\,\eta^2$ over $\Psi_{2.}^D$ subject to $\omega_{c,\varepsilon} \le b$.

(c) Given any $a \in \mathbb{H}_{\bullet}$, $A \in \mathbb{R}^{1\times k}$, invoking $\sigma' = \sigma'_{a,A}$, $\sigma'' = \sigma''_{a,A}$ from (94), and $Y = A\Lambda - a$, the following bound holds for all $\eta \in \Psi_{2.}^D$,

$$\begin{aligned} &\mathrm{E}\,\varepsilon \sup_{\bullet} \eta \vee (-\mathrm{E}\,\varepsilon \inf_{\bullet} \eta) \\ &\quad\ge (\sigma'' \,\mathrm{E}\,\mathrm{E}_{\bullet} Y^+ \sup_{\bullet} \eta) \vee (-\sigma' \,\mathrm{E}\,\mathrm{E}_{\bullet} Y^- \inf_{\bullet} \eta) \\ &\quad\ge (\sigma'' \,\mathrm{E}\,\eta Y^+) \vee (-\sigma' \,\mathrm{E}\,\eta Y^-) \end{aligned} \tag{101}$$

Hence

$$(\sigma' + \sigma'')\,\omega_{c,\varepsilon;0}^{\min} \ge \sigma'\sigma''\,(\mathrm{E}\,\eta Y^+ - \mathrm{E}\,\eta Y^-) = \sigma'\sigma'' A D' \tag{102}$$

from which $\ge$ follows in (95). To show $\le$, consider the convex problem

$$\omega_{c,\varepsilon}(\eta) = \min! \qquad \eta \in \mathbb{H}, \ \mathrm{E}_{\bullet}\eta = 0, \ \mathrm{E}\,\eta\Lambda' = D \tag{103}$$

which is well-posed, since $\mathrm{E}\,\varepsilon^2 < \infty$), and has a finite value, as $\omega_{c,\varepsilon}$ is finite on $\Psi_{\infty.}^D$ due to $\mathrm{E}\,\varepsilon < \infty$. Thus, whether $\omega_{c,\varepsilon}^{\min}$ is attained or not, there exist multipliers $a \in \mathbb{H}_{\bullet}$ and $A \in \mathbb{R}^{1\times k}$ such that, setting $Y = A\Lambda - a$,

$$\omega_{c,\varepsilon}^{\min} - AD' = \inf\{\, \omega_{c,\varepsilon}(\eta) - \mathrm{E}\,\eta Y \mid \eta \in \mathbb{H} \,\} \tag{104}$$

An approximation argument shows that this inf may without change be restricted to $\mathbb{L}$, and then to functions $\eta \in \mathbb{L}$ such that $\eta = \inf_{\bullet} \eta$ if $Y < 0$, and $\eta = \sup_{\bullet} \eta$ if $Y > 0$. Thus,

$$\omega_{c,\varepsilon}^{\min} - AD' = \inf\{\, \mathrm{E}\,\varepsilon c'' \vee (-\mathrm{E}\,\varepsilon c') - \mathrm{E}\,c'' \mathrm{E}_{\bullet} Y^+ + \mathrm{E}\,c' \mathrm{E}_{\bullet} Y^- \,\} \tag{105}$$

where this inf is taken over all functions $c', c'' \in \mathbb{M}_{\bullet}$, $c' \le c''$, that are bounded, or (again by approximation) at least are square integrable. Now either $\mathrm{E}\,\varepsilon c'' \geq -\mathrm{E}\,\varepsilon c'$ or $\mathrm{E}\,\varepsilon c'' \leq -\mathrm{E}\,\varepsilon c'$. Thus the preceding inf may be calculated under, say, the second constraint:

$$-\mathrm{E}\,\varepsilon c' - \mathrm{E}\,c''\,\mathrm{E}_{\bullet}\,Y^{+} + \mathrm{E}\,c'\,\mathrm{E}_{\bullet}\,Y^{-} = \min! \qquad \begin{array}{c} c', c'' \in \mathbb{H}_{\bullet},\ c' \leq c'' \\ \mathrm{E}\,\varepsilon(c' + c'') \leq 0 \end{array} \tag{106}$$

This problem is again convex and well-posed (if, without restriction, ε is not identically 0). So there exists some $\beta \in [0, \infty)$ such that

$$\omega_{c,\varepsilon}^{\min} - AD' \leq \inf\bigl\{ \mathrm{E}\,c''\,\bigl(\beta\varepsilon - \mathrm{E}_{\bullet}\,Y^{+}\bigr) - \mathrm{E}\,c'\,\bigl((1-\beta)\varepsilon - \mathrm{E}_{\bullet}\,Y^{-}\bigr) \bigr\} \tag{107}$$

where this inf is now taken over all functions $c', c'' \in \mathbb{M}_{\bullet}$, $c' \leq c''$, such that $\mathrm{E}\,|c'|^2 + \mathrm{E}\,|c''|^2 < \infty$. Since $\omega_{c,\varepsilon}^{\min} - AD' > -\infty$, we conclude that

$$\beta\varepsilon \geq \mathrm{E}_{\bullet}\,Y^{+}, \qquad (1-\beta)\varepsilon \geq \mathrm{E}_{\bullet}\,Y^{-} \tag{108}$$

and then we see that the inf in (107) equals 0. Under the first constraint, the result is the same, only with β renamed to $1-\beta$. Therefore, equality is achieved in (107), so that $\omega_{c,\varepsilon}^{\min} = AD'$. But

$$\omega_{c,\varepsilon}^{\min} = AD' \leq \frac{AD'}{(1/\sigma') + (1/\sigma'')} \tag{109}$$

since $1/\sigma' \leq \beta$ and $1/\sigma'' \leq 1-\beta$ due to (108). Thus (95) is proved. And in case $\omega_{c,\varepsilon}^{\min} > 0$, our multipliers $a \in \mathbb{H}_{\bullet}$, $A \in \mathbb{R}^{1\times k}$, and $\beta \in [0, \infty)$ must verify

$$\beta = \frac{1}{\sigma'_{a,A}} = 1 - \frac{1}{\sigma''_{a,A}} \tag{110}$$

Now suppose that $\omega_{c,\varepsilon}^{\min} > 0$, and consider any $a \in \mathbb{H}_{\bullet}$, $A \in \mathbb{R}^{1\times k}$ attaining the max in (95). Then, if $\bar{\eta} \in \Psi_{2\bullet}^{D}$ achieves $\omega_{c,\varepsilon}^{\min}$, equalities must hold throughout in (101) and (102). Obviously, these are realized iff $\bar{\eta}$ is of form (96). Under assumption (46), such $\bar{\eta} \in \Psi_{2\bullet}^{D}$ is improved by conditional averaging over the event $\{A\Lambda = a\}$, instead of $\{\Lambda_f = m\}$ in the proof to Lemma 7.4.6; we only need that $Ax\Lambda_f = a(x)$ determines Λ_f uniquely a.e. $K(dx)$, for the second line of (42) [p 281]. ////

$* = c,\ t = \alpha = 1,\ s = 0, e$

The only difference between problems $O_{c,1}^{\mathrm{tr}}(b)$ and $O_{c,0}^{\mathrm{tr}}(b)$ is the stronger side condition $\mathrm{E}_{\bullet} = 0$ (vs. $\mathrm{E} = 0$), which results in a centering function (instead of a centering vector) and also creates well-posedness complications since we are in the case $1 = \alpha < 2$ [cf. p 276]. Then the extra assumption

$$\operatorname{med} \Lambda_f(F) \ \text{ unique} \tag{111}$$

is required in Theorem 7.4.13 a—unless K has finite support (so that $\mathbb{H}_{\bullet}$ becomes finite-dimensional), or F verifies the symmetry condition 7.5(27) below (so that the constraint $\mathrm{E}_{\bullet} = 0$ may be removed); under the latter condition, see also Theorem 7.5.15.

Theorem 7.4.13 [Problem $O^{\mathrm{tr}}_{c,1;s}(b)$; $s = 0, e$]

(a) *In case* $\omega^{\min}_{c,1;s} < b < \omega_{c,1;s}(\eta_h)$, *under assumption* (111), *there exist a Borel measurable function* $a\colon \mathbb{R}^k \to \mathbb{R}^p$, *and some matrix* $A \in \mathbb{R}^{p\times k}$, *such that*

$$\tilde\eta = (A\Lambda - a)\,w\,, \qquad w = \min\Bigl\{1,\ \frac{b}{|A\Lambda - a|}\Bigr\} \tag{112}$$

is the solution, and

$$\omega_{c,1;s}(\tilde\eta) = b \tag{113}$$

Conversely, if some $\tilde\eta \in \Psi^D_{2\bullet}$ *has form* (112), *for any function* $a\colon \mathbb{R}^k \to \mathbb{R}^p$, *any matrix* $A \in \mathbb{R}^{p\times k}$, *and bound* $b \in (0,\infty]$, *then* $\tilde\eta$ *solves* $O^{\mathrm{tr}}_{c,1;s}(b)$.

(b) *Given* $\beta \in [0,\infty)$, *the unique solution* $\tilde\eta \in \Psi^D_{2\bullet}$ *to the mean square error problem* $O^{\mathrm{ms}}_{c,1;s}(\beta)$,

$$\mathrm{E}\,|\eta|^2 + \beta\,\omega^2_{c,1;s}(\eta) = \min! \qquad \eta \in \Psi^D_{2\bullet} \tag{114}$$

is of sufficient and, under assumption (111), *also of necessary form* (112), *with* $b \in [0,\infty]$ *and* β *related as in* 5.5(80); *that is,*

$$\beta b = \mathrm{E}\,\bigl(|A\Lambda - a| - b\bigr)^+ \tag{115}$$

(c) *With any* $m = \operatorname{med}\Lambda_f(F)$ *we have*

$$\omega^{\min}_{c,1;s} = \max\Bigl\{\frac{\operatorname{tr} AD'}{\mathrm{E}\,|Ax|\,\mathrm{E}\,|\Lambda_f - m|}\ \Bigm|\ A \in \mathbb{R}^{p\times k}\setminus\{0\}\Bigr\} \tag{116}$$

There exist $A \in \mathbb{R}^{p\times k}\setminus\{0\}$ *and* $\bar\eta \in \Psi^D_2$ *achieving* $\omega^{\min}_{c,1;s} = b$, *respectively. And then necessarily*

$$\bar\eta = b\,\frac{Ax}{|Ax|}\,\operatorname{sign}(\Lambda_f - m) \qquad \text{on } \bigl\{Ax(\Lambda_f - m) \neq 0\bigr\} \tag{117}$$

Under condition (46), *the solution is*

$$\tilde\eta = b\,\frac{Ax}{|Ax|}\,\mathbf{I}(Ax \neq 0)\bigl(\operatorname{sign}(\Lambda_f - m) + \gamma\,\mathbf{I}(\Lambda_f = m)\bigr) \tag{118}$$

with $\gamma \in [-1,1]$ *determined from* $\mathrm{E}\operatorname{sign}(\Lambda_f - m) + \gamma P(\Lambda_f = m) = 0$.

Remark 7.4.14 By iterating the argument for partial influence curves of lower case form (117) on the remaining part $\{Axu = 0\}$ of the sample space, assuming normal errors $F = \mathcal{N}(0,1)$, Müller (1987) has been able to determine the general lower bias case solution, which thus has additional branches of lower case form and a final branch of main case form (112), each with its own multipliers A_m and matrices D_m that add up to D. While our arguments still apply in case $\operatorname{rk} D_m \le p$, the possible singularity of Fisher information due to the successive restriction of the sample space requires explicit reparametrizations (respectively, generalized inverses) [cf. also Kurotschka and Müller (1992)]. ////

PROOF Recall that $\omega_{c,1;0}(\eta) = \omega_{c,1;e}(\eta) = \sup_P |\eta| = \omega_{c,1}(\eta)$.

(a) Keeping all side conditions except $\mathrm{E}\,\eta'\Lambda = D$ back in the domain, the convex problem

$$\mathrm{E}\,|\eta|^2 = \min! \qquad \eta \in \mathbb{H},\ \mathrm{E}_{\bullet}\,\eta = 0,\ \mathrm{E}\,\eta\Lambda' = D,\ \omega_{c,1}(\eta) \le b \tag{119}$$

stays well-posed [for any $\psi \in \Psi_{\infty\bullet}$ such that $\omega_{c,1}(D\psi) < b$ consider $\eta = A\psi$ for $A \approx D$]. Thus there exists some multiplier $A \in \mathbb{R}^{p\times k}$ such that the solution $\tilde\eta$ minimizes the Lagrangian

$$L(\eta) = \mathrm{E}\,|\eta|^2 - 2\,\mathrm{E}\,\eta'A\Lambda = \mathrm{E}\,\mathrm{E}_{\bullet}\,|\eta|^2 - 2\,\mathrm{E}_{\bullet}\,\eta'A\Lambda \tag{120}$$

among all $\eta \in \mathbb{H}$ satisfying $\mathrm{E}_{\bullet}\,\eta = 0$, $\omega_{c,1}(\eta) \le b$. This Lagrangian can be minimized by minimization of the conditional expectation inside, at each point x, subject to $\mathrm{E}_{\bullet}\,\eta = 0$ and $\sup_{\bullet} |\eta| \le b$. The conditional problem given any $x \in \mathbb{R}^k$ refers to the space $L_2^p(F)$ with (ordinary) expectation under F; it is well-posed as $b > 0$. Thus there exists some vector $a(x) \in \mathbb{R}^p$ such that the x section of the solution $\tilde\eta$ may tentatively be found by minimization of the conditional Lagrangian

$$\begin{aligned} L_{\bullet}(\eta) &= \mathrm{E}_{\bullet}\,|\eta|^2 - 2\,\mathrm{E}_{\bullet}\,\eta'A\Lambda + 2\,a(x)'\,\mathrm{E}_{\bullet}\,\eta \\ &= \mathrm{E}_{\bullet}\,\bigl|\eta - \bigl(Ax\Lambda_f - a(x)\bigr)\bigr|^2 + \text{const} \end{aligned} \tag{121}$$

over all $\eta(x,.) \in L_2^p(F)$ such that $\sup_{\bullet} |\eta| \le b$. Pointwise minimization of the integrand implies the asserted form (112) of the optimal section $\tilde\eta(x,.)$.

However, not only to ensure that $\tilde\eta$ of form (112) is (product) measurable, but already to justify the minimization of expectations by pointwise minimization of integrands—the way we have calculated $\tilde\eta(x,.)$—the function a should turn out measurable: Spell out $\mathrm{E}_F\,\tilde\eta(x,.) = 0$ to obtain form (55) of $a(x) = \vartheta(x)Ax$, with $c(x) = b$ (so that $\mathrm{E}_{\bullet}\,w > 0$ always) and $\vartheta(x)$ determined from the equation

$$0 = \int \bigl(\Lambda_f(u) - \vartheta(x)\bigr) \min\Bigl\{1, \frac{r(x)}{|\Lambda_f(u) - \vartheta(x)|}\Bigr\}\, F(du) \tag{122}$$

with $r(x) = b/|Ax|$. Thus, under assumption (111), Lemma C.2.4 ensures us that $\vartheta(x)$ is a continuous function of $|Ax|$ where $Ax \ne 0$ [$\vartheta(x) = 0$ if $Ax = 0$]. This proves measurability of a.

Conversely, let us assume some $\tilde\eta \in \Psi_{2\bullet}^D$ of this form (112). Then (K almost) all x sections minimize the corresponding conditional Lagrangians $L_{\bullet}$ over $L_2^p(F)$ subject to $\sup_{\bullet} |\eta| \le b$, hence $\mathrm{E}_{\bullet}\,|\eta|^2 - 2\,\mathrm{E}_{\bullet}\,\eta'A\Lambda$ among all $\eta \in \mathbb{H}$ subject to $\mathrm{E}_{\bullet}\,\eta = 0$, $\sup_{\bullet} |\eta| \le b$. On integration w.r.t. $K(dx)$, then $\tilde\eta$ minimizes the Lagrangian L in (120) over $\mathbb{H}$ subject to $\mathrm{E}_{\bullet} = 0$ and $\omega_{c,1} \le b$. As $L(\eta) = \mathrm{E}\,|\eta|^2 - 2\,\mathrm{tr}\,AD'$ for $\eta \in \Psi_{2\bullet}^D$, $\tilde\eta$ solves $O_{c,1}^{\mathrm{tr}}(b)$.

(c) For every $A \in \mathbb{R}^{p\times k}$, $\vartheta \in \mathbb{R}$, and all $\eta \in \Psi_{2\bullet}^D$ we have the bound,

$$\mathrm{tr}\,AD' = \mathrm{E}\,\eta'A\Lambda = \mathrm{E}\,\eta'Ax(\Lambda_f - \vartheta) \le \omega_{c,1}(\eta)\,\mathrm{E}\,|Ax|\,\mathrm{E}\,|\Lambda_f - \vartheta| \tag{123}$$

This proves $\geq$ in (116). To verify attainability, let us consider the problem

$$\sup_{\bullet} |\eta| = \min! \qquad \eta \in \mathbb{H}, \ \mathrm{E}_{\bullet}\, \eta = 0, \ \mathrm{E}\, \eta \Lambda' = D \tag{124}$$

where $\omega_{c,1}^{\min} \in (0,\infty)$ is attained [pp 198, 277]. Keeping the $\mathrm{E}_{\bullet}$ constraint in the domain, this convex problem is well-posed (for any $\psi \in \Psi_{\infty \bullet}$ consider $A\psi$ with $A \in \mathbb{R}^{p\times k}$). Thus there exists some $A \in \mathbb{R}^{p\times k}$ such that

$$\begin{aligned} \omega_{c,1}^{\min} - \operatorname{tr} AD' &= \min\{\, \omega_{c,1}(\eta) - \mathrm{E}\, \eta' A \Lambda \mid \eta \in \mathbb{H}, \ \mathrm{E}_{\bullet}\, \eta = 0 \,\} \\ &= \min\{\, \sup_P |\eta| - \mathrm{E}\, \eta' A x (\Lambda_f - m) \mid \eta \in \mathbb{H}, \ \mathrm{E}_{\bullet}\, \eta = 0 \,\} \end{aligned} \tag{125}$$

Insert the special functions $\eta = c\, e_{Ax} \big(\operatorname{sign}(\Lambda_f - m) + \gamma\, \mathbf{I}(\Lambda_f = m)\big)$, with any $c \in \mathbb{M}_{\bullet}^{+}$ such that $\mathrm{E}\, c^2 < \infty$, and $\gamma \in [-1,1]$ determined in such a way that $\gamma P(\Lambda_f = m) = -\,\mathrm{E} \operatorname{sign}(\Lambda_f - m)$. Thus

$$\begin{aligned} \omega_{c,1}^{\min} - \operatorname{tr} AD' &\leq \inf\{\, \sup_K c - \mathrm{E}\, c |Ax|\, \mathrm{E}\, |\Lambda_f - m| \mid c \in \mathbb{M}_{\bullet}^{+},\ \mathrm{E}\, c^2 < \infty \,\} \\ &= \inf\{\, \kappa \big(1 - \mathrm{E}\, |Ax|\, \mathrm{E}\, |\Lambda_f - m|\big) \mid \kappa \in [0,\infty) \,\} \end{aligned} \tag{126}$$

where $\kappa = \sup_K c$. We conclude:

$$1 \geq \mathrm{E}\, |Ax|\, \mathrm{E}\, |\Lambda_f - m| \tag{127}$$

hence

$$\omega_{c,1}^{\min} \leq \operatorname{tr} AD' \leq \frac{\operatorname{tr} AD'}{\mathrm{E}\, |Ax|\, \mathrm{E}\, |\Lambda_f - m|} \tag{128}$$

and therefore,

$$1 = \mathrm{E}\, |Ax|\, \mathrm{E}\, |\Lambda_f - m|, \qquad \omega_{c,1}^{\min} = \operatorname{tr} AD' = \frac{\operatorname{tr} AD'}{\mathrm{E}\, |Ax|\, \mathrm{E}\, |\Lambda_f - m|} \tag{129}$$

For $\bar\eta \in \Psi_{2\bullet}$ and such a matrix $A \neq 0$ that both achieve $\omega_{c,1}^{\min} = b$, respectively, bound (123) turns into an equality iff $\bar\eta$ is of form (117). Under condition (46), such an $\bar\eta$ may be improved according to Lemma 7.4.6 and Remark 7.4.7. Then (41) applies with constant $c = \sup_{\bullet} |\bar\eta| = b$ and yields the improved version $\hat\eta = \bar\eta\, \mathbf{I}(Y \neq 0) + \bar a\, \mathbf{I}(Y = 0)$ with $Y = Ax(\Lambda_f - m)$, which is of form (118). And this must already be the unique minimum norm solution, since

$$\mathrm{E}\, |\hat\eta|^2 \underset{(46)}{=} b^2 \big(P(\Lambda_f \neq m) + \gamma^2 P(\Lambda_f = m)\big) \tag{130}$$

does not depend on the matrix A. ////

$$\boxed{\boldsymbol{* = c,\ t = \alpha = 2,\ s = e}}$$

While the constraint $\|\varepsilon\|_1 \leq 1$ leads to bounded partial influence curves of *Hampel-Krasker* type, the seemingly similar and only slightly stronger constraint $\|\varepsilon\|_2 \leq 1$, results in partial influence curves of *Huber* type that only bound the errors/residuals but not the regressor. In Theorem 7.4.15 a, this different kind of solution becomes explicit here at least under the unique median assumption (111); under the symmetry condition 7.5(27) on F, see also Corollary 7.5.14.

Theorem 7.4.15 [Problem $O^{\mathrm{tr}}_{c,2;e}(b)$]

(a) *In case* $\omega^{\min}_{c,2;e} < b < \omega_{c,2;e}(\eta_h)$, *there exist some multipliers* $a \in \mathbb{H}_{\bullet}$, $A \in \mathbb{R}^{p\times k}$, $c \in \mathbb{M}^{+}_{\bullet}$, *and* $\beta \in (0,\infty)$, *such that the solution is given by*

$$\tilde{\eta} = (A\Lambda - a)w\,, \qquad w = \min\Bigl\{1,\, \frac{c}{|A\Lambda - a|}\Bigr\} \tag{131}$$

$$\mathrm{E}_{\bullet}\bigl(|A\Lambda - a| - c\bigr)^{+} = \beta c \tag{132}$$

$$\mathrm{E}\, c^2 = b^2 \tag{133}$$

Conversely for any $a \in \mathbb{H}_{\bullet}$, $A \in \mathbb{R}^{p\times k}$, *and* $\beta \in [0,\infty)$, *some* $c \in \mathbb{M}^{+}_{\bullet}$ *is defined by* (132), *and if any* $\tilde{\eta} \in \Psi^{D}_{2\bullet}$ *is of form* (131), *then* $\tilde{\eta}$ *is the solution to problem* $O^{\mathrm{tr}}_{c,2;e}(b)$ *with bound* b *given by* (133).

Under assumption (111), *there exist constants* $\vartheta \in \mathbb{R}$ *and* $r, \rho \in (0,\infty)$ *such that the solution attains the following necessary and sufficient form,*

$$\tilde{\eta} = \rho D\mathcal{K}^{-1}x(\Lambda_f - \vartheta)w\,, \qquad w = \min\Bigl\{1,\, \frac{r}{|\Lambda_f - \vartheta|}\Bigr\} \tag{134}$$

where

$$\mathrm{E}\,(\Lambda_f - \vartheta)w = 0 \tag{135}$$

$$\rho\,\mathrm{E}\,(\Lambda_f - \vartheta)^2 w = 1 \tag{136}$$

$$r^2\rho^2\,\mathrm{tr}\, D\mathcal{K}^{-1}D' = b^2 \tag{137}$$

(b) *Given* $\beta \in [0,\infty)$, *the unique solution* $\tilde{\eta} \in \Psi^{D}_{2\bullet}$ *to the mean square error problem* $O^{\mathrm{ms}}_{c,2;e}(\beta)$,

$$\mathrm{E}\,|\eta|^2 + \beta\,\omega^2_{c,2;e}(\eta) = \min! \qquad \eta \in \Psi^{D}_{2\bullet} \tag{138}$$

is of necessary and sufficient form (131) *and* (132).

(c) *With any* $m = \mathrm{med}\,\Lambda_f(F)$, *the minimal bias is*

$$\omega^{\min}_{c,2;e} = \frac{\sqrt{\mathrm{tr}\, D\mathcal{K}^{-1}D'}}{\mathrm{E}\,|\Lambda_f - m|} \tag{139}$$

and

$$\tilde{\eta} = \frac{D\mathcal{K}^{-1}x}{\mathrm{E}\,|\Lambda_f - m|}\bigl(\mathrm{sign}(\Lambda_f - m) + \gamma\,\mathbf{I}(\Lambda_f = m)\bigr) \tag{140}$$

with $\gamma \in [-1,1]$ *determined from* $\mathrm{E}\,\mathrm{sign}(\Lambda_f - m) + \gamma P(\Lambda_f = m) = 0$, *is the solution in case* $b = \omega^{\min}_{c,2;e}$.

PROOF Recall that $\omega^2_{c,2;e}(\eta) = \mathrm{E}\sup_{\bullet}|\eta|^2$.

(a) As the following problem is convex and well-posed ($\tilde{\alpha} \le 2!$),

$$\mathrm{E}\,|\eta|^2 = \min! \qquad \eta \in \mathbb{H},\ \mathrm{E}_{\bullet}\,\eta = 0,\ \mathrm{E}\,\eta\Lambda' = D,\ \omega^2_{c,2;e}(\eta) \le b^2 \tag{141}$$

there exist multipliers $a \in \mathbb{H}_{\bullet}$, $A \in \mathbb{R}^{p\times k}$, $\beta \in [0,\infty)$ such that the solution $\tilde{\eta}$, in comparison with all $\eta \in \mathbb{H}$ of finite $\omega_{c,2;e}(\eta)$, satisfies

$$L(\eta) = \mathrm{E}\,|\eta - Y|^2 + \beta\,\omega^2_{c,2;e}(\eta) \ge \mathrm{E}\,|\tilde{\eta}|^2 - 2\,\mathrm{tr}\,AD' + \mathrm{E}\,|Y|^2 + \beta b^2 \tag{142}$$

for $Y = A\Lambda - a$. On approximating η_h by $\Psi^D_{\infty\cdot}$, $\beta > 0$ as $b < \omega_{c,2;e}(\eta_h)$.

By Lemma 7.4.3 a, $\tilde\eta$ is some $\chi_c = Y \min\{1, c/|Y|\}$ with $c \in \mathbb{M}^+_\cdot$ that may be determined by minimization of the convex Lagrangian

$$L_1(c) = L(\chi_c) = \mathrm{E}\Bigl(\mathrm{E}_\cdot\, \bigl|(|Y| - c)^+\bigr|^2 + \beta c^2\Bigr) \tag{143}$$

Differentiation w.r.t. $c(x)$ [Lemma C.2.3] yields (132).

Conversely, form (131) and (132) of some $\tilde\eta \in \Psi^D_{2\cdot}$, for any multipliers $a \in \mathbb{H}_\cdot$, $A \in \mathbb{R}^{p\times k}$, $b, \beta \in [0,\infty)$ and $c \in \mathbb{M}^+_\cdot$, means that $\tilde\eta$ minimizes the corresponding Lagrangian L_1 on $\mathbb{M}^+_\cdot$, hence L on $\mathbb{H}$, and therefore solves problem $O^{\mathrm{tr}}_{c,2;e}(b)$ with its own bias b. Concerning c, observe that

$$\begin{aligned} c = \inf\{\, \tau \in [0,\infty] \mid \mathrm{E}_\cdot\, \bigl(|A\Lambda - a| - \tau\bigr)^+ \le \beta\tau \,\} \\ c > \tau \iff \mathrm{E}_\cdot\, \bigl(|A\Lambda - a| - \tau\bigr)^+ > \beta\tau \end{aligned} \tag{144}$$

where the functions $\mathrm{E}_\cdot\, \bigl(|A\Lambda - a| - \tau\bigr)^+$, given the fact that $a \in \mathbb{M}_\cdot$, are all measurable functions of x. This implies that $c \in \mathbb{M}^+_\cdot$.

In view of the relations (55), if $Ax = 0$ but $c > 0$, it would follow that $a = \vartheta Ax = 0$, hence $Y = 0$ in (132) so that $0 = \mathrm{E}_\cdot\, c^- = \beta c > 0$, which is a contradiction. Therefore, introducing the residuals clipping function $r = c/|Ax|$ for $Ax \ne 0$, and zero else, equations (131)–(133) read

$$\tilde\eta = Ax(\Lambda_f - \vartheta)w, \qquad w = \min\Bigl\{1, \frac{r}{|\Lambda_f - \vartheta|}\Bigr\} \tag{145}$$

$$\mathrm{E}_\cdot\, \bigl(|\Lambda_f - \vartheta| - r\bigr)^+ = \beta r \tag{146}$$

$$\mathrm{E}\, |Ax|^2 r^2 = b^2 \tag{147}$$

But in general it is not visible, whether the functions $\vartheta(x)$ and $r(x)$ are in fact constants. This may be shown under assumption (111), with the help of Lemma C.2.4: Put $\bar r = \sup_F |\Lambda_f|$. For any $r \in (0, \bar r)$ let $\vartheta_r \in \mathbb{R}$ and then $\rho_r \in (0,\infty)$ be defined by

$$0 = \int (\Lambda_f - \vartheta_r) \min\Bigl\{1, \frac{r}{|\Lambda_f - \vartheta_r|}\Bigr\}\, dF \tag{148}$$

$$1 = \rho_r \int (\Lambda_f - \vartheta_r)^2 \min\Bigl\{1, \frac{r}{|\Lambda_f - \vartheta_r|}\Bigr\}\, dF \tag{149}$$

Then

$$\tilde\eta_r = \rho_r DK^{-1} x (\Lambda_f - \vartheta_r) \min\Bigl\{1, \frac{r}{|\Lambda_f - \vartheta_r|}\Bigr\} \tag{150}$$

defines a partial influence curve $\tilde\eta_r \in \Psi^D_{2\cdot}$ with bias

$$b_r^2 = \mathrm{E} \sup_\cdot |\tilde\eta_r|^2 = r^2 \rho_r^2 \operatorname{tr} DK^{-1}D' \tag{151}$$

By the sufficiency of this form, $\tilde\eta_r$ sure is the solution to $O^{\mathrm{tr}}_{c,2;e}(b_r)$.

The bias b_r, like ϑ_r and hence ρ_r (dominated convergence), is a continuous function of $r \in (0,\infty)$. As $\lim_{r\to\bar r} \vartheta_r = 0$ and $\lim_{r\to\bar r} \rho_r = 1/\mathcal{I}_f$,

$$\lim_{r\to\bar r} b_r^2 = \frac{\bar r^2}{\mathcal{I}_f^2}\,\mathrm{tr}\,D\mathcal{K}^{-1}D' = \omega_{c,2;e}^2(\eta_h) \tag{152}$$

Writing

$$r\rho_r \int |\Lambda_f - \vartheta_r| \min\Bigl\{1, \frac{|\Lambda_f - \vartheta_r|}{r}\Bigr\}\, dF = 1 \tag{153}$$

we get

$$\lim_{r\to 0} r\rho_r = \delta, \quad \text{where } \delta^{-1} = \mathrm{E}\,|\Lambda_f - m| \tag{154}$$

Therefore

$$\lim_{r\to 0} b_r^2 = \delta^2\,\mathrm{tr}\,D\mathcal{K}^{-1}D' \tag{155}$$

The intermediate value theorem now gives the result.

(c) Note that $\omega_{c,2;e}^{\min}$ is finite and attained [argument p 277].

For every $a \in \mathbb{H}_.$ and $A \in \mathbb{R}^{p\times k}$, setting $Y = A\Lambda - a$, the following bound is true for all $\eta \in \Psi_{2.}^D$,

$$\begin{aligned}\mathrm{E}\sup_.|\eta|^2 &= \mathrm{E}\,|Y|\sup_.|\eta| - \mathrm{E}\bigl(|Y| - \sup_.|\eta|\bigr)\sup_.|\eta| \\ &\ge \mathrm{E}\,|Y||\eta| - \mathrm{E}\,(|Y|-c)c, \qquad c = \sup_.|\eta| \\ &\ge \mathrm{E}\,Y'\eta - \mathrm{E}\,(|Y|-c)c = \mathrm{tr}\,AD' + \mathrm{E}\,c\bigl(c - \mathrm{E}_.|Y|\bigr)\end{aligned} \tag{156}$$

Unconstrained minimization of the lower bound w.r.t. $c(x)$ yields

$$c = \tfrac{1}{2}\mathrm{E}_.|Y| \tag{157}$$

and thus

$$\mathrm{E}\sup_.|\eta|^2 \ge \mathrm{tr}\,AD' - \tfrac{1}{4}\mathrm{E}\,\bigl(\mathrm{E}_.|Y|\bigr)^2 \tag{158}$$

Therefore

$$\mathrm{E}\sup_.|\eta|^2 \ge \mathrm{tr}\,AD' - \tfrac{1}{4}\mathrm{E}\,|Ax|^2\bigl(\mathrm{E}\,|\Lambda_f - m|\bigr)^2 \tag{159}$$

since $\mathrm{E}_.|Y| \ge |Ax|\,\mathrm{E}_.|\Lambda_f - m|$ as in (86).

Given any $A \in \mathbb{R}^{p\times k}$, some $\bar\eta \in \Psi_{2.}^D$ achieves the lower bound iff

$$\begin{aligned}\bar\eta &= c\,e_Y &&\text{on } \{Y \ne 0\} \\ Y &= Ax(\Lambda_f - m), && c = \sup_.|\bar\eta| = \frac{1}{2}|Ax|\,\mathrm{E}\,|\Lambda_f - m|\end{aligned} \tag{160}$$

In particular, $\bar\eta = 0$ for $Ax = 0$. Thus, on the one hand, $\bar\eta$ may be improved to $\hat\eta$ of form (36) [Lemma 7.4.6]. This $\hat\eta$ already is the solution $\tilde\eta$ stated in (140) since, on the other hand, the matrix A verifies the relation (53), which in view of (160) yields that, with $1/\delta = \mathrm{E}\,|\Lambda_f - m|$, necessarily

$$A = 2\delta^2 D\mathcal{K}^{-1} \tag{161}$$

Finally, for this A and the corresponding $\bar{\eta}$ of form (160), we check that

$$\mathrm{E}\sup_{\bullet}|\bar{\eta}|^2=\delta^2\operatorname{tr}D\mathcal{K}^{-1}D'=\operatorname{tr}AD'-\frac{1}{4}\mathrm{E}\,|Ax|^2\big(\mathrm{E}\,|\Lambda_f-m|\big)^2 \tag{162}$$

which means equalities throughout (156)–(159). ////

7.4.4 Total Variation Optimality

The total variation oscillations $\omega_{v,\varepsilon;0}$ and $\omega_{v,1;0}$ can be handled exactly, in one dimension $p=1$ and for general dimension $p\geq 1$, respectively.

$\boldsymbol{*=v,\ t=\varepsilon,\ s=0;\ p=1}$

Theorem 7.4.16 [Problem $O^{\mathrm{tr}}_{v,\varepsilon;0}(b)$; $p=1$] *Assume* (9): $\mathrm{E}\,\varepsilon<\infty$.

(a) *Then, in case* $\omega^{\min}_{v,\varepsilon;0}<b<\omega_{v,\varepsilon;0}(\eta_h)$, *there exist some multipliers* $A\in\mathbb{R}^{1\times k}$, $c'\in\mathbb{M}^-_{\bullet}$, $c''\in\mathbb{M}^+_{\bullet}$, *and* $\beta\in(0,\infty)$, *such that the solution is given by*

$$\tilde{\eta}=c'\vee A\Lambda\wedge c''=A\Lambda w\,,\qquad w=1\wedge\max\Big\{\frac{c'}{A\Lambda},\frac{c''}{A\Lambda}\Big\} \tag{163}$$

where

$$\begin{array}{rl}c'=c''=0 & \text{if } |Ax|\,\mathrm{E}\,\Lambda_f^+\leq\beta\varepsilon\\ \mathrm{E}_{\bullet}(A\Lambda-c'')^+=\mathrm{E}_{\bullet}(c'-A\Lambda)^+=\beta\varepsilon & \text{if } |Ax|\,\mathrm{E}\,\Lambda_f^+>\beta\varepsilon\end{array} \tag{164}$$

and

$$\mathrm{E}\,\tilde{\eta}\Lambda'=D \tag{165}$$

$$\mathrm{E}\,\varepsilon(c''-c')=b \tag{166}$$

Conversely, for any $A\in\mathbb{R}^{1\times k}$ *and* $\beta\in[0,\infty)$, *some functions* $c'\in\mathbb{M}^-_{\bullet}$ *and* $c''\in\mathbb{M}^+_{\bullet}$ *are defined by* (164), *and if some* $\tilde{\eta}$ *is of form* (163)–(165), *then* $\tilde{\eta}\in\Psi^D_{2\bullet}$ *and* $\tilde{\eta}$ *is the solution to* $O^{\mathrm{tr}}_{v,\varepsilon;0}(b)$ *with* b *given by* (166).

(b) *Given* $\beta\in[0,\infty)$, *the unique solution* $\tilde{\eta}$ *to the problem*

$$\mathrm{E}\,\eta^2+2\beta\,\omega_{v,\varepsilon;0}(\eta)=\min!\qquad \eta\in\Psi^D_{2\bullet} \tag{167}$$

is of necessary and sufficient form (163)–(165).

(c) *We have*

$$\omega^{\min}_{v,\varepsilon;0}=\max\Big\{\frac{\sigma_A\,AD'}{\mathrm{E}\,\Lambda_f^+}\;\Big|\;A\in\mathbb{R}^{1\times k}\Big\} \tag{168}$$

where $\sigma_A=\sup\{\tau\in[0,\infty)\mid K(\varepsilon\geq\tau|Ax|)=1\}$ *as in* (73).

Suppose that $\omega^{\min}_{v,\varepsilon;0}>0$, *and* $\omega^{\min}_{v,\varepsilon;0}$ *is achieved by some* $\bar{\eta}\in\Psi^D_{2\bullet}$. *Choose any matrix* $A\in\mathbb{R}^{1\times k}$ *attaining the max in* (168). *Then necessarily*

$$\bar{\eta}=\begin{cases}c'\,\mathbf{I}(A\Lambda<0)+c''\,\mathbf{I}(A\Lambda>0) & \text{if}\quad A\Lambda\neq 0\\ 0 & \text{if}\ \ \varepsilon>\sigma_A|Ax|\end{cases} \tag{169}$$

where $c' = \inf_{.} \tilde\eta \in \mathbb{M}_{.}^{-}$, $c'' = \sup_{.} \tilde\eta \in \mathbb{M}_{.}^{+}$. *Under condition* (46), *the solution* $\tilde\eta$ *in addition coincides with some function* $\bar a \in \mathbb{M}_{.}$ *on* $\{\Lambda_f = 0\}$. *In case* $k = 1$, *under the assumption* (46): $K(x = 0) = 0$, *the solution is*

$$\tilde\eta = A|x| \operatorname{E} \Lambda_f^+ \left(\frac{\mathbf{I}(x\Lambda_f > 0)}{F(x\Lambda_f > 0)} - \frac{\mathbf{I}(x\Lambda_f < 0)}{F(x\Lambda_f < 0)} \right) \mathbf{I}(\varepsilon = \sigma_1 |x|) \tag{170}$$

with the scalar $A \in \mathbb{R}$ *determined by* $\operatorname{E} \tilde\eta \Lambda = D$.

PROOF Recall that $\omega_{v,\varepsilon;0}(\eta) = \omega_{v,\varepsilon}(\eta) = \operatorname{E} \varepsilon (\sup_{.} \eta - \inf_{.} \eta)$.

(a) With the constraint $\operatorname{E}_{.} = 0$ removed, the problem

$$\operatorname{E} \eta^2 = \min! \qquad \eta \in \mathbb{H},\ \operatorname{E} \eta \Lambda' = D,\ \operatorname{E} \varepsilon (\sup_{.} \eta - \inf_{.} \eta) \le b \tag{171}$$

is convex and well-posed. Thus there exist some multipliers $A \in \mathbb{R}^{1\times k}$ and $\beta \in [0, \infty)$ such that the solution $\tilde\eta$, in comparison with all other $\eta \in \mathbb{H}$ of finite $\omega_{v,\varepsilon}(\eta)$, satisfies

$$\begin{aligned} L(\eta) &= \operatorname{E} (\eta - A\Lambda)^2 + 2\beta \operatorname{E} \varepsilon (\sup_{.} \eta - \inf_{.} \eta) \\ &\ge \operatorname{E} \tilde\eta^2 - 2AD' + \operatorname{E} (A\Lambda)^2 + 2\beta b = L(\tilde\eta) \end{aligned} \tag{172}$$

Approximating η_h by $\mathbb{L}$, on which $\omega_{v,\varepsilon}$ is finite due to $\operatorname{E} \varepsilon < \infty$, it follows that $\beta > 0$, hence (166). And that $2AD' \ge \operatorname{E} \tilde\eta^2 + 2\beta b > 0$ follows by insertion of $\eta = 0$.

By Lemma 7.4.4 a, the solution $\tilde\eta$ must be some $\eta_c = c' \vee A\Lambda \wedge c''$ with $c' \in \mathbb{M}_{.}^{-}$ and $c'' \in \mathbb{M}_{.}^{+}$ (namely, its own essential extrema). These functions may be optimized by minimization of the Lagrangian $L_1(c) = L(\eta_c)$,

$$L_1(c) = \operatorname{E} \Big(\operatorname{E}_{.} \big|(A\Lambda - c'')^+\big|^2 + \operatorname{E}_{.} \big|(c' - A\Lambda)^+\big|^2 + 2\beta\varepsilon(c'' - c') \Big) \tag{173}$$

Pointwise minimization w.r.t. $c' \le 0 \le c''$ by differentiation of the convex integrand [Lemma C.2.3] yields (164).

Given any $A \in \mathbb{R}^{1\times k}$ and $\beta \in [0, \infty)$, we have

$$\begin{aligned} c' &= \sup\{\, \tau \in [-\infty, 0] \mid \operatorname{E}_{.} (\tau - A\Lambda)^+ \le \beta\varepsilon \,\} \\ c'' &= \inf\{\, \tau \in [0, \infty] \mid \operatorname{E}_{.} (A\Lambda - \tau)^+ \le \beta\varepsilon \,\} \end{aligned} \tag{174}$$

Hence $c' \in \mathbb{M}_{.}^{-}$ and $c'' \in \mathbb{M}_{.}^{+}$. Since moreover (164) entails that $\operatorname{E}_{.} \tilde\eta = 0$ for $\tilde\eta$ of form (163)–(165), the converse follows.

(c) Given any $A \in \mathbb{R}^{1\times k}$, and defining σ_A as in (73), the following lower bound is true for all $\eta \in \Psi_{2.}^{D}$,

$$\begin{aligned} &\operatorname{E} \varepsilon (\sup_{.} \eta - \inf_{.} \eta) \\ &\quad \ge \sigma_A \operatorname{E} |Ax| (\sup_{.} \eta - \inf_{.} \eta) \\ &\quad = \frac{\sigma_A}{\operatorname{E} \Lambda_f^+} \operatorname{E} |Ax| (\sup_{.} \eta - \inf_{.} \eta) \operatorname{E} \Lambda_f^+ \end{aligned} \tag{175}$$

where $|Ax|\,\mathrm{E}\,\Lambda_f^+ = (Ax)^+\,\mathrm{E}\,\Lambda_f^+ + (Ax)^-\,\mathrm{E}\,\Lambda_f^- = \mathrm{E}_.(A\Lambda)^+ = \mathrm{E}_.(A\Lambda)^-$. So (175) may be continued by

$$\begin{aligned} &= \frac{\sigma_A}{\mathrm{E}\,\Lambda_f^+}\,\mathrm{E}\Big((\sup_.\eta)\,\mathrm{E}_.(A\Lambda)^+ - (\inf_.\eta)\,\mathrm{E}_.(A\Lambda)^-\Big) \\ &= \frac{\sigma_A}{\mathrm{E}\,\Lambda_f^+}\,\mathrm{E}\Big((\sup_.\eta)(A\Lambda)^+ - (\inf_.\eta)(A\Lambda)^-\Big) \\ &\geq \frac{\sigma_A}{\mathrm{E}\,\Lambda_f^+}\,\mathrm{E}\left(\eta(A\Lambda)^+ - \eta(A\Lambda)^-\right) = \frac{\sigma_A}{\mathrm{E}\,\Lambda_f^+}\,\mathrm{E}\,\eta A\Lambda = \frac{\sigma_A\,AD'}{\mathrm{E}\,\Lambda_f^+} \end{aligned} \tag{176}$$

This proves $\geq$ in (168). To prove $\leq$ and attainability of the max in (168), consider the problem

$$\mathrm{E}\,\varepsilon(\sup_.\eta - \inf_.\eta) = \min! \qquad \eta\in\mathbb{H},\ \mathrm{E}\,\eta\Lambda' = D \tag{177}$$

which is convex and well-posed [argument p 197], and has a finite value because $\Psi_\infty^D \neq \emptyset$ and $\mathrm{E}\,\varepsilon < \infty$. Thus there exists some multiplier $A\in\mathbb{R}^{1\times k}$ such that, whether $\omega_{v,\varepsilon}^{\min}$ is attained or not,

$$\omega_{v,\varepsilon}^{\min} - AD' = \inf\{\,\omega_{v,\varepsilon}(\eta) - \mathrm{E}\,\eta A\Lambda\,\} \tag{178}$$

where the inf may be taken over $\mathbb{H}$ subject to $\omega_{v,\varepsilon}$ finite, or over $\mathbb{H}$ unrestricted, or only over $\mathbb{L}$. By Lemma 7.4.4 b, we conclude that

$$\omega_{v,\varepsilon}^{\min} - AD' = \inf\Big\{\,\mathrm{E}\,c''\left(\varepsilon - \mathrm{E}_.\,(A\Lambda)^+\right) - \mathrm{E}\,c'\left(\varepsilon - \mathrm{E}_.\,(A\Lambda)^-\right)\Big\} \tag{179}$$

where the inf may be taken over all functions $c', c''\in\mathbb{M}_.$, $c'\leq c''$, that are bounded, or are square integrable (argue by approximation). It follows that

$$\varepsilon \geq \mathrm{E}_.\,(A\Lambda)^+ = \mathrm{E}_.\,(A\Lambda)^- = |Ax|\,\mathrm{E}\,\Lambda_f^+ \tag{180}$$

Thus the inf is 0, and $\sigma_A \geq \mathrm{E}\,\Lambda_f^+$, so that

$$\omega_{v,\varepsilon}^{\min} = AD' \leq \frac{\sigma_A\,AD'}{\mathrm{E}\,\Lambda_f^+} \tag{181}$$

This completes the proof of (168).

Now suppose that $\omega_{v,\varepsilon}^{\min} > 0$, and that some $\bar\eta\in\Psi_{2.}^D$ achieves $\omega_{v,\varepsilon}^{\min}$. Let $A\in\mathbb{R}^{1\times k}$ be any matrix attaining the max in (168). Then the bounds (175) and (177) with $\sigma_A > 0$ must become equalities, and these are realized iff $\bar\eta$ is of form (169); especially $K(\varepsilon = \sigma_A|Ax|) > 0$ must hold. Under condition (46), every $\bar\eta\in\Psi_{2.}^D$ of form (169) can be improved by conditional averaging over the event $\{\Lambda_f = 0\}$ [instead of $\{\Lambda_f = m\}$ in the proof to Lemma 7.4.6].

Moreover, every $\bar\eta \in \Psi_{2\bullet}^{D}$ of form (169) verifies the bound

$$\begin{aligned}\mathrm{E}\,\bar\eta^2 - 2AD' &\ge \mathrm{E}\,\bar\eta^2\,\mathbf{I}(A\Lambda \neq 0) - 2\,\mathrm{E}\,\bar\eta A\Lambda \\ &= \mathrm{E}\,(c'')^2\,\mathrm{E}_{\bullet}\,\mathbf{I}(A\Lambda > 0) + (c')^2\,\mathrm{E}_{\bullet}\,\mathbf{I}(A\Lambda < 0) \\ &\quad - 2\,\mathrm{E}\,(c'' - c')|Ax|\,\mathrm{E}\,\Lambda_f^+\end{aligned} \tag{182}$$

Pointwise minimization of the integrand on the event $\{\varepsilon = \sigma_A|Ax|\}$ by differentiation w.r.t. c' and c'' yields that, on $\{\varepsilon = \sigma_A|Ax|\}$,

$$\tilde c' = -\frac{|Ax|\,\mathrm{E}\,\Lambda_f^+}{\mathrm{E}_{\bullet}\,\mathbf{I}(A\Lambda < 0)}\,, \qquad \tilde c'' = \frac{|Ax|\,\mathrm{E}\,\Lambda_f^+}{\mathrm{E}_{\bullet}\,\mathbf{I}(A\Lambda > 0)} \tag{183}$$

where $\bar\eta = 0$ outside $\{\varepsilon = \sigma_A|Ax|\}$ anyway. Thus we try

$$\tilde\eta = |Ax|\,\mathrm{E}\,\Lambda_f^+ \Big(\frac{\mathbf{I}(A\Lambda > 0)}{\mathrm{E}_{\bullet}\,\mathbf{I}(A\Lambda > 0)} - \frac{\mathbf{I}(A\Lambda < 0)}{\mathrm{E}_{\bullet}\,\mathbf{I}(A\Lambda < 0)}\Big)\,\mathbf{I}\big(\varepsilon = \sigma_A|Ax|\big) \tag{184}$$

Then $\tilde\eta = 0$ on $\{A\Lambda = 0\}$, so that nothing would be lost in the first line of (182), and $\mathrm{E}_{\bullet}\,\tilde\eta = 0$ holds. It remains to verify that $\mathrm{E}\,\tilde\eta\Lambda' = D$. In view of the relations

$$\begin{aligned}\mathrm{E}_{\bullet}\,\Lambda_f\,\mathbf{I}(A\Lambda > 0) &= \operatorname{sign}(Ax)\,\mathrm{E}\,\Lambda_f^+ = -\,\mathrm{E}_{\bullet}\,\Lambda_f\,\mathbf{I}(A\Lambda < 0) \\ \mathrm{E}_{\bullet}\,\mathbf{I}(A\Lambda > 0) &= F(Ax\Lambda_f > 0) = F\big(\operatorname{sign}(Ax)\Lambda_f > 0\big)\end{aligned}$$

and similar relations with $\{A\Lambda < 0\}$, this condition reads

$$D = \big(\mathrm{E}\,\Lambda_f^+\big)^2\Big(\frac{1}{F(\Lambda_f > 0)} + \frac{1}{F(\Lambda_f < 0)}\Big)\,A\,\mathrm{E}\,xx'\,\mathbf{I}\big(\varepsilon = \sigma_A|Ax|\big) \tag{185}$$

Note that the existence of such an $A \in \mathbb{R}^{1\times k}$ implies that A automatically achieves the max in (168), since $\tilde\eta$ defined by (184) to this A would be a partial influence curve achieving equalities in (175), (177).

Under assumption (46), the situation looks promising if only the event $\{\varepsilon = \sigma_A|Ax|\}$ would not depend on A (yet to be found). Since no analogue to Lemma 7.4.10 is available, we retreat to dimension $k = 1$, where we have $\sigma_A = \sigma_1/|A|$, and so

$$\omega_{v,\varepsilon}^{\min} = \frac{\sigma_1|D|}{\mathrm{E}\,\Lambda_f^+} \tag{186}$$

Then (185) can be solved for A, and the minimum norm (variance) solution $\tilde\eta$ given by (170) is the function $\tilde\eta$ defined by (184) to this A. ////

$* = v,\ t = \alpha = 1,\ s = 0$

Theorem 7.4.17 [Problem $O^{\mathrm{tr}}_{v,1;0}(b)$]

(a) *In case* $\omega^{\min}_{v,1;0} < b < \omega_{v,1;0}(\eta_h)$, *there exist some matrix* $A \in \mathbb{R}^{p\times k}$ *and some function* $c \in \mathbb{M}^-_{\bullet}$ *such that the solution to* $O^{\mathrm{tr}}_{v,1;0}(b)$ *is*

$$\tilde\eta = Ax\Bigl(\frac{c}{|Ax|} \vee \Lambda_f \wedge \frac{c+b}{|Ax|}\Bigr) = A\Lambda w \qquad w = 1 \wedge \max\Bigl\{\frac{c}{|Ax|\Lambda_f}\,,\,\frac{c+b}{|Ax|\Lambda_f}\Bigr\} \tag{187}$$

where

$$\mathrm{E}_{\bullet}\,(c - |Ax|\Lambda_f)^+ = \mathrm{E}_{\bullet}\,\bigl(|Ax|\Lambda_f - c - b\bigr)^+ \tag{188}$$

$$D = A\,\mathrm{E}\,xx'\,\mathrm{E}_{\bullet}\,\Lambda_f^2 w \tag{189}$$

$$\omega_{v,1;0}(\tilde\eta) = b \tag{190}$$

Conversely, for any $A \in \mathbb{R}^{p\times k}$ *and* $b \in (0,\infty)$, (188) *defines some* $c \in \mathbb{M}^-_{\bullet}$. *And if some* $\tilde\eta$ *is of form* (187)–(189), *then* $\tilde\eta \in \Psi^D_{2\bullet}$ *and* $\tilde\eta$ *is the solution.*

(b) *Given* $\beta \in [0,\infty)$, *the unique solution* $\tilde\eta$ *to problem* $O^{\mathrm{ms}}_{v,1;0}(\beta)$,

$$\mathrm{E}\,|\eta|^2 + \beta\omega^2_{v,1;0}(\eta) = \min! \qquad \eta \in \Psi^D_{2\bullet} \tag{191}$$

is of necessary and sufficient form (187)–(189), *with* $b \in [0,\infty]$ *and* β *related by*

$$\beta b = \mathrm{E}\,\bigl(|Ax|\Lambda_f - c - b\bigr)^+ \tag{192}$$

(c) *It holds that*

$$\omega^{\min}_{v,1;0} = \max\Bigl\{\frac{\operatorname{tr} AD'}{\mathrm{E}\,|Ax|\,\mathrm{E}\,\Lambda_f^+}\;\Big|\; A \in \mathbb{R}^{p\times k}\setminus\{0\}\Bigr\} \tag{193}$$

Assume (46). *Then there exists some matrix* $A \in \mathbb{R}^{p\times k}$ *such that the solution in the case* $b = \omega^{\min}_{v,1;0}$ *is, for* $Ax \neq 0$, *given by*

$$\tilde\eta = b\frac{Ax}{|Ax|}\Bigl(\frac{P(\Lambda_f<0)}{P(\Lambda_f\neq 0)}\,\mathbf{I}(\Lambda_f>0) - \frac{P(\Lambda_f>0)}{P(\Lambda_f\neq 0)}\,\mathbf{I}(\Lambda_f<0)\Bigr) \tag{194}$$

and

$$D = b\,\mathrm{E}\,\frac{Axx'}{|Ax|}\,\mathbf{I}(Ax\neq 0)\,\mathrm{E}\,\Lambda_f^+ \tag{195}$$

Conversely, if $A \in \mathbb{R}^{p\times k}$ *is any matrix verifying* (195) *with* $b = \omega^{\min}_{v,1;0}$, *then* $\tilde\eta$ *defined by* (194) *for* $Ax \neq 0$ *and zero else, is the solution.*

PROOF Recall that

$$\begin{aligned}\omega_{v,1;0}(\eta) &= \sup\nolimits_{|e|=1}\sup\nolimits_{K(dx)}\bigl(\sup_{\bullet} e'\eta - \inf_{\bullet} e'\eta\bigr)\\ &= \sup\nolimits_{K(dx)}\sup\nolimits_{|e|=1}\bigl(\sup_{\bullet} e'\eta - \inf_{\bullet} e'\eta\bigr) = \omega_{v,1}(\eta)\end{aligned} \tag{196}$$

where the interchange holds by the continuity for any fixed x, of $\inf_{\bullet} e'\eta$ and $\sup_{\bullet} e'\eta$ as functions of $e \in \mathbb{R}^p$, arguing similarly to 5.3(32) [p175].

(a) On removing the constraint $\mathrm{E}_{\bullet} = 0$, our convex, well-posed problem reads

$$\mathrm{E}\,|\eta|^2 = \min! \qquad \eta \in \mathbb{H},\ \mathrm{E}\,\eta\Lambda' = D,\ \omega_{v,1}(\eta) \le b \tag{197}$$

Thus, for some multiplier $A \in \mathbb{R}^{p\times k}$, the solution to $O^{\mathrm{tr}}_{v,1}(b)$ also solves problem

$$L(\eta) = \mathrm{E}\,|\eta - A\Lambda|^2 = \min! \qquad \eta \in \mathbb{H},\ \omega_{v,1}(\eta) \le b \tag{198}$$

For every $\eta \in \mathbb{H}$ such that $\omega_{v,1}(\eta) \le b$, the following lower bound holds,

$$\begin{aligned} L(\eta) &= \mathrm{E}\,\mathrm{E}_{\bullet} \sup\nolimits_{|e|=1} (e'\eta - e'A\Lambda)^2 \ge \mathrm{E}\,\mathrm{E}_{\bullet} \big(e'_{Ax}\eta - |Ax|\Lambda_f\big)^2 \\ &\ge \mathrm{E}\Big(\mathrm{E}_{\bullet}\big|(c - |Ax|\Lambda_f)^+\big|^2 + \mathrm{E}_{\bullet}\big|(|Ax|\Lambda_f - c - b)^+\big|^2\Big) \end{aligned} \tag{199}$$

where $c = \inf_{\bullet} e'_{Ax}\eta$. Minimization of the convex integrand by differentiation w.r.t. $c(x)$ [Lemma C.2.3] yields equation (188), which for $Ax \ne 0$ determines $c(x)$ uniquely (if $Ax = 0$ then $\tilde\eta = 0$ anyway).

The corresponding function $\tilde\eta = e_{Ax}\big(c \vee |Ax|\Lambda_f \wedge (c+b)\big)$ defined by (187) is in $\mathbb{H}$ (as $|\tilde\eta| \le |A\Lambda|$), fulfills $\omega_{v,1}(\tilde\eta) \le b$ (as $\sup_{\bullet} e'\tilde\eta \le c + b$ and $\inf_{\bullet} e'\tilde\eta \ge c$ for all $e \in \mathbb{R}^p$, $|e| = 1$, with equalities for $e = e_{Ax}$), and achieves equality in (199) as

$$\begin{aligned} L(\tilde\eta) &= \mathrm{E}\,\mathrm{E}_{\bullet} \big|e_{Ax}\big(c \vee |Ax|\Lambda_f \wedge (c+b)\big) - Ax\Lambda_f\big|^2 \\ &= \mathrm{E}\,\mathrm{E}_{\bullet} \big|\big(c \vee |Ax|\Lambda_f \wedge (c+b)\big) - |Ax|\Lambda_f\big|^2 \\ &= \mathrm{E}\,\mathrm{E}_{\bullet} \big|(c - |Ax|\Lambda_f)^+\big|^2 + \big|(|Ax|\Lambda_f - c - b)^+\big|^2 \end{aligned} \tag{200}$$

Hence, this $\tilde\eta$ solves problem (198), which by the strict convexity of L can have only one solution; namely, the solution of $O^{\mathrm{tr}}_{v,1}(b)$.

Conversely, any function $\tilde\eta$ of form (187)–(189) is in $\Psi^D_{2\bullet}$, solves the corresponding problem (198), hence also problem $O^{\mathrm{tr}}_{v,1}(b)$.

(c) Given any $A \in \mathbb{R}^{p\times k}$, the following bound holds for all $\eta \in \Psi^D_{2\bullet}$,

$$\begin{aligned} \operatorname{tr} AD' &= \mathrm{E}\,\eta' A\Lambda = \mathrm{E}\,\mathrm{E}_{\bullet}\, \eta' Ax\Lambda_f = \mathrm{E}\,|Ax|\,\mathrm{E}_{\bullet}\, e'_{Ax}\eta\Lambda_f \\ &\le \mathrm{E}\,|Ax|\,\mathrm{E}_{\bullet}\big(\sup\nolimits_{\bullet}(e'_{Ax}\eta)\Lambda_f^+ - \inf\nolimits_{\bullet}(e'_{Ax}\eta)\Lambda_f^-\big) \\ &= \mathrm{E}\,|Ax|\big(\sup\nolimits_{\bullet}(e'_{Ax}\eta) - \inf\nolimits_{\bullet}(e'_{Ax}\eta)\big)\,\mathrm{E}\,\Lambda_f^+ \\ &\overset{(196)}{\le} \omega_{v,1}(\eta)\,\mathrm{E}\,|Ax|\,\mathrm{E}\,\Lambda_f^+ \end{aligned} \tag{201}$$

This proves $\ge$ in (193). To verify $\le$ and attainability, consider the problem

$$\omega_{v,1}(\eta) = \min! \qquad \eta \in \mathbb{H},\ \mathrm{E}\,\eta\Lambda' = D \tag{202}$$

(with $\mathrm{E}_{\bullet} = 0$ removed). It is convex, well-posed, and $\omega^{\min}_{v,1} \in (0,\infty)$ is attained [p 277]. Thus there exists some matrix $A \in \mathbb{R}^{p\times k}$ such that

$$\begin{aligned} \omega^{\min}_{v,1} - \operatorname{tr} AD' &= \min\{\, \omega_{v,1}(\eta) - \mathrm{E}\,\eta' A\Lambda \mid \eta \in \mathbb{H} \,\} \\ &\le \inf\{\, 2c\big(1 - \mathrm{E}\,|Ax|\,\mathrm{E}\,\Lambda_f^+\big) \mid c \in [0,\infty) \,\} \end{aligned} \tag{203}$$

where the upper bound is obtained for $\eta_c = c\,e_{Ax}\,\mathrm{sign}\,\Lambda_f$. Hence

$$\mathrm{E}\,|Ax|\,\mathrm{E}\,\Lambda_f^+ \le 1 \tag{204}$$

$$\omega_{v,1}^{\min} \le \mathrm{tr}\,AD' \le \frac{\mathrm{tr}\,AD'}{\mathrm{E}\,|Ax|\,\mathrm{E}\,\Lambda_f^+} \tag{205}$$

and A must achieve the max in (193). If A is any such matrix, and $\bar\eta \in \Psi_{2\bullet}^D$ achieves $b = \omega_{v,1}^{\min}$, respectively, then bound (201) entails that necessarily

$$e_{Ax}'\bar\eta = c\,\mathbf{I}(\Lambda_f \ne 0) + b\,\mathbf{I}(\Lambda_f > 0) \qquad \text{on } \{A\Lambda \ne 0\} \tag{206}$$

namely, with $c = \inf_{\bullet} e_{Ax}'\bar\eta$. Therefore,

$$\begin{aligned} \mathrm{E}\,|\bar\eta|^2 &\ge \mathrm{E}\,(e_{Ax}'\bar\eta)^2\,\mathbf{I}(A\Lambda \ne 0) \\ &= \mathrm{E}\,\mathbf{I}(Ax \ne 0)\big(c^2(x)P(\Lambda_f < 0) + |c(x) + b|^2 P(\Lambda_f > 0)\big) \end{aligned} \tag{207}$$

and equality would be achieved in (207) by

$$\hat\eta = e_{Ax}\big(c\,\mathbf{I}(\Lambda_f \ne 0) + b\,\mathbf{I}(\Lambda_f > 0)\big) \tag{208}$$

Minimization of the lower bound in (207) w.r.t. $c(x)$ leads to the constant choice

$$c = -b\,\frac{P(\Lambda_f > 0)}{P(\Lambda_f \ne 0)} \tag{209}$$

and thus to $\tilde\eta$ of form (194), which certainly satisfies $\mathrm{E}_{\bullet}\,\tilde\eta = 0$. It remains to choose the matrix A such that, in addition to achieving the max in (193), also (195): $\mathrm{E}\,\tilde\eta\Lambda' = D$, holds. Then the corresponding $\tilde\eta$ is the solution.

For this purpose, we invoke Lemma 7.4.10: Under condition (46), there exists some matrix $A \in \mathbb{R}^{p\times k}$ such that

$$D = \tilde b\,\mathrm{E}\,\frac{Axx'}{|Ax|}\,\mathbf{I}(Ax \ne 0)\,\mathrm{E}\,\Lambda_f^+, \qquad \tilde b = \frac{\mathrm{tr}\,AD'}{\mathrm{E}\,|Ax|\,\mathrm{E}\,\Lambda_f^+} \tag{210}$$

This is $\mathrm{E}\,\tilde\eta\Lambda' = D$ for the function

$$\tilde\eta = \tilde b\,e_{Ax}\Big(\frac{P(\Lambda_f < 0)}{P(\Lambda_f \ne 0)}\,\mathbf{I}(\Lambda_f > 0) - \frac{P(\Lambda_f > 0)}{P(\Lambda_f \ne 0)}\,\mathbf{I}(\Lambda_f < 0)\Big) \tag{211}$$

which moreover satisfies $\mathrm{E}_{\bullet}\,\tilde\eta = 0$, so that we have $\tilde\eta \in \Psi_{2\bullet}^D$. And because

$$\tilde b = \omega_{v,1}(\tilde\eta) \ge b \ge \frac{\mathrm{tr}\,AD'}{\mathrm{E}\,|Ax|\,\mathrm{E}\,\Lambda_f^+} = \tilde b \tag{212}$$

the matrix A also achieves the max in (193), and $\tilde\eta$ is the solution.

Conversely, if $A \in \mathbb{R}^{p\times k}$ is any matrix verifying (195) with $b = \omega_{v,1}^{\min}$ then

$$\mathrm{tr}\,AD' = b\,\mathrm{E}\,|Ax|\,\mathrm{E}\,\Lambda_f^+ \tag{213}$$

Hence A achieves the max in (193). The corresponding $\tilde\eta$ defined by (194) is in $\Psi_{2\bullet}^D$ and achieves $\omega_{v,1}^{\min}$. It also is of form (208) to achieve equality in (207) with the minimizing choice (209) of c, hence is the solution. ////

7.4.5 Hellinger Optimality

$$\boxed{* = h,\ t = \varepsilon,\ s = e}$$

In view of 7.3(25), the problems $O^{\mathrm{tr}}_{h,\varepsilon;0}(b)$ and $O^{\mathrm{tr}}_{h,\varepsilon;e}(b)$ coincide if $p = 1$. Thus problem $O^{\mathrm{tr}}_{h,\varepsilon;e}(b)$ also provides the solution (coordinatewise) for problem $O^{\mathrm{tr}}_{h,\varepsilon;\infty}(b)$ in case $p \ge 1$.

Theorem 7.4.18 [Problem $O^{\mathrm{tr}}_{h,\varepsilon;e}(b)$] *Assume* (9): $\mathrm{E}\,\varepsilon < \infty$.

(a) *Then, in case* $\omega^{\min}_{h,\varepsilon;e} < b < \omega_{h,\varepsilon;e}(\eta_h)$, *there exist some* $A \in \mathbb{R}^{p\times k}$ *and* $\beta \in (0,\infty)$ *such that the solution is of the form*

$$\tilde\eta(x,u) = Ax\Lambda_f(u)h_\varepsilon(x) \tag{214}$$

$$h_\varepsilon(x) = \Bigl(1 - \beta\frac{\varepsilon(x)}{|Ax|}\Bigr)^+ \tag{215}$$

$$D = \mathcal{I}_f A\,\mathrm{E}\,xx'h_\varepsilon(x) \tag{216}$$

$$b = \sqrt{8\mathcal{I}_f}\ \mathrm{E}\,\varepsilon(x)\bigl(|Ax| - \beta\varepsilon(x)\bigr)^+ \tag{217}$$

Conversely, form (214)–(217) implies that $\tilde\eta$ *is the solution to* $O^{\mathrm{tr}}_{h,\varepsilon;e}(b)$.

(b) *Given* $\beta \in [0,\infty)$, *the unique solution* $\tilde\eta$ *to the problem*

$$\mathrm{E}\,|\eta|^2 + \sqrt{\frac{\mathcal{I}_f}{2}}\,\beta\,\omega_{h,\varepsilon;e}(\eta) = \min! \qquad \eta \in \Psi_{2\cdot}^D \tag{218}$$

is of necessary and sufficient form (214)–(216).

(c) *Setting* $\sigma_A = \sup\bigl\{\tau \in [0,\infty) \bigm| K(\varepsilon \ge \tau|Ax|) = 1\bigr\}$ *as in (73), we have*

$$\omega^{\min}_{h,\varepsilon;e} = \sqrt{8}\,\max\Bigl\{\frac{\sigma_A \operatorname{tr} AD'}{\sqrt{\mathcal{I}_f}} \Bigm| A \in \mathbb{R}^{p\times k}\Bigr\} \tag{219}$$

Suppose that $\omega^{\min}_{h,\varepsilon;e} > 0$, *and* $\omega^{\min}_{h,\varepsilon;e}$ *is achieved by some* $\bar\eta \in \Psi_{2\cdot}^D$. *Choose any matrix* $A \in \mathbb{R}^{p\times k}$ *that attains the max in (219). Then necessarily*

$$\bar\eta = Ax\Lambda_f(u)h(x)\,\mathbf{I}\bigl(\varepsilon = \sigma_A|Ax|\bigr) \tag{220}$$

for some $h \in \mathbb{M}_\cdot^+$. *In case* $k = 1$, *under assumption (46), the solution is*

$$\tilde\eta = \frac{D}{\mathcal{I}_f\mathcal{K}_1}\,x\Lambda_f\,\mathbf{I}(\varepsilon = \sigma_1|x|), \qquad \mathcal{K}_1 = \mathrm{E}\,x^2\,\mathbf{I}(\varepsilon = \sigma_1|x|) \tag{221}$$

We note that

$$h_\varepsilon(x) = \begin{cases} 0 & \text{if } \beta\varepsilon(x) \ge |Ax| \\ 1 & \text{if } \beta\varepsilon(x) = 0 \end{cases} \tag{222}$$

PROOF Recall that $\omega_{h,\varepsilon;e}(\eta) = \sqrt{8}\,\mathrm{E}\,\varepsilon\bigl(\mathrm{E}_\cdot\,|\eta|^2\bigr)^{1/2}$.

(a) Upon elimination of the constraint $\mathrm{E}_{\bullet} = 0$, we solve the problem

$$\mathrm{E}\,|\eta|^2 = \min! \qquad \eta \in \mathbb{H},\ \mathrm{E}\,\eta\Lambda' = D,\ \omega_{h,\varepsilon;e}(\eta) \le b \tag{223}$$

which is convex, and well-posed as $\mathrm{E}\,\varepsilon < \infty$ [arguments pp 197, 276]. Thus there exist multipliers $A \in \mathbb{R}^{p\times k}$ and $\beta_0 \in [0,\infty)$ such that the solution $\tilde{\eta}$, in comparison with all other $\eta \in \mathbb{H}$ of finite $\omega_{h,\varepsilon;e}(\eta)$, satisfies

$$L(\eta) = \mathrm{E}\,|\eta|^2 - 2\,\mathrm{E}\,\eta' A\Lambda + \beta_0\,\omega_{h,\varepsilon;e}(\eta) \ge \mathrm{E}\,|\tilde{\eta}|^2 - 2\,\mathrm{tr}\,AD' + \beta_0 b \tag{224}$$

As $b < \omega_{h,\varepsilon;e}(\eta_h)$, necessarily $\beta_0 > 0$; otherwise, as η_h is in the closure of $\Psi_\infty^D \subset \mathbb{H}$, (224) would imply that $\mathrm{E}\,|\tilde{\eta}|^2 \le \mathrm{E}\,|\eta_h|^2$, hence $\tilde{\eta} = \eta_h$, which is a contradiction. Therefore $\omega_{h,\varepsilon;e}(\tilde{\eta}) = b$ is achieved.

For each $\eta \in \mathbb{H}$, Lemma 7.4.5 a applies with $h^2 = \mathrm{E}_{\bullet}\,|\eta|^2$, $Y = A\Lambda$. Since $\mathrm{E}\,|\tilde{\chi} - Y|^2 \le \mathrm{E}\,|\eta - Y|^2$ and $\omega_{h,\varepsilon;e}(\tilde{\chi}) \le \omega_{h,\varepsilon;e}(\eta)$, the solution $\tilde{\chi}$ improves on η, hence $\tilde{\eta}$ itself is of this form.

To determine the optimal $h \in \mathbb{M}_{\bullet}^{+}$ in $\chi_h = Y\min\{1,h\}$, subsuming $\mathrm{E}_{\bullet}\,|Y|^2$ into h^2, we minimize the Lagrangian $L_1(h) = L(\chi_h) + \mathrm{E}\,|Y|^2$, essentially,

$$L_1(h) = \mathcal{I}_f\,\mathrm{E}\,|Ax|^2\big|(1-h)^+\big|^2 + \beta_0\sqrt{8\mathcal{I}_f}\,\mathrm{E}\,\varepsilon|Ax|h \tag{225}$$

by minimizing the integrand, which is convex in $h(x)$, pointwise. Setting $\beta = \beta_0\sqrt{2/\mathcal{I}_f}$, this yields $|Ax|(1-h_\varepsilon)^+ = \beta\varepsilon$, and hence the asserted form of the weight $1 \wedge h_\varepsilon$ in (215).

Conversely, if $\tilde{\eta}$ is of form (214)–(217) for any contamination curve ε, any $A \in \mathbb{R}^{p\times k}$, and $\beta, b \in [0,\infty)$, then $\tilde{\eta}$ minimizes the corresponding Lagrangian L_1, hence L. Since $\tilde{\eta}$ also is in $\Psi_{2\bullet}^D$, it solves $O_{h,\varepsilon;e}^{\mathrm{tr}}(b)$.

(c) Given any $A \in \mathbb{R}^{p\times k}$, the following bound is true for all $\eta \in \Psi_{2\bullet}^D$,

$$\begin{aligned}
\mathrm{E}\,\varepsilon\big(\mathrm{E}_{\bullet}\,|\eta|^2\big)^{1/2} &\ge \sigma_A\,\mathrm{E}\,|Ax|\big(\mathrm{E}_{\bullet}\,|\eta|^2\big)^{1/2} \\
&\ge \frac{\sigma_A}{\sqrt{\mathcal{I}_f}}\,\mathrm{E}\,\mathrm{E}_{\bullet}\,|Ax\Lambda_f||\eta| = \frac{\sigma_A}{\sqrt{\mathcal{I}_f}}\,\mathrm{E}\,|A\Lambda||\eta| \\
&\ge \frac{\sigma_A}{\sqrt{\mathcal{I}_f}}\,\mathrm{E}\,\eta' A\Lambda = \frac{\sigma_A}{\sqrt{\mathcal{I}_f}}\,\mathrm{tr}\,AD'
\end{aligned} \tag{226}$$

which proves $\ge$ in (219). To get $\le$ and attainability, let us solve

$$\sqrt{8}\,\mathrm{E}\,\varepsilon\big(\mathrm{E}_{\bullet}\,|\eta|^2\big)^{1/2} = \min! \qquad \eta \in \mathbb{H},\ \mathrm{E}\,\eta\Lambda' = D \tag{227}$$

Since $\mathrm{E}\,\varepsilon < \infty$, the value $\omega_{h,\varepsilon;e}^{\min}$ is finite, and the problem well-posed (consider $\eta = A\psi$ for any $\psi \in \Psi_\infty$ and $A \approx D$). Thus, whether the min is attained or not, there exists some multiplier $A \in \mathbb{R}^{p\times k}$ such that

$$\begin{aligned}
\omega_{h,\varepsilon;e}^{\min} - \mathrm{tr}\,AD' &= \inf\big\{\,\sqrt{8}\,\mathrm{E}\,\varepsilon\big(\mathrm{E}_{\bullet}\,|\eta|^2\big)^{1/2} - \mathrm{E}\,\eta' A\Lambda \bigm| \eta \in \mathbb{L}\,\big\} \\
&= \inf\big\{\,\sqrt{\mathcal{I}_f}\,\mathrm{E}\,|Ax|h\big(\sqrt{8}\,\varepsilon - |Ax|\sqrt{\mathcal{I}_f}\,\big)\,\big\}
\end{aligned} \tag{228}$$

by a restriction to $\eta_h = A\Lambda h$ with square integrable weight $h \in \mathbb{M}_{\bullet}^{+}$, according to Lemma 7.4.5 b. It follows that

$$\sqrt{8}\,\varepsilon \geq \sqrt{\mathcal{I}_f}\,|Ax| \tag{229}$$

Hence

$$\omega_{h,\varepsilon;e}^{\min} = \operatorname{tr} AD' \leq \sqrt{8}\,\frac{\sigma_A \operatorname{tr} AD'}{\sqrt{\mathcal{I}_f}} \tag{230}$$

Now suppose that $\omega_{h,\varepsilon;e}^{\min} > 0$, and that some $\bar{\eta} \in \Psi_{2\bullet}^{D}$ achieves $\omega_{h,\varepsilon;e}^{\min}$. Choose any matrix $A \in \mathbb{R}^{p\times k}$ attaining the max in (219). Then equality in bound (226) is realized iff $\bar{\eta}$ is of form (220).

Under condition (46), a constant weight $h = h_A$ in (220) leads to the solution $\tilde{\eta}$, which is optimal in the positive definite sense, by the converse part of Lemma 7.4.8 with the event $\{\varepsilon = 0\}$ replaced by $\{\varepsilon = \sigma_A |Ax|\}$, provided that this latter event, hence the weight h, does not depend on A. For dimension $k = 1$, this is actually the case since $\sigma_A = \sigma_1/|A|$ if $k = 1$. Especially,

$$\omega_{h,\varepsilon;0}^{\min} = \sqrt{8}\,\frac{\sigma_1 |D|}{\sqrt{\mathcal{I}_f}} \tag{231}$$

is the minimal bias in dimension $k = 1$. ////

$* = h,\ t = \alpha = 1,\ s = 0, e$

Problems $O_{h,1;0}^{\mathrm{tr}}(b)$ and $O_{h,1;e}^{\mathrm{tr}}(b)$ have the same solution; in particular, the lower and upper bias bounds $\omega_{h,1;s}^{\min}$ and $\omega_{h,1;s}(\eta_h)$ agree for $s = 0$ and $s = e$, respectively. However, contrary to what Bickel (1984, p 1356) suggests in the subcase $p = k = 1$, the solution is definitely not η_h.

Theorem 7.4.19 [Problem $O_{h,1;s}^{\mathrm{tr}}(b)$; $s = 0, e$]

(a) *In case $\omega_{h,1;s}^{\min} < b < \omega_{h,1;s}(\eta_h)$, there exists some matrix $A \in \mathbb{R}^{p\times k}$ such that the solution $\tilde{\eta}$ to $O_{h,1;s}^{\mathrm{tr}}(b)$ attains the form*

$$\tilde{\eta}(x,u) = Ax\Lambda_f(u)h_1(x) \tag{232}$$

$$h_1(x) = \min\Bigl\{1,\ \frac{b}{\sqrt{8\mathcal{I}_f}\,|Ax|}\Bigr\} \tag{233}$$

$$D = \mathcal{I}_f A \operatorname{E} xx' h_1(x) \tag{234}$$

Conversely, form (232)–(234), for any $A \in \mathbb{R}^{p\times k}$ and $b \in (0,\infty]$, entails that $\tilde{\eta}$ is the solution to $O_{h,1;s}^{\mathrm{tr}}(b)$.

(b) *Given $\beta \in [0,\infty)$, the unique solution $\tilde{\eta}$ to problem $O_{h,1;s}^{\mathrm{ms}}(\beta)$,*

$$\operatorname{E}|\eta|^2 + \beta\omega_{h,1;s}^2(\eta) = \min! \qquad \eta \in \Psi_{2\bullet}^{D} \tag{235}$$

is of necessary and sufficient form (232)–(234), with $b \in (0,\infty]$ and β related by

$$8\beta b = \operatorname{E}\bigl(\sqrt{8\mathcal{I}_f}\,|Ax| - b\bigr)^{+} \tag{236}$$

(c) *We have*

$$\omega_{h,1;s}^{\min} = \sqrt{8}\,\max\Bigl\{\,\frac{\operatorname{tr} AD'}{\sqrt{\mathcal{I}_f}\;\mathrm{E}\,|Ax|}\;\Big|\; A \in \mathbb{R}^{p\times k}\setminus\{0\}\Bigr\} \tag{237}$$

Assume (46). Then there exists some matrix $A \in \mathbb{R}^{p\times k}$ such that the solution $\tilde{\eta}$ to $O_{h,1;s}^{\mathrm{tr}}(b)$ in the lower case $b = \omega_{h,1;s}^{\min}$ is given by

$$\tilde{\eta} = \frac{b}{\sqrt{8\mathcal{I}_f}}\,\frac{Ax}{|Ax|}\,\mathbf{I}(Ax \neq 0)\,\Lambda_f(u) \tag{238}$$

where

$$D = \frac{b}{\sqrt{8}}\sqrt{\mathcal{I}_f}\;\mathrm{E}\,\frac{Axx'}{|Ax|}\,\mathbf{I}(Ax \neq 0) \tag{239}$$

Conversely, if $A \in \mathbb{R}^{p\times k}$ is any matrix verifying (239) with $b = \omega_{h,1;s}^{\min}$, then $\tilde{\eta}$ defined by (238) is the solution to $O_{h,1;s}^{\mathrm{tr}}(b)$.

PROOF Recall that

$$\begin{aligned}\omega_{h,1;0}^2(\eta) &= 8\,\sup\nolimits_{|e|=1}\sup\nolimits_K \mathrm{E}_{\bullet}(e'\eta)^2 \\ &= 8\,\sup\nolimits_K\sup\nolimits_{|e|=1}\mathrm{E}_{\bullet}(e'\eta)^2 \le 8\,\sup\nolimits_K \mathrm{E}_{\bullet}\,|\eta|^2 = \omega_{h,1;e}^2(\eta)\end{aligned} \tag{240}$$

where the interchange, as in (196), holds by the continuity for any fixed x, of $\mathrm{E}_{\bullet}(e'\eta)^2$ as a function of $e \in \mathbb{R}^p$, arguing similarly to 5.3(32) [p 175].

(a) On removing the constraint $\mathrm{E}_{\bullet} = 0$, the convex, well-posed problem reads

$$\mathrm{E}\,|\eta|^2 = \min! \qquad \eta \in \mathbb{H},\ \mathrm{E}\,\eta\Lambda' = D,\ \omega_{h,1;0}(\eta) \le b \tag{241}$$

Thus, for some multiplier $A \in \mathbb{R}^{p\times k}$, the solution to $O_{h,1;0}^{\mathrm{tr}}(b)$ also solves

$$L(\eta) = \mathrm{E}\,|\eta - A\Lambda|^2 = \min! \qquad \eta \in \mathbb{H},\ \omega_{h,1;0}(\eta) \le b \tag{242}$$

For every $\eta \in \mathbb{H}$ such that $\omega_{h,1}(\eta) \le b$, the following bound holds

$$L(\eta) = \mathrm{E}\,\mathrm{E}_{\bullet}\,\sup\nolimits_{|e|=1}\bigl(e'\eta - e'A\Lambda\bigr)^2 \ge \mathrm{E}\,\bigl(e_{Ax}'\eta - |Ax|\Lambda_f\bigr)^2 \tag{243}$$

where $\chi = e_{Ax}'\eta \in L_2(P)$ and $8\,\mathrm{E}_{\bullet}\,\chi^2 \le b^2$. Applying Lemma 7.4.5 a in the case $p = 1$, the solution to the auxiliary problem

$$\mathrm{E}\,\bigl(\chi - |Ax|\Lambda_f\bigr)^2 = \min! \qquad \chi \in L_2(P),\ 8\,\mathrm{E}_{\bullet}\,\chi^2 \le b^2 \tag{244}$$

is

$$\tilde{\chi} = e_{Ax}'\tilde{\eta} = |Ax|\Lambda_f h_1(x) \tag{245}$$

with $\tilde{\eta}$ and h_1 given by (232), (233). Then $8\,\mathrm{E}_{\bullet}\,\tilde{\eta}^2 \le 8\mathcal{I}_f|Ax|^2h_1^2 \le b^2$, and $\tilde{\eta}$ achieves equality in (243) since, for each fixed x, $\bigl|e'(\eta - A\Lambda)\bigr|$ is maximized by $e = e_{Ax}$. Hence $\tilde{\eta}$ solves problem (242), which, by the strict convexity of L, has only one solution, namely, the solution of $O_{h,1;0}^{\mathrm{tr}}(b)$.

Conversely, any function $\bar{\eta}$ of form (232)–(234) is in $\Psi_{2\bullet}^D$, and solves the corresponding problem (242), hence $O_{h,1;0}^{\rm tr}(b)$. In view of (240), it is easy to check that $\omega_{h,1;0}(\bar{\eta}) = \omega_{h,1;e}(\bar{\eta})$, and since $\omega_{h,1;e} \ge \omega_{h,1;0}$ in general, this $\bar{\eta}$ also solves problem $O_{h,1;e}^{\rm tr}(b)$.

(c) Given any $A \in \mathbb{R}^{p\times k}$, the following bound holds for all $\eta \in \Psi_{2\bullet}^D$,

$$\begin{aligned}\operatorname{tr} AD' &= \mathrm{E}\,\eta' A\Lambda = \mathrm{E}\,\mathrm{E}_{\bullet}\,\eta' Ax\Lambda_f = \mathrm{E}\,|Ax|\,\mathrm{E}_{\bullet}\,e'_{Ax}\eta\Lambda_f \\ &\le \sqrt{\mathcal{I}_f}\;\mathrm{E}\,|Ax|\big|\mathrm{E}_{\bullet}(e'_{Ax}\eta)^2\big|^{1/2} \le \sqrt{\mathcal{I}_f}\;\mathrm{E}\,|Ax|\sup_{|e|=1}\big|\mathrm{E}_{\bullet}(e'\eta)^2\big|^{1/2} \\ &\le \sqrt{\mathcal{I}_f}\;\mathrm{E}\,|Ax|\sup_K\sup_{|e|=1}\big|\mathrm{E}_{\bullet}(e'\eta)^2\big|^{1/2} \end{aligned} \tag{246}$$

In view of (240), this proves $\ge$ in (237). To verify $\le$ and attainability, we solve

$$\omega_{h,1;e}(\eta) = \min! \qquad \eta \in \mathbb{H},\ \mathrm{E}\,\eta\Lambda' = D \tag{247}$$

(with $\mathrm{E}_{\bullet} = 0$ removed). This problem is convex, well-posed, and its value $\omega_{h,1;e}^{\min} \in (0,\infty)$ is attained [p 277]. Thus there is some $A \in \mathbb{R}^{p\times k}$ such that

$$\begin{aligned}\omega_{h,1;e}^{\min} - \operatorname{tr} AD' &= \min\{\,\omega_{h,1;e}(\eta) - \mathrm{E}\,\eta' A\Lambda \mid \eta \in \mathbb{H}\,\} \\ &= \inf\{\,\sqrt{8\mathcal{I}_f}\,\sup_K |Ax|h - \mathcal{I}_f\,\mathrm{E}\,|Ax|^2 h \mid |Ax|h \in L_2(K)\,\} \\ &= \inf\{\,\kappa\big(\sqrt{8\mathcal{I}_f} - \mathcal{I}_f\,\mathrm{E}\,|Ax|\big) \mid \kappa \in [0,\infty)\,\}\end{aligned} \tag{248}$$

The first reduction on functions $\eta = A\Lambda h$ is feasible by Lemma 7.4.5 b, the second puts $\kappa = \sup_K |Ax|h$ and then chooses $|Ax|h$ constant κ least favorably. Thus (237) is completely proved by concluding that

$$\sqrt{8\mathcal{I}_f} \ge \mathcal{I}_f\,\mathrm{E}\,|Ax| \tag{249}$$

$$\omega_{h,1;e}^{\min} = \operatorname{tr} AD' \le \sqrt{8}\,\frac{\operatorname{tr} AD'}{\sqrt{\mathcal{I}_f}\;\mathrm{E}\,|Ax|} \tag{250}$$

Now let $A \in \mathbb{R}^{p\times k}$ be any matrix achieving the max in (237), and consider any $\bar{\eta} \in \Psi_{2\bullet}^D$ such that $\omega_{h,1;0}(\bar{\eta}) = b = \omega_{h,1;s}^{\min}$. Then equalities in (246) hold iff, first, $e'_{Ax}\bar{\eta} = \kappa(x)\Lambda_f$, so that $\mathrm{E}_{\bullet}(e'_{Ax}\bar{\eta})^2 = \kappa^2(x)\,\mathcal{I}_f$ which, secondly, has always to be maximal. Hence $\kappa(x)$ must be constant some $\kappa \in (0,\infty)$, which can be determined from $\operatorname{tr} AD' = \mathrm{E}\,\bar{\eta}' A\Lambda = \kappa\mathcal{I}_f\,\mathrm{E}\,|Ax|$, where it holds that $\sqrt{8}\,\operatorname{tr} AD' = b\sqrt{\mathcal{I}_f}\;\mathrm{E}\,|Ax|$. Thus we obtain that

$$e'_{Ax}\bar{\eta} = \kappa\Lambda_f\,, \qquad \kappa = \frac{b}{\sqrt{8\mathcal{I}_f}} \tag{251}$$

Consequentially,

$$\mathrm{E}\,|\bar{\eta}|^2 \ge \mathrm{E}\,(e'_{Ax}\bar{\eta})^2 = \kappa^2\,\mathcal{I}_f = \frac{b^2}{8} \tag{252}$$

To achieve this bound, we invoke Lemma 7.4.10: Under condition (46), there exists some matrix $A \in \mathbb{R}^{p\times k}$ such that

$$D = \tilde{\kappa}\mathcal{I}_f\,\mathrm{E}\,\frac{Axx'}{|Ax|}\,\mathbf{I}(Ax \ne 0)\,, \qquad \tilde{\kappa} = \frac{\operatorname{tr} AD'}{\mathcal{I}_f\,\mathrm{E}\,|Ax|} \tag{253}$$

This is $\mathrm{E}\,\tilde{\eta}\Lambda' = D$ for the function

$$\tilde{\eta} = \tilde{\kappa}\, e_{Ax}\, \Lambda_f \tag{254}$$

As also $\mathrm{E}_{\bullet}\, \tilde{\eta} = 0$, we have $\tilde{\eta} \in \Psi_{2\bullet}^D$. And because

$$8\,\tilde{\kappa}^2\, \mathcal{I}_f = \omega_{h,1;e}^2(\tilde{\eta}) \ge b^2 \ge 8\, \frac{(\operatorname{tr} AD')^2}{\mathcal{I}_f\big(\mathrm{E}\,|Ax|\big)^2} = 8\,\tilde{\kappa}^2\, \mathcal{I}_f \tag{255}$$

actually $\omega_{h,1;e}(\tilde{\eta}) = b$ and $\tilde{\kappa} = \kappa$. Hence (238) and (239) are implied by (254) and (253), respectively. Achieving $\mathrm{E}\,|\tilde{\eta}|^2 = b^2/8$, $\tilde{\eta}$ is the solution.

Conversely, if $A \in \mathbb{R}^{p\times k}$ is any matrix verifying (239) with $b = \omega_{h,1;s}^{\min}$, then

$$\operatorname{tr} AD' = \frac{b}{\sqrt{8}}\, \sqrt{\mathcal{I}_f}\; \mathrm{E}\,|Ax| \tag{256}$$

Hence A achieves the max in (237). The corresponding $\tilde{\eta}$ defined by (238) is in $\Psi_{2\bullet}^D$, attains $\omega_{h,1;e}(\tilde{\eta}) = b$, and achieves equality in (252). ////

Remark 7.4.20 Similar optimization problems arise in ARMA time series models with (conditionally) contaminated transition probabilities, where the regressor variable x is replaced by some suitable function of the past of the process [Staab (1984)], Hummel (1992)], and in generalized linear models [Schlather (1994)]. ////

7.5 Least Favorable Contamination Curves

Several optimality problems can be summarized by Huber's (1983) idea, not only to derive certain robust regression estimators as minimax in suitable average contamination models, but also to look for least favorable contamination curves; these describe the situations against which minimax estimators safeguard most. We essentially follow Rieder (1987 b).

Saddle Points

Definition 7.5.1 *For problem $O_{*,\alpha;s}^{\mathrm{tr}}(b)$ of exponent $\alpha \in [1,\infty]$, the contamination curve $\tilde{\varepsilon} \in \mathbb{M}_{\bullet}^{+}$ is called* least favorable, *if $\|\tilde{\varepsilon}\,\|_\alpha \le 1$ and if the solution $\tilde{\eta}$ to problem $O_{*,\tilde{\varepsilon};s}^{\mathrm{tr}}(b)$ satisfies*

$$\omega_{*,\tilde{\varepsilon};s}(\tilde{\eta}) = \omega_{*,\alpha;s}(\tilde{\eta}) \tag{1}$$

Then $\tilde{\eta}$ is called minimax *and the pair $(\tilde{\eta},\tilde{\varepsilon}\,)$* a saddle point.

The notions refer to a *game*: Nature may choose her $\varepsilon \in \mathbb{M}_{\bullet}^{+}$ arbitrarily subject to $\|\varepsilon\,\|_\alpha \le 1$. For each such ε, the statistician's strategies consist in those as. linear functionals (respectively, unbiased as. linear estimators) whose partial influence curves η obey the bias constraint $\omega_{*,\varepsilon;s}(\eta) \le b$. Under these restrictions, the statistician aims to minimize $\mathrm{E}\,|\eta|^2$.

Remark 7.5.2

(a) Condition (1), in view of 7.2(3), means that $\omega_{*,\varepsilon;s}(\tilde{\eta}) \leq \omega_{*,\tilde{\varepsilon};s}(\tilde{\eta})$ for all other contamination curves $\varepsilon \in \mathbb{M}_{\bullet}^{+}$ such that $\|\varepsilon\|_{\alpha} \leq 1$.

(b) It is intuitively obvious that $\tilde{\varepsilon}$ must satisfy $\|\tilde{\varepsilon}\|_{\alpha} = 1$. It is also plausible that the constant $\varepsilon_1 = 1$ is least favorable in case $\alpha = \infty$.

(c) If $(\tilde{\eta}, \tilde{\varepsilon})$ is a saddle point for problem $O^{\mathrm{tr}}_{*,\alpha;s}(b)$, then $\tilde{\eta}$ not only solves problem $O^{\mathrm{tr}}_{*,\alpha;s}(b)$ but minimizes $\mathrm{E}\,|\eta|^2$ among all $\eta \in \Psi_{2\bullet}^{D}$ subject to the possibly weaker bound $\omega_{*,\tilde{\varepsilon};s}(\eta) \leq b$.

(d) The saddle point definition suits constrained optimization: Nature's choice of ε has no effect on the objective function but only on the side condition, and thus on the available strategies. Saddle points for particular combinations of covariance and bias as objective function, like mean square error, and fixed width confidence probabilities, and with $\Psi_{2\bullet}^{D}$ as constant set of strategies, can be handled this way [Rieder (1987 a)].

(e) Saddle points will be determined by first finding the solution $\tilde{\eta}$ of problem $O^{\mathrm{tr}}_{*,\alpha;s}(b)$, secondly solving for $\tilde{\varepsilon}$ in (1), and thirdly proving $\tilde{\eta}$ to be the solution also to problem $O^{\mathrm{tr}}_{*,\tilde{\varepsilon};s}(b)$. ////

In this subsection, for the existence results on Lagrange multipliers (only), we make a restriction to the main cases; that is, we suppose bounds b such that, respectively,

$$\omega^{\min}_{*,t;s} < b < \omega_{*,t;s}(\eta_h) \tag{2}$$

thus avoiding the (awkward) lower bias cases $b = \omega^{\min}_{*,t;s}$ by assumption. And we require integrability 7.4(9): $\|\varepsilon\|_1 = \mathrm{E}\,\varepsilon < \infty$, in the cases $t = \varepsilon$. Then the optimization problems will be well-posed, and the multiplier β positive. Moreover, the neighborhoods are restricted to types $* = c, h$ (contamination, Hellinger), and the conditional bias version is set to $s = e$.

7.5.1 Hellinger Saddle Points

$\boldsymbol{* = h,\ t = \varepsilon}$

Recall the main case solution to problem $O^{\mathrm{tr}}_{h,\varepsilon;e}(b)$ from Theorem 7.4.18 a.

$\boldsymbol{* = h,\ \alpha = \infty}$

Theorem 7.5.3 [Problem $O^{\mathrm{tr}}_{h,\infty;e}(b)$] *Assume* (2).

(a) *Then there exist* $A \in \mathbb{R}^{p \times k}$, $\beta \in (0, \infty)$ *such that the solution is*

$$\tilde{\eta}(x,u) = A x \Lambda_f(u) h_\infty(x) \tag{3}$$

$$h_\infty(x) = \Bigl(1 - \frac{\beta}{|Ax|}\Bigr)^{+} \tag{4}$$

$$D = \mathcal{I}_f\, A\, \mathrm{E}\, x x' h_\infty(x) \tag{5}$$

$$b = \sqrt{8 \mathcal{I}_f}\ \mathrm{E}\,\bigl(|Ax| - \beta\bigr)^{+} \tag{6}$$

Conversely, form (3)–(6) implies that $\tilde\eta$ *is the solution to* $O^{\mathrm{tr}}_{h,\infty;e}(b)$.

(b) *A contamination curve* $\tilde\varepsilon$ *is least favorable iff*

$$1 \ge \tilde\varepsilon(x) \ge \min\{1, \gamma|Ax|\} \tag{7}$$

with $\gamma = 1/\beta$, *where* $\min\{1,\gamma|Ax|\} = \gamma|Ax|(1-h_\infty)$.

Remark 7.5.4

(a) As $h_\infty(x) = 0$ for $|Ax| \le \beta$, observations at small Ax have no influence, and $\tilde\varepsilon(x) = 1$ where $h_\infty(x) \ne 0$.

(b) For $b \ge \omega_{h,\infty;e}(\eta_h)$ problem $O^{\mathrm{tr}}_{h,\infty;e}(b)$ has solution η_h, and a contamination curve $\tilde\varepsilon \le 1$ is least favorable iff $\tilde\varepsilon(x) = 1$ for $Ax \ne 0$. ////

PROOF As $\omega_{h,\infty;e} = \omega_{h,\varepsilon_1;e}$ for $\varepsilon_1 = 1$, part (a) is actually a special case of Theorem 7.4.18 a, so that $h_\infty = h_{\varepsilon_1}$, and only (b) must be checked.

But an $\tilde\varepsilon \le 1$ achieves $\mathrm{E}\,\tilde\varepsilon|Ax|h_\infty = \mathrm{E}\,|Ax|h_\infty$ iff $|Ax| > \beta$ implies that $\tilde\varepsilon = 1$. And then, whenever $h_\infty > 0$,

$$h_\infty = \Bigl(1 - \frac{\beta}{|Ax|}\Bigr)^+ = \Bigl(1 - \beta\frac{\tilde\varepsilon}{|Ax|}\Bigr)^+ = h_{\tilde\varepsilon} \tag{8}$$

Moreover, $h_\infty = 0$ entails $h_{\tilde\varepsilon} = 0$ iff $|Ax| \le \beta\tilde\varepsilon$ when $|Ax| \le \beta$. ////

$\boldsymbol{* = h,\ \alpha \in (1,\infty)}$

Theorem 7.5.5 [Problem $O^{\mathrm{tr}}_{h,\alpha;e}(b)$] *Let* $\alpha, \tilde\alpha \in (1,\infty)$ *be conjugate exponents, and assume (2).*

(a) *Then there exist some* $A \in \mathbb{R}^{p\times k}$, $\beta \in (0,\infty)$, *and* $h_\alpha \in \mathcal{M}^+_{\textstyle .}$ *such that the solution is of the form*

$$\tilde\eta(x,u) = Ax\Lambda_f(u)h_\alpha(x) \tag{9}$$

$$|Ax|\bigl(1 - h_\alpha(x)\bigr) = \beta\bigl(|Ax|h_\alpha(x)\bigr)^{\tilde\alpha-1} \tag{10}$$

$$D = \mathcal{I}_f A\,\mathrm{E}\,xx'h_\alpha(x) \tag{11}$$

$$b = \sqrt{8\mathcal{I}_f}\,\Bigl|\mathrm{E}\,\bigl(|Ax|h_\alpha(x)\bigr)^{\tilde\alpha}\Bigr|^{1/\tilde\alpha} \tag{12}$$

Conversely, form (9)–(12) implies that $\tilde\eta$ *is the solution to* $O^{\mathrm{tr}}_{h,\alpha;e}(b)$.

(b) *The unique least favorable contamination curve is*

$$\tilde\varepsilon(x) = \gamma|Ax|\bigl(1 - h_\alpha(x)\bigr) \tag{13}$$

with $\gamma \in (0,\infty)$ *determined by* $\|\tilde\varepsilon\|_\alpha = 1$.

Remark 7.5.6

(a) The weight $0 < h_\alpha(x) < 1$ is uniquely determined by (10) if $Ax \neq 0$; its value for $Ax = 0$ being irrelevant. It depends on x only through $|Ax|$, as does the least favorable contamination curve $\tilde{\varepsilon}(x)$. When standardized by the linear rate, a small ratio $\tilde{\varepsilon}(x)/|Ax|$ allows $h_\alpha(x) \approx 1$, while a large ratio $\tilde{\varepsilon}(x)/|Ax|$ forces $h_\alpha(x) \approx 0$.

(b) From (10) it follows that

$$\lim_{|Ax|\to 0} h_\alpha(x) = \left\{ \begin{matrix} 1 & \text{if } \alpha < 2 & 0 \\ 0 & \text{if } \alpha > 2 & 1 \end{matrix} \right\} = \lim_{|Ax|\to\infty} h_\alpha(x) \tag{14}$$

In view of 7.4(233), (4), this also holds for the marginal cases $\alpha = 1, \infty$.

(c) For $b \geq \omega_{h,\alpha;e}(\eta_h)$, problem $O^{\mathrm{tr}}_{h,\alpha;e}(b)$ has solution η_h [also the solution to $O^{\mathrm{tr}}_{h,0;s}(b)$ for $s = 0, 2, \infty$], and the unique least favorable contamination curve, in view of (21) below, is

$$\tilde{\varepsilon}(x) \propto \left|D\mathcal{K}^{-1}x\right|^{\tilde{\alpha}-1}, \qquad \|\tilde{\varepsilon}\|_\alpha = 1 \tag{15}$$

This especially holds for exponent $\alpha = 2$, in which case $\omega^{\min}_{h,2;e} = \omega_{h,2;e}(\eta_h)$ and the least favorable $\tilde{\varepsilon}(x) \propto |D\mathcal{K}^{-1}x|$ is of linear order.

(d) Given $\beta \in [0, \infty)$, the unique solution $\tilde{\eta}$ to the problem

$$\mathrm{E}\,|\eta|^2 + \frac{1}{\tilde{\alpha}\, 2^{\tilde{\alpha}}\, (2\mathcal{I}_f)^{\tilde{\alpha}/2-1}}\, \beta\, \omega^{\tilde{\alpha}}_{h,\alpha;e}(\eta) = \min! \qquad \eta \in \Psi^D_{2\bullet} \tag{16}$$

is of necessary and sufficient form (9)–(11). ////

PROOF [Theorem 7.5.5]

(a) Upon elimination of the constraint $\mathrm{E}_\bullet = 0$, we solve the problem

$$\mathrm{E}\,|\eta|^2 = \min! \qquad \eta \in \mathbb{H},\ \mathrm{E}\,\eta\Lambda' = D,\ \omega_{h,\alpha;e}(\eta) \leq b \tag{17}$$

which is convex and well-posed [arguments pp 197, 276]. Thus there exist multipliers $A \in \mathbb{R}^{p\times k}$, $\beta_0 \in [0, \infty)$ such that for the solution $\tilde{\eta}$, and all other $\eta \in \mathbb{H}$ of finite $\omega_{h,\alpha;e}(\eta)$,

$$L(\eta) = \mathrm{E}\,|\eta|^2 - 2\,\mathrm{E}\,\eta' A\Lambda + \beta_0\, \omega^{\tilde{\alpha}}_{h,\alpha;e}(\eta) \geq \mathrm{E}\,|\tilde{\eta}|^2 - 2\,\mathrm{tr}\, AD' + \beta_0 b^{\tilde{\alpha}} \tag{18}$$

Again $\beta_0 > 0$ as $b < \omega_{h,\alpha;e}(\eta_h)$, hence $\omega_{h,\alpha;e}(\tilde{\eta}) = b$. Lemma 7.4.5 a allows a restriction to all $\chi_h = Y \min\{1, h\}$ with $h \in \mathbb{M}^+_\bullet$, $Y = A\Lambda$. Then minimization of the Lagrangian $L_1(h) = L(\chi_h) + \mathrm{E}\,|Y|^2$, essentially,

$$L_1(h) = \mathcal{I}_f\, \mathrm{E}\,|Ax|^2 \left|(1-h)^+\right|^2 + \beta_0\, (8\mathcal{I}_f)^{\tilde{\alpha}/2}\, \mathrm{E}\,(|Ax|h)^{\tilde{\alpha}} \tag{19}$$

by pointwise minimization of the integrand, which is convex in $h(x)$, on setting $2\beta\mathcal{I}_f = \tilde{\alpha}\beta_0\, (8\mathcal{I}_f)^{\tilde{\alpha}/2}$ yields the weight h_α determined by (10).

Conversely, if $\tilde{\eta}$ is of form (9)–(12) for any $A \in \mathbb{R}^{p\times k}$, $h_\alpha \in \mathbb{M}^+_\bullet$ and $\beta \in [0, \infty)$, then $\tilde{\eta}$ minimizes the corresponding Lagrangian L_1, hence L. Since $\tilde{\eta}$ also is in $\Psi^D_{2\bullet}$, it solves problem $O^{\mathrm{tr}}_{h,\alpha;e}(b)$ with its own bias b.

(b) For contamination curves $\|\varepsilon\|_\alpha \le 1$, Hölder's inequality says that

$$\mathrm{E}\,\varepsilon|Ax|h_\alpha \le \bigl|\mathrm{E}\,(|Ax|h_\alpha)^{\tilde\alpha}\bigr|^{1/\tilde\alpha} \tag{20}$$

Equality is achieved by $\tilde\varepsilon$ iff $\|\tilde\varepsilon\|_\alpha = 1$ and $\tilde\varepsilon^\alpha \propto (|Ax|h_\alpha)^{\tilde\alpha}$; that is,

$$\tilde\varepsilon \propto (|Ax|h_\alpha)^{\tilde\alpha/\alpha} \propto \beta(|Ax|h_\alpha)^{\tilde\alpha-1} \underset{(10)}{=} |Ax|(1-h_\alpha) \tag{21}$$

And then

$$h_\alpha = \Bigl(1 - \beta\frac{\tilde\varepsilon}{|Ax|}\Bigr)^+ = h_{\tilde\varepsilon} \tag{22}$$

is realized with $\beta = 1/\gamma$ chosen in 7.4(215). ////

$\boldsymbol{* = h,\ \alpha = 1}$

Recall the main case solution to problem $O^{\mathrm{tr}}_{h,1;e}(b)$ from Theorem 7.4.19 a.

Theorem 7.5.7 [Problem $O^{\mathrm{tr}}_{h,1;e}(b)$] *Assume* (2).

(a) *The solution has the necessary and sufficient form* 7.4(232)–7.4(234).

(b) *A least favorable contamination curve is given by*

$$\tilde\varepsilon(x) = \gamma|Ax|\bigl(1 - h_1(x)\bigr) = \gamma\Bigl(|Ax| - \frac{b}{\sqrt{8\mathcal{I}_f}}\Bigr)^+ \tag{23}$$

with $\gamma \in (0,\infty)$ *such that* $\|\tilde\varepsilon\|_1 = 1$.

Remark 7.5.8

(a) This least favorable contamination curve shifts all contamination to positions $|Ax| > b/\sqrt{8\mathcal{I}_f}$, and $h_1(x) = 1$ where $\tilde\varepsilon(x) = 0$.

(b) If the multiplier matrix A is unique, so is the least favorable contamination curve $\tilde\varepsilon$. ////

PROOF Although Theorem 7.4.19 a applies, we give another proof of (a) using the oscillation version $\omega_{h,1;e}$ directly.

(a) For the convex, well-posed minimum norm problem

$$\mathrm{E}\,|\eta|^2 = \min! \qquad \eta \in \mathbb{H},\ \mathrm{E}\,\eta\Lambda' = D,\ \omega_{h,1;e}(\eta) \le b \tag{24}$$

there exists a multiplier $A \in \mathbb{R}^{p\times k}$ such that the solution $\tilde\eta$ also solves

$$\mathrm{E}\,|\eta - A\Lambda|^2 = \min! \qquad \eta \in \mathbb{H},\ 8\,\mathrm{E}_.\,|\eta|^2 \le b^2 \tag{25}$$

Using Lemma 7.4.5 a, we get form 7.4(232)–7.4(234) of $\tilde\eta$, and conversely.

(b) A contamination curve $\|\tilde\varepsilon\|_1 \le 1$ achieves $\sqrt{8\mathcal{I}_f}\,\mathrm{E}\,\tilde\varepsilon|Ax|h_1 = b$ iff $\|\tilde\varepsilon\|_1 = 1$, and $\tilde\varepsilon = 0$ holds whenever $b > \sqrt{8\mathcal{I}_f}\,|Ax|$. The $\tilde\varepsilon$ defined by (23) is such. And then also

$$h_1 = \min\Bigl\{1, \frac{b}{\sqrt{8\mathcal{I}_f}\,|Ax|}\Bigr\} = \Bigl(1 - \beta\frac{\tilde\varepsilon}{|Ax|}\Bigr)^+ = h_{\tilde\varepsilon} \tag{26}$$

is realized with $\beta = 1/\gamma$ chosen in 7.4(215). ////

7.5.2 Contamination Saddle Points

In addition to the general regression model assumptions, we impose the following kind of symmetry condition on the law $\Lambda_f(F)$ of Λ_f under F,

$$\int \Lambda_f(u) \min\Bigl\{1, \frac{b}{|\Lambda_f(u)|}\Bigr\} F(du) = 0\,, \qquad b \in (0,\infty) \tag{27}$$

By this simplifying assumption, we get rid of the constraint $\mathrm{E}_{.} = 0$ also in case $* = c$. Like in case $* = h$ then $\mathrm{E}_{.}\,\tilde\eta = 0$ will be satisfied automatically by the solutions $\tilde\eta$ of the problems $O^{\mathrm{tr}}_{c,t;e}(b)$ for $t = \varepsilon, \alpha$ that are derived under neglect of the constraint $\mathrm{E}_{.} = 0$.

We introduce the (decreasing) functions $\Gamma, \Gamma^{-1}\colon [0,\infty] \to [0,\infty]$,

$$\Gamma(\tau) = \mathrm{E}\,\bigl(|\Lambda_f| - \tau\bigr)^+\,, \qquad \Gamma^{-1}(\gamma) = \inf\{\,\tau \ge 0 \mid \Gamma(\tau) \le \gamma\,\} \tag{28}$$

so that $\Gamma^{-1}(\gamma) = 0$ for $\gamma > \Gamma(0)$, and $\Gamma^{-1}\bigl(\Gamma(\tau)\bigr) = \tau$ for $\tau < \bar\tau$, and, in principle, $\Gamma^{-1}(0)$ may be any value larger or equal $\bar\tau = \sup_F |\Lambda_f|$. For example,

$$\Gamma(\tau) = 2\bigl(\varphi(\tau) - \tau\,\Phi(-\tau)\bigr) \tag{29}$$

in the case of standard normal errors $F = \mathcal{N}(0,1) = \varphi\, d\lambda = \Phi$.

The solutions in case $* = c$ and for $t = \varepsilon, \alpha$ will be given in terms of a weight function

$$w_t(x,u) = \min\Bigl\{1, \frac{r_t(x)}{|\Lambda_f(u)|}\Bigr\} \tag{30}$$

representing the clipping of (transformed) residuals $\Lambda_f(u)$ at $\pm r_t(x)$.

$\boxed{* = c,\ t = \varepsilon}$

Theorem 7.5.9 [Problem $O^{\mathrm{tr}}_{c,\varepsilon;e}(b)$] *Assume* (2), (27), *and* $\mathrm{E}\,\varepsilon < \infty$. *Then there exist some matrix* $A \in \mathbb{R}^{p\times k}$ *and constant* $\beta \in (0,\infty)$ *such that the solution is of the form*

$$\tilde\eta(x,u) = Ax\Lambda_f(u)\,w_\varepsilon(x,u) \tag{31}$$

$$r_\varepsilon(x) = \Gamma^{-1}\Bigl(\beta\,\frac{\varepsilon(x)}{|Ax|}\Bigr) \tag{32}$$

$$D = A\,\mathrm{E}\,xx'\,\mathrm{E}_x\,\Lambda_f^2(u)\,w_\varepsilon(x,u) \tag{33}$$

$$b = \mathrm{E}\,\varepsilon(x)|Ax|r_\varepsilon(x) \tag{34}$$

Conversely, form (31)–(34) *implies that* $\tilde\eta$ *is the solution to* $O^{\mathrm{tr}}_{c,\varepsilon;e}(b)$.

PROOF Theorem 7.4.11 a with the function a set equal to 0 in $\mathbb{H}_{.}$, and the residuals clipping function $r_\varepsilon = c/|Ax|$. ////

Note that

$$w_\varepsilon(x,u) = \begin{cases} 0 & \text{if } \ \beta\varepsilon(x) \ge |Ax|\Gamma(0) \\ 1 & \text{if } \ \beta\varepsilon(x) = 0 \end{cases} \tag{35}$$

$* = c, \ \alpha = \infty$

Theorem 7.5.10 [Problem $O^{\mathrm{tr}}_{c,\infty;e}(b)$] *Assume* (2) *and* (27).

(a) *Then there exist some matrix* $A \in \mathbb{R}^{p\times k}$ *and constant* $\beta \in (0,\infty)$ *such that the solution is of the form*

$$\tilde\eta(x,u) = Ax\Lambda_f(u)w_\infty(x,u) \tag{36}$$

$$r_\infty(x) = \Gamma^{-1}\Bigl(\frac{\beta}{|Ax|}\Bigr) \tag{37}$$

$$D = A\,\mathrm{E}\,xx'\,\mathrm{E}_x\,\Lambda_f^2(u)w_\infty(x,u) \tag{38}$$

$$b = \mathrm{E}\,|Ax|r_\infty(x) \tag{39}$$

Conversely, form (36)–(39) *implies that* $\tilde\eta$ *is the solution to* $O^{\mathrm{tr}}_{c,\infty;e}(b)$.

(b) *A contamination curve* $\tilde\varepsilon$ *is least favorable iff*

$$1 \ge \tilde\varepsilon(x) \ge \min\bigl\{1, \gamma|Ax|\Gamma(0)\bigr\} \tag{40}$$

with $\gamma = 1/\beta$, *where* $\min\bigl\{1,\gamma|Ax|\Gamma(0)\bigr\} = \gamma|Ax|\Gamma(r_\infty)$.

Remark 7.5.11

(a) As $r_\infty(x) = 0$ if $|Ax|\Gamma(0) \le \beta$, observations at small Ax have no influence, and $\tilde\varepsilon(x) = 1$ where $r_\infty(x) \neq 0$.

(b) For bound $b \ge \omega_{c,\infty;e}(\eta_h)$, problem $O^{\mathrm{tr}}_{c,\infty;e}(b)$ has solution η_h, and in case $\bar r$ is finite, a contamination curve $\tilde\varepsilon \le 1$ is least favorable iff $\tilde\varepsilon(x) = 1$ whenever $Ax \neq 0$. ////

PROOF As $\omega_{c,\infty;e} = \omega_{c,\varepsilon_1;e}$ for $\varepsilon_1 = 1$, part (a) is in fact a special case of Theorem 7.5.9, so that $r_\infty = r_{\varepsilon_1}$, and only (b) must be checked.

But an $\tilde\varepsilon \le 1$ achieves $\mathrm{E}\,\tilde\varepsilon|Ax|r_\infty = \mathrm{E}\,|Ax|r_\infty$ iff $\tilde\varepsilon = 1$ whenever $|Ax|r_\infty > 0$, that is $|Ax|\Gamma(0) > \beta$. And then, whenever $r_\infty > 0$,

$$r_\infty = \Gamma^{-1}\Bigl(\frac{\beta}{|Ax|}\Bigr) = \Gamma^{-1}\Bigl(\beta\frac{\tilde\varepsilon}{|Ax|}\Bigr) = r_{\tilde\varepsilon} \tag{41}$$

Moreover, $r_\infty = 0$ entails $r_{\tilde\varepsilon} = 0$ iff $|Ax|\Gamma(0) \le \beta\tilde\varepsilon$ when $|Ax|\Gamma(0) \le \beta$. ////

$* = c, \ \alpha \in (1,\infty)$

Theorem 7.5.12 [Problem $O^{\mathrm{tr}}_{c,\alpha;e}(b)$] *Assume* (2), (27), *and conjugate exponents* $\alpha, \tilde\alpha \in (1,\infty)$.

(a) *Then there exist some* $A \in \mathbb{R}^{p\times k}$, $\beta \in (0,\infty)$, *and* $r_\alpha \in \mathbb{M}_\cdot^+$, *such that the solution is of the form*

$$\tilde\eta(x,u) = Ax\Lambda_f(u)w_\alpha(x,u) \tag{42}$$

$$|Ax|\Gamma\bigl(r_\alpha(x)\bigr) = \beta\bigl(|Ax|r_\alpha(x)\bigr)^{\tilde\alpha-1} \tag{43}$$

$$D = A\,\mathrm{E}\,xx'\,\mathrm{E}_x\,\Lambda_f^2(u)w_\alpha(x,u) \tag{44}$$

$$b = \Bigl|\mathrm{E}\,\bigl(|Ax|r_\alpha(x)\bigr)^{\tilde\alpha}\Bigr|^{1/\tilde\alpha} \tag{45}$$

Conversely, form (42)–(45) implies that $\tilde{\eta}$ is the solution to $O^{\mathrm{tr}}_{c,\alpha;e}(b)$.

(b) *The unique least favorable contamination curve is*

$$\tilde{\varepsilon}(x) = \gamma |Ax| \Gamma\bigl(r_\alpha(x)\bigr) \tag{46}$$

with $\gamma \in (0,\infty)$ determined by $\|\tilde{\varepsilon}\|_\alpha = 1$.

Remark 7.5.13 Let $\bar{r} = \sup_F |\Lambda_f|$.

(a) The clipping value $0 < r_\alpha(x) < \bar{r}$ is uniquely determined by (43) if $Ax \neq 0$; its value for $Ax = 0$ being irrelevant. It depends on x only through $|Ax|$, as does the least favorable contamination curve $\tilde{\varepsilon}(x)$. Standardized by the linear rate, a small ratio $\tilde{\varepsilon}(x)/|Ax|$ allows $r_\alpha(x) \approx \bar{r}$, while a large ratio $\tilde{\varepsilon}(x)/|Ax|$ forces $r_\alpha(x) \approx 0$.

(b) From (43) it follows that,

$$\lim_{|Ax|\to 0} r_\alpha(x) = \left\{\begin{matrix} \bar{r} & \text{if } \alpha < 2 & 0 \\ 0 & \text{if } \alpha > 2 & \bar{r} \end{matrix}\right\} = \lim_{|Ax|\to\infty} r_\alpha(x) \tag{47}$$

In view of (37), this also holds for $\alpha = \infty$, and by (60) similarly for $\alpha = 1$.

(c) For $b \geq \omega_{c,\alpha;e}(\eta_h)$, problem $O^{\mathrm{tr}}_{c,\alpha;e}(b)$ has solution η_h, and in view of (53), provided that $\bar{r}$ is finite,

$$\tilde{\varepsilon}(x) \propto \bigl|D\mathcal{K}^{-1}x\bigr|^{\tilde{\alpha}-1}, \qquad \|\tilde{\varepsilon}\|_\alpha = 1 \tag{48}$$

is the unique least favorable contamination curve.

(d) Given $\beta \in [0,\infty)$, the unique solution $\tilde{\eta}$ to the problem

$$\mathrm{E}\,|\eta|^2 + \frac{2}{\tilde{\alpha}}\beta\,\omega^{\tilde{\alpha}}_{c,\alpha;e}(\eta) = \min! \qquad \eta \in \Psi^D_{2\boldsymbol{.}} \tag{49}$$

is of necessary and sufficient form (42)–(44). ////

PROOF [Theorem 7.5.12]

(a) Omitting the condition $\mathrm{E}_{\boldsymbol{.}} = 0$ on account of (27), and by the familiar arguments (convexity, well-posedness), the solution $\tilde{\eta}$ satisfies

$$L(\eta) = \mathrm{E}\,|\eta|^2 - 2\,\mathrm{E}\,\eta' A\Lambda + \beta_0\,\omega^{\tilde{\alpha}}_{c,\alpha;e}(\eta) \geq \mathrm{E}\,|\tilde{\eta}|^2 - 2\,\mathrm{tr}\,AD' + \beta_0 b^{\tilde{\alpha}} \tag{50}$$

for some $A \in \mathbb{R}^{p\times k}$, $\beta_0 \in (0,\infty)$, among all $\eta \in \mathbb{H}$ of finite $\omega_{c,\alpha;e}(\eta)$. Lemma 7.4.3 a justifies the restriction to all $\chi_r = Y\min\{1, r/|\Lambda_f|\}$ with $r \in \mathbb{M}^+_{\boldsymbol{.}}$ and $Y = A\Lambda$. Thus we arrive at the Lagrangian $L(\chi_r) + \mathrm{E}\,|Y|^2$, essentially,

$$L_1(r) = \mathrm{E}\,|Ax|^2\,\mathrm{E}_{\boldsymbol{.}}\,\bigl|(|\Lambda_f| - r)^+\bigr|^2 + \beta_0\,\mathrm{E}\,(|Ax|r)^{\tilde{\alpha}} \tag{51}$$

Upon setting $2\beta = \tilde{\alpha}\beta_0$, pointwise minimization by differentiation yields the residuals clipping function r_α determined by (43).

Conversely, form (42)–(45) of $\tilde{\eta}$, for any $A \in \mathbb{R}^{p\times k}$, $\beta, b \in [0,\infty)$, implies that $\tilde{\eta}$ minimizes the corresponding Lagrangian L_1, hence L. Since also $\tilde{\eta}$ is in $\Psi^D_{2\boldsymbol{.}}$, it solves problem $O^{\mathrm{tr}}_{c,\alpha;e}(b)$.

(b) For contamination curves $\|\varepsilon\|_\alpha \le 1$, Hölder's inequality says that

$$\mathrm{E}\,\varepsilon|Ax|r_\alpha \le \bigl|\mathrm{E}\,(|Ax|r_\alpha)^{\tilde\alpha}\bigr|^{1/\tilde\alpha} \tag{52}$$

and equality is achieved by $\tilde\varepsilon$ iff $\|\tilde\varepsilon\|_\alpha = 1$ and $\tilde\varepsilon^\alpha \propto (|Ax|r_\alpha)^{\tilde\alpha}$; that is,

$$\tilde\varepsilon \propto (|Ax|r_\alpha)^{\tilde\alpha/\alpha} \propto \beta(|Ax|r_\alpha)^{\tilde\alpha-1} \underset{(43)}{=} |Ax|\Gamma(r_\alpha) \tag{53}$$

And then

$$r_\alpha = \Gamma^{-1}(\Gamma(r_\alpha)) = \Gamma^{-1}\Bigl(\beta\frac{\tilde\varepsilon}{|Ax|}\Bigr) = r_{\tilde\varepsilon} \tag{54}$$

is realized with $\beta = 1/\gamma$ chosen in (32). ////

$* = \mathrm{c},\ \alpha = 2$

For $\alpha = 2$, the factor $|Ax|$ drops out in (43), so that r_2 is constant; that is, the truncation of residuals is done irrespectively of the position (Huber M estimates) [cf. Theorem 7.4.15 a if (27) fails]. The least favorable contamination curve is of linear order, and actually the same as for the corresponding square conditional Hellinger model.

Corollary 7.5.14 [Problem $O^{\mathrm{tr}}_{\mathrm{c},2;e}(b)$] *Assume (2) and (27).*

(a) *Then the solution is given by*

$$\tilde\eta(x,u) = \rho D\mathcal{K}^{-1}x\Lambda_f(u)w_2(u) \tag{55}$$

$$r_2 = \frac{b}{\rho\sqrt{\operatorname{tr} D\mathcal{K}^{-1}D'}} \tag{56}$$

$$1 = \rho\,\mathrm{E}\,\Lambda_f^2(u)w_2(u) \tag{57}$$

(b) *The unique least favorable contamination curve is*

$$\tilde\varepsilon(x) \propto |D\mathcal{K}^{-1}x|, \qquad \|\tilde\varepsilon\|_2 = 1 \tag{58}$$

$* = \mathrm{c},\ \alpha = 1$

For mean conditional contamination, bounded (Hampel–Krasker) partial influence curves [also the solutions to the unconditional problems $O^{\mathrm{tr}}_{\mathrm{c},0;0}(b)$] emerge as minimax [cf. Theorem 7.4.13 a if (27) fails], with some least favorable contamination curve.

Theorem 7.5.15 [Problem $O^{\mathrm{tr}}_{\mathrm{c},1;e}(b)$] *Assume (2) and (27).*

(a) *Then there exists some $A \in \mathbb{R}^{p\times k}$ such that the solution is*

$$\tilde\eta(x,u) = Ax\Lambda_f(u)w_1(x,u) \tag{59}$$

$$r_1(x) = \frac{b}{|Ax|} \tag{60}$$

$$D = A\,\mathrm{E}\,xx'\,\mathrm{E}_x\,\Lambda_f^2(u)w_1(x,u) \tag{61}$$

Conversely, form (59)–(61) implies that $\tilde{\eta}$ is the solution.

(b) *A least favorable contamination curve is given by*

$$\tilde{\varepsilon}(x) = \gamma|Ax|\Gamma\bigl(r_1(x)\bigr) = \gamma|Ax|\Gamma\Bigl(\frac{b}{|Ax|}\Bigr) \tag{62}$$

with $\gamma \in (0,\infty)$ determined by $\|\tilde{\varepsilon}\|_1 = 1$.

Remark 7.5.16

(a) The least favorable contamination curve $\tilde{\varepsilon}$ given by (62) shifts all contamination to positions $|Ax| > b/\bar{r}$, and $w_1(x,u) = 1$ where $\tilde{\varepsilon}(x) = 0$.

(b) If the multiplier matrix A is unique, so is the least favorable contamination curve $\tilde{\varepsilon}$. ////

PROOF

(a) After dropping the constraint $\mathrm{E}_{\boldsymbol{.}} = 0$ on account of (27), and in view of 7.3(29), we are back in problem $O^{\mathrm{tr}}_{c,0;0}(b)$ solved by Theorem 5.5.1.

(b) Some contamination curve $\|\tilde{\varepsilon}\|_1 \le 1$ achieves $\mathrm{E}\,\tilde{\varepsilon}|Ax|r_1 = b$ iff $\|\tilde{\varepsilon}\|_1 = 1$ and $\tilde{\varepsilon} = 0$ whenever $b > |Ax|\bar{r}$. The $\tilde{\varepsilon}$ defined by (62) is such. And then, since $r_1 > 0$ for $Ax \ne 0$, also

$$r_1 = \Gamma^{-1}\bigl(\Gamma(r_1)\bigr) = \Gamma^{-1}\Bigl(\beta\frac{\tilde{\varepsilon}}{|Ax|}\Bigr) = r_{\tilde{\varepsilon}} \tag{63}$$

is realized with $\beta = 1/\gamma$ chosen in (32). ////

Remark 7.5.17

(a) The Huber estimator, when compared with the Hampel–Krasker estimator, is minimax in a similar though slightly smaller model: square, as opposed to mean, conditional contamination. The difference is reflected by their least favorable contamination curves, but only after standardization by the linear rate: Thus one gets a constant, respectively $\Gamma\bigl(b/|Ax|\bigr)$ increasing from 0 to its finite maximum $\Gamma(0) = \mathrm{E}\,|\Lambda_f|$ as $|Ax|$ goes from 0 to ∞. In this relative sense, Hampel–Krasker safeguards against contamination at leverage points, more than Huber does.

(b) The minimax solutions in the average conditional contamination $(* = c)$ and Hellinger $(* = h)$ models differ principally in that the transformed residuals $\Lambda_f(u)$ are truncated in absolute value at some $r_\alpha(x)$, respectively downweighted uniformly by some factor $h_\alpha(x)$. Nevertheless, there is this formal relationship between the clipping and weight functions:

$$(1 - h_\alpha)\Gamma(0) \simeq \Gamma(r_\alpha) \tag{64}$$

by which also the least favorable contamination curves in corresponding contamination and Hellinger models are connected. These generally show quite similar features. For the exponents $\alpha = 2, \infty$, despite of different minimax solutions, the least favorable contamination curves coincide. ////

Remark 7.5.18 By invariance of $\omega_{*,\varepsilon;e}$ under orthogonal projection onto the tangent subspace $W_{2;*,\varepsilon}$ given by 7.2(32), the saddle points $(\tilde{\eta},\tilde{\varepsilon})$ just derived satisfy 7.4(18): $\tilde{\eta}\in W^p_{2;*,\tilde{\varepsilon}}$; that is, $\tilde{\eta} = Ax\Lambda_f$ on $\{\tilde{\varepsilon}=0\}$. Actually, except in the case of exponent $\alpha=1$, they even satisfy

$$\tilde{\eta}\in V^p_{2;*,\tilde{\varepsilon}} \tag{65}$$

where is given by 7.2(31); that is, $\tilde{\eta}=0$ if $\tilde{\varepsilon}=0$. Even in case $\alpha=1$, however, the value of the game won't be decreased by projection onto $V_{2;*,\tilde{\varepsilon}}$ [cf. Proposition 7.3.5]. ////

7.6 Equivariance Under Basis Change

The general theory of equivariance under reparametrizations, developed in Subsection 5.5.4, applies to the group of linear transformations

$$g(\theta)=B\theta\,,\qquad B\in\mathbb{R}^{k\times k},\ \det B\neq 0 \tag{1}$$

of the regression parameter $\theta\in\mathbb{R}^k$. Such parameter transformations may equally be interpreted as change of basis in design space, since

$$y=x'B\theta+u=(B'x)'\theta+u \tag{2}$$

under $P^g_\theta=P_{g(\theta)}$. The reparametrized model $\mathcal{P}^g$ is just another linear regression model of type 2.4(36)–2.4(40) with the following error and regressor distributions,

$$F^g=F\,,\qquad K^g=B'K\ \text{(image measure)} \tag{3}$$

so that

$$\Lambda^g_f=\Lambda_f\,,\qquad \mathcal{I}^g_f=\mathcal{I}_f\,,\qquad \mathcal{K}^g=B'\mathcal{K}B \tag{4}$$

Denoting by $i(\theta)=B^{-1}\theta$ the inverse of g, we have $P^g_{i(\theta)}=P_\theta$. More generally, using in 7.2(3): $Q(dx,dy)=Q(dy|x)\,Q(dx)$, the basis change

$$y=x'\theta+u=(B'x)'B^{-1}\theta+u \tag{5}$$

any probability Q on $\mathbb{B}^{k+1}$ may be rewritten as $Q=Q^g$, where

$$\begin{gathered}Q^g(dx,dy)=Q^g(dy|x)\,Q^g(dx)\\ Q^g(dx)=B'Q(dx)\,,\qquad Q^g(dy|B'x)=Q(dy|x)\end{gathered} \tag{6}$$

Thus, the contamination curves ε^g and ε of one and the same conditional neighborhood $U^g_{*,\varepsilon^g}(i(\theta),r)=U_{*,\varepsilon}(\theta,r)$ about $P^g_{i(\theta)}=P_\theta$ in the two models, respectively, are connected by

$$\varepsilon^g(B'x)=\varepsilon(x) \tag{7}$$

In view of 5.5(106), the L_2 derivatives and Fisher informations of $\mathcal{P}^g$ and $\mathcal{P}$ at $i(\theta)$ and θ, respectively, are related by

$$\Lambda^g_{i(\theta)} = B'\Lambda_\theta = B'x\Lambda_f\,, \qquad \mathcal{I}^g_{i(\theta)} = B'\mathcal{I}_\theta B = \mathcal{I}_f B'\mathcal{K}B = \mathcal{I}_f\mathcal{K}^g \tag{8}$$

The equivariance condition 5.5(107) reads

$$\psi^g_{i(\theta)} = B^{-1}\psi_\theta \tag{9}$$

This correspondence between the two classes of influence curves $\Psi^g_2(i(\theta))$ and $\Psi_2(\theta)$, within the same $L^k_2(P_\theta)$, also relates the two subclasses of conditionally centered influence curves $\Psi^g_{2\bullet}(i(\theta))$ and $\Psi_{2\bullet}(\theta)$ to each other.

Remark 7.6.1 Bickel (1984), indicating the dependence on model $\mathcal{P}^g$ by the regressor distribution $B'K$ [cf. his formula (3.15)], writes

$$\psi^g_{i(\theta)} = \psi^{B'K}_{B^{-1}\theta}(B'x, y) \tag{10}$$

Assuming (10), the equivariance relation (9) reads

$$\psi^K_\theta(x, y) = B\psi^{B'K}_{B^{-1}\theta}(B'x, y) \tag{11}$$

This notion of equivariant influence curve suggests that ψ should depend, in some more or less explicit way, (solely) on the regressor distribution. For the solutions to our optimality problems, representation (10) actually turns out to be true—but rather as a result of the proofs than by definition.

Assuming (11), the zeros $S^{B'K} = (S^{B'K}_n)$ and $S^K = (S^K_n)$ of the M equations based on $\psi^{B'K}$ and ψ^K, respectively, are related by

$$B\,S^{B'K}_n\big((B'x_i, y_i)\big) = S^K_n\big((x_i, y_i)\big) \tag{12}$$

But Bickel (1984) now pretends that $S^{B'K}$ and S^K are one and the same as. estimator S.

The situation is different for Huber regression M estimates, including LSE, which have influence curves of the form

$$\psi_\theta(x, y) = \mathcal{K}^{-1}x\,\varphi(y - x'\theta) \tag{13}$$

with $\varphi \in \Psi_2(F)$ any influence curve in the location model induced by F [cf. Subsection 5.2]. Already by (5): $x'\theta = (B'x)'B^{-1}\theta$, the zero $S = (S_n)$ to such an M equation verifies

$$B\,S_n\big((B'x_i, y_i)\big) = S_n\big((x_i, y_i)\big) \tag{14}$$

which is the classical definition of equivariant regression estimator. This implies the following representation of the estimator S^g in terms of S,

$$S^g_n\big((x_i, y_i)\big) \underset{\text{(M)}}{\overset{(9)}{=}} B^{-1}S_n\big((x_i, y_i)\big) \overset{(14)}{=} S_n\big((B'x_i, y_i)\big) \tag{15}$$

which, without justification, is assumed by Bickel (1984) in general. ////

In the following, it is no restriction to set $\theta = 0$, and then to drop the fixed parameter value, and also $i(\theta) = B^{-1}\theta = 0$, from notation.

7.6.1 Unstandardized Solutions

Denoting by $\tilde{\varrho}$ the solution to problem $O^{\rm tr}_{*,t;s}(b)$ w.r.t. model $\mathcal{P}$ $(t=\varepsilon,\alpha)$, we shall check whether $B^{-1}\tilde{\varrho}$ coincides with the solution $\tilde{\varrho}^g$ to problem $O^{g,\rm tr}_{*,t;s}(b^g)$ w.r.t. model $\mathcal{P}^g$ (for $t=\varepsilon^g,\alpha$; respectively). Since unstandardized bias is not invariant, the bias bound b will naturally have to be adapted to the (known) parametrization, by setting

$$b^g=\omega_{*,t;s}(B^{-1}\tilde{\varrho}) \tag{16}$$

Equivariance (9) of the solutions to the problems $O^{\rm tr}_{*,t;s}(b)$ considered, for $*=c,h$ and $s=e$, will be achieved only in the case $t=\alpha=2$ (square conditional contamination). The other unstandardized solutions in our list are not equivariant, essentially by the previous argument [see p 214], since a different clipping norm cannot be compensated for.

For simplicity, we only consider the main bias cases; in the lower bias cases, similar arguments apply and the results are effectively the same. We are assuming $p=k$ and $D=\mathbb{I}_k$. In the case $*=c$, we use representation 7.4(56) of the solutions 7.4(54), by means of residuals conditional centering and clipping functions $\vartheta\in\mathbb{M}_{\bullet}$ and $r\in\mathbb{M}_{\bullet}^{+}$. In the case $*=h$, the solutions are of form 7.4(57), 7.4(58) and essentially determined by a residuals weight function $h\in\mathbb{M}_{\bullet}^{+}$. In addition to ε^g and b^g defined by (7) and (16), given the nonsingular linear transform $g(\theta)=B\theta$, we employ the matrix notation

$$A^g=B^{-1}A\,(B')^{-1} \tag{17}$$

so that $A^gB'=B^{-1}A$. In general, the ratio $|A^gB'x|/|Ax|$ is not constant.

$\boxed{*=c,\ t=\varepsilon,\ s=e}$ (not equivariant) In view of Theorem 7.4.11 a, the main case solution $\tilde{\varrho}$ to problem $O^{\rm tr}_{c,\varepsilon;e}(b)$ is of the form

$$\tilde{\varrho}=Ax(\Lambda_f-\vartheta)\min\Bigl\{1,\ \frac{r}{|\Lambda_f-\vartheta|}\Bigr\} \tag{18}$$

where

$$\begin{aligned} r=0 &\qquad \text{if } |Ax|\,\mathrm{E}_{\bullet}|\Lambda_f-\vartheta|\le\beta\varepsilon \\ |Ax|\,\mathrm{E}_{\bullet}\bigl(|\Lambda_f-\vartheta|-r\bigr)^+=\beta\varepsilon &\qquad \text{if } |Ax|\,\mathrm{E}_{\bullet}|\Lambda_f-\vartheta|>\beta\varepsilon \end{aligned} \tag{19}$$

and

$$\mathrm{E}\,\varepsilon|Ax|r=b \tag{20}$$

Thus

$$B^{-1}\tilde{\varrho}=A^gB'x(\Lambda_f-\vartheta)\min\Bigl\{1,\ \frac{r}{|\Lambda_f-\vartheta|}\Bigr\} \tag{21}$$

with still the same residuals clipping function $r\in\mathbb{M}_{\bullet}^{+}$,

$$\begin{aligned} r=0 &\qquad \text{if } |A^gB'x|\,\mathrm{E}_{\bullet}|\Lambda_f-\vartheta|\le\beta\tilde{\varepsilon}^g \\ |A^gB'x|\,\mathrm{E}_{\bullet}\bigl(|\Lambda_f-\vartheta|-r\bigr)^+=\beta\tilde{\varepsilon}^g &\qquad \text{if } |A^gB'x|\,\mathrm{E}_{\bullet}|\Lambda_f-\vartheta|>\beta\tilde{\varepsilon}^g \end{aligned} \tag{22}$$

where

$$\tilde{\varepsilon}^g(B'x) = \varepsilon(x)\,\frac{|A^g B'x|}{|Ax|} \tag{23}$$

Setting

$$\tilde{b}^g = \mathrm{E}\,\tilde{\varepsilon}^g |A^g B'x| r \tag{24}$$

it is true that $B^{-1}\tilde{\varrho}$ solves problem $O^{g,\mathrm{tr}}_{c,\tilde{\varepsilon}^g;e}(\tilde{b}^g)$, but it does not solve problem $O^{g,\mathrm{tr}}_{c,\varepsilon^g;e}(b^g)$, since $\tilde{\varepsilon}^g$ given by (23) differs from ε^g defined by (7).

$\boxed{* = c,\ t = \alpha = \infty,\ s = e}$ (not equivariant) As a consequence of the preceding case, since the contamination curve $|A^g B'x|/|Ax|$ connected via (23) to ε_1 constant 1 is not a constant again, the solution to $O^{\mathrm{tr}}_{c,\infty;e}(b)$ is not equivariant.

$\boxed{* = c,\ t = \alpha = 1,\ s = 0, e}$ (not equivariant) Theorem 7.4.13 a, under the unique median assumption 7.4(111), yields the following form of the main case solution $\tilde{\varrho}$ to problem $O^{\mathrm{tr}}_{c,1;s}(b)$:

$$\tilde{\varrho} = Ax(\Lambda_f - \vartheta)\min\Bigl\{1, \frac{b}{|Ax(\Lambda_f - \vartheta)|}\Bigr\} \tag{25}$$

Thus

$$B^{-1}\tilde{\varrho} = A^g B'x(\Lambda_f - \vartheta)\min\Bigl\{1, \frac{b}{|Ax(\Lambda_f - \vartheta)|}\Bigr\} \tag{26}$$

As $b|A^g B'x|/|Ax|$ is not constant, this $B^{-1}\tilde{\varrho}$ does not agree with the solution $\tilde{\varrho}^g$ to problem $O^{g,\mathrm{tr}}_{c,1;s}(b^g)$.

$\boxed{* = c,\ t = \alpha = 2,\ s = e}$ (equivariant) In view of Theorem 7.4.15 a, the main case solution $\tilde{\varrho}$ to problem $O^{\mathrm{tr}}_{c,2;e}(b)$ is of the form

$$\tilde{\varrho} = Ax(\Lambda_f - \vartheta)\min\Bigl\{1, \frac{r}{|\Lambda_f - \vartheta|}\Bigr\} \tag{27}$$

where

$$\mathrm{E}_{.}\,\bigl(|\Lambda_f - \vartheta| - r\bigr)^+ = \beta r \tag{28}$$

and

$$\mathrm{E}\,|Ax|^2 r^2 = b^2 \tag{29}$$

Thus

$$B^{-1}\tilde{\varrho} = A^g B'x(\Lambda_f - \vartheta)\min\Bigl\{1, \frac{r}{|\Lambda_f - \vartheta|}\Bigr\} \tag{30}$$

with still the same residuals clipping function $r \in \mathbb{M}_{\cdot}^+$. By (16) we have

$$(b^g)^2 = \mathrm{E}\,|A^g B'x|^2 r^2 \tag{31}$$

Thus $B^{-1}\tilde{\varrho}$ attains the form of the solution $\tilde{\varrho}^g$ to problem $O^{g,\mathrm{tr}}_{c,2;e}(b^g)$ [whether or not ϑ and r are constants].

Under the unique median assumption 7.4(111) on F, and in the lower bias case, verification of (9) just amounts to $B^{-1}\mathcal{K}^{-1}x = (\mathcal{K}^g)^{-1}B'x$ and the modification

$$(b^g)^2 = b^2\,\frac{\operatorname{tr}(\mathcal{K}^g)^{-1}}{\operatorname{tr}\mathcal{K}^{-1}} \tag{32}$$

of the bias bound; in particular, of the minimal bias, by the last factor.

$\boxed{* = c,\ t = \alpha \in (1,\infty)\setminus\{2\},\ s = e}$ (not equivariant) Under the symmetry condition 7.5(27) on F, and employing the map Γ from 7.5(28), the main case solution $\tilde{\varrho}$ to $O^{\mathrm{tr}}_{c,\alpha;e}(b)$ given by Theorem 7.5.12 is of the form

$$\tilde{\varrho} = Ax\Lambda_f\,\min\Bigl\{1,\ \frac{r_\alpha}{|\Lambda_f|}\Bigr\} \tag{33}$$

where

$$\beta\bigl(|Ax|r_\alpha\bigr)^{\tilde{\alpha}-1} = |Ax|\,\mathrm{E}_{\bullet}\bigl(|\Lambda_f| - r_\alpha\bigr)^+ = |Ax|\Gamma(r_\alpha) \tag{34}$$

Thus

$$B^{-1}\tilde{\varrho} = A^gB'x\Lambda_f\,\min\Bigl\{1,\ \frac{r_\alpha}{|\Lambda_f|}\Bigr\} \tag{35}$$

But since

$$|A^gB'x|\Gamma(r_\alpha) = \beta\bigl(|A^gB'x|r_\alpha\bigr)^{\tilde{\alpha}-1}\,\frac{|Ax|^{\tilde{\alpha}-2}}{|A^gB'x|^{\tilde{\alpha}-2}} \tag{36}$$

contains the RHS correction factor, this $B^{-1}\tilde{\varrho}$ does not agree with the solution $\tilde{\varrho}^g$ to problem $O^{g,\mathrm{tr}}_{c,\alpha;e}(b^g)$ [unless the conjugate exponent $\tilde{\alpha} = 2$].

$\boxed{* = v,\ t = \alpha = 1,\ s = 0}$ (not equivariant) The main case solution $\tilde{\varrho}$ to problem $O^{\mathrm{tr}}_{v,1;0}(b)$, which is given by Theorem 7.4.17 a, is of the form

$$\tilde{\varrho} = Ax\bigl(r \vee \Lambda_f \wedge (r+\delta)\bigr) \tag{37}$$

where

$$\mathrm{E}_{\bullet}\,(r - \Lambda_f)^+ = \mathrm{E}_{\bullet}\,(\Lambda_f - r - \delta)^+, \qquad \delta = \frac{b}{|Ax|} \tag{38}$$

Thus

$$B^{-1}\tilde{\varrho} = A^gB'x\bigl(r \vee \Lambda_f \wedge (r+\delta)\bigr) \tag{39}$$

As $b|A^gB'x|/|Ax|$ is not constant, $B^{-1}\tilde{\varrho}$ cannot agree with the solution $\tilde{\varrho}^g$ to problem $O^{g,\mathrm{tr}}_{v,1;0}(b^g)$.

$\boxed{* = h,\ t = \varepsilon,\ s = e}$ (not equivariant) The main case solution $\tilde{\varrho}$ to problem $O^{\mathrm{tr}}_{h,\varepsilon;e}(b)$, in view of Theorem 7.4.18 a, is of the form

$$\tilde{\varrho} = Ax\Bigl(1 - \beta\,\frac{\varepsilon}{|Ax|}\Bigr)^+\Lambda_f \tag{40}$$

where

$$b = \sqrt{8\mathcal{I}_f}\,\mathrm{E}\bigl(|Ax| - \beta\varepsilon\bigr)^+\varepsilon \tag{41}$$

Thus

$$B^{-1}\tilde{\varrho} = A^g B'x\Big(1 - \beta\frac{\tilde{\varepsilon}^g}{|A^g B'x|}\Big)^+\Lambda_f \tag{42}$$

with $\tilde{\varepsilon}^g$ defined by (23). Setting

$$\tilde{b}^g = \sqrt{8\mathcal{I}_f}\;\mathrm{E}\left(|A^g B'x| - \beta\tilde{\varepsilon}^g\right)^+\tilde{\varepsilon}^g \tag{43}$$

it is true that $B^{-1}\tilde{\varrho}$ solves problem $O^{g,\mathrm{tr}}_{h,\tilde{\varepsilon}^g;e}(\tilde{b}^g)$, but it does not solve the problem $O^{g,\mathrm{tr}}_{h,\varepsilon^g;e}(b^g)$, since $\tilde{\varepsilon}^g$ given by (23) differs from ε^g defined by (7).

$\boxed{* = h,\ t = \alpha = \infty,\ s = e}$ (not equivariant) As a consequence of the preceding case, since the contamination curve $|A^g B'x|/|Ax|$ connected via (23) to ε_1 constant 1 is not a constant again, the solution to $O^{\mathrm{tr}}_{h,\infty;e}(b)$ is not equivariant.

$\boxed{* = h,\ t = \alpha = 1,\ s = 0, e}$ (not equivariant) The solution $\tilde{\varrho}$ to problem $O^{\mathrm{tr}}_{h,1;s}(b)$ in the main case, by Theorem 7.4.19 a, is of the form

$$\tilde{\varrho} = Ax\Lambda_f\,\min\Big\{1,\ \frac{b}{\sqrt{8\mathcal{I}_f}\,|Ax|}\Big\} \tag{44}$$

Thus

$$B^{-1}\tilde{\varrho} = A^g B'x\Lambda_f\,\min\Big\{1,\ \frac{b}{\sqrt{8\mathcal{I}_f}\,|Ax|}\Big\} \tag{45}$$

As $b|A^g B'x|/|Ax|$ is not constant, this $B^{-1}\tilde{\varrho}$ cannot agree with the solution $\tilde{\varrho}^g$ to problem $O^{g,\mathrm{tr}}_{h,1;s}(b^g)$.

$\boxed{* = h,\ t = \alpha = 2,\ s = 0, e}$ (equivariant) The classical scores, satisfying $B^{-1}\psi_h = B^{-1}\mathcal{I}_f^{-1}\mathcal{K}^{-1}x\Lambda_f = \mathcal{I}_f^{-1}(\mathcal{K}^g)^{-1}B'x\Lambda_f = \psi_h^g$, are equivariant.

$\boxed{* = h,\ t = \alpha \in (1,\infty)\setminus\{2\},\ s = e}$ (not equivariant) The main case solution $\tilde{\varrho}$ to problem $O^{\mathrm{tr}}_{h,\alpha;e}(b)$, in view of Theorem 7.5.5, attains the form

$$\tilde{\varrho} = Ax\Lambda_f h_\alpha\,, \qquad |Ax|(1 - h_\alpha) = \beta\left(|Ax|h_\alpha\right)^{\tilde{\alpha}-1} \tag{46}$$

Thus

$$B^{-1}\tilde{\varrho} = A^g B'x\Lambda_f h_\alpha \tag{47}$$

But since

$$|A^g B'x|(1 - h_\alpha) = \beta\left(|A^g B'x|h_\alpha\right)^{\tilde{\alpha}-1}\frac{|Ax|^{\tilde{\alpha}-2}}{|A^g B'x|^{\tilde{\alpha}-2}} \tag{48}$$

contains the RHS correction factor, this $B^{-1}\tilde{\varrho}$ cannot agree with the solution $\tilde{\varrho}^g$ to problem $O^{g,\mathrm{tr}}_{h,\alpha;e}(b^g)$ [unless $\tilde{\alpha} = 2$].

7.6.2 M Standardized Equivariant Solutions

According to Subsection 5.5.4, equivariance may be enforced by employing invariant objective functions and bias versions that are based on non-Euclidean norms standardized by model dependent matrices $M = M' > 0$ such that

$$B'MB = M^g \tag{49}$$

in view of 5.5(116). We are still assuming $\theta = 0$. Now let $\psi^g \in \Psi^g_{2(\bullet)}$ and $\psi \in \Psi_{2(\bullet)}$ be any two influence curves (conditionally centered, respectively) that are related by (9): $\psi^g = B^{-1}\psi$. Then,

$$(\psi^g)'M^g\psi^g = \psi'M\psi \tag{50}$$

This, first, implies invariance of the M standardized covariance,

$$\mathcal{C}^g\bigl((M^g)^{1/2}\psi^g\bigr) = \mathrm{E}^g\,(\psi^g)'M^g\psi^g = \mathrm{E}\,\psi'M\psi = \mathcal{C}\,(M^{1/2}\psi) \tag{51}$$

and hence of the trace. Moreover, in view of the definitions 7.2(21), 7.2(27), and 7.2(28), the corresponding tangent classes $\mathcal{G}^g_{*,t}$ and $\mathcal{G}_{*,t}$ coincide; for $t = 0$, $t = \alpha$, and $t = \varepsilon^g, \varepsilon$, respectively, where ε^g is related to ε via (7). Thus the corresponding M standardized oscillation stays invariant,

$$\omega^{M,g}_{*,t;s}(\psi^g) = \omega^g_{*,t;s}\bigl((M^g)^{1/2}\psi^g\bigr) = \omega_{*,t;s}\bigl(M^{1/2}\psi\bigr) = \omega^M_{*,t;s}(\psi) \tag{52}$$

As for the exact version ($s = 0$), see 7.4(121). The approximate version ($s = e$), by definition, is based on $|M^{1/2}\psi|$, and (52) follows from (50).

Therefore, by the general argument given in Subsection 5.5.4 [p215], the solution $\tilde{\varrho}$ to the standardized problem $O^{M,\mathrm{tr}}_{*,t;s}(b)$ $[\,t = 0, \varepsilon, \alpha;\ s = 0, e\,]$ is equivariant, being automatically related by (9) to the solution $\tilde{\varrho}^g$ to the corresponding standardized problem $O^{M,g,\mathrm{tr}}_{*,t;s}(b)$ $[\,t = 0, \varepsilon^g, \alpha;\ s = 0, e\,]$.

Due to bias invariance, the bound $b = b^g$ need not be adjusted to the parametrization. Minimal standardized bias is denoted by

$$\omega^{M,\min}_{*,t;s} = \inf\bigl\{\,\omega^M_{*,t;s}(\psi) \bigm| \psi \in \Psi_2\,\bigr\}, \qquad \omega^M_{*,t;s}(\psi) = \omega_{*,t;s}(M^{1/2}\psi) \tag{53}$$

Then the explicit solutions to $O^{M,\mathrm{tr}}_{*,t;s}(b)$ [M standardized, equivariant] may be derived from the corresponding unstandardized ones, for $p = k$, by:

$$\eta \rightsquigarrow M^{1/2}\eta, \qquad D \rightsquigarrow M^{1/2}D$$

the formal substitutions 5.5(127), and then equating $D = \mathbb{I}_k$.

Thus, the solutions will be of forms 7.4(54) and 7.4(57), respectively, expressed by some multiplier $\beta \in (0,\infty)$, some matrix $A \in \mathbb{R}^{k\times k}$, some conditional centering term $a \in \mathbb{H}_\bullet$ (zero, in case $* = h$), some clipping function $c \in \mathbb{M}^+_\bullet$ and corresponding weight w in case $* = c$, respectively some weight function $h \in \mathbb{M}^+_\bullet$ in case $* = h$. Representations 7.4(55) and 7.4(58) still apply with the modified weights w and h. In case $* = c$, therefore, form 7.4(56) is available which is based on residuals conditional centering and clipping functions $\vartheta \in \mathbb{M}_.$ and $r \in \mathbb{M}_.^+$.

We only sketch the essential form of the standardized solutions. As for the explicit formulation of the theorems, see the corresponding unstandardized results or, for example, Theorem 5.5.12.

$\boxed{* = c,\ t = \varepsilon,\ s = e}$ In view of Theorem 7.4.11, and (18)–(20), the main case solution to problem $O^{M,\mathrm{tr}}_{c,\varepsilon;e}(b)$ is of the form

$$\tilde{\varrho} = Ax(\Lambda_f - \vartheta) \min\Bigl\{1, \frac{r}{|\Lambda_f - \vartheta|}\Bigr\} \tag{54}$$

where

$$\begin{aligned} r = 0 \qquad & \text{if } |M^{1/2}Ax| \,\mathrm{E}_.\, |\Lambda_f - \vartheta| \le \beta\varepsilon \\ |M^{1/2}Ax| \,\mathrm{E}_.\, \bigl(|\Lambda_f - \vartheta| - r\bigr)^+ = \beta\varepsilon \qquad & \text{if } |M^{1/2}Ax| \,\mathrm{E}_.\, |\Lambda_f - \vartheta| > \beta\varepsilon \end{aligned} \tag{55}$$

and

$$\mathrm{E}\, \varepsilon |M^{1/2}Ax| r = b \tag{56}$$

The lower bias, with any $m = \operatorname{med} \Lambda_f(F)$, and $\sigma_{M^{1/2}A}$ defined by 7.4(73), is

$$\omega^{M,\min}_{c,\varepsilon;e} = \max\Bigl\{ \frac{\sigma_{M^{1/2}A} \operatorname{tr} AM}{\mathrm{E}\, |\Lambda_f - m|} \Bigm| A \in \mathbb{R}^{k \times k} \Bigr\} \tag{57}$$

The lower case forms 7.4(75) and 7.4(76) of $\bar{\varrho}$ with $c = \sup_. |M^{1/2}\bar{\varrho}|$ and the lower case solution $\tilde{\varrho}$, under assumption 7.4(46), both get $|M^{1/2}Ax|$ in the denominator and vanish if $\varepsilon > \sigma_{M^{1/2}A} |M^{1/2}Ax|$, where A is any matrix achieving the max in (57). Other things remain equal.

$\boxed{* = c,\ t = \alpha = \infty,\ s = e}$ The standardized solution in this case is obtained from the preceding result by inserting $\varepsilon = \varepsilon_1$ constant 1.

$\boxed{* = c,\ t = \alpha = 1,\ s = 0, e}$ In view of Theorem 7.4.13, and (25), under the unique median assumption 7.4(111), the main case solution to problem $O^{M,\mathrm{tr}}_{c,1;s}(b)$ is of the form

$$\tilde{\varrho} = Ax(\Lambda_f - \vartheta) \min\Bigl\{1, \frac{b}{|M^{1/2}Ax(\Lambda_f - \vartheta)|}\Bigr\} \tag{58}$$

The standardized lower bias, with any $m = \operatorname{med} \Lambda_f(F)$, is

$$\omega^{M,\min}_{c,1;0} = \max\Bigl\{ \frac{\operatorname{tr} AM}{\mathrm{E}\, |M^{1/2}Ax| \,\mathrm{E}\, |\Lambda_f - m|} \Bigm| A \in \mathbb{R}^{k \times k} \setminus \{0\} \Bigr\} \tag{59}$$

The lower case form 7.4(117) and solution 7.4(118), under the continuous design assumption 7.4(46), get $|M^{1/2}Ax|$ in the denominator (instead of $|Ax|$). Other things are kept and stay equal, respectively.

$\boxed{* = c,\ t = \alpha = 2,\ s = e}$ In view of Theorem 7.4.15, and (27)–(29), since $|M^{1/2}Ax|$ drops out from 7.4(132), as $|Ax|$ did from (28), the standardized problem $O^{M,\mathrm{tr}}_{c,2;e}(b)$ and the unstandardized problem $O^{\mathrm{tr}}_{c,2;e}(b')$, have the same solution, where the respective bias bounds are defined in terms of the optimal (in both senses) influence curve via

$$(b')^2 = \mathrm{E}\,|Ax|^2 r^2\,, \qquad b^2 = \mathrm{E}\,|M^{1/2}Ax|^2 r^2 \tag{60}$$

Especially, under the unique median assumption 7.4(111), we have

$$b^2 = (b')^2\,\frac{\operatorname{tr} M\mathcal{K}^{-1}}{\operatorname{tr}\mathcal{K}^{-1}} \tag{61}$$

and by this factor, also the lower standardized oscillation

$$\omega^{M,\min}_{c,2;e} = \frac{\sqrt{\operatorname{tr} M\mathcal{K}^{-1}}}{\mathrm{E}\,|\Lambda_f - m|} \tag{62}$$

is related to the lower unstandardized oscillation.

$\boxed{* = c,\ t = \alpha \in (1,\infty)\setminus\{2\},\ s = e}$ According to Theorem 7.5.12, under the symmetry condition 7.5(27) on F, and in view of (33), (34), the main case solution to problem $O^{M,\mathrm{tr}}_{c,\alpha;e}(b)$ is of the following form,

$$\tilde{\varrho} = Ax\Lambda_f w_\alpha\,, \qquad w_\alpha = \min\Bigl\{1,\ \frac{r_\alpha}{|\Lambda_f|}\Bigr\} \tag{63}$$

where

$$|M^{1/2}Ax|\,\mathrm{E}_\cdot\,\bigl(|\Lambda_f| - r_\alpha\bigr)^+ = \beta\bigl(|M^{1/2}Ax|r_\alpha\bigr)^{\tilde{\alpha}-1} \tag{64}$$

$$\mathbb{I}_k = \mathrm{E}\,xx'\,\mathrm{E}_\cdot\,\Lambda_f^2 w_\alpha\,, \qquad b = \Bigl|\mathrm{E}\,\bigl(|M^{1/2}Ax|r_\alpha\bigr)^{\tilde{\alpha}}\Bigr|^{1/\tilde{\alpha}} \tag{65}$$

which is both necessary and sufficient.

$\boxed{* = v,\ t = \alpha = 1,\ s = 0}$ In view of Theorem 7.4.17 a, and (37), (38), the main case solution to problem $O^{M,\mathrm{tr}}_{v,1;0}(b)$ is of the form

$$\tilde{\varrho} = Ax\bigl(r \vee \Lambda_f \wedge (r+\delta)\bigr) \tag{66}$$

where

$$\mathrm{E}_\cdot\,(r - \Lambda_f)^+ = \mathrm{E}_\cdot\,(\Lambda_f - r - \delta)^+\,, \qquad \delta = \frac{b}{|M^{1/2}Ax|} \tag{67}$$

In view of Theorem 7.4.17 c, the minimum standardized bias is given by

$$\omega^{M,\min}_{v,1;0} = \max\Bigl\{\frac{\operatorname{tr} AM}{\mathrm{E}\,|M^{1/2}Ax|\,\mathrm{E}\,\Lambda_f^+}\ \Bigm|\ A \in \mathbb{R}^{k\times k}\setminus\{0\}\Bigr\} \tag{68}$$

The lower case solution of form 7.4(194), under the continuous design assumption 7.4(46), gets $|M^{1/2}Ax|$ in the denominator (instead of $|Ax|$). Other things are kept and stay equal, respectively.

$\boxed{* = h,\ t = \varepsilon,\ s = e}$ By Theorem 7.4.18 a, and in view of (40), (41), the main case solution to problem $O^{M,\mathrm{tr}}_{h,\varepsilon;e}(b)$ attains the following necessary and sufficient form,

$$\tilde{\varrho} = Ax\Lambda_f h_\varepsilon\,, \qquad h_\varepsilon = \Bigl(1 - \beta\,\frac{\varepsilon}{|M^{1/2}Ax|}\Bigr)^+ \tag{69}$$

$$\mathbb{I}_k = \mathcal{I}_f\,\mathrm{E}\,xx'h_\varepsilon\,, \qquad b = \sqrt{8\mathcal{I}_f}\;\mathrm{E}\,\bigl(|M^{1/2}Ax| - \beta\varepsilon\bigr)^+\varepsilon \tag{70}$$

In view of Theorem 7.4.18 c, with $\sigma_{M^{1/2}A}$ defined by 7.4(73), the lower standardized bias is

$$\omega^{M,\min}_{h,\varepsilon;e} = \sqrt{8}\,\max\Bigl\{\,\frac{\sigma_{M^{1/2}A}\,\mathrm{tr}\,AM}{\sqrt{\mathcal{I}_f}}\Bigm| A\in\mathbb{R}^{k\times k}\Bigr\} \tag{71}$$

The influence curves of lower case form 7.4(220) now vanish on the event $\varepsilon > \sigma_{M^{1/2}A}|M^{1/2}Ax|$. Other things remain equal. In particular, the lower case solution of form 7.4(221), for dimension $k = 1$, under the continuous design assumption 7.4(46), stays the same.

$\boxed{* = h,\ t = \alpha = \infty,\ s = e}$ The standardized solution in this case is obtained from the preceding result by inserting $\varepsilon = \varepsilon_1$ constant 1.

$\boxed{* = h,\ t = \alpha = 1,\ s = 0, e}$ In view of Theorem 7.4.19 a, and (44), the main case solution to the problems $O^{M,\mathrm{tr}}_{h,1;s}(b)$ is of the following necessary and sufficient form,

$$\tilde{\varrho} = Ax\Lambda_f h_1\,, \qquad h_1 = \min\Bigl\{1,\ \frac{b}{\sqrt{8\mathcal{I}_f}\,|M^{1/2}Ax|}\Bigr\} \tag{72}$$

where $\mathbb{I}_k = \mathcal{I}_f\,\mathrm{E}\,xx'h_1$. By Theorem 7.4.19 c, the minimal standardized bias is

$$\omega^{M,\min}_{h,1;s} = \sqrt{8}\,\max\Bigl\{\,\frac{\mathrm{tr}\,AM}{\sqrt{\mathcal{I}_f}\;\mathrm{E}\,|M^{1/2}Ax|}\Bigm| A\in\mathbb{R}^{k\times k}\setminus\{0\}\Bigr\} \tag{73}$$

The lower case solution, under assumption 7.4(46), is of form 7.4(238) but, in the denominator, $|Ax|$ replaced by $|M^{1/2}Ax|$. Other things stay equal.

$\boxed{* = h,\ t = \alpha = 2,\ s = 0, e}$ The equivariant solution to the two problems $O^{M,\mathrm{tr}}_{h,2;s}(b)$ is the classical scores $\psi_h = \mathcal{I}_f^{-1}\mathcal{K}^{-1}x\Lambda_f$, with its own standardized bias $b = \omega^M_{h,2;s}(\psi_h)$, respectively, as the bound.

$\boxed{* = h,\ t = \alpha \in (1,\infty)\setminus\{2\},\ s = e}$ The main case solution to problem $O^{M,\mathrm{tr}}_{h,\alpha;e}(b)$, in view of Theorem 7.5.5, and (46), is of the following necessary and sufficient form

$$\tilde{\varrho} = Ax\Lambda_f h_\alpha\,, \qquad |M^{1/2}Ax|(1-h_\alpha) = \beta\bigl(|M^{1/2}Ax|h_\alpha\bigr)^{\tilde{\alpha}-1} \tag{74}$$

$$\mathbb{I}_k = \mathcal{I}_f\,\mathrm{E}\,xx'h_\alpha\,, \qquad b = \sqrt{8\mathcal{I}_f}\,\Bigl|\mathrm{E}\,\bigl(|M^{1/2}Ax|h_\alpha\bigr)^{\tilde{\alpha}}\Bigr|^{1/\tilde{\alpha}} \tag{75}$$

Self-Standardization

Self-standardization, as developed in Subsection 5.5.4 in the general parameter case, applies literally to linear regression with unconditional neighborhoods. In particular, the self-standardized bounded influence curve given by 5.5(140),

$$\hat{\varrho} = Y \min\Bigl\{1, \frac{b}{|Y|_{\mathcal{C}(\hat{\varrho})}}\Bigr\}, \qquad Y = A\Lambda - a$$

with some vector $a \in \mathbb{R}^k$ and matrix $A \in \mathbb{R}^{k \times k}$, is available in the regression situation $* = c$, $t = 0$, $s = 0$, providing an equivariant influence curve, which is optimally robust solving the minimum trace problem 5.5(142) standardized by its own covariance $\mathcal{C}(\hat{\varrho})$ [cf. p217].

If $\hat{\varrho}$ happens to be conditionally centered [e.g, under symmetry 7.5(27) of the error distribution F], it also supplies an equivariant, optimal self-standardized solution in the regression model $* = c$, $t = \alpha = 1$, $s = 0, e$. By 7.4(55): $a = \vartheta A x$, and $\operatorname{rk} \mathcal{K} = k$, necessarily $a = 0$ in case $\mathrm{E}_{\cdot}\, \hat{\varrho} = 0$.

In the subcase that, moreover, the regressor distribution K is spherically symmetric, $\hat{\varrho}$ coincides with the solution $\tilde{\varrho}$ to the unstandardized problem $O^{\mathrm{tr}}_{c,t;s}(b)$ [$t = 0, 1$; $s = 0, e$]. This may be seen from the representation 7.4(55), which, on right/left multiplication by orthogonal matrices, shows that then the matrix A, like $\mathcal{K}$, is a multiple of the identity $D = \mathbb{I}_k$. Hence this is also true for the covariance $\mathcal{C}(\tilde{\varrho})$.

7.6.3 Robust Prediction

Suppose that

$$M = \int x x' \, \widetilde{K}(dx) \in \mathbb{R}^{k \times k}, \qquad \operatorname{rk} M = k \tag{76}$$

for any measure $\widetilde{K} \in \mathcal{M}(\mathbb{B}^k)$. Then M is a norming matrix satisfying (49): $M^g = B'MB$, if in the reparametrized model $\mathcal{P}^g$ as regressor weight the image measure $\widetilde{K}^g = B'\widetilde{K}$ is employed.

Remark 7.6.2 The standardized MSE problem $O^{M,\mathrm{ms}}_{c,0;0}(1)$ with norming matrix M of kind (76), has also been studied by Samarov (1985). In view of (49), the finite-sample M standardized MSE stays invariant under the basis change $g(\theta) = B'\theta$,

$$\mathrm{E}\,(S_n - \theta)' M (S_n - \theta) = \mathrm{E}\,(S^g_n - B^{-1}\theta)' M^g (S^g_n - B^{-1}\theta) \tag{77}$$

if in model $\mathcal{P}^g$ the estimator $S^g_n = B^{-1} S_n$ is used. This equivariance does not yet imply the representations (15) or (14),

$$S^g_n\bigl((x_i, y_i)\bigr) = S_n\bigl((B'x_i, y_i)\bigr), \qquad B S_n\bigl((B'x_i, y_i)\bigr) = S_n\bigl((x_i, y_i)\bigr)$$

Even though $P^g_{i(\theta)} = P_\theta$, model $\mathcal{P}^g$ must be distinguished from $\mathcal{P}$. ////

Invariant MSE standardized by a matrix M of form (76) may be related to robust prediction: Suppose that $\sqrt{n}\,(S_n-\theta)\sim\mathcal{N}(b,\mathcal{C}(\psi))$, in distribution, asymptotically, with some bias $b\in\mathbb{R}$. The next observation, which is made at $x_{n+1}\in\mathbb{R}^k$, is $y_{n+1}=x_{n+1}{}'\theta+u_{n+1}$ and may reasonably be predicted by the estimate

$$\widehat{y}_{n+1}=x_{n+1}{}'S_n \tag{78}$$

Then, due to stochastic independence of (x_i,u_i) for $i=1,\ldots,n,n+1$, we have

$$\mathrm{E}\,|\widehat{y}_{n+1}-y_{n+1}|^2=\mathrm{E}\,\big|x_{n+1}{}'(S_n-\theta)\big|^2+\mathrm{E}\,u_{n+1}^2 \tag{79}$$

Therefore, under the assumption of a finite error variance,

$$\int u^2\,F(du)<\infty \tag{80}$$

the suitably rescaled MSE of prediction is, essentially,

$$\mathrm{E}\,\big|x_{n+1}{}'\sqrt{n}\,(S_n-\theta)\big|^2 \tag{81}$$

which, justifying the interchange as in 5.5(79), asymptotically amounts to

$$\mathrm{E}\,\big|\mathcal{N}(x'b,x'\mathcal{C}(\psi)x)\big|^2=x'bb'x+x'\mathcal{C}(\psi)x \tag{82}$$

Integration w.r.t. $\widetilde{K}(dx)$ gives the expected as. MSE of prediction,

$$\int\Big(x'bb'x+x'\mathcal{C}(\psi)x\Big)\,\widetilde{K}(dx)=b'Mb+\operatorname{tr}M\mathcal{C}(\psi) \tag{83}$$

Maximization over the biasses b finally leads to the invariant MSE objective function,

$$\operatorname{tr}M\mathcal{C}(\psi)+\big|\omega_{*,t;s}^{M}(\psi)\big|^2 \tag{84}$$

This expected as. MSE of prediction/invariant MSE, with the conditional regression bias terms, can be minimized as in Theorems 5.5.7 and 5.5.12 b, or by applying the unstandardized MSE type results in Section 7.4 upon the formal substitutions 5.5(127).

Both ways, the same type of solutions are obtained as for the corresponding standardized minimum trace problems—with multiplier $\beta=1$, provided that the bias terms occur quadratic in the Lagrangians of the corresponding minimum trace problems. But possibly $\beta\neq1$, if the bias terms in the previous Lagrangians occur there to a power other than 2, or if the bias weight β in the MSE objective function (84), as in 5.5(78) and 5.5(125), is chosen more general anyway.

Appendix A

Weak Convergence of Measures

A.1 Basic Notions

We collect some basic notions and facts about weak convergence of finite measures.

Definitions

A sequence of finite measures $Q_n \in \mathcal{M}_b(\mathcal{A}_n)$ on some general measurable spaces $(\Omega_n, \mathcal{A}_n)$ is called *bounded* if

$$\limsup_{n\to\infty} Q_n(\Omega_n) < \infty \tag{1}$$

Let Ξ be a metric space with distance d and Borel σ algebra $\mathcal{B}$, and denote by $\mathcal{C}(\Xi)$ the set of all bounded continuous functions $f\colon \Xi \to \mathbb{R}$. A sequence $Q_n \in \mathcal{M}_b(\mathcal{B})$ is called *tight* if for every $\varepsilon \in (0,\infty)$ there exists a compact subset $K \subset \Xi$ such that

$$\limsup_{n\to\infty} Q_n(\Xi \setminus K) < \varepsilon \tag{2}$$

A sequence $Q_n \in \mathcal{M}_b(\mathcal{B})$ is said to *converge weakly* to some $Q_0 \in \mathcal{M}_b(\mathcal{B})$ if

$$\lim_{n\to\infty} \int f\,dQ_n = \int f\,dQ_0 \tag{3}$$

for all $f \in \mathcal{C}(\Xi)$; notation: $Q_n \xrightarrow{\;\mathrm{w}\;} Q_0$.

Continuous Mapping Theorem

By definition, weak convergence is compatible with continuous transformations. The continuity assumption can even be alleviated. For any function h denote by D_h the set of its discontinuities.

Proposition A.1.1 *Let $(\Xi, \mathcal{B})$ and $(\widetilde{\Xi}, \widetilde{\mathcal{B}})$ be two metric sample spaces. Consider a sequence $Q_0, Q_1, \dots$ in $\mathcal{M}_b(\mathcal{B})$ and a Borel measurable transformation $h: (\Xi, \mathcal{B}) \to (\widetilde{\Xi}, \widetilde{\mathcal{B}})$ such that*

$$Q_0(D_h) = 0 \tag{4}$$

(a) *Then $Q_n \xrightarrow{w} Q_0$ implies weak convergence of the image measures,*

$$h(Q_n) \xrightarrow{w} h(Q_0) \tag{5}$$

(b) *Let $(\widetilde{\Xi}, \widetilde{\mathcal{B}}) = (\mathbb{R}, \mathbb{B})$ and h in addition be bounded. Then*

$$\lim_{n\to\infty} \int h\, dQ_n = \int h\, dQ_0 \tag{6}$$

PROOF Billingsley (1968; Theorems 5.1 and 5.2, pp 30, 31). ////

Prokhorov's Theorem

Weak sequential compactness is connected with tightness via Prokhorov's theorem. A topological space is *polish* if it is separable (a countable dense subset) and metrizable to become complete (Cauchy sequences converge).

Proposition A.1.2 *Let $Q_n \in \mathcal{M}_b(\mathcal{B})$ be a sequence of finite measures on the metric space Ξ with Borel σ algebra $\mathcal{B}$.*

(a) [direct half] *Boundedness and tightness of (Q_n) imply its weak sequential compactness: Every subsequence of (Q_n) has a subsequence that converges weakly.*

(b) [converse half] *If (Q_n) is weakly sequentially compact, and the space Ξ is polish, then (Q_n) is necessarily bounded and tight.*

PROOF Billingsley (1968; Theorems 6.1 and 6.2) or Billingsley (1971; Theorem 4.1). ////

Fourier Transforms

Suppose a finite-dimensional Euclidean sample space $(\Xi, \mathcal{B}) = (\mathbb{R}^k, \mathbb{B}^k)$. Then the Fourier transform, or characteristic function, $\widehat{Q}: \mathbb{R}^k \to \mathbb{C}$ of any finite measure $Q \in \mathcal{M}_b(\mathbb{B}^k)$ is defined by

$$\widehat{Q}(t) = \int \exp(it'x)\, Q(dx), \qquad t \in \mathbb{R}^k \tag{7}$$

The continuity theorem describes weak convergence by pointwise convergence of Fourier transforms.

Proposition A.1.3 *Let $Q_1, Q_2, \ldots$ be a sequence in $\mathcal{M}_b(\mathbb{B}^k)$.*

(a) *If $Q_n \xrightarrow{\mathrm{w}} Q_0$ for some $Q_0 \in \mathcal{M}_b(\mathbb{B}^k)$ then $\lim_{n\to\infty} \widehat{Q}_n(t) = \widehat{Q}_0(t)$ for all $t \in \mathbb{R}^k$; even uniformly on compacts.*

(b) *If $\varphi\colon \mathbb{R}^k \to \mathbb{C}$ is continuous at $t = 0$ and $\lim_{n\to\infty} \widehat{Q}_n(t) = \varphi(t)$ for all $t \in \mathbb{R}^k$ then $\varphi = \widehat{Q}_0$ for some $Q_0 \in \mathcal{M}_b(\mathbb{B}^k)$ and $Q_n \xrightarrow{\mathrm{w}} Q_0$.*

PROOF Billingsley (1968; Theorem 7.6, p46). ////

The Extended Real Line

In the context of log likelihoods, a modification of tightness is useful: A bounded sequence $R_n \in \mathcal{M}_b(\bar{\mathbb{B}})$ is called *tight to the right* if for every $\varepsilon \in (0, \infty)$ there is some $t \in \mathbb{R}$ such that

$$\limsup_{n\to\infty} R_n\big((t, \infty]\big) < \varepsilon \tag{8}$$

Moreover, the space $(\Xi, \mathcal{B}) = (\bar{\mathbb{R}}, \bar{\mathbb{B}})$, the extended real line equipped with its Borel σ algebra, occurs naturally. We consider $\bar{\mathbb{R}} = [-\infty, \infty]$ isometric to the closed Euclidean interval $[-1, 1]$ via the map $u \mapsto u/(1 + |u|)$. The usual arithmetic in $\bar{\mathbb{R}}$ consisting of certain conventions about addition and multiplication by $\pm\infty$, is employed [Rudin (1974; 1.22)].

Viewed through the same isometry, the real line $\mathbb{R}$ is the Euclidean interval $(-1, 1)$. Weak convergence on $(\bar{\mathbb{R}}, \bar{\mathbb{B}})$ is connected with weak convergence of the measures restricted to $(\mathbb{R}, \mathbb{B})$, the real line equipped with its Borel σ algebra; of course, the one-point measures $\mathbf{I}_n$ at $n \in \mathbb{N}$ converge weakly in $\bar{\mathbb{R}}$, but their restrictions to $\mathbb{R}$ do not.

Proposition A.1.4 *For $n \geq 0$ let $Q_n \in \mathcal{M}_b(\bar{\mathbb{B}})$ such that*

$$\lim_{n\to\infty} Q_n(\{-\infty\}) = Q_0(\{-\infty\}), \qquad \lim_{n\to\infty} Q_n(\{\infty\}) = Q_0(\{\infty\}) \tag{9}$$

Then, denoting the restrictions onto $\mathbb{B}$ by Q_n', we have

$$Q_n \xrightarrow{\mathrm{w}} Q_0 \iff Q_n' \xrightarrow{\mathrm{w}} Q_0' \tag{10}$$

PROOF Every continuous, bounded function on $\bar{\mathbb{R}}$ is continuous, bounded on $\mathbb{R}$. So we directly see from the definition of weak convergence that, under assumption (9), the RHS of (10) implies the LHS of (10). As not every continuous, bounded function on $\mathbb{R}$ has a continuous extension to $\bar{\mathbb{R}}$, the converse is not so obvious. However, we can, as for the intervals $[-1, 1]$ and $(-1, 1)$, invoke the distribution function criterion: Denoting the distribution functions evaluated at $t \in \mathbb{R}$ by

$$Q_n(t) = Q_n([-\infty, t]), \qquad Q_n'(t) = Q_n(t) - Q_n(\{-\infty\})$$

and by C_0 the set of all $t \in \mathbb{R}$ such that Q_0 is continuous at t, then

$$Q_n \xrightarrow{\mathrm{w}} Q_0 \iff \begin{cases} Q_n(\bar{\mathbb{R}}) \longrightarrow Q_0(\bar{\mathbb{R}}) \\ Q_n(t) \longrightarrow Q_0(t), \quad \forall t \in C_0 \end{cases} \tag{11}$$

$$Q_n' \xrightarrow{\mathrm{w}} Q_0' \iff \begin{cases} Q_n(\mathbb{R}) \longrightarrow Q_0(\mathbb{R}) \\ Q_n'(t) \longrightarrow Q_0'(t), \quad \forall t \in C_0 \end{cases} \tag{12}$$

From this the converse is immediate. ////

Sequences of Random Variables

The notions of tightness and weak convergence apply in particular when the measures Q_n are the image measures of $P_n \in \mathcal{M}_b(\mathcal{A}_n)$ under random variables $X_n \colon (\Omega_n, \mathcal{A}_n) \to (\Xi, \mathcal{B})$ [i.e., the laws of X_n under P_n],

$$Q_n = X_n(P_n) \tag{13}$$

In this situation we write $X_n \xrightarrow{\mathrm{w}} X_0$ for $X_n(P_n) \xrightarrow{\mathrm{w}} X_0(P_0)$.

Lemma A.1.5 *Consider three sequences of random variables $X_{m,n}$, X_m, and Y_n, with values in some separable metric sample space $(\Xi, \mathcal{B}, d)$, such that*

$$X_{m,n} \xrightarrow{\mathrm{w}} X_m \xrightarrow{\mathrm{w}} X_0 \tag{14}$$

as, first, $n \to \infty$ and then, secondly, $m \to \infty$. Moreover, for all $\varepsilon \in (0,1)$,

$$\lim_{m\to\infty} \limsup_{n\to\infty} \Pr\bigl(d(X_{m,n}, Y_n) > \varepsilon\bigr) = 0 \tag{15}$$

Then

$$Y_n \xrightarrow{\mathrm{w}} X_0 \tag{16}$$

PROOF Billingsley (1968; Theorem 4.2). Using any metric to metrize weak convergence, this is just the triangle inequality. ////

For a sequence of random variables X_n that are defined on the same probability space $(\Omega_0, \mathcal{A}_0, P_0)$, *almost sure convergence* to some X_0 means

$$\lim_{n\to\infty} X_n(\omega) = X_0(\omega) \qquad \text{a.e. } P_0(d\omega) \tag{17}$$

With possibly varying domains $(\Omega_n, \mathcal{A}_n)$, *convergence in probability* of X_n to some point $x \in \Xi$ is defined by the condition that for all $\varepsilon \in (0,1)$,

$$\lim_{n\to\infty} P_n\bigl(d(X_n, x) > \varepsilon\bigr) = 0 \tag{18}$$

If Ξ is separable, and the Ξ-valued random variables X_n again have the same probability space $(\Omega_0, \mathcal{A}_0, P_0)$ as domain, convergence of X_n in probability to X_0 is defined by

$$\lim_{n\to\infty} P_0\bigl(d(X_n, X_0) > \varepsilon\bigr) = 0 \tag{19}$$

for all $\varepsilon \in (0,1)$. Corresponding notations for stochastic convergence are:

$$X_n \xrightarrow{P_n} x\,, \quad X_n = x + \mathrm{o}_{P_n}(n^0)\,, \quad X_n \xrightarrow{P_0} X_0\,, \quad X_n = X_0 + \mathrm{o}_{P_0}(n^0)$$

Skorokhod Representation

Convergence almost surely implies convergence in probability. Convergence in probability implies weak convergence, and convergence almost surely along some subsequence [Chung (1974; Theorem 4.2.3)]. The Skorokhod representation sometimes allows a substitution of weak convergence by almost sure convergence. Let $\lambda_0 = R(0,1)$, the rectangular on $(0,1)$.

Proposition A.1.6 *Suppose random variables X_n with values in some polish sample space $(\Xi, \mathcal{B})$ that converge weakly,*

$$X_n(P_n) \longrightarrow_{\mathrm{w}} X_0(P_0)\,, \qquad \text{as } n \to \infty \tag{20}$$

Then there exist random variables Y_n on $\big([0,1], \mathbb{B} \cap [0,1], \lambda_0\big)$ such that

$$Y_n(\lambda_0) = X_n(P_n)\,, \qquad n \geq 0 \tag{21}$$

and

$$\lim_{n\to\infty} Y_n = Y_0 \qquad \text{a.e. } \lambda_0 \tag{22}$$

PROOF Billingsley (1971; Theorem 3.3). ////

Cramér–Wold Device

This is an immediate consequence of the continuity theorem, which reduces weak convergence of $\mathbb{R}^k$-valued random variables to weak convergence on $\mathbb{R}$.

Proposition A.1.7 *Let $X_0, X_1, \ldots$ be random variables with values in some finite-dimensional Euclidean sample space $(\Xi, \mathcal{B}) = (\mathbb{R}^k, \mathbb{B}^k)$.*
Then $X_n \longrightarrow_{\mathrm{w}} X_0$ iff $t'X_n \longrightarrow_{\mathrm{w}} t'X_0$ for every $t \in \mathbb{R}^k$.

Subsequence Argument

The useful subsequence argument, although proven indirectly itself, allows direct proofs otherwise, which usually give more insight.

Lemma A.1.8 *A sequence of real numbers converges to a limit iff every subsequence has a further subsequence that converges to this limit.*

The argument extends to notions of convergence that, by definition, are based on the convergence of real numbers; for example, weak convergence of measures, and contiguity.

A.2 Convergence of Integrals

As for the monotone and dominated convergence theorems, we refer to Rudin (1974; 1.26 and 1.34).

Fatou's Lemma

Formulated under the assumption of weak convergence, Fatou's lemma for nonnegative, lower semicontinuous integrands has been the basic convergence tool in Chapter 3. Again Ξ denotes a general metric space equipped with Borel σ algebra $\mathcal{B}$.

Lemma A.2.1 *If $g\colon \Xi \to [0,\infty]$ is l.s.c. and $Q_n \xrightarrow{\mathrm{w}} Q$ in $\mathcal{M}_b(\mathcal{B})$, then*

$$\int g\,dQ \le \liminf_{n\to\infty} \int g\,dQ_n \tag{1}$$

PROOF For any $\varepsilon \in (0,1)$ and any closed subset $F \subset \Xi$, the function

$$f(x) = 0 \vee \Bigl(1 - \frac{d(x,F)}{\varepsilon}\Bigr) \wedge 1 \tag{2}$$

of the distance $d(x,F)$ from $x \in \Xi$ to F is bounded (uniformly) continuous, and

$$f(x) = \begin{cases} 1 & \text{if } x \in F \\ 0 & \text{if } x \notin F^\varepsilon \end{cases} \tag{3}$$

since $x \in F^\varepsilon$ iff $d(x,F) \le \varepsilon$. It follows that

$$Q_n(F) \le \int f\,dQ_n \overset{\text{A.1(3)}}{\longrightarrow} \int f\,dQ \le Q(F^\varepsilon)$$

hence

$$\limsup_{n\to\infty} Q_n(F) \le Q(F) \tag{4}$$

as $\varepsilon \downarrow 0$ since $F^\varepsilon \downarrow F$ (closed) and $Q(F^\varepsilon) \downarrow Q(F)$. Using $Q_n(\Xi) \to Q(\Xi)$ and taking complements we see that for every open set $G \subset \Xi$,

$$\liminf_{n\to\infty} Q_n(G) \ge Q(G) \tag{5}$$

The assertion is that this is true for more general nonnegative l.s.c. functions g than just for indicator functions $\mathbf{I}_G$ of open sets $G \subset \Xi$.

It suffices to prove the result for bounded $0 \le g \le b < \infty$ since

$$\liminf_{n\to\infty} \int g\,dQ_n \ge \liminf_{n\to\infty} \int b \wedge g\,dQ_n \ge \int b \wedge g\,dQ \uparrow \int g\,dQ$$

as $b \uparrow \infty$ [monotone convergence]. By a linear transformation and the convergence of total masses we thus may assume that $0 < g < 1$. Then

$$g \le g_m = \sum_{i=1}^m \frac{i}{m}\,\mathbf{I}\Bigl(\frac{i-1}{m} < g \le \frac{i}{m}\Bigr) = \frac{1}{m}\sum_{i=1}^m \mathbf{I}\Bigl(g > \frac{i-1}{m}\Bigr) \le g + \frac{1}{m}$$

for every $m \in \mathbb{N}$, where by l.s.c. the sets $\{g > i/m\}$ are open. Thus,

$$\int g\,dQ \le \int g_m\,dQ = \frac{1}{m}\sum_{i=0}^{m-1} Q\Big(g > \frac{i}{m}\Big) \overset{(5)}{\le} \frac{1}{m}\sum_{i=0}^{m-1} \liminf_{n\to\infty} Q_n\Big(g > \frac{i}{m}\Big)$$
$$\le \liminf_{n\to\infty} \frac{1}{m}\sum_{i=0}^{m-1} Q_n\Big(g > \frac{i}{m}\Big) = \liminf_{n\to\infty} \int g_m\,dQ_n$$
$$\le \liminf_{n\to\infty} \int g\,dQ_n + \frac{1}{m}Q(\Xi)$$

The auxiliary fact that $\liminf_n a_n + \liminf_n b_n \le \liminf_n (a_n + b_n)$ for two sequences in $\mathbb{R}$ is a primitive form of Fatou's lemma. Then let $m \to \infty$. ////

The more common version of Fatou's lemma [Rudin (1974; 1.28)] states that, for random variables $X_n \ge 0$ on some measure space $(\Omega, \mathcal{A}, P)$,

$$\mathrm{E} \liminf_{n\to\infty} X_n \le \liminf_{n\to\infty} \mathrm{E}\, X_n \tag{6}$$

Employing the variables

$$Y_n = \inf_{m\ge n} X_m\,, \qquad X_n \ge Y_n \longrightarrow X = \liminf_{n\to\infty} X_n \quad \text{a.e.}$$

and using the identifications

$$Q_n = \mathcal{L}\,Y_n\,, \quad Q = \mathcal{L}\,X\,, \qquad g = \mathrm{id}_{[0,\infty]}$$

this is a consequence of the version just proved (at least, if P is finite).

Uniform Integrability

Some further convergence results in Sections 2.4 and 3.2 use uniform integrability. Let

$$X_n\colon (\Omega_n, \mathcal{A}_n, P_n) \longrightarrow (\mathbb{R}^k, \mathbb{B}^k)\,, \qquad n \ge 0$$

be random variables taking values in some finite-dimensional Euclidean space. Then $X_1, X_2, \ldots$ are called *uniformly integrable* if

$$\lim_{c\to\infty} \sup_{n\ge 1} \int_{\{|X_n|>c\}} |X_n|\,dP_n = 0 \tag{7}$$

The following proposition is known as Vitali's theorem.

Proposition A.2.2 *Assume the sequence of random variables X_n are defined on the same probability space $(\Omega_0, \mathcal{A}_0, P_0)$, such that*

$$X_n \xrightarrow{P_0} X_0 \qquad \text{as } n \to \infty \tag{8}$$

and, for some $r \in (0, \infty)$,

$$\mathrm{E}\,|X_n|^r < \infty, \qquad n \geq 1 \tag{9}$$

Then the following statements (10)–(13) *are pairwise equivalent:*

$$|X_1|^r, |X_2|^r, \ldots \text{ are uniformly integrable} \tag{10}$$

$$\lim_{n\to\infty} \mathrm{E}\,|X_n - X_0|^r = 0 \tag{11}$$

$$\lim_{n\to\infty} \mathrm{E}\,|X_n|^r = \mathrm{E}\,|X_0|^r < \infty \tag{12}$$

$$\limsup_{n\to\infty} \mathrm{E}\,|X_n|^r \leq \mathrm{E}\,|X_0|^r < \infty \tag{13}$$

If any one of the conditions (10)–(13) *is satisfied, and if* $r \geq 1$, *then*

$$\lim_{n\to\infty} \mathrm{E}\,X_n = \mathrm{E}\,X_0 \tag{14}$$

PROOF Chung (1974, Theorem 4.5.4). This result states pairwise equivalence of (10)–(12), and also holds for dimension $k \geq 1$. (13) implies (12) by Fatou's lemma. The following bound using Hölder's inequality for $r \geq 1$,

$$\big|\mathrm{E}\,X_n - \mathrm{E}\,X_0\big| \leq \mathrm{E}\,|X_n - X_0| \leq \big(\mathrm{E}\,|X_n - X_0|^r\big)^{1/r} \xrightarrow[(11)]{} 0$$

gives us (14). ////

Conditions (7)–(14), except for (8) and (11), depend only on the single laws $\mathcal{L}\,X_n$. Therefore, using the Skorokhod representation, the assumption (8) of convergence in probability can be weakened to convergence in law, if one cancels (11).

Corollary A.2.3 *Suppose* A.1(20), *and* (9) *for some* $r \in (0, \infty)$. *Then*

$$(10) \iff (12) \iff (13)$$

If any one of these conditions is satisfied, and $r \geq 1$, *then* (14) *holds.*

Scheffé's Lemma

If densities converge almost surely, then in mean.

Lemma A.2.4 *On some measure space* $(\Omega, \mathcal{A}, \mu)$, *consider a sequence of measurable functions* $q_n\colon (\Omega, \mathcal{A}) \to (\mathbb{R}, \mathbb{B})$ *such that, as* $n \to \infty$,

$$0 \leq q_n \longrightarrow q \quad \text{a.e. } \mu\,, \qquad \int q_n\,d\mu \longrightarrow \int q\,d\mu < \infty \tag{15}$$

Then

$$\int |q_n - q|\,d\mu \longrightarrow 0 \tag{16}$$

PROOF By the assumptions, we have

$$q \geq (q - q_n)^+ \longrightarrow 0 \qquad \text{a.e. } \mu \tag{17}$$

and q has a finite μ integral, which will be denoted by E in this proof. Thus, the dominated convergence theorem yields

$$\mathrm{E}\,(q - q_n)^+ \longrightarrow 0$$

But

$$\mathrm{E}\,(q - q_n)^- = \mathrm{E}\,(q - q_n)^+ - \mathrm{E}\,(q - q_n) \longrightarrow 0$$

hence also

$$\mathrm{E}\,|q_n - q| = \mathrm{E}\,(q - q_n)^+ + \mathrm{E}\,(q - q_n)^- \longrightarrow 0$$

which proves the asserted L_1 convergence. ////

A.3 Smooth Empirical Process

In Sections 1.5 and 1.6 some results on the empirical process in $\mathcal{C}[0,1]$ and $\mathcal{D}[0,1]$ have been used. Let $u_i \sim F_0$ be i.i.d. random variables with rectangular distribution function $F_0 = \mathrm{id}_{[0,1]}$ on the unit interval. The corresponding order statistics at sample size n are denoted by

$$u_{n:1} \leq \cdots \leq u_{n:n}$$

and

$$\hat{F}_n(u_1, \ldots, u_n; s) = \hat{F}_{u,n}(s) = \frac{1}{n} \sum_{i=1}^{n} \mathbf{I}(u_i \leq s) \tag{1}$$

is the rectangular empirical distribution function $\hat{F}_n \colon [0,1]^n \to \mathcal{D}[0,1]$.

Donsker's Theorem

The smoothed rectangular empirical distribution function $\bar{F}_{u,n}$ is given by

$$(n+1)\bar{F}_{u,n}(s) = \begin{cases} 0 & \text{if } s = 0 \\ i & \text{if } s = u_{n:i}\,, \quad i = 1, \ldots, n \\ n+1 & \text{if } s = 1 \\ \text{linear} & \text{in between} \end{cases} \tag{2}$$

where the underlying observations $u_1, \ldots, u_n$ may indeed be considered pairwise distinct with F_0^n probability 1.

Proposition A.3.1 *For i.i.d. random variables u_i with rectangular distribution function F_0 on $[0,1]$, the smoothed rectangular empirical process defined via (2) converges weakly in $\mathcal{C}[0,1]$,*

$$\sqrt{n}\,(\bar{F}_{u,n} - F_0)(F_0^n) \xrightarrow{\ \mathrm{w}\ } B \tag{3}$$

to the Brownian Bridge B on $\mathcal{C}[0,1]$.

PROOF Billingsley (1968; Theorem 13.1, p 105). ////

Now let F be an arbitrary distribution function on $\mathbb{R}$. The left continuous pseudoinverse $F^{-1}\colon [0,1] \to \bar{\mathbb{R}}$ defined by

$$F^{-1}(s) = \inf\bigl\{x \in \mathbb{R} \bigm| F(x) \geq s\bigr\} \tag{4}$$

satisfies

$$F^{-1}(s) \leq x \iff s \leq F(x) \tag{5}$$

Therefore, starting with i.i.d. rectangular variables $u_i \sim F_0$, the inverse probability transformation produces an i.i.d. sequence $x_i \sim F$,

$$x_i = F^{-1}(u_i) \tag{6}$$

As F^{-1} is monotone, the corresponding order statistics are related by

$$x_{n:i} = F^{-1}(u_{n:i})$$

The empirical distribution function $\hat{F}_{x,n}\colon \mathbb{R}^n \to \mathcal{D}(\mathbb{R})$ based on $x_1, \ldots, x_n$, because of (5), is linked to the rectangular empirical via

$$\begin{aligned} \hat{F}_{x,n}(y) = \hat{F}_n(x_1, \ldots, x_n; y) &= \frac{1}{n} \sum_{i=1}^{n} \mathbf{I}(x_i \leq y) \\ &= \hat{F}_n(u_1, \ldots, u_n; F(y)) = \hat{F}_{u,n}(F(y)) \end{aligned} \tag{7}$$

This relation cannot so easily be achieved for the smoothed versions.

Randomization

Therefore, we prefer the following construction that defines the rectangulars u_i by means of the x_i and some randomization, subject to (6) a.e..

Consider i.i.d. observations $x_i \sim F$. In addition, let $v_i \sim F_0$ be an i.i.d. sequence of rectangular variables stochastically independent of the sequence (x_i). Randomizing over the jump heights, define

$$u_i = F(x_i - 0) + v_i \bigl(F(x_i) - F(x_i - 0)\bigr) \tag{8}$$

Then we have for all $s \in (0,1)$ and $t = F^{-1}(s)$,

$$\begin{aligned} \Pr(u_i < s) &= \Pr\bigl(F(x_i) < s\bigr) + \Pr\bigl(F(x_i - 0) < s \leq F(x_i),\ u_i < s\bigr) \\ &= \Pr(x_i < t) + \Pr\bigl(x_i = t,\ v_i F(\{t\}) < s - F(t-0)\bigr) \\ &= F(t-0) + F(\{t\}) \frac{s - F(t-0)}{F(\{t\})} = s \end{aligned} \tag{9}$$

So in fact $u_i \sim F_0$. Since $F(x_i) \geq u_i$ by construction, it follows that

$$x_i \geq F^{-1}(u_i) \sim F$$

Hence, for u_i constructed from $x_i \sim F$ via (8), equality (6) must hold except on an event of probability 0, which may in the sequel be neglected.

Using construction (8) the randomized smoothed empirical distribution function $\bar{F}_{x,n}$ based on $x_1, \ldots, x_n$ is, by definition,

$$\bar{F}_{x,n}(y) = \bar{F}_n(x_1, \ldots, x_n; y) = \bar{F}_n(u_1, \ldots, u_n; F(y)) = \bar{F}_{u,n}(F(y)) \qquad (10)$$

where the randomization variables $v_1, \ldots, v_n$ are suppressed notationally.

Extending the spaces $\mathcal{C}[0,1]$ and $\mathcal{D}[0,1]$, let $\mathcal{C}(\mathbb{R})$ and $\mathcal{D}(\mathbb{R})$ be the spaces of bounded functions from $\mathbb{R}$ to $\mathbb{R}$ that are continuous or, respectively, right continuous with left limits, both equipped with the sup norm $\|.\|$. Then the $\mathcal{D}$ spaces become nonseparable and, relative to these spaces, the empirical $\hat{F}_n$ non-Borel measurable. Equipped with the Skorokhod topology, however, the $\mathcal{D}$ spaces would not be topological vector spaces [Billingsley (1968; §18, p 150; Problem 3, p 123)].

Corollary A.3.2 *For i.i.d. random variables x_i with arbitrary distribution function F on $\mathbb{R}$, the randomized smoothed empirical process defined by (10) converges weakly in $\mathcal{D}(\mathbb{R})$,*

$$\sqrt{n}\,(\bar{F}_{x,n} - F)(F^n) \xrightarrow{\mathrm{w}} B \circ F \qquad (11)$$

to the Brownian Bridge on $\mathcal{C}[0,1]$ composed with F. If F is continuous, the convergence takes place in $\mathcal{C}(\mathbb{R})$.

PROOF The composition with F is a continuous linear transformation from $\mathcal{C}[0,1]$ to $\mathcal{D}(\mathbb{R})$, mapping process (2) into process (10). By the continuous mapping theorem [Proposition A.1.1], the result is a consequence of Proposition A.3.1. ////

Sup Norm Metrics

Weak convergence in the $\mathcal{D}$ spaces does not automatically supply the compacts required for compact differentiability, as the converse half of Prokhorov's theorem [Proposition A.1.2 b] is not available for nonseparable spaces; because of this, smoothing and the $\mathcal{C}$ spaces are required for compact differentiability. For bounded differentiability, the following boundedness of the empirical process in sup norm suffices; actually, more general sup norms have been used in Section 1.6.

Let $\mathcal{Q}'$ denote the set of all measurable functions $q\colon [0,1] \to [0,\infty]$ such that

$$\int_0^1 \frac{d\lambda}{q^2} < \infty \qquad (12)$$

and which for some $0 < \varepsilon < 1/2$ increase on $(0,\varepsilon)$, are bounded away from 0 on $(\varepsilon, 1-\varepsilon)$, and decrease on $(1-\varepsilon, 1)$. Define

$$\mathcal{Q} = \bigl\{\, q \in \mathcal{D}[0,1] \bigm| \exists\, q' \in \mathcal{Q}'\colon q' \le q \,\bigr\} \qquad (13)$$

If $0 < \delta \leq 1/2$, then

$$q(s) = \big(s(1-s)\big)^{1/2-\delta} \tag{14}$$

is such a function $q \in \mathcal{Q}$; in particular, the constant $q = 1$ is in $\mathcal{Q}$.

Given $q \in \mathcal{Q}$ define

$$\mathcal{D}_q[0,1] = \big\{\, f \in \mathcal{D}[0,1] \mid \|f\|_q < \infty \,\big\}, \qquad \|f\|_q = \sup\left|\frac{f}{q}\right| \tag{15}$$

The space $\mathcal{D}_q[0,1]$ is still nonseparable and the empirical distribution function non-Borel measurable. This deficiency to deal with the empirical process in its original form and space, has given rise to several modified definitions of weak convergence for nonmeasurable maps in the literature.

Definition 2.1 of Pyke and Shorack (1968) may be formulated this way: A sequence of functions X_n with values in some metric space Ξ converges weakly to some function X if, for all continuous real-valued functions f on Ξ that make the subsequent compositions measurable, the laws converge weakly in the classical sense,

$$f \circ X_n \xrightarrow{\mathrm{w}} f \circ X \tag{16}$$

The definition obviously coincides with the classical definition in case X_n and X are measurable. A continuous mapping theorem is immediate.

It might seem surprising that nonmeasurability should add to weak convergence, but in specific examples one may hope that at least the interesting functions f make measurability happen. In the case of the empirical process, the sup norm $\|\,.\,\|_q$ is such a function, because it can be evaluated over a countable dense subset of $[0,1]$, thus being the supremum over a countable number of projections, which each are Borel measurable.

Proposition A.3.3 *For $q \in \mathcal{Q}$ the rectangular empirical process converges weakly in $\mathcal{D}_q[0,1]$,*

$$\sqrt{n}\,(\hat{F}_{u,n} - F_0)(F_0^n) \xrightarrow{\mathrm{w}} B \tag{17}$$

to the Brownian Bridge B on $\mathcal{C}[0,1]$; in particular, the sequence of its sup norms $\sqrt{n}\,\|\hat{F}_{u,n} - F_0\|_q(F_0^n)$ is tight on $\mathbb{R}$.

PROOF Pyke and Shorack (1968; Theorem 2.1), and (16) for $f = \|\,.\,\|_q$. ////

Remark A.3.4 O'Reilly (1974; Theorem 2) was able to replace assumption (12) by a somewhat weaker integrability condition that is both necessary and sufficient for (17). ////

For general $F \in \mathcal{M}_1(\mathbb{B})$ we divide by the composition $q \circ F$ of $q \in \mathcal{Q}$ with the distribution function F on $\mathbb{R}$ to obtain the sup norm $\|\,.\,\|_{qF}$. The empirical process is bounded in these norms, uniformly in F.

Proposition A.3.5 *For* $q \in \mathcal{Q}$ *defined by* (12)–(13),

$$\lim_{M\to\infty} \limsup_{n\to\infty} \sup\bigl\{ F^n\bigl(\sqrt{n}\,\|\hat{F}_n - F\|_{qF} > M\bigr) \bigm| F \in \mathcal{M}_1(\mathbb{B}) \bigr\} = 0 \tag{18}$$

PROOF First assume $F = F_0$ and let $\varepsilon \in (0,1)$. Then, for q constant 1, Proposition A.3.1 gives us a compact subset K of $\mathcal{C}[0,1]$ such that

$$\liminf_{n\to\infty} F_0^n\bigl(\sqrt{n}\,(\bar{F}_n - F_0) \in K\bigr) \ge 1 - \varepsilon \tag{19}$$

Invoking the finite norm bound of this compact, choose

$$M = 1 + \sup_{f\in K} \|f\|$$

Then

$$\limsup_{n\to\infty} F_0^n\bigl(\sqrt{n}\,\|\bar{F}_n - F_0\| > M - 1\bigr) \le \varepsilon$$

and as $\|\hat{F}_n - \bar{F}_n\| \le 1/n$,

$$\limsup_{n\to\infty} F_0^n\bigl(\sqrt{n}\,\|\hat{F}_n - F_0\| > M\bigr) \le \varepsilon \tag{20}$$

For general $q \in \mathcal{Q}$ we conclude this directly from Proposition A.3.3. If also the distribution function F is arbitrary, use representation (6). Then,

$$\begin{aligned} &F^n\bigl(\|\hat{F}_n - F\|_{qF} > M\bigr) \\ &= F_0^n\left(\sup_{x\in\mathbb{R}} \frac{1}{q(F(x))} \left| \frac{1}{n}\sum_{i=1}^n \mathbf{I}(F^{-1}(u_i) \le x) - F(x) \right| > M \right) \\ &= F_0^n\left(\sup_{x\in\mathbb{R}} \frac{1}{q(F(x))} \left| \frac{1}{n}\sum_{i=1}^n \mathbf{I}(u_i \le F(x)) - F(x) \right| > M \right) \\ &\le F_0^n\left(\sup_{s\in[0,1]} \frac{1}{q(s)} \left| \frac{1}{n}\sum_{i=1}^n \mathbf{I}(u_i \le s) - s \right| > M \right) \le \varepsilon \end{aligned} \tag{21}$$

because $F(\mathbb{R}) \subset [0,1]$; a one-point measure being the most favorable case. The least favorable case occurs if F is continuous (e.g., if $F = F_0$). ////

Independent, Non-I.I.D. Multivariate Case

The basic result for q constant 1 has an extension to independent non-i.i.d. and m variate observations, which is used in Subsection 6.3.2 to prove locally uniform $\sqrt{n}$ consistency of the Kolmogorov MD estimate: Assume a finite-dimensional Euclidean sample space $(\mathbb{R}^m, \mathbb{B}^m)$. Given an array of probabilities $P_{n,1}, \ldots, P_{n,n} \in \mathcal{M}_1(\mathbb{B}^m)$ let

$$\hat{P}_n = \frac{1}{n}\sum_{i=1}^n \mathbf{I}_{x_i}\,, \qquad \bar{P}_n = \frac{1}{n}\sum_{i=1}^n P_{n,i}\,, \qquad P_n^{(n)} = \bigotimes_{i=1}^n P_{n,i} \tag{22}$$

denote the empirical measure at sample size n, the average and the product measures, respectively, and Kolmogorov distance by d_κ.

Proposition A.3.6 *It holds that*

$$\lim_{M\to\infty} \sup_{n\geq 1} \sup\big\{ P_n^{(n)}\big(\sqrt{n}\, d_\kappa(\hat{P}_n, \bar{P}_n) > M\big) \bigm| P_{n,i} \in \mathcal{M}_1(\mathbb{B}^m) \big\} = 0 \tag{23}$$

PROOF This follows from Bretagnolle's (1980) extension to the independent non-i.i.d. and m variate case of the Dvoretzky–Kiefer–Wolfowitz exponential bound, and from LeCam (1982) who, by a different technique, proved that

$$P_n^{(n)}\Big(\sqrt{n}\, d_\kappa(\hat{P}_n, \bar{P}_n) > M + \frac{1}{\sqrt{n}}\Big) \leq 2^{m+6} \exp\Big(-\frac{M^2}{48}\Big) \tag{24}$$

for all arrays $P_{n,i} \in \mathcal{M}_1(\mathbb{B}^m)$, all $n \geq 1$, and all $M \in (0, \infty)$. ////

A.4 Square Integrable Empirical Process

Let $(\mathbb{R}^m, \mathbb{B}^m)$ be some finite-dimensional Euclidean sample space. Given an array of probabilities $Q_{n,i} \in \mathcal{M}_1(\mathbb{B}^m)$, the empirical process based on stochastically independent observations $x_i \sim Q_{n,i}$ at sample size n is

$$Y_n(x_1, \dots, x_n; y) = \frac{1}{\sqrt{n}} \sum_{i=1}^n \big(\mathbf{I}(x_i \leq y) - Q_{n,i}(y)\big) = \sqrt{n}\,(\hat{P}_n - \bar{Q}_n) \tag{1}$$

where $\hat{P}_n$ denotes the empirical distribution function,

$$Q_{n,i}(y) = Q_{n,i}\big(\{\, x \in \mathbb{R}^m \mid x \leq y \,\}\big), \qquad \bar{Q}_n = \frac{1}{n} \sum_{i=1}^n Q_{n,i}$$

and $\leq$ is meant coordinatewise.

Let $\mu \in \mathcal{M}_\sigma(\mathbb{B}^m)$ be some σ finite weight. We shall study weak convergence of the empirical process Y_n in the Hilbert space $L_2(\mu)$. This, on the one hand, is a large function space but, on the other hand, admits few continuous functions. The space $L_2(\mu)$ is separable [Lemma C.2.5], hence polish. Given any ONB (e_α), which must necessarily be countable, define the sequence of Fourier coefficients

$$\pi_\alpha(z) = \langle z | e_\alpha \rangle = \int z(y) e_\alpha(y)\, \mu(dy) \tag{2}$$

Using Bessel's equality,

$$\|z\|^2 = \sum_{\alpha=1}^\infty \langle z | e_\alpha \rangle^2 \tag{3}$$

we can show that the set algebra $\mathcal{A}$ induced by (π_α),

$$\mathcal{A} = \big\{\, \pi_A^{-1}(B) \bigm| A \subset \mathbb{N} \text{ finite}, B \in \mathbb{B}^A \,\big\} \tag{4}$$

where $\pi_A = (\pi_\alpha)_{\alpha \in A}$, generates the Borel σ algebra $\mathcal{B}$ on $L_2(\mu)$. Thus $\mathcal{A}$ is measure determining: Any two finite measures P and Q on $\mathcal{B}$ such that $\pi_A(Q) = \pi_A(P)$ for all finite subsets $A \subset \mathbb{N}$ agree on $\mathcal{B}$. But $\mathcal{A}$ is not convergence determining: For example, as $e_\alpha \not\to 0$ the Dirac measures $P_\alpha = \mathbf{I}_{e_\alpha}$ do not tend weakly to $P_0 = \mathbf{I}_0$ although $\pi_A(P_\alpha) = \mathbf{I}_0 = \pi_A(P_0)$ holds on $\mathbb{B}^A$ for every finite subset $A \subset \mathbb{N}$, as soon as $\alpha > \max A$.

Tightness

In the light of Prokhorov's theorem [Proposition A.1.2] tightness is essential for weak convergence in $L_2(\mu)$. The following criterion employs the truncated norms

$$\|z\|_\nu^2 = \sum_{\alpha > \nu} \langle z | e_\alpha \rangle^2 \tag{5}$$

which are defined for all $\nu \in \mathbb{N}$, given the ONB (e_α).

Proposition A.4.1 *Let $\Pi \subset \mathcal{M}_b(\mathcal{B})$ be a subset, and (Q_n) a sequence, of finite measures on $(L_2(\mu), \mathcal{B})$.*

(a) *The set Π is tight iff*

$$\lim_{M \to \infty} \sup_{Q \in \Pi} Q\big(\|z\| > M\big) = 0 \tag{6}$$

and for all $\varepsilon \in (0,1)$,

$$\lim_{\nu \to \infty} \sup_{Q \in \Pi} Q\big(\|z\|_\nu > \varepsilon\big) = 0 \tag{7}$$

(b) *The sequence (Q_n) is tight iff*

$$\lim_{M \to \infty} \limsup_{n \to \infty} Q_n\big(\|z\| > M\big) = 0 \tag{8}$$

and for all $\varepsilon \in (0,1)$,

$$\lim_{\nu \to \infty} \limsup_{n \to \infty} Q_n\big(\|z\|_\nu > \varepsilon\big) = 0 \tag{9}$$

Remark A.4.2 The tightness condition on the norm cannot be dropped [Prokhorov (1956; Theorem 1.13), Parthasaraty (1967; VI Theorem 2.2)]. For example, the sequence of Dirac probabilities $Q_n = \mathbf{I}_{ne_1}$ satisfies condition (7), without being tight. ////

PROOF Assume Π tight. That is, for every $\delta \in (0,1)$ there exists a compact subset K of $L_2(\mu)$ such that

$$\sup_{Q \in \Pi} Q\big(L_2(\mu) \setminus K\big) < \delta \tag{10}$$

The norm being continuous, the family of image measures $\|\,.\,\|(Q)$ is necessarily tight, hence fulfill (6). Given $\varepsilon, \delta \in (0,1)$ choose the compact K according to (10) and consider the cover by open balls of radius $\varepsilon/2$.

By compactness there are finitely many $z_1, \ldots, z_r \in K$ such that for every $z \in K$ there is some $j = 1, \ldots, r$ such that $\|z - z_j\| < \varepsilon/2$ hence $\|z - z_j\|_\nu < \varepsilon/2$ for all $\nu \geq 1$. We can choose $\nu_0 \in \mathbb{N}$ so that $\|z_j\|_\nu < \varepsilon/2$ for all $j = 1, \ldots, r$ and $\nu \geq \nu_0$; thus $\|z\|_\nu \leq \|z - z_j\|_\nu + \|z_j\|_\nu < \varepsilon$ holds for all $z \in K$. It follows that

$$\sup_{Q \in \Pi} Q\bigl(\|z\|_\nu > \varepsilon\bigr) \leq \sup_{Q \in \Pi} Q\bigl(L_2(\mu) \setminus K\bigr) < \delta \tag{11}$$

for all $\nu \geq \nu_0$, which proves (7).

Conversely, given any $\delta \in (0,1)$, there exists some $M \in (0, \infty)$ by (6), and for each $j \in \mathbb{N}$ we can choose some $\nu_j \in \mathbb{N}$ by (7), such that

$$\sup_{Q \in \Pi} Q\bigl(\|z\| > M\bigr) < \frac{\delta}{2} \tag{12}$$

respectively,

$$\sup_{Q \in \Pi} Q\Bigl(\, \|z\|_{\nu_j} > \frac{1}{j}\Bigr) < \frac{\delta}{2^{j+1}} \tag{13}$$

Then the closed set

$$K = \bigl\{\|z\| \leq M\bigr\} \cap \bigcap_{j=1}^{\infty} \bigl\{\, \|z\|_{\nu_j} \leq 1/j\,\bigr\} \tag{14}$$

satisfies (10) due to (12) and (13). As $L_2(\mu)$ is complete, it remains to show that K is totally bounded [Billingsley (1968; Appendix I, p 217)]: Given any $\varepsilon \in (0,1)$ choose some $j \in \mathbb{N}$ such that $j^2 > 2/\varepsilon^2$ and then any finite $\varepsilon/\sqrt{2\nu_j}$ net Γ of the compact $[-M, M\,] \subset \mathbb{R}$. Then the set of all elements

$$w = \gamma_1 e_1 + \cdots + \gamma_{\nu_j} e_{\nu_j}\,, \qquad \gamma_1, \ldots, \gamma_{\nu_j} \in \Gamma$$

turns out to be a finite ε net for K. Indeed, for every $z \in K$ we have

$$\|z - w\|^2 = \sum_{\alpha=1}^{\nu_j} \bigl|\langle z|e_\alpha\rangle - \gamma_\alpha\bigr|^2 + \|z\|_{\nu_j}^2 \tag{15}$$

where $\|z\|_{\nu_j}^2 \leq 1/j^2 < \varepsilon^2/2$ and the first sum, since $\bigl|\langle z|e_\alpha\rangle\bigr| \leq \|z\| \leq M$, can by suitable choice of γ_α be made smaller than $\nu_j \varepsilon^2/(2\nu_j) = \varepsilon^2/2$.

This completes the proof of (a). Obviously, (b) is implied by (a). ////

Some Bounds

To justify Fubini's theorem in subsequent arguments, some bounds are needed. Consider any two distribution functions $P, Q \in \mathcal{M}_1(\mathbb{B}^m)$ such that

$$\int Q(1-Q)\, d\mu + \int |Q - P|^2\, d\mu < \infty \tag{16}$$

Then, for every $e \in L_2(\mu)$, the following bound holds by Cauchy–Schwarz,

$$\begin{aligned}
&\iint \big|\mathbf{I}(x \le y) - P(y)\big|\big|e(y)\big|\, Q(dx)\, \mu(dy) \\
&\le \iint \big|\mathbf{I}(x \le y) - Q(y)\big|\big|e(y)\big|\, Q(dx)\, \mu(dy) + \int |Q - P||e|\, d\mu \qquad (17)\\
&\le \langle \sqrt{Q(1-Q)}\, |e| \rangle + \langle |Q-P||e| \rangle \le \|e\| \,\big(\big\|\sqrt{Q(1-Q)}\,\big\| + \|Q-P\| \big)
\end{aligned}$$

The upper bound is finite by (16). In particular, with $Q = P$, this bound applies to every distribution function P such that

$$\int P(1-P)\, d\mu < \infty \tag{18}$$

And given such a P, another distribution function Q fulfills (16) if

$$\int |Q - P|\, d\mu + \int |Q - P|^2\, d\mu < \infty \tag{19}$$

Furthermore, for $Q = P \in \mathcal{M}_1(\mathbb{B}^m)$ satisfying (18) and $e \in L_2(\mu)$ we have

$$\begin{aligned}
&\iiint \big|\mathbf{I}(x \le y) - Q(y)\big|\big|\mathbf{I}(x \le z) - Q(z)\big|\big|e(y)\big|\big|e(z)\big|\, Q(dx)\, \mu(dy)\, \mu(dz) \\
&\le \iint \sqrt{Q(y)(1-Q(y))}\, \sqrt{Q(z)(1-Q(z))}\, |e(y)||e(z)|\, \mu(dy)\, \mu(dz) \\
&= \Big| \int |e| \sqrt{Q(1-Q)}\, d\mu \Big|^2 \le \|e\|^2 \int Q(1-Q)\, d\mu < \infty \qquad (20)
\end{aligned}$$

For $Q \in \mathcal{M}_1(\mathbb{B}^m)$ such that $\int Q(1-Q)\, d\mu < \infty$ and $e \in L_2(\mu)$ we define

$$V(Q,e) = \iint \big(Q(y \wedge z) - Q(y)Q(z)\big) e(y) e(z)\, \mu(dy)\, \mu(dz) \tag{21}$$

taking the minimum coordinatewise. Then, by bound (20) and Fubini,

$$0 \le V(Q,e) = \int Z^2\, dQ \le \|e\|^2 \int Q(1-Q)\, d\mu \tag{22}$$

with

$$Z = \int \big(\mathbf{I}(x \le y) - Q(y)\big) e(y)\, \mu(dy) \tag{23}$$

The function Z, by bound (17) and Fubini's theorem, is a random variable of expectation 0 under Q.

The Empirical Process

This is now applied to the empirical process Y_n under the joint distribution $Q_n^{(n)} = Q_{n,1} \otimes \cdots \otimes Q_{n,n}$ of the independent observations $x_i \sim Q_{n,i}$.

Lemma A.4.3 *Given $Q_{n,1}, \ldots, Q_{n,n} \in \mathcal{M}_1(\mathbb{B}^m)$ such that*

$$\max_{i=1,\ldots,n} \int Q_{n,i}(1 - Q_{n,i})\, d\mu < \infty \tag{24}$$

there exists a Borel set $D_n \in \mathbb{B}^{mn}$ of measure $Q_n^{(n)}(D_n) = 1$ such that $Y_n \colon D_n \to L_2(\mu)$ is Borel measurable. Moreover, for every $e \in L_2(\mu)$,

$$\int \langle Y_n | e \rangle^2\, dQ_n^{(n)} = \frac{1}{n} \sum_{i=1}^n V(Q_{n,i}, e) \tag{25}$$

and

$$\int \|Y_n\|^2\, dQ_n^{(n)} = \frac{1}{n} \sum_{i=1}^n \int Q_{n,i}(1 - Q_{n,i})\, d\mu \tag{26}$$

PROOF Introduce the following functions, which are product measurable,

$$Y_{n,i} = \mathbf{I}(x_i \le y) - Q_{n,i}(y) \tag{27}$$

Fubini's theorem for $Y_{n,i}^2 \ge 0$ ensures that $\int Y_{n,i}^2\, d\mu$ is measurable and

$$\int \|Y_{n,i}\|^2\, dQ_{n,i} = \iint Y_{n,i}^2\, d\mu\, dQ_{n,i} = \int Q_{n,i}(1 - Q_{n,i})\, d\mu \tag{28}$$

Putting $D_{n,i} = \bigl\{ \|Y_{n,i}\| < \infty \bigr\}$ it follows from (24) that $Q_{n,i}(D_{n,i}) = 1$.

Let $e \in L_2(\mu)$. Then, restricted onto $D_{n,i}$, the scalar product

$$Z_{n,i} = \langle Y_{n,i} | e \rangle = \int \bigl(\mathbf{I}(x_i \le y) - Q_{n,i}(y) \bigr) e(y)\, \mu(dy) \tag{29}$$

is finite by Cauchy–Schwarz, and measurable by Fubini.

Set $D_n = D_{n,1} \times \cdots \times D_{n,n}$. Then $Q_n^{(n)}(D_n) = 1$. On D_n, by the triangle inequality, $\sqrt{n}\, \|Y_n\| \le \sum_i \|Y_{n,i}\| < \infty$, hence $Y_n(D_n) \subset L_2(\mu)$. On D_n, moreover,

$$\langle Y_n | e \rangle = \frac{1}{\sqrt{n}} \sum\nolimits_i Z_{n,i} \tag{30}$$

hence $Y_n \colon D_n \to L_2(\mu)$ is Borel measurable.

The variables $Z_{n,i}$ are stochastically independent and of expectation zero [Fubini]. Thus (25) follows from (22) and the Bienaymé equality.

By the Cauchy–Schwarz inequality, we have

$$\begin{aligned} \iint |Y_{n,i}||Y_{n,j}|\, dQ_n^{(n)}\, d\mu &\le \int \sqrt{Q_{n,i}(1 - Q_{n,i})}\, \sqrt{Q_{n,j}(1 - Q_{n,j})}\, d\mu \\ &\le \Bigl| \int Q_{n,i}(1 - Q_{n,i})\, d\mu \Bigr|^{1/2} \Bigl| \int Q_{n,j}(1 - Q_{n,j})\, d\mu \Bigr|^{1/2} \underset{(24)}{<} \infty \end{aligned} \tag{31}$$

Therefore, Fubini's theorem is also enforced in

$$\int \|Y_n\|^2 \, dQ_n^{(n)} = \frac{1}{n} \sum_{i,j=1}^{n} \iint Y_{n,i} Y_{n,j} \, dQ_n^{(n)} \, d\mu \tag{32}$$

Thus (26) follows since, for fixed y, the binomial variables $Y_{n,i}$ are stochastically independent and of expectations zero. ////

Theorem A.4.4 *Given an array* $Q_{n,i} \in \mathcal{M}_1(\mathbb{B}^m)$ *and* $P \in \mathcal{M}_1(\mathbb{B}^m)$.

(a) [norm boundedness] *Then, under the condition that*

$$\limsup_{n\to\infty} \frac{1}{n} \sum_{i=1}^{n} \int Q_{n,i}(1 - Q_{n,i}) \, d\mu < \infty \tag{33}$$

the sequence $\|Y_n\|(Q_n^{(n)})$ *is tight on* $\mathbb{R}$.

(b) [weak convergence of $\langle Y_n|e\rangle$] *Suppose* (18), (33), *and*

$$\lim_{n\to\infty} \frac{1}{n} \sum_{i=1}^{n} V(Q_{n,i}, e) = V(P, e) \tag{34}$$

for all $e \in L_2(\mu)$. *Then, for all* $e \in L_2(\mu)$,

$$\langle Y_n|e\rangle \, (Q_n^{(n)}) \xrightarrow{\mathrm{w}} \mathcal{N}(0, V(P, e)) \tag{35}$$

(c) [weak convergence of Y_n] *Assume* (18), (34), *and*

$$\lim_{n\to\infty} \frac{1}{n} \sum_{i=1}^{n} \int Q_{n,i}(1 - Q_{n,i}) \, d\mu = \int P(1 - P) \, d\mu \tag{36}$$

Then

$$Y_n(Q_n^{(n)}) \xrightarrow{\mathrm{w}} Y \tag{37}$$

where Y *is the Gaussian process on* $L_2(\mu)$ *such that for all* $e \in L_2(\mu)$,

$$\mathcal{L}\,\langle Y|e\rangle = \mathcal{N}(0, V(P, e)) \tag{38}$$

PROOF

(a) By the Chebyshev inequality, for every $M \in (0, \infty)$,

$$Q_n^{(n)}(\|Y_n\| > M) \overset{(26)}{\leq} \frac{1}{M^2 n} \sum\nolimits_i \int Q_{n,i}(1 - Q_{n,i}) \, d\mu \tag{39}$$

Thus tightness follows by assumption (33) if we let $M \to \infty$.

(b) For any $e \in L_2(\mu)$, Lemma C.2.6 supplies a sequence $e_n \in L_2(\mu)$ such that

$$\|e_n - e\| \longrightarrow 0, \qquad \mu(e_n \neq 0) \sup |e_n| = \mathrm{o}(\sqrt{n}\,) \tag{40}$$

Then we have $\sqrt{n}\,\langle Y_n|e_n\rangle = \sum_i \tilde{Z}_{n,i}$ with the following random variables

$$\tilde{Z}_{n,i} = \int \bigl(\mathbf{I}(x_i \le y) - Q_{n,i}(y)\bigr) e_n(y)\,\mu(dy) \tag{41}$$

These are bounded by $\max_i |\tilde{Z}_{n,i}| \le \mu(e_n \neq 0) \sup |e_n| = \mathrm{o}(\sqrt{n}\,)$.

As for the second moments, we have

$$\begin{aligned} &\left| \left| \frac{1}{n} \sum\nolimits_i \int \tilde{Z}_{n,i}^2 \, dQ_{n,i} \right|^{1/2} - \left| \frac{1}{n} \sum\nolimits_i \int Z_{n,i}^2 \, dQ_{n,i} \right|^{1/2} \right| \\ &\le \left| \frac{1}{n} \sum_{i=1}^n \int \bigl(\tilde{Z}_{n,i} - Z_{n,i}\bigr)^2 \, dQ_{n,i} \right|^{1/2} \\ &\le \left| \frac{1}{n} \sum_{i=1}^n \iint \bigl(\mathbf{I}(x_i \le y) - Q_{n,i}(y)\bigr)^2 \mu(dy)\, \|e_n - e\|^2 \, Q_{n,i}(dx_i) \right|^{1/2} \\ &= \|e_n - e\| \left| \frac{1}{n} \sum_{i=1}^n \int Q_{n,i}(1 - Q_{n,i}) \, d\mu \right|^{1/2} \xrightarrow[(40)]{(33)} 0 \end{aligned} \tag{42}$$

Thus, in view of (22), (34) implies that

$$\lim_{n\to\infty} \frac{1}{n} \sum\nolimits_i \int \tilde{Z}_{n,i}^2 \, dQ_{n,i} = V(P,e) \tag{43}$$

If $V(P,e) > 0$, the Lindeberg–Feller theorem [Proposition 6.2.1 c, with ψ_n substituted by $\tilde{Z}_{n,i}$] yields that

$$\langle Y_n|e_n\rangle(Q_n^{(n)}) \xrightarrow{\mathrm{w}} \mathcal{N}(0, V(P,e)) \tag{44}$$

But by assumption (33), according to (a), the sequence $\|Y_n\|(Q_n^{(n)})$ is tight, hence by (40),

$$\bigl|\langle Y_n|e_n - e\rangle\bigr| \le \|Y_n\|\,\|e_n - e\| \xrightarrow{Q_n^{(n)}} 0 \tag{45}$$

Thus (35) follows. If $V(P,e) = 0$ then for all $\varepsilon \in (0,1)$ by Chebyshev,

$$Q_n^{(n)}\bigl(\bigl|\langle Y_n|e\rangle\bigr| > \varepsilon\bigr) \le \frac{1}{\varepsilon^2 n} \sum\nolimits_i \int Z_{n,i}^2 \, dQ_{n,i} \xrightarrow[(22)]{(34)} 0 \tag{46}$$

so that (35) holds in this case as well.

(c) Conditions (36) and (18) imply (33), hence by (a) ensure the norm boundedness required for (8). To verify condition (9), we introduce the auxiliary process

$$Y_0(x;y) = \mathbf{I}(x \le y) - P(y) \tag{47}$$

which, under (18), is well defined on the probability space $(\mathbb{R}^m, \mathbb{B}^m, P)$ with values in $(L_2(\mu), \mathcal{B})$. Then

$$\int \|Y_0\|^2 \, dP = \int P(1-P) \, d\mu \tag{48}$$

and

$$\int \langle Y_0 | e_\alpha \rangle^2 \, dP = V(P, e_\alpha) \tag{49}$$

for all $\alpha \geq 1$. In view of (25) and (26), assumptions (34) and (36) now read

$$\lim_{n\to\infty} \int \|Y_n\|^2 \, dQ_n^{(n)} = \int \|Y_0\|^2 \, dP \tag{50}$$

and

$$\lim_{n\to\infty} \int \langle Y_n | e_\alpha \rangle^2 \, dQ_n^{(n)} = \int \langle Y_0 | e_\alpha \rangle^2 \, dP \tag{51}$$

for all $\alpha \geq 1$. Subtracting the terms number $\alpha = 1, \ldots, \nu$, it follows that

$$\lim_{n\to\infty} \int \|Y_n\|_\nu^2 \, dQ_n^{(n)} = \int \|Y_0\|_\nu^2 \, dP \tag{52}$$

for all $\nu \geq 1$. However, $\|Y_0\| < \infty$ a.e. P by assumption (18) and (48). Therefore, $\|Y_0\|_\nu \to 0$ a.e. P as $\nu \to \infty$. At the same time, $\|Y_0\|_\nu^2$ is dominated by $\|Y_0\|^2 \in L_1(P)$. It now follows that

$$\begin{aligned} 0 = \lim_{\nu\to\infty} \int \|Y_0\|_\nu^2 \, dP &= \lim_{\nu\to\infty} \lim_{n\to\infty} \int \|Y_n\|_\nu^2 \, dQ_n^{(n)} \\ &\geq \varepsilon^2 \lim_{\nu\to\infty} \lim_{n\to\infty} Q_n^{(n)}\big(\|Y_n\|_\nu > \varepsilon\big) \end{aligned} \tag{53}$$

for all $\varepsilon \in (0,1)$. By Proposition A.4.1 b, the sequence $Y_n(Q_n^{(n)})$ is tight, hence has weak cluster points $\mathcal{L}Y$ [Prokhorov]. Then (38) must hold as a consequence of (35) [(18) and (36) implying (33), part (b) is in force]. By Cramér–Wold, if finite real linear combinations $e = \gamma_1 e_1 + \cdots + \gamma_r e_r$ of the ONB vectors are inserted, we see that $\mathcal{L}Y$ is uniquely determined on the algebra $\mathcal{A}$ induced by (π_α). Hence $\mathcal{L}Y$ is unique. Thus (c) is proved.

The proof also shows that $\mathcal{L}\langle Y|e\rangle$ is the normal distribution with the same first and second moments as $\langle Y_0|e\rangle(P)$, for all $e \in L_2(\mu)$. Moreover, the finite-dimensional distributions are multivariate normal,

$$\big(\langle Y|h_1\rangle, \ldots \langle Y|h_r\rangle\big)' \sim \mathcal{N}(0, C(r)) \tag{54}$$

for arbitrary functions $h_1, \ldots, h_r \in L_2(\mu)$, with the covariance entries

$$\begin{aligned} \mathrm{E}\, \langle Y|h_\alpha\rangle \langle Y|h_\beta\rangle &= \int \langle Y_0|h_\alpha\rangle \langle Y_0|h_\beta\rangle \, dP \\ &= \iint \big(P(y \wedge z) - P(y)P(z)\big)\, h_\alpha(y) h_\beta(z)\, \mu(dy)\, \mu(dz) \end{aligned} \tag{55}$$

number $\alpha, \beta = 1, \ldots, r < \infty$. ////

Remark A.4.5

(a) Conditions (18), (24) and (33) hold automatically if μ is finite.

(b) Actually, only the statements (a) and (b) of Theorem A.4.4 are needed when in Subsection 6.3.2 this result is applied to the MD estimate S_μ. For example, X_α constant e_α defines a norm bounded sequence such that, for every $e \in L_2(\mu)$, we have $\lim_\alpha \langle X_\alpha | e\rangle = \langle 0|e\rangle$. But, obviously, (X_α) does not converge weakly (to the only possible limit 0). ////

Neighborhood Type Conditions

The following lemma ensures the assumptions made in Theorem A.4.4 by conditions that the distribution functions $Q_{n,i}$ are suitably close to P.

Lemma A.4.6 *Assume* (18).

(a) *Then the conditions* (33) *and* (36) *are fulfilled if, respectively,*

$$\limsup_{n\to\infty} \max_{i=1,\dots,n} \int |Q_{n,i} - P|\, d\mu < \infty \tag{56}$$

$$\lim_{n\to\infty} \max_{i=1,\dots,n} \int |Q_{n,i} - P|\, d\mu = 0 \tag{57}$$

(b) *Condition* (34) *is fulfilled if* (56) *holds and if for all* $\varepsilon \in (0,1)$ *and all* $B \in \mathbb{B}^m$ *such that* $\mu(B) < \infty$,

$$\lim_{n\to\infty} \max_{i=1,\dots,n} \mu\big(B \cap \{|Q_{n,i} - P| > \varepsilon\}\big) = 0 \tag{58}$$

and if, in case $m > 1$, *in addition one of the assumptions* (59)–(61) *is made,*

$$\lim_{n\to\infty} \max_{i=1,\dots,n} d_\kappa(Q_{n,i}, P) = 0 \tag{59}$$

$$\operatorname{support} \mu = \mathbb{R}^m \quad \text{and the distribution function } P \text{ continuous} \tag{60}$$

$$\mu = \mu_1 \otimes \cdots \otimes \mu_m \quad \text{a finite product measure} \tag{61}$$

Remark A.4.7 Condition (58) is implied by $\lim_n \max_i d_\mu(Q_{n,i}, P) = 0$. In turn, if μ is finite, (58) implies (57). ////

PROOF Since $\big|Q_{n,i}(1 - Q_{n,i}) - P(1-P)\big| \le |Q_{n,i} - P|$, part (a) follows. For the proof of (b), assumption (58) allows a passage to subsequences that converge a.e. μ [Bauer (1974; Satz 19.6)]; that is, we can assume a sequence of distribution functions $Q_n = Q_{n,i_n}$ that converge to P a.e. μ. In any case then, a.e. $\mu \otimes \mu(dy, dz)$,

$$\lim_{n\to\infty} Q_n(y) Q_n(z) = P(y) P(z) \tag{62}$$

In case $m = 1$, moreover $Q_n(y \wedge z) = Q_n(y) \wedge Q_n(z)$, and likewise for P, hence a.e. $\mu \otimes \mu(dy, dz)$,

$$\lim_{n\to\infty} Q_n(y \wedge z) = P(y \wedge z) \tag{63}$$

so that a.e. $\mu \otimes \mu(dy, dz)$,

$$\lim_{n\to\infty} Q_n(y \wedge z) - Q_n(y)Q_n(z) = P(y \wedge z) - P(y)P(z) \tag{64}$$

The convergence (63), hence (64), also obviously hold in case $m > 1$ under the additional assumption (59), or under (60), in which case $Q_n(y) \to P(y)$ on a dense subset hence, by continuity of P, at all $y \in \mathbb{R}^m$ and even in d_κ.

Assumptions (18), (56) imply (24). Therefore, given any $e, \tilde{e} \in L_2(\mu)$, the variables

$$\begin{aligned} Z_n &= \int \big(\mathbf{I}(x \le y) - Q_n(y)\big) e(y)\, \mu(dy) \\ \tilde{Z}_n &= \int \big(\mathbf{I}(x \le y) - Q_n(y)\big) \tilde{e}(y)\, \mu(dy) \end{aligned} \tag{65}$$

are well defined under $x \sim Q_n$. In the manner of (42) we obtain

$$\begin{aligned} \Big|\sqrt{V(Q_n, \tilde{e})} - \sqrt{V(Q_n, e)}\Big| &\le \|\tilde{e} - e\| \, \Big| \int Q_n(1 - Q_n)\, d\mu \Big|^{1/2} \\ &\le \|\tilde{e} - e\| \, \Big| \int P(1 - P)\, d\mu + \int |Q_n - P|\, d\mu \Big|^{1/2} \end{aligned} \tag{66}$$

and

$$\Big|\sqrt{V(P, \tilde{e})} - \sqrt{V(P, e)}\Big| \le \|\tilde{e} - e\| \, \Big| \int P(1 - P)\, d\mu \Big|^{1/2} \tag{67}$$

In view of these bounds, and by assumptions (18) and (56), it suffices to prove

$$\begin{aligned} &\iint \big(Q_n(y \wedge z) - Q_n(y)Q_n(z)\big) \tilde{e}(y)\tilde{e}(z)\, \mu(dy)\, \mu(dz) \\ &\qquad \longrightarrow \iint \big(P(y \wedge z) - P(y)P(z)\big) \tilde{e}(y)\tilde{e}(z)\, \mu(dy)\, \mu(dz) \end{aligned} \tag{68}$$

for suitable $\tilde{e}$ arbitrarily close to e. As such choose $\tilde{e}$ bounded measurable so that $\mu(\tilde{e} \ne 0) < \infty$ [Lemma C.2.6]. Then $\tilde{e} \in L_1(\mu)$, and (68) follows from (64) by dominated convergence.

In case $m > 1$, under the additional assumption (61), $L_2(\mu) \subset L_1(\mu)$ as μ is finite. Thus, by dominated convergence, (62) implies that

$$\lim_{n\to\infty} \iint \big(Q_n(y)Q_n(z) - P(y)P(z)\big) e(y)e(z)\, \mu(dy)\, \mu(dz) = 0 \tag{69}$$

and it remains to show that

$$\lim_{n\to\infty} \iint \big(Q_n(y \wedge z) - P(y \wedge z)\big) e(y)e(z)\, \mu(dy)\, \mu(dz) = 0 \tag{70}$$

It is no restriction to assume μ a product probability. The distinction in each coordinate $j = 1, \dots, m$ whether $y_j \le z_j$ or $y_j > z_j$ defines us a partition (D_r) of $\mathbb{R}^{2m}$. Let $w_j = y_j$ in case $y_j \le z_j$ on D_r, and $w_j = z_j$ in case $y_j > z_j$ on D_r. Then

$$\begin{aligned}
&\iint \left|Q_n(y \wedge z) - P(y \wedge z)\right|^2 \mu(dy)\,\mu(dz) \\
&\quad = \sum_{r=1}^{2^m} \int \!\cdots\! \int_{D_r} \left|Q_n(w) - P(w)\right|^2 \bigotimes_{j=1}^{n} \mu_j(dy_j) \bigotimes_{j=1}^{n} \mu_j(dz_j) \\
&\quad \le \sum_{r=1}^{2^m} \int \!\cdots\! \int \left|Q_n(w) - P(w)\right|^2 \bigotimes_{j=1}^{n} \mu_j(dy_j) \bigotimes_{j=1}^{n} \mu_j(dz_j) \\
&\quad = \sum_{r=1}^{2^m} \int \!\cdots\! \int \left|Q_n(w) - P(w)\right|^2 \bigotimes_{j=1}^{n} \mu_j(dw_j) = 2^m \int |Q_n - P|^2 \, d\mu
\end{aligned} \tag{71}$$

The upper bound tends to 0 since $1 \ge |Q_n - P| \to 0$ a.e. μ and μ is finite. On applying the Cauchy–Schwarz inequality, (70) follows. ////

Example A.4.8 Contrary to dimension $m = 1$ the additional assumptions (59) or (60) or (61) cannot be dispensed with in Theorem A.4.4 b, c.

For example, let $m = 2$ and consider the probability weight

$$\mu = \frac{1}{2}\left(\mathbf{I}_a + \mathbf{I}_b\right), \qquad a = \begin{pmatrix} 0 \\ 1 \end{pmatrix}, \; b = \begin{pmatrix} 1 \\ 0 \end{pmatrix} \tag{72}$$

Counting the four open unit quadrants I, II, III, IV counterclockwise, there are probabilities P and Q such that

$$\begin{aligned}
&P(\mathrm{I}) = 0, \quad P(\mathrm{II}) = \tfrac{1}{3}, \quad P(\mathrm{III}) = \tfrac{1}{3}, \quad P(\mathrm{IV}) = \tfrac{1}{3} \\
&Q(\mathrm{I}) = \tfrac{1}{3}, \quad Q(\mathrm{II}) = 0, \quad Q(\mathrm{III}) = \tfrac{2}{3}, \quad Q(\mathrm{IV}) = 0
\end{aligned} \tag{73}$$

The corresponding distribution functions satisfy

$$\begin{aligned}
Q(a) &= Q(\mathrm{II}) + Q(\mathrm{III}) = \tfrac{2}{3} = P(a) \\
Q(b) &= Q(\mathrm{III}) + Q(\mathrm{IV}) = \tfrac{2}{3} = P(b)
\end{aligned} \tag{74}$$

Hence $Q = P$ a.e. μ, whereas

$$Q(a \wedge b) = Q(\mathrm{III}) = \tfrac{2}{3} \neq \tfrac{1}{3} = P(a \wedge b) \tag{75}$$

Theorem A.4.4 applied to the constant arrays $Q_{n,i} = P$ and $Q_{n,i} = Q$, respectively, yields the following as. normality of the empirical process,

$$\sqrt{n}\,(\hat{P}_n - P)(P^n) \xrightarrow{\;\mathrm{w}\;} \mathcal{N}(0, C_P), \quad \sqrt{n}\,(\hat{P}_n - Q)(Q^n) \xrightarrow{\;\mathrm{w}\;} \mathcal{N}(0, C_Q) \tag{76}$$

The different as. covariances may be computed as in (55); namely,

$$C_P = \frac{1}{36}\begin{pmatrix} 2 & -1 \\ -1 & 2 \end{pmatrix}, \qquad C_Q = \frac{1}{36}\begin{pmatrix} 2 & 2 \\ 2 & 2 \end{pmatrix} \tag{77}$$

In this example, the indicator variables $e_1 = \mathbf{I}_a$, $e_2 = \mathbf{I}_b$ make an ONB for $L_2(\mu)$, which may be identified with $\mathbb{R}^2$ via $h = h(a)\,\mathbf{I}_a + h(b)\,\mathbf{I}_b$. ////

Appendix B

Some Functional Analysis

B.1 A Few Facts

We recall a few facts from functional analysis that are needed for optimization and Lagrange multiplier theorems.

Lemma B.1.1 *Every nonempty, closed, convex subset of a Hilbert space contains a unique element of smallest norm.*

PROOF Rudin (1974; Theorem 4.10). ////

For a (real) linear topological space X, the topological dual X^* is defined as the space of all continuous linear functionals $x^*: X \to \mathbb{R}$. The weak topology on X is the initial topology generated by X^*. A subset A of X is called *weakly sequentially compact* if every sequence in A has a subsequence that converges weakly to some limit in X (may be outside A).

Lemma B.1.2 *Let X be a locally convex linear topological space.*

(a) *Then a subset of X is weakly bounded iff it is bounded.*

(b) *A convex subset of X is weakly closed iff it is closed.*

(c) *Suppose X is a Hilbert space. Then a subset of X is weakly sequentially compact iff it is bounded.*

PROOF

(a) Rudin (1973; Theorem 3.18).

(b) Dunford and Schwartz (1957; Vol. I, Theorem V 3.13).

(c) Dunford and Schwartz (1957; Vol. I, Corollary IV 4.7). ////

Proposition B.1.3 *If A and B are disjoint convex subsets of a real linear topological space X, and $A^\circ \neq \emptyset$, there exists some $x^* \in X^*$, $x^* \neq 0$, such that $x^*a \geq x^*b$ for all $a \in A$, $b \in B$.*

PROOF Dunford and Schwartz (1957; Vol. I, Theorem V 2.8), Rudin (1973; Theorem 3.4). ////

B.2 Lagrange Multipliers

This section derives the Lagrange multipliers theorems needed in Chapter 5. We introduce the following objects:

$$
\begin{array}{ll}
X, Y, Z & \text{three real topological vector spaces} \\
A & \text{a convex subset of } X \\
C & \text{a convex cone in } Z \text{ with vertex at } 0 \\
 & \text{and nonempty interior } C^{\circ} \neq \emptyset \\
f\colon A \longrightarrow \mathbb{R} & \text{a convex function} \\
G\colon A \longrightarrow Z & \text{a convex map} \\
H\colon X \longrightarrow Y & \text{a linear operator}
\end{array}
\tag{1}
$$

The set C being a convex cone with vertex at 0 means that $sw + tz \in C$ whenever $w, z \in C$ and $s, t \in [0, \infty)$. C is called a positive cone in Z as it induces a partial order according to: $w \leq z$ iff $z - w \in C$. It is this ordering to which convexity of G refers. C induces a partial order also in the topological dual Z^* via: $0 \leq z^*$ iff $z^* z \geq 0$ for all $z \in C$. The interior point assumption $C^{\circ} \neq \emptyset$ is needed for the separation theorem [Proposition B.1.3], which will provide us with Lagrange multipliers.

The optimization problems considered are of the following convex linear type,

$$f(x) = \min! \qquad x \in A,\ G(x) \leq z_0,\ Hx = y_0 \tag{2}$$

where $y_0 \in Y$ and $z_0 \in Z$ are fixed elements. To avoid trivialities, the value m is without further mention assumed finite,

$$-\infty < m = \inf\{\, f(x) \mid x \in A,\ G(x) \leq z_0,\ Hx = y_0 \,\} < \infty \tag{3}$$

Dealing with the different constraints separately, we first study the purely convex problem,

$$f(x) = \min! \qquad x \in A,\ G(x) \leq z_0 \tag{4}$$

which can be subsumed under (2) on identifying $Y = \{0\}$, $H = 0$.

The following result resembles the Kuhn–Tucker theorems of Luenberger (1969; §8.3, Theorem 1) and Joffe, Tichomirov (1979; §1.1, Theorem 2).

Theorem B.2.1 [Problem (4)] *There exist* $r \in [0, \infty)$, $0 \leq z^* \in Z^*$, *not both zero, such that*

$$rm + z^* z_0 = \inf\{\, r f(x) + z^* G(x) \mid x \in A \,\} \tag{5}$$

(a) *If* $x_0 \in A$ *satisfying* $G(x_0) \leq z_0$ *achieves the infimum in* (3), *then also in* (5), *and*

$$z^* G(x_0) = z^* z_0 \tag{21}$$

(b) *We have* $r > 0$ *if there exists some* $x_1 \in A$ *such that*

$$z_0 - G(x_1) \in C^{\circ} \tag{6}$$

PROOF Introduce the set

$$K = \{ (t,z) \in \mathbb{R} \times Z \mid \exists x \in A : f(x) < t,\, G(x) \leq z \}$$

A, f, and G being convex, so is K. $K^{\circ} \neq \emptyset$ as $(m,\infty) \times (z_0 + C) \subset K$, and $(m, z_0) \notin K$ by the definition of m. Thus the separation theorem supplies some $r \in \mathbb{R}$, $z^* \in Z^*$, not both zero, such that for all $(t,z) \in K$,

$$rm + z^* z_0 \leq rt + z^* z \tag{7}$$

Insert $z = z_0$ and $m < t \uparrow \infty$ to get $r \geq 0$. Also $z^* \geq 0$ follows from (7) if we insert $z = z_0 + w$ with any $w \in C$ and let $t \downarrow m$.

For $x \in A$ put $z = G(x)$ and let $t \downarrow f(x)$ in (7). Thus, for all $x \in A$,

$$rm + z^* z_0 \leq r f(x) + z^* G(x) \tag{8}$$

As $z^* \geq 0$ it follows that for all $x \in A$ satisfying $G(x) \leq z_0$,

$$rm + z^* z_0 \leq r f(x) + z^* G(x) \leq r f(x) + z^* z_0 \tag{9}$$

In this relation, $f(x)$ may approach m. Thus (8) and (9) imply (5).

(a) If $x_0 \in A$, $G(x_0) \leq z_0$, $f(x_0) = m$, then equalities hold in (9) for $x = x_0$. Thus $r f(x_0) + z^* G(x_0) = rm + z^* z_0$ and so $z^* G(x_0) = z^* z_0$.

(b) Suppose that $z_0 - G(x_1) \in C^{\circ}$ for some $x_1 \in A$, and $r = 0$. Then $z^*(z_0 - G(x_1)) = 0$ follows from (9). So the cone C is a neighborhood of a zero of z^*, on which $z^* \geq 0$ attains only nonnegative values. But this enforces $z^* = 0$, and so both multipliers would vanish. ////

We next turn to convex linear problems that have only a linear constraint,

$$f(x) = \min! \qquad x \in A,\, Hx = y_0 \tag{10}$$

which may be subsumed under (2) on identifying $Z = \{0\}$, $G = 0$.

The remaining convex constraint $x \in A$ may formally be removed by passing to the convex extension $\bar{f} = f\,\mathbf{I}_A + \infty\,\mathbf{I}_{X \setminus A}$ of f. As H is linear, a smooth Lagrange multiplier analogue [Luenberger (1969; §9.3, Theorem 1)] seems tempting that, instead of the inverse function theorem, invokes the notion of *subdifferential* for $\bar{f}$ [Joffe and Tichomirov (1979; §0.3.2)].

By definition, the subdifferential $\partial \bar{f}(x_0)$ of $\bar{f}$ at $x_0 \in X$ consists of all continuous linear functionals $u^* \in X^*$ such that for all $x \in A$,

$$\bar{f}(x_0) + u^*(x - x_0) \leq \bar{f}(x) \tag{11}$$

The convex indicator of the set $\{H = y_0\}$ is denoted by χ; that is, $\chi(x) = 0$ if $Hx = y_0$, and $\chi(x) = \infty$ if $Hx \neq y_0$.

Proposition B.2.2 [Problem (10)] *Let X and Y be Banach spaces, and assume the linear operator H is continuous and onto; $HX = Y$. Suppose there exists an $x \in A^{\circ}$ such that $Hx = y_0$ and f is continuous at x.*

Then any $x_0 \in A$, $Hx_0 = y_0$, solves (10) iff there is some $y^* \in Y^*$ such that

$$f(x_0) + y^* Hx_0 \leq f(x) + y^* Hx, \qquad x \in A \tag{12}$$

PROOF If $x_0 \in X$ solves (10) it minimizes $g = \bar{f} + \chi$ (convex) on X, and $g(x_0)$ is finite as $m < \infty$. Thus $0 \in \partial g(x_0)$ by the definition of subdifferential. Then the Moreau–Rockafellar theorem [Joffe and Tichomirov (1979), Theorem 0.3.3)] tells us that

$$\partial g(x_0) = \partial \bar{f}(x_0) + \partial \chi(x_0)$$

It is easy to check, and actually follows from the open mapping theorem [Luenberger (1969; §6.6, Theorem 2)], that in X^*,

$$\partial \chi(x_0) = (\ker H)^{\perp} = \operatorname{im} H^*$$

where H^* denotes the adjoint of H. Thus there exists some $y^* \in Y^*$ such that

$$u^* = -H^* y^* = -y^* H \in \partial \bar{f}(x_0)$$

This proves the only nontrivial direction. ////

This version turns out to be only of limited, theoretical interest since the oscillation terms we have in mind are convex and weakly l.s.c.; however, except in the Hellinger case, they may be discontinuous everywhere (so that $A^{\circ} = \emptyset$ in these applications). The alternative assumption of the Moreau–Rockafellar theorem, namely, $A \cap \{H = y_0\}^{\circ} \neq \emptyset$, is not acceptable either in our setup as it would entail that $H = 0$. Therefore, the following result is obtained by once more using the separation theorem directly.

Theorem B.2.3 [Problem (10)] *If there is some subset $V \subset A$ such that*

$$(HV)^{\circ} \neq \emptyset \tag{18}$$

$$\sup\{ f(x) \mid x \in V \} < \infty \tag{13}$$

then there exist $r \in [0, \infty)$, $y^* \in Y^*$, not both zero, such that

$$rm + y^* y_0 = \inf\{ rf(x) + y^* Hx \bigm| x \in A \} \tag{14}$$

(a) *If $x_0 \in A$, $Hx_0 = y_0$, achieves the infimum in* (3) *then also in* (14).

(b) *We have $r > 0$ if*

$$y_0 \in (HV)^{\circ} \tag{22}$$

Remark B.2.4 In case $\dim Y = k < \infty$, condition (13) can be cancelled:

For (18) ensures that HV includes a simplex B of nonempty interior, which is the convex hull of its $k+1$ vertices $y_i = Hx_i$ with $x_i \in V$. Then f (convex, finite-valued) is bounded by $\max_{i=0,\dots,k} f(x_i) < \infty$ on the convex hull $\widetilde{V}$ of the x_i's, while still $H\widetilde{V} = B$ is achieved. ////

Example B.2.5 In case $\dim Y = \infty$, condition (13) cannot be dispensed with, as shown by this counterexample:

$$\begin{array}{ll} A = X = Y, & H = \mathrm{id}_X\,, \quad y_0 = 0 \\ h\colon Y \longrightarrow \mathbb{R} & \text{linear, discontinuous} \\ g\colon \mathbb{R} \longrightarrow \mathbb{R} & \text{convex,} \quad \inf g < g(0) \\ f = g \circ h & \end{array}$$

For every infinite-dimensional Banach space Y such h exist [Dunford and Schwartz (1957; Vol. I, Exercise V.7.2, p436)].

In this situation we have $m = g(0)$, whereas the infimum in (14) equals $r \inf g$ if $y^* = 0$, and $-\infty$ if $y^* \neq 0$. Clearly, condition (13) is violated since h is unbounded on every neighborhood of 0 and $\sup g = \infty$ [thus the interval $(\bar{v}, \infty)$ in the following proof would be empty].

This example shows that Joffe and Tichomirov (1979; §1.1, Theorem 5) are wrong in case $\dim Y = \infty$. ////

PROOF [Theorem B.2.3] Introduce the convex set

$$K = \{\, (t, y) \in \mathbb{R} \times Y \mid \exists x \in A\colon f(x) < t,\ Hx = y \,\}$$

which does not contain (m, y_0). Denoting $\bar{v} = \sup\{ f(x) \mid x \in V \} < \infty$, we have $(\bar{v}, \infty) \times HV \subset K$ hence $K^\circ \neq \emptyset$. The separation theorem supplies multipliers $r \in \mathbb{R}$, $y^* \in Y$, not both zero, such that for all $(t, y) \in K$,

$$rm + y^* y_0 \leq rt + y^* y \tag{15}$$

As $(m, \infty) \times \{y_0\} \subset K$, it follows that $r \geq 0$.

For $x \in A$ let $t \downarrow f(x)$ in (15). Thus for all $x \in A$,

$$rm + y^* y_0 \leq r f(x) + y^* Hx \tag{16}$$

In (16) let $f(x) \downarrow m$ on $x \in A$, $Hx = y_0$. Thus (14) is proved.

(a) If $x_0 \in A$, $Hx_0 = y_0$, achieves $f(x_0) = m$ then equality holds in (16) for $x = x_0$. This proves a).

(b) If $y_0 \in (HV)^\circ$ and $r = 0$, (15) implies that $y^* y_0 \leq y^* y$ for all y in the neighborhood HV of y_0, which would enforce that also $y^* = 0$. ////

To settle the general convex linear problem (2) we only need to apply Theorem B.2.1 to the triple f, G, $A \cap \{H = y_0\}$, which gives us multipliers r_1, z_1^*, and then Theorem B.2.3 to the triple $r_1 f + z_1^* G$, H, A, providing multipliers r_2 and y^*. Thus, the following result obtains with the final multipliers $r = r_1 r_2$, $z^* = r_2 z_1^*$ and y^*.

Theorem B.2.6 [Problem (2)] *There exist* $r_1 \in [0, \infty)$, $0 \le z_1^* \in Z^*$, *not both zero, such that*

$$r_1 m + z_1^* z_0 = \inf\{\, r_1 f(x) + z_1^* G(x) \mid x \in A,\, Hx = y_0 \,\} \tag{17}$$

Assume there is some subset $V \subset A$ *such that*

$$(HV)^\circ \ne \emptyset \tag{18}$$

$$\sup\{\, r_1 f(x) + z_1^* G(x) \bigm| x \in V \,\} < \infty \tag{19}$$

then there exist $r \in [0, \infty)$, $0 \le z^* \in Z^*$, $y^* \in Y^*$, *not all three zero, so that*

$$rm + z^* z_0 + y^* y_0 = \inf\{\, r f(x) + z^* G(x) + y^* Hx \mid x \in A \} \tag{20}$$

(a) *If some* $x_0 \in A$ *satisfying* $G(x_0) \le z_0$ *and* $Hx_0 = y_0$ *achieves the infimum in* (3), *then also in* (17) *and in* (20), *and it holds that*

$$z^* G(x_0) = z^* z_0 \tag{21}$$

(b) *If*

$$y_0 \in (HV)^\circ \tag{22}$$

not both r *and* z^* *can be zero. We have* $r > 0$ *under condition* (22) *and if there is some* $x_1 \in A$ *such that*

$$z_0 - G(x_1) \in C^\circ, \qquad Hx_1 = y_0 \tag{23}$$

Remark B.2.7

(a) In case $\dim Y < \infty$, assumption (19) can be dispensed with, which follows if we apply the argument given in Remark B.2.4 to the convex Lagrangian $r_1 f + z_1^* G$. As demonstrated by counterexample, the cancellation of condition (19) is not feasible in case $\dim Y = \infty$.

(b) Suppose $r = 0$ in Theorem B.2.6 a, and consider any $x \in A$ satisfying the constraints $G(x) \le z_0$, $Hx = y_0$. Then

$$L(x) = z^* G(x) + y^* Hx \le z^* z_0 + y^* y_0 = L(x_0) = \min L(A) \tag{24}$$

Hence x too minimizes the Lagrangian L on the unrestricted domain A and necessarily achieves the equality $z^* G(x) = z^* z_0$. Thus, the optimization seems to be determined exclusively by the side conditions, which in fact cannot distinguish any particular solution. ////

For the sake of completeness we add the degenerate case, when condition (18) cannot be fulfilled. Denote by M the linear span of the image HA in Y, and by $\overline{M}$ the (topological) closure of M in Y.

Proposition B.2.8 [Problem (2)] *Let the space* Y *be locally convex, and suppose that* $\overline{M} \ne Y$. *Then there exist* $r \in [0, \infty)$, $0 \le z^* \in Z^*$, $y^* \in Y^*$, *not all three zero, such that* (20) *holds; namely,* $r = 0$, $z^* = 0$, *and any nonzero* $y^* \perp \overline{M}$. *With this choice, every* $x_0 \in A$ *attains the infimum in* (20), *which is zero, and fulfills* (21).

Definition B.2.9 *Problem* (2) *is called* well-posed *if there exist* $x_1 \in A$ *and* $V \subset A$ *such that conditions* (19), (22), *and* (23) *are fulfilled.*

We note that in the first subproblems ($Y = \{0\}$, $H = 0$), well-posedness reduces to condition (6) [put $V = \{x\}$ for any $x \in A$], while in the second subproblems ($Z = \{0\}$, $G = 0$) well-posedness reduces to (13) and (22) [as $Z^* = \{0\} \implies r_1 > 0$]. In the well-posed case, w.l.o.g. $r = 1$.

Remark B.2.10

(a) Like the separating hyperplanes, , the Lagrange multipliers r, y^*, z^* need not be unique. But every solution x_0 to problem (2) minimizes all corresponding Lagrangians $L = rf + z^*G + y^*H$ of the problem. The explicit forms of x_0 derived under possibly different sets of Lagrange multipliers express the same x_0.

(b) If f or G are extended-valued, with values in $\mathbb{R} \cup \{\infty\}$, respectively, $(\mathbb{R} \cup \{\infty\})^k$, for some $k < \infty$, the convexity of these maps refers to the usual arithmetic and ordering in $\mathbb{R} \cup \{\infty\}$, in $(\mathbb{R} \cup \{\infty\})^k$ coordinatewise so. If G is such a map, it shall nevertheless be assumed that $z_0 \in \mathbb{R}^k$, and $\mathbb{R}^k$ is taken for Z. Then all arguments and results of this section carry over with the replacement

$$A \rightsquigarrow A \cap \{f < \infty\} \cap \{G \in \mathbb{R}^k\} \tag{25}$$

This substitution does not affect the value m in (3), and in some instances, notation is lightened again [e.g., $G(x) \le z_0 \in \mathbb{R}^k$ entails that $G(x) \in \mathbb{R}^k$]. Concerning the infimum of the Lagrangian $L = rf + z^*G + y^*H$, the restrictions $f < \infty$ and $G \in \mathbb{R}^k$, respectively, can be removed if $r > 0$, respectively, if $z^* > 0$ in $\mathbb{R}^k$. ////

B.2.1 Neyman–Pearson Lemma

We demonstrate Theorem B.2.6 by deriving the classical Neyman–Pearson and fundamental lemmas concerning tests φ between probability measures on some sample space $(\Omega, \mathcal{A})$. [The extension to 2-alternating capacities, however, requires other techniques; Huber (1965, 1968, 1969); Huber and Strassen (1973); Rieder (1977); Bednarski (1981, 1982); Buja (1984–1986)].

Proposition B.2.11 *The simple testing problem between two probabilities* $P, Q \in \mathcal{M}_1(\mathcal{A})$, *at level* $\alpha \in [0, 1]$,

$$\int \varphi\, dQ = \max! \qquad \varphi \text{ test}, \int \varphi\, dP \le \alpha \tag{26}$$

has a solution. There exist numbers $r, z \in [0, \infty)$, *not both zero, such that every solution* $\varphi^\star$ *to* (26) *is of the form*

$$\varphi^\star = \begin{cases} 1 & \text{if } r\, dQ > z\, dP \\ 0 & \text{if } r\, dQ < z\, dP \end{cases} \tag{27}$$

and satisfies

$$z \int \varphi^\star \, dP = z\alpha \tag{28}$$

Moreover,

$$\int \varphi^\star \, dP < \alpha \implies \int \varphi^\star \, dQ = 1 \tag{29}$$

and

$$\alpha > 0 \implies r > 0 \tag{30}$$

PROOF A solution exists as the tests on $(\Omega, \mathcal{A})$ are weakly sequentially compact [Lehmann (1986; Theorem 3, p 576), Noelle and Plachky (1968), Witting (1985; Satz 2.14, p 205)]. For this result and the following application of Theorem B.2.1, choose any dominating measure $\mu \in \mathcal{M}_\sigma(\mathcal{A})$ such that $dP = p\,d\mu$ and $dQ = q\,d\mu$, and regard tests as elements of $L_\infty(\mu)$. Denote expectation under μ by E_μ. Then define

$$\begin{aligned} X &= L_\infty(\Omega, \mathcal{A}, \mu), \quad A = \{\, \varphi \in X \mid 0 \le \varphi \le 1 \text{ a.e. } \mu \,\} \\ Z &= \mathbb{R}, \quad z_0 = \alpha, \quad f(\varphi) = -\,\mathrm{E}_\mu\, \varphi q, \quad G(\varphi) = \mathrm{E}_\mu\, \varphi p \end{aligned} \tag{31}$$

As $-1 \le m \le 0$ and $C^\circ = (0, \infty)$, Theorem B.2.1 is in force: There exist multipliers $r, z \in [0, \infty)$, not both of them zero, such that

$$\begin{aligned} -r\,\mathrm{E}_\mu\, \varphi^\star q + z\alpha &= \inf\{\, -r\,\mathrm{E}_\mu\, \varphi q + z\,\mathrm{E}_\mu\, \varphi p \mid \varphi \in A \,\} \\ &= \inf\{\, \mathrm{E}_\mu\, \varphi(zp - rq) \mid 0 \le \varphi \le 1 \,\} \end{aligned}$$

Pointwise minimization of the integrand leads to (27), which function is indeed measurable on the event $\{rq \ne zp\}$. By Theorem B.2.1 a, relation (21) holds which, in view of the identifications (31), is (28).

If $\varphi^\star$ does not exhaust level α, then $z = 0$ by (28), hence $r > 0$ and $\varphi^\star = 1$ a.e. Q; which proves (29). If $\alpha > 0$, condition (6) is fulfilled by the zero test, thus $r > 0$ holds according to Theorem B.2.1 b. ////

Remark B.2.12 The Neyman–Pearson lemma can be proved more constructively, using the following critical value and randomization $\gamma \in [0, 1]$,

$$c = \inf\{\, u \in [0, \infty] \mid P(q > up) \le \alpha \,\} \tag{32}$$

$$\gamma\, P(q = cp) = \alpha - P(q > cp) \tag{33}$$

Then by construction

$$\mathrm{E}_\mu(\varphi^\star - \varphi)(q - cp) \ge 0 \tag{34}$$

for the test

$$\varphi^\star = \mathbf{I}(q > cp) + \gamma\, \mathbf{I}(q = cp) \tag{35}$$

and any other test φ of level α, from which the assertions follow. ////

More generally, testing power is to be maximized subject to a finite number of level constraints, $k \geq 0$ inequalities and $n \geq 0$ equalities, which are defined relative to some σ finite measure μ, using levels $\alpha_1, \ldots, \alpha_{k+n} \in [0,1]$ and any integrands $q_0, q_1, \ldots, q_{k+n} \in L_1(\mu)$. The testing problem then, which for example leads to the classical unbiased two-sided tests in exponential families, attains the following form:

$$\mathrm{E}_\mu \varphi q_0 = \max! \tag{36}$$

among all tests φ on $(\Omega, \mathcal{A})$ subject to the side conditions

$$\mathrm{E}_\mu \varphi q_i \leq \alpha_i \quad (i \leq k), \qquad \mathrm{E}_\mu \varphi q_i = \alpha_i \quad (i > k) \tag{37}$$

Proposition B.2.13 *Suppose there exists a test φ satisfying (37).*

(a) *Then problem (36), (37) has a solution.*

(b) *If*

$$\big\{ (\mathrm{E}_\mu \varphi q_{k+1}, \ldots, \mathrm{E}_\mu \varphi q_{k+n}) \bigm| \varphi \text{ test} \big\}^\circ \neq \emptyset \tag{38}$$

then there exist multipliers $r, z_1, \ldots, z_k \in [0, \infty)$, $y_{k+1}, \ldots, y_{k+n} \in \mathbb{R}$, not all of them zero, such that every solution $\varphi^\star$ to (36), (37) is of the form

$$\varphi^\star = \begin{cases} 1 & \text{if } \; r q_0 > \sum_{i \leq k} z_i q_i + \sum_{i>k} y_i q_i \\ 0 & \text{if } \; r q_0 < \sum_{i \leq k} z_i q_i + \sum_{i>k} y_i q_i \end{cases} \tag{39}$$

and satisfies

$$z_i \, \mathrm{E}_\mu \varphi^\star q_i = z_i \alpha_i \qquad (i \leq k) \tag{40}$$

If

$$(\alpha_{k+1}, \ldots, \alpha_{k+n}) \in \big\{ (\mathrm{E}_\mu \varphi q_{k+1}, \ldots, \mathrm{E}_\mu \varphi q_{k+n}) \bigm| \varphi \text{ test} \big\}^\circ \tag{41}$$

then $r, z_1, \ldots, z_k$ are not all zero. We have $r > 0$, under condition (41) and if there exists a test φ such that

$$\mathrm{E}_\mu \varphi q_i < \alpha_i \quad (i \leq k), \qquad \mathrm{E}_\mu \varphi q_i = \alpha_i \quad (i > k) \tag{42}$$

(c) *If a test $\varphi^\star$ satisfies side condition (37) and is of form (39) and (40), for any numbers $r \in (0, \infty)$, $z_1, \ldots, z_k \in [0, \infty)$, and $y_{k+1}, \ldots, y_{k+n} \in \mathbb{R}$, then $\varphi^\star$ solves problem (36), (37).*

PROOF

(a) By weak sequential compactness of the tests on $(\Omega, \mathcal{A})$.

(b) Make the following identifications:

$$\begin{gathered} X = L_\infty(\Omega, \mathcal{A}, \mu), \quad A = \{ \varphi \in X \mid 0 \leq \varphi \leq 1 \text{ a.e. } \mu \} \\ Y = \mathbb{R}^n, \quad Z = \mathbb{R}^k, \quad C = [0, \infty)^k \\ z_0 = (\alpha_1, \ldots, \alpha_k)', \quad y_0 = (\alpha_{k+1}, \ldots, \alpha_{k+n})' \\ f(\varphi) = -\, \mathrm{E}_\mu \varphi q_0, \quad G(\varphi) = (\mathrm{E}_\mu \varphi q_1, \ldots, \mathrm{E}_\mu \varphi q_k)' \\ H(\varphi) = (\mathrm{E}_\mu \varphi q_{k+1}, \ldots, \mathrm{E}_\mu \varphi q_{k+n})' \end{gathered} \tag{43}$$

Then $C^{\circ} = (0,\infty)^k \neq \emptyset$. By the assumption of a test satisfying the side conditions, the value m is $< \infty$, and also $m > -\infty$ as $f \geq -\mathrm{E}_\mu q_0^+$. Assumption (38) ensures condition (18) with $V = A$. Condition (19) is automatic since $\dim Y < \infty$ [Remark B.2.7 a]. Thus Theorem B.2.6 provides multipliers $r, z_1, \ldots, z_k \in [0,\infty)$ and $y_{k+1}, \ldots, y_{k+n} \in \mathbb{R}$, not all of them zero, such that, by Theorem B.2.6 a, every solution $\varphi^\star$ to problem (36) and (37) minimizes the corresponding Lagrangian,

$$\begin{aligned} -r\,\mathrm{E}_\mu \varphi^\star q_0 &+ \sum\nolimits_{i\leq k} z_i\alpha_i + \sum\nolimits_{i>k} y_i\alpha_i \\ &= \inf\Big\{ \int \varphi\Big(-r q_0 + \sum\nolimits_{i\leq k} z_i q_i + \sum\nolimits_{i>k} y_i q_i\Big)\, d\mu \;\Big|\; \varphi \in A \Big\} \\ &= -\int \Big(-r q_0 + \sum\nolimits_{i\leq k} z_i q_i + \sum\nolimits_{i>k} y_i q_i\Big)^- \, d\mu \end{aligned} \tag{44}$$

and satisfies

$$\sum\nolimits_{i\leq k} z_i\,\mathrm{E}_\mu \varphi^\star q_i = \sum\nolimits_{i\leq k} z_i\alpha_i \tag{45}$$

This implies (39) and (40). That (41) and the existence of a test satisfying (42) enforces $r > 0$, follows from Theorem B.2.6 b.

(c) Form (39) implies that $\varphi^\star$ minimizes the Lagrangian

$$\begin{aligned} L(\varphi) &= -r\,\mathrm{E}_\mu \varphi q_0 + \sum\nolimits_{i\leq k} z_i\,\mathrm{E}_\mu \varphi q_i + \sum\nolimits_{i>k} y_i\,\mathrm{E}_\mu \varphi q_i \\ &= \int \varphi\Big(-r q_0 + \sum\nolimits_{i\leq k} z_i q_i + \sum\nolimits_{i>k} y_i q_i\Big)\, d\mu \\ &\geq -\int \Big(r q_0 - \sum\nolimits_{i\leq k} z_i q_i - \sum\nolimits_{i>k} y_i q_i\Big)^+ \, d\mu = L(\varphi^\star) \end{aligned} \tag{46}$$

among all tests φ on $(\Omega, \mathcal{A})$. Moreover, $\varphi^\star$ is assumed to satisfy (37) and (40). If then also the test φ meets the side conditions (37), it follows that

$$\begin{aligned} z_i\,\mathrm{E}_\mu \varphi q_i \leq z_i\alpha_i &\overset{(40)}{=} z_i\,\mathrm{E}_\mu \varphi^\star q_i \qquad (i \leq k) \\ y_i\,\mathrm{E}_\mu \varphi q_i = y_i\alpha_i &\underset{(37)}{=} y_i\,\mathrm{E}_\mu \varphi^\star q_i \qquad (i > k) \end{aligned} \tag{47}$$

Therefore,

$$\begin{aligned} L(\varphi^\star) &= -r\,\mathrm{E}_\mu \varphi^\star q_0 + \sum\nolimits_{i\leq k} z_i\,\mathrm{E}_\mu \varphi^\star q_i + \sum\nolimits_{i>k} y_i\,\mathrm{E}_\mu \varphi^\star q_i \\ &\overset{(46)}{\leq} -r\,\mathrm{E}_\mu \varphi q_0 + \sum\nolimits_{i\leq k} z_i\,\mathrm{E}_\mu \varphi q_i + \sum\nolimits_{i>k} y_i\,\mathrm{E}_\mu \varphi q_i \\ &\overset{(47)}{\leq} -r\,\mathrm{E}_\mu \varphi q_0 + \sum\nolimits_{i\leq k} z_i\,\mathrm{E}_\mu \varphi^\star q_i + \sum\nolimits_{i>k} y_i\,\mathrm{E}_\mu \varphi^\star q_i \end{aligned}$$

Since $r > 0$ by assumption, this implies that

$$\mathrm{E}_\mu \varphi^\star q_0 \geq \mathrm{E}_\mu \varphi q_0$$

which is the assertion. ////

Appendix C

Complements

C.1 Parametric Finite-Sample Results

Some classical optimality results from parametric statistics are occasionally needed, or are alluded to, for illustrating the different views and techniques of robust statistics.

Neyman's Criterion

This result, associated with the names of Halmos, Savage, and Neyman, characterizes sufficiency in the dominated case.

Proposition C.1.1 *Given a family $\mathcal{P} = \{ P_\theta \mid \theta \in \Theta \}$ of probability measures on a measurable space $(\Omega, \mathcal{A})$ that is dominated by any $\mu \in \mathcal{M}_\sigma(\mathcal{A})$.*

Then a sub-σ algebra $\mathcal{B} \subset \mathcal{A}$ is sufficient for $\mathcal{P}$ iff there is some measurable function $h: (\Omega, \mathcal{A}) \to (\mathbb{R}, \mathbb{B})$ and for every $\theta \in \Theta$ there exists some $\mathcal{B}$ measurable function $p_\theta: (\Omega, \mathcal{B}) \to (\mathbb{R}, \mathbb{B})$ such that

$$dP_\theta = p_\theta h \, d\mu \tag{1}$$

PROOF Lehmann (1986; 2.6 Corollary 1, p 55). ////

Rao–Blackwell Theorem

Proposition C.1.2 *Given a family $\mathcal{P} = \{ P_\theta \mid \theta \in \Theta \}$ of probability measures on some measurable space $(\Omega, \mathcal{A})$ and a sub-σ algebra $\mathcal{B} \subset \mathcal{A}$ that is sufficient for $\mathcal{P}$. For $p \in \mathbb{N}$ and $\theta \in \Theta$ let $\ell_\theta: \mathbb{R}^p \to [0, \infty)$ be a convex function. Then for every estimator $S: (\Omega, \mathcal{A}) \to (\mathbb{R}^p, \mathbb{B}^p)$ having finite expectations $\mathrm{E}_\theta S$ under P_θ for all $\theta \in \Theta$, there exists an estimator $S_{\bullet}: (\Omega, \mathcal{B}) \to (\mathbb{R}^p, \mathbb{B}^p)$ of uniformly smaller risk,*

$$\mathrm{E}_\theta \, \ell_\theta(S_{\bullet}) \le \mathrm{E}_\theta \, \ell_\theta(S) \tag{2}$$

and same expectations,

$$\mathrm{E}_\theta\, S_{\bullet} = \mathrm{E}_\theta\, S \tag{3}$$

namely, a θ-free version of the conditional expectation of S given the sufficient σ algebra $\mathcal{B}$,

$$S_{\bullet} = \mathrm{E}_{\bullet}(S|\mathcal{B}) \tag{4}$$

If ℓ_θ is strictly convex and equality holds in (2), *then $S = S_{\bullet}$* a.e. P_θ.

PROOF By convexity of ℓ_θ and Jensen's inequality for conditional expectations, we have

$$\ell_\theta \circ \mathrm{E}_{\bullet}(S|\mathcal{B}) \le \mathrm{E}_\theta\big(\ell_\theta(S)\big|\mathcal{B}\big) \qquad \text{a.e. } P_\theta \tag{5}$$

and (2) follows by taking expectations under P_θ. Equality in (2) entails equality a.e. P_θ in (5), and then, for ℓ_θ strictly convex, the uniqueness statement of Jensen's inequality applies. ////

Lehmann–Scheffè Theorem

The Rao–Blackwell theorem does not rely on, although it is compatible with, the concept of (expectation) unbiasedness. Then the Lehmann–Scheffè theorem is the following uniqueness corollary to the Rao–Blackwell result, making essential use of prescribed expectations.

Corollary C.1.3 *Assume the family of restrictions of P_θ onto the sufficient sub-σ algebra $\mathcal{B}$ of $\mathcal{A}$ is complete, and consider any two estimators $\tilde{S}, S\colon (\Omega, \mathcal{A}) \to (\mathbb{R}^p, \mathbb{B}^p)$ that have finite and identical expectations under all $\theta \in \Theta$,*

$$\mathrm{E}_\theta\, \tilde{S} = \mathrm{E}_\theta\, S \tag{6}$$

Then for all $\theta \in \Theta$,

$$\tilde{S}_{\bullet} = S_{\bullet} \qquad \text{a.e. } P_\theta \tag{7}$$

PROOF By (3), (6), and the very definition of completeness. ////

The completeness of exponential families is the uniqueness theorem for Laplace transforms.

Lemma C.1.4 *Every exponential family of probability measures on some finite-dimensional $(\mathbb{R}^k, \mathbb{B}^k)$,*

$$P_\zeta(dx) = c_\zeta\, e^{\zeta' x}\, \nu(dx), \qquad x \in \mathbb{R}^k,\ \zeta \in Z \tag{8}$$

whose parameter set $Z \subset \mathbb{R}^k$ has nonempty interior $Z^\circ \neq \emptyset$, is complete.

PROOF By reparametrization of P_ζ and rescaling of ν we may achieve that $0 \in Z^\circ$. Then assume some real-valued Borel measurable function h on $\mathbb{R}^k$ such that

$$\int h\, dP_\zeta = 0, \qquad \zeta \in Z \tag{9}$$

which means that the expectations of the positive and negative parts of h are finite and the same. Introducing the finite measures

$$d\mu^+ = h^+\,d\nu\,, \qquad d\mu^- = h^-\,d\nu \tag{10}$$

we obtain

$$\int e^{\zeta' x}\,\mu^+(dx) = \int e^{\zeta' x}\,\mu^-(dx)\,, \qquad \zeta \in Z \tag{11}$$

That is, the Laplace transforms of μ^+ and μ^- are finite, and coincide on a set which contains some nonempty open ball $|\zeta| < \delta$. Now either appeal to the uniqueness theorem for Laplace transforms. Or extend the Laplace transforms holomorphically in each variable to the strip $|\Re\xi| < \delta$ in $\mathbb{C}^k$. By the uniqueness theorem from complex analysis, since these extensions agree on a set with accumulation points, they must be the same. In particular, they show identical values for purely imaginary arguments. In other words, the Fourier transforms of μ^+ and μ^- coincide. Thus the uniqueness theorem for Fourier transforms applies. Both ways, $\mu^+ = \mu^-$ follows. Therefore,

$$h^+ = h^- = 0 \qquad \text{a.e. } \nu \tag{12}$$

and hence $h = 0$ a.e. ν. ////

Gauss–Markov Theorem

The Gauss–Markov theorem neither uses sufficiency nor completeness, but instead is restricted to (expectation) unbiased estimators that moreover are linear in the observations.

For $k, n \in \mathbb{N}$, $k \le n$, consider the linear model

$$Y = X\theta + U \tag{13}$$

with regression parameter $\theta \in \mathbb{R}^k$, design matrix $X \in \mathbb{R}^{n\times k}$ of rank

$$\operatorname{rk} X = r \le k \tag{14}$$

and error distribution of unknown scale $\sigma \in (0,\infty)$ such that

$$\mathrm{E}\,U = 0, \qquad \operatorname{Cov} U = \sigma^2\,\mathbb{I}_n \tag{15}$$

Denote by $C(X) \subset \mathbb{R}^n$ the column space of X. Choosing $\{e_1,\dots,e_r\}$ any ONB of the column space $C(X)$, the symmetric idempotent matrix

$$\begin{aligned}\Pi &= (e_1,\dots,e_r)(e_1,\dots,e_r)' \\ &= X(X'X)^{-1}X'\,, \qquad \text{if } r = k\end{aligned} \tag{16}$$

defines the orthogonal projection of $\mathbb{R}^n$ onto $C(X)$.

Proposition C.1.5 *Assume the linear model* (13)–(15) *and, for* $p \in \mathbb{N}$, *let* $B \in \mathbb{R}^{p\times k}$ *be some matrix defining the parameter of interest* $B\theta$.

(a) *Then a linear estimator*

$$S = AY \tag{17}$$

based on any matrix $A \in \mathbb{R}^{p\times n}$ *is unbiased for* $B\theta$ *iff*

$$AX = B \tag{18}$$

(b) *Suppose the linear estimator* (17) *is unbiased for* $B\theta$. *Then the linear estimator*

$$\widehat{S} = A\Pi Y \tag{19}$$

based on least squares is also unbiased for $B\theta$, *and*

$$\operatorname{Cov}\widehat{S} \le \operatorname{Cov} S \tag{20}$$

Equality is achieved in (20) *iff* $A = A\Pi$.

(c) *If* $A, A'' \in \mathbb{R}^{p\times n}$ *are any two matrices verifying* (18), *then*

$$A''\Pi = A\Pi \tag{21}$$

PROOF

(a) $B\theta = \mathrm{E}\, S = A\,\mathrm{E}\, Y = AX\theta$ for all $\theta \in \mathbb{R}^k$, since $\mathrm{E}\, U = 0$.

(b) $\mathrm{E}\,\widehat{S} = A\Pi X\theta = AX\theta = B\theta$, by (18) and $X\theta \in C(X)$.

Using this unbiasedness and $\operatorname{Cov} U = \mathbb{I}_n$ in this proof,

$$0 \le \operatorname{Cov}(S - \widehat{S}) = A(\mathbb{I}_n - \Pi)A' = \operatorname{Cov} S - \operatorname{Cov}\widehat{S} \tag{22}$$

since also $\mathbb{I}_n - \Pi$ is symmetric idempotent.

(c) For every $y \in \mathbb{R}^n$ there is some $\theta \in \mathbb{R}^k$ (if $r < k$ nonunique) such that $\Pi y = X\theta$. Then $A\Pi y = AX\theta = B\theta = A''X\theta = A''\Pi y$, by (18). ////

C.2 Some Technical Results

We collect some useful auxiliary technical results.

C.2.1 Calculus

Integration by Parts

Integration by parts can be linked up with Fubini's theorem. For the proof, just integrate $\mathbf{I}(x < y)$ w.r.t. $(\mu \otimes \nu)(dx, dy)$ and apply Fubini's theorem.

Lemma C.2.1 *Let μ and ν be two finite real measures on the Borel σ algebra $\Omega \cap \mathbb{B}$ of some measurable subset $\Omega \in \mathbb{B}$ of the real line. Then*

$$\begin{aligned} \int_\Omega \mu\{x \in \Omega \mid x < y\}\, \nu(dy) &= \int_\Omega \nu\{y \in \Omega \mid y > x\}\, \mu(dx) \\ &= \mu(\Omega)\nu(\Omega) - \int_\Omega \nu\{y \in \Omega \mid y \le x\}\, \mu(dx) \end{aligned} \tag{1}$$

Hájek's Lemma

Under a suitable integrability condition, even non-Lipschitz transforms of absolutely continuous functions may be absolutely continuous. The next such lemma due to Hájek (1972) has been the stepping stone to the proof of L_2 differentiability of location models (with finite Fisher information).

Lemma C.2.2 *Suppose that the function $f\colon \mathbb{R} \to \mathbb{R}$, $f \ge 0$ a.e. λ, is absolutely continuous on bounded intervals, with derivative f', and assume that*

$$\int_a^b |\partial\sqrt{f}\,|\, d\lambda < \infty, \qquad -\infty < a < b < \infty \tag{2}$$

for the function

$$\partial\sqrt{f} = \frac{f'}{2\sqrt{f}}\, \mathbf{I}(f \ne 0) \tag{3}$$

Then $\sqrt{f}$ is absolutely continuous on bounded intervals, and its derivative is the function $\partial\sqrt{f}$.

PROOF Being absolutely continous on bounded intervals, the function f is differentiable a.e. λ. As there can be only a countable number of points u such that $f(u) = 0$ and f has derivative $f'(u) \ne 0$ it follows that

$$\lambda(f = 0, f' \ne 0) = 0 \tag{4}$$

The square root function being differentiable and Lipschitz bounded on every interval with positive endpoints, $\sqrt{f}$ is absolutely continous with derivative $\partial\sqrt{f}$ [chain rule], on every interval where f is bounded away from 0. Thus, the representation

$$\sqrt{f}(b) - \sqrt{f}(a) = \int_a^b \partial\sqrt{f}\, d\lambda \tag{5}$$

holds for all $a, b \in \mathbb{R}$ such that f is strictly positive on $[a, b]$. By the continuity of f and dominated convergence, using the integrability assumption (2), representation (5) extends to all $a, b \in \mathbb{R}$ such that f is strictly positive on (a, b).

For arbitrary $a, b \in \mathbb{R}$, the open set $(a, b) \cap \{f > 0\}$ is the countable union of pairwise disjoint intervals (a_i, b_i) such that $f(a_i) = 0$ if $a_i > a$, and $f(b_i) = 0$ if $b_i < b$. By the definition of $\partial\sqrt{f}$ it holds that $\partial\sqrt{f} = 0$ on $(a,b) \cap \{f = 0\}$. We thus obtain that

$$\int_a^b \partial\sqrt{f}\, d\lambda \overset{(2)}{=} \sum_{i=1}^{\infty} \int_{a_i}^{b_i} \partial\sqrt{f}\, d\lambda \overset{(5)}{=} \sum_{i=1}^{\infty} \sqrt{f}(b_i) - \sqrt{f}(a_i) = \sqrt{f}(b) - \sqrt{f}(a)$$

which proves the assertion. ////

Differentiable Lagrangians and Continuous Multipliers

The following lemma provides the differentiability for the minimization of certain Lagrangians in Sections 5.5, 7.4, and 7.5.

Lemma C.2.3 *Given a probability P and some $Y \in L_2(P)$. Then the functions $f, g \colon \mathbb{R} \to [0,\infty)$ defined by*

$$f(t) = \int |(t-Y)^+|^2\, dP, \qquad g(t) = \int |(Y-t)^+|^2\, dP \tag{6}$$

are differentiable on $\mathbb{R}$ and have the following derivatives,

$$f'(t) = 2\int (t-Y)^+\, dP, \qquad g'(t) = -2\int (Y-t)^+\, dP \tag{7}$$

Proof Without restriction we only deal with f. Then, for $h \downarrow 0$,

$$\begin{aligned} f(t+h) - f(t) &= \int_{\{Y<t+h\}} (t+h-Y)^2\, dP - \int_{\{Y<t\}} (t-Y)^2\, dP \\ &= 2h \int_{\{Y<t\}} (t-Y)\, dP + h^2 P(Y<t) + \int_{\{t\le Y<t+h\}} (t+h-Y)^2\, dP \\ &= 2h \int (t-Y)^+\, dP + \mathrm{O}(h^2) \end{aligned}$$

and likewise,

$$\begin{aligned} f(t) - f(t-h) &= \int_{\{Y<t\}} (t-Y)^2\, dP - \int_{\{Y<t-h\}} (t-h-Y)^2\, dP \\ &= 2h \int_{\{Y<t-h\}} (t-Y)\, dP - h^2 P(Y<t-h) + \int_{\{t-h\le Y<t\}} (t-Y)^2\, dP \\ &= 2h \int (t-Y)^+\, dP + \mathrm{O}(h^2) \end{aligned}$$

since $0 \le \int_{\{t-h\le Y<t\}} (t-Y)\, dP \le h \downarrow 0$. ////

The next kind of implicit function lemma enables us in Theorem 7.4.13 a to sidestep the well-posedness condition B.2(19), and in Theorem 7.4.15 a to derive a more explicit solution than merely by Lagrange multipliers.

Lemma C.2.4 *Given a real-valued random variable Y and a probablity P on some sample space, let the function $M: \mathbb{R} \times (0, \infty) \to \mathbb{R}$ be defined by*

$$M(\vartheta, r) = \int (Y - \vartheta) \min\Bigl\{1, \frac{r}{|Y - \vartheta|}\Bigr\}\, dP \tag{8}$$

Then for every $r \in (0, \infty)$ there exists some $\vartheta(r) \in \mathbb{R}$ such that

$$M\bigl(\vartheta(r), r\bigr) = 0 \tag{9}$$

If Y has finite expectation $\mathrm{E}\, Y$ under P, and $\bar{r} = \sup_P |Y|$, then

$$\lim_{r \to \bar{r}} \vartheta(r) = \mathrm{E}\, Y \tag{10}$$

Suppose the median $m = \operatorname{med} Y(P)$ is unique. Then the function $\vartheta(.)$ is uniquely determined, continuous on $(0, \infty)$, and has limit

$$\lim_{r \to 0} \vartheta(r) = m \tag{11}$$

PROOF Fix any $r \in (0, \infty)$. The existence of a zero $\vartheta(r)$ is a consequence of the intermediate value theorem since, by continuity of the integrand and dominated convergence, $M(.\,, r)$ is continuous in the first argument and has the limits

$$\lim_{\vartheta \to \mp\infty} M(\vartheta, r) = \pm r \tag{12}$$

For $\vartheta_1 < \vartheta_2$, the difference $\Delta M(.\,, r) = M(\vartheta_1, r) - M(\vartheta_2, r)$ can in the case that $\vartheta_1 + r < \vartheta_2 - r$ be calculated to

$$\begin{aligned} \Delta M(.\,, r) &= 2r\, P(\vartheta_1 + r \le Y \le \vartheta_2 - r) \\ &\quad + \int_{\{|Y - \vartheta_1| < r\}} (Y - \vartheta_1 + r)\, dP \\ &\quad - \int_{\{|Y - \vartheta_2| < r\}} (Y - \vartheta_2 - r)\, dP \end{aligned} \tag{13}$$

and to

$$\begin{aligned} \Delta M(.\,, r) &= (\vartheta_2 - \vartheta_1)\, P(\vartheta_2 - r < Y < \vartheta_1 + r) \\ &\quad + \int_{\{\vartheta_1 - r < Y \le \vartheta_2 - r\}} (Y - \vartheta_1 + r)\, dP \\ &\quad - \int_{\{\vartheta_1 + r \le Y < \vartheta_2 + r\}} (Y - \vartheta_2 - r)\, dP \end{aligned} \tag{14}$$

in the case $\vartheta_1 + r \ge \vartheta_2 - r$. Both times $\Delta M(.\,, r) \ge 0$, so that the function $M(.\,, r)$ is decreasing, and $\Delta M(.\,, r) = 0$ is achieved iff

$$P(\vartheta_1 - r < Y < \vartheta_2 + r) = 0 \tag{15}$$

Assume $\mathrm{E}\,Y$ finite, and let $r_n \to \bar{r}$. Suppose that $\vartheta_n = \vartheta(r_n) \leq \mathrm{E}\,Y - \delta$ infinitely often, for some $\delta \in (0,1)$. Then the monotonicity in the first argument as shown, and continuity of M in the second argument [dominated convergence] would imply that along such n,

$$0 = M(\vartheta_n, r_n) \geq M(\mathrm{E}\,Y - \delta, r_n) \longrightarrow M(\mathrm{E}\,Y - \delta, \bar{r}) = \delta \tag{16}$$

which is a contradiction. On likewise ruling out that $\vartheta_n \geq \mathrm{E}\,Y + \delta$ occurs infinitely often, (10) is proved.

Now suppose $\operatorname{med} Y(P)$ is unique. Then (15) for $\vartheta_1 \leq \vartheta = \vartheta(r) \leq \vartheta_2$ would imply that

$$P(\vartheta - r < Y < \vartheta + r) = 0 \tag{17}$$

hence

$$0 = \frac{M(\vartheta, r)}{r} = P(Y \geq \vartheta + r) - P(Y \leq \vartheta - r) \tag{18}$$

and therefore

$$P(Y \geq \vartheta + r) = P(Y \leq \vartheta - r) = \tfrac{1}{2} \tag{19}$$

So the entire interval $(\vartheta - r, \vartheta + r)$ would be medians. Thus $\vartheta(r)$ is unique. To prove continuity let $r_n \to r \in (0,\infty)$ and assume $\vartheta_n = \vartheta(r_n) \leq \vartheta - \delta$ infinitely often, for some $\delta \in (0,1)$ and $\vartheta = \vartheta(r)$. Then the monotonicity in the first argument, and continuity of M in the second argument [dominated convergence] would imply that along such n,

$$0 = M(\vartheta_n, r_n) \geq M(\vartheta - \delta, r_n) \longrightarrow M(\vartheta - \delta, r) > M(\vartheta, r) = 0 \tag{20}$$

a contradiction. That $\vartheta_n \geq \vartheta + \delta$ infinitely often, may be ruled out likewise. On writing

$$\frac{M(\vartheta, r)}{r} = \int \operatorname{sign}(Y - \vartheta) \min\Big\{1, \frac{|Y - \vartheta|}{r}\Big\}\, dP \tag{21}$$

we obtain that

$$\lim_{r \to 0} \frac{M(\vartheta, r)}{r} = \int \operatorname{sign}(Y - \vartheta)\, dP \tag{22}$$

for every $\vartheta \in \mathbb{R}$. Thus (11) follows by the previous indirect argument. ////

C.2.2 Topology

Separability of $L_2(\mu)$

The weak convergence theory in Section A.4 uses the separability of $L_2(\mu)$.

Lemma C.2.5 *Let $\mu \in \mathcal{M}_\sigma(\mathcal{B})$ be some σ finite measure on the Borel σ algebra $\mathcal{B}$ of some separable metric space Ξ. Then, for $1 \leq r < \infty$, the integration spaces $L_r(\mu) = L_r(\Xi, \mathcal{B}, \mu)$ are separable.*

The space $L_\infty(\mu)$ is, of course, not separable in general.

PROOF If Ξ_0 is countable dense in Ξ and $\mathcal{E}_0$ denotes the countable system of all balls with centers in Ξ_0 and rational radii, then $\sigma(\mathcal{E}_0) = \mathcal{B}$. Enlarge $\mathcal{E}_0$ to the system $\mathcal{E} = \mathcal{E}_0 \cup \mathcal{E}_0^c$ of all sets that themselves, or whose complements, are in $\mathcal{E}_0$. Then the system $\bigcap_{\mathrm{f}} \mathcal{E}$ of all intersections of finite subfamilies of $\mathcal{E}$ is still countable, as well as the algebra $\alpha(\mathcal{E})$ generated by $\mathcal{E}$, which is the system of all unions of finite subfamilies of $\bigcap_{\mathrm{f}} \mathcal{E}$,

$$\alpha(\mathcal{E}) = \bigcup\nolimits_{\mathrm{f}} \bigcap\nolimits_{\mathrm{f}} \mathcal{E}$$

Let us first assume $\mu \in \mathcal{M}_b(\mathcal{B})$ finite, and introduce the system $\mathcal{G}$ of all sets $B \in \mathcal{B}$ with the property that for every $\varepsilon \in (0,1)$ there exists an $A \in \alpha(\mathcal{E})$ such that $\mu(A \triangle B) < \varepsilon$. Then $\mathcal{G}$ is a monotone class and includes $\alpha(\mathcal{E})$; therefore $\mathcal{G} = \mathcal{B}$. This proves the assertion of the lemma in the case of indicator variables $X = \mathbf{I}_B$. General variables $X \in L_r(\mu)$ can be approximated arbitrarily closely by finite linear combinations with rational coefficients of indicators of events in $\alpha(\mathcal{E})$, which thus constitute a countable dense subset of $L_r(\mu)$.

In case $\mu \in \mathcal{M}_\sigma(\mathcal{B})$ there exists a function $g \in L_1(\mu)$ such that $g > 0$ everywhere. Let $\nu \in \mathcal{M}_b(\mathcal{B})$ be defined by $d\nu = g\,d\mu$. Then $L_r(\nu)$ is separable, and isomorphic to $L_r(\mu)$ via the isometry $f \mapsto \sqrt[r]{g}\, f$. ////

Elementary Functions Dense

The following lemma has been used for the inclusion of directions, which are approximately least favorable to the MD functional T_μ, into suitable neighborhoods, and for the as. distribution theory of the MD estimate S_μ, in the case of a σ finite weight μ.

Lemma C.2.6 *Let $(\Omega, \mathcal{B}, \mu)$ be any measure space, and $r \in [1, \infty)$. Then the measurable functions $e\colon (\Omega, \mathcal{B}) \to (\mathbb{R}, \mathbb{B})$ such that*

$$\#e(\Omega) < \infty, \qquad \mu(e \neq 0) < \infty \tag{23}$$

are dense in $L_r(\mu) = L_r(\Omega, \mathcal{B}, \mu)$.

PROOF Rudin (1974; Theorem 3.13, p 70). ////

C.2.3 Matrices

Singular Value Decomposition

In Subsection 5.3.2 we have used that for every matrix $A \in \mathbb{R}^{p \times k}$ of finite dimensions $p \le k$,

$$\max\mathrm{ev}\, AA' = \max\mathrm{ev}\, A'A \tag{24}$$

In fact, the larger $A'A$ has the same eigenvalues as AA' plus additional zeros. This is true because the eigenvalues of AA' are $d_1^2, \ldots, d_p^2$, and those of $A'A$ are $d_1^2, \ldots, d_p^2, 0, \ldots, 0$, where $\mathrm{diag}\,(d_1, \ldots, d_p)$ denotes the diagonal matrix D in the following singular value decomposition of $C = A'$.

Proposition C.2.7 *For every matrix* $C \in \mathbb{R}^{k\times p}$ *of finite dimensions* $k \geq p$ *there exist two orthogonal matrices* $U \in \mathbb{R}^{k\times k}$ *and* $V \in \mathbb{R}^{p\times p}$ *such that*

$$C = U \begin{pmatrix} D \\ 0 \end{pmatrix} V' \tag{25}$$

where $D = \operatorname{diag}(d_1, \ldots, d_p)$ *with elements* $d_1 \geq d_2 \geq \cdots \geq d_p \geq 0$.

PROOF Golub, van Loan (1983; Theorem 2.3-1, p 16). ////

By the way, if $0 = d_p = \cdots = d_{r+1} < d_r$ and $\widetilde{D} = \operatorname{diag}(d_1, \ldots, d_r)$, then

$$C^- = V \begin{pmatrix} \widetilde{D}^{-1} & 0 \\ 0 & 0 \end{pmatrix} U' \tag{26}$$

defines a *generalized inverse* of C satisfying $CC^-C = C$.

Robust Covariances

Assuming a finite-dimensional Euclidean sample space $\mathbb{R}^k$, robust location and scatter functionals that are equivariant under affine transformations have been defined as solutions $(\vartheta, V) \in \mathbb{R}^k \times \mathbb{R}^{k\times k}$, $V = V' > 0$, to a set of equations,

$$\begin{aligned} \int u_1\bigl(|y-\vartheta|_V\bigr)\,(y-\vartheta)\,dQ &= 0 \\ \int u_2\bigl(|y-\vartheta|_V^2\bigr)\,(y-\vartheta)(y-\vartheta)'\,dQ &= V \end{aligned} \tag{27}$$

where $|y-\vartheta|_V^2 = (y-\vartheta)'V^{-1}(y-\vartheta)$, under certain conditions on the pair of functions $u_1, u_2\colon [0,\infty) \to [0,\infty)$ and on the measure $Q \in \mathcal{M}_1(\mathbb{B}^k)$.

The existence result in this context due to Maronna (1976) will suffice to obtain a self-standardized influence curve in Subsection 5.5.4; confer also Hampel et al. (1986; Theorem 3, p 246). For a more general exposition on robust covariances themselves, see Huber (1981; Chapter 8).

We now take up the notation of Subsection 5.5.4, but drop the fixed value of the parameter $\theta \in \Theta$. In particular, Λ is the L_2 derivative of the parametric model at θ, and E denotes expectation, and $\mathcal{C}$ the covariance, under the fixed parametric measure $P = P_\theta$.

Proposition C.2.8 *Let* $b \in \bigl(\sqrt{k}, \infty\bigr)$, *and assume there is some* $\alpha \in (0,1)$ *such that for all* $(k-1)$*-dimensional hyperplanes* $H \subset \mathbb{R}^k$,

$$P(\Lambda \in H) \leq 1 - k/b^2 - \alpha \tag{28}$$

Then there exist some matrix $A \in \mathbb{R}^{k\times k}$ *and vector* $a \in \mathbb{R}^k$ *such that the function*

$$\varrho = (A\Lambda - a)\min\Bigl\{1, \frac{b}{|A\Lambda - a|_{\mathcal{C}(\varrho)}}\Bigr\} \tag{29}$$

is an influence curve at P; *that is,* $\varrho \in L_2^p(P)$ *and* $\mathrm{E}\,\varrho = 0$, $\mathrm{E}\,\varrho\Lambda' = \mathbb{I}_k$.

PROOF Identify the measure Q and the pair of functions u_1, u_2 as follows,

$$Q = \Lambda(P), \qquad u_1(s) = \min\{1, b/s\}, \qquad u_2(s) = \min\{1, b^2/s\} \tag{30}$$

Then

$$\psi_1(s) = s \wedge b, \qquad \psi_2(s) = s \wedge b^2, \qquad K_1 = b, \qquad K_2 = b^2 \tag{31}$$

in Maronna's (1976) notation, and his conditions (A), (B), and (C) are fulfilled; (D) too as $b^2 > k$. Condition (E), in the present setup, is just (28).

Thus Maronna (1976; Theorem 2) and Schönholzer (1979; Satz 1), invoking Brouwer's fixed point theorem, supply a solution (ϑ, V) to (27). This set of equations now means that

$$\mathrm{E}\,\chi = 0, \qquad \mathcal{C}(\chi) = \mathrm{E}\,\chi\chi' = \mathbb{I}_k \tag{32}$$

for the function

$$\chi = V^{-1/2}(\Lambda - \vartheta)\min\Bigl\{1, \frac{b}{|\Lambda - \vartheta|_V}\Bigr\} \tag{33}$$

where $B = \mathrm{E}\,\chi\Lambda'$ is nonsinglar. With $A = B^{-1}V^{-1/2}$ then define

$$\varrho = B^{-1}\chi = A(\Lambda - \vartheta)\min\Bigl\{1, \frac{b}{|A(\Lambda - \vartheta)|_{(B'B)^{-1}}}\Bigr\} \tag{34}$$

which, since $(B'B)^{-1} = \mathcal{C}(\varrho)$, is the desired influence curve. ////

Bibliography

Anderson, T.W. (1955): The integral of a symmetric unimodal function over a symmetric convex set and some probability inequalities. *Proc. Amer. Math. Soc.* **6** 170–176.

Averbukh, V.I. and Smolyanov, O.G. (1967): The theory of differentiation in linear topological spaces. *Russian Math. Surveys* **22** 201–258.

Averbukh, V.I. and Smolyanov, O.G. (1968): The various definitions of the derivative in linear topological spaces. *Russian Math. Surveys* **23** 67–113.

Bahadur, R.R. (1966): A note on quantiles in large samples. *Ann. Math. Stat.* **37** 577–580.

Bauer, H. (1974): *Wahrscheinlichkeitstheorie und Grundzüge der Maßtheorie* 2. Auflage. W. de Gruyter, Berlin.

Bednarski, T. (1981): On solutions of minimax test problems for special capacities. *Z. Wahrsch. verw. Gebiete* **58** 397–405.

Bednarski, T. (1982): Binary experiments, minimax tests, and 2-alternating capacities. *Ann. Statist.* **10** 226–232.

Begun, J.M., Hall, W.J., Huang, W.M. and Wellner, J.A. (1983): Information and asymptotic efficiency in parametric-nonparametric models. *Ann. Statist.* **11** 432–452.

Beran, R.J. (1977): Minimum Hellinger distance estimates for parametric models. *Ann. Statist.* **5** 445–463.

Beran, R.J. (1981 a): Efficient robust estimates in parametric models. *Z. Wahrsch. verw. Gebiete* **55** 91–108.

Beran, R.J. (1981 b): Efficient and robust tests in parametric models. *Z. Wahrsch. verw. Gebiete* **57** 73–86.

Beran, R.J. (1982): Robust estimation in models for independent nonidentically distributed data. *Ann. Statist.* **10** 415–428.

Beran, R.J. (1984): Minimum Distance Procedures. In *Handbook of Statistics* Vol. 4 (P.R. Krishnaiah and P.K. Sen, eds.), 741–754. Elsevier, New York.

Bickel, P.J. (1973): On some analogues to linear combinations of order statistics in the linear model. *Ann. Statist.* **1** 597–616.

Bickel, P.J. (1975): One-step Huber estimates in the linear model. *J. Amer. Statist. Assoc.* **70** 428–434.

Bickel, P.J. (1976): Another look at robustness—a review of reviews and some new developments. *Scand. J. Statist.* **3** 145–168.

Bickel, P.J. (1981): Quelques aspects de la statistique robuste. In *Ecole d'Eté de Probabilités de Saint Flour IX 1979* (P.L. Hennequin, ed.), 1–72. Lecture Notes in Mathematics #876. Springer-Verlag, Berlin.

Bickel, P.J. (1984): Robust regression based on infinitesimal neighborhoods. *Ann. Statist.* **12** 1349–1368.

Bickel, P.J. and Lehmann, E.L. (1975): Descriptive statistics for nonparametric models. II. Location. *Ann. Statist.* **3** 1045–1069.

Billingsley, P. (1968): *Convergence of Probability Measures*. Wiley, New York.

Billingsley, P. (1971): *Weak Convergence of Measures: Applications in Probability*. SIAM, Philadelphia.

Blyth, C.R. (1951): On minimax statistical decision procedures and their admissibility. *Ann. Math. Stat.* **22** 22–42.

Boos, D. (1979): A differential for L statistics. *Ann. Statist.* **7** 955–959.

Bretagnolle, J. (1980): Statistique de Kolmogorov–Smirnov pour un échantillon non-équireparti. In *Aspects Statistiques et Aspects Physiques des Processus Gaussiens* 39–44. Centre National de la Recherche Scientifique, Saint Flour.

Buja, A. (1984): Simultaneously least favorable experiments I: upper standard functionals and sufficiency. *Z. Wahrsch. verw. Gebiete* **65** 367–384.

Buja, A. (1985): Simultaneously least favorable experiments II: upper standard loss functions and their applications. *Z. Wahrsch. verw. Gebiete* **69** 387–420.

Buja, A. (1986): On the Huber–Strassen theorem. *Probab. Th. Rel. Fields* **73** 149–152.

Chung, K.L. (1974): *A Course in Probability Theory*, 2nd ed. Academic Press, New York.

Dieudonné, J. (1960): *Foundations of Modern Analysis*. Academic Press, New York.

Donoho, D.L. and Liu, R.C. (1988 a): The 'automatic' robustness of minimum distance functionals. *Ann. Statist.* **16** 552–586.

Donoho, D.L. and Liu, R.C. (1988 b): Pathologies of some minimum distance estimators. *Ann. Statist.* **16** 587–608.

Droste, W. and Wefelmeyer, W. (1984): On Hájek's convolution theorem. *Statistics and Decisions* **2** 131–144.

Dunford, N. and Schwartz, J.T. (1957): *Linear Operators I—General Theory*. Wiley-Interscience, New York.

Feller, W. (1966): *An Introduction to Probability Theory and Its Applications* Vol. II. Wiley, New York.

Ferguson, Th.S. (1967): *Mathematical Statistics—A Decision-Theoretic Approach*. Academic Press, New York.

Fernholz, L.T. (1979): *von Mises Calculus for Statistical Functionals*. Lecture Notes in Statistics #19. Springer-Verlag, New York.

Fréchet, M. (1937): Sur la notion de différentielle dans l'analyse générale. *J. Math. Pures Appl.* **16** 233–250.

Golub, G.H. and van Loan, C.F. (1983): *Matrix Computations*. The Johns Hopkins University Press, Baltimore, Maryland.

Hájek, J. (1968): Asymptotic normality of simple linear rank statistics under alternatives. *Ann. Math. Stat.* **39** 325–354.

Hájek, J. (1970): A characterization of limiting distributions of regular estimates. *Z. Wahrsch. verw. Gebiete* **14** 323–330.

Hájek, J. (1972): Local asymptotic minimax and admissibility in estimation. *Proc. Sixth Berkeley Symp. Math. Stat. Prob.* **1** 175–194. Univ. California Press, Berkeley.

Hájek, J. and Šidák, Z. (1967): *Theory of Rank Tests.* Academic Press, New York.

Hampel, F.R. (1968): *Contributions to the theory of robust estimation.* Ph.D. Thesis, University of California, Berkeley.

Hampel, F.R. (1971): A general qualitative definition of robustness. *Ann. Math. Stat.* **42** 1887–1896.

Hampel, F.R. (1974): The influence curve and its role in robust estimation. *J. Amer. Statist. Assoc.* **69** 383–393.

Hampel, F.R., Ronchetti, E.M., Rousseeuw, P.J. and Stahel, W.A. (1986): *Robust Statistics—The Approach Based on Influence Functions.* Wiley, New York.

Hodges, J.L. and Lehmann, E.L. (1950): Some applications of the Cramèr–Rao inequality. *Proc. Second Berkeley Symp. Math. Stat. Prob.* **1** 13–22. Univ. California Press, Berkeley.

Huber, P.J. (1964): Robust estimation of a location parameter. *Ann. Math. Stat.* **35** 73–101.

Huber, P.J. (1965): A robust version of the probability ratio test. *Ann. Math. Stat.* **36** 1753–1758.

Huber, P.J. (1966): Strict efficiency excludes superefficiency (abstract). *Ann. Math. Stat.* **37** 1425.

Huber, P.J. (1967): The behavior of maximum likelihood estimates under non-standard conditions. *Proc. Fifth Berkeley Symp. Math. Stat. Prob.* **1** 221–233. Univ. California Press, Berkeley.

Huber, P.J. (1968): Robust confidence limits. *Z. Wahrsch. verw. Gebiete* **10** 269–278.

Huber, P.J. (1969): *Théorie de l'Inférence Statistique Robuste.* Les Presses de l'Université de Montréal.

Huber, P.J. (1972): Robust statistics: A review. *Ann. Math. Stat.* **43** 1041–1067.

Huber, P.J. (1973): Robust regression: Asymptotics, conjectures, and Monte Carlo. *Ann. Statist.* **1** 799–821.

Huber, P.J. (1977): Robust methods of estimation of regression coefficients. *Math. Operationsforschung Statist. Ser. Statist.* **8** 41–53.

Huber, P.J. (1981): *Robust Statistics.* Wiley, New York.

Huber, P.J. (1983): Minimax aspects of bounded influence regression. *J. Amer. Statist. Assoc.* **78** 66–80.

Huber, P.J. (1991): Between robustness and diagnostics. In *Directions in Robust Statistics and Diagnostics* (W. Stahel and S. Weisberg, eds.), Part I, 121–130. The IMA Volumes in Mathematics and Its Applications #33. Springer-Verlag, New York.

Huber, P.J. and Strassen, V. (1973): Minimax tests and the Neyman–Pearson lemma for capacities. *Ann. Statist.* **1** 251–263.

Huber-Carol, C. (1970): *Étude asymptotique de tests robustes.* Thèse de Doctorat, ETH Zürich.

Hummel, T. (1992): *Robustes Testen in Zeitreihenmodellen.* Diplomarbeit, Universität Bayreuth.

Jaeckel, L.A. (1972): Estimating regression coefficients by minimizing the dispersion of the residuals. *Ann. Math. Stat.* **43** 1449–1458.

Jain, N.C. and Marcus, M.B. (1975): Central limit theorems for $C(S)$-valued random variables. *J. Funct. Anal.* **19** 216–231.

James, W. and Stein, C. (1961): Estimation under quadratic loss. *Proc. Fourth Berkeley Symp. Math. Stat. Prob.* **1** 361–379. Univ. California Press, Berkeley.

Joffe, A.D. and Tichomirov, V.M. (1979): *Theorie der Extremalaufgaben*. VEB Deutscher Verlag der Wissenschaften, Berlin.

Jurečková, J. (1969): Asymptotic linearity of a rank statistic in the regression parameter. *Ann. Math. Stat.* **40** 1889–1900.

Jurečková, J. (1971): Nonparametric estimates of regression coefficients. *Ann. Math. Stat.* **42** 1328–1338.

Keller, H.H. (1974): *Differential Calculus in Locally Convex Spaces*. Lecture Notes in Mathematics #417. Springer-Verlag, Berlin.

Koshevnik, Yu.A. and Levit, B.Ya. (1976): On a nonparametric analogue of the information matrix. *Theory Probab. Appl.* **21** 738–753.

Koul, H.L. (1985): Minimum distance estimation in multiple linear regression. *Sankhyā* **A 47** 57–74.

Krasker, W.S. (1980): Estimation in linear regression with disparate data points. *Econometrica* **48** 1333–1346.

Kurotschka, V. and Müller, C. (1992): Optimum robust estimation of linear aspects in conditionally contaminated linear models. *Ann. Statist.* **20** 331–350.

LeCam, L. (1953): On some asymptotic properties of maximum likelihood estimates and related Bayes estimates. *Univ. of California Publ. in Statistics* **1** 277–330.

LeCam, L. (1960): Locally asymptotically normal families of distributions. *Univ. of California Publ. in Statistics* **3** 37–98.

LeCam, L. (1969): *Théorie Asymptotique de la Décision Statistique*. Les Presses de l'Université de Montréal.

LeCam, L. (1972): Limits of Experiments. *Proc. Sixth Berkeley Symp. Math. Stat. Prob.* **1** 245–261. Univ. California Press, Berkeley.

LeCam, L. (1979): On a theorem of Hájek. In *Contributions to Statistics: J. Hájek Memorial Volume* (J. Jurečková, ed.), 119–137. Academia, Prague.

LeCam, L. (1982): Limit theorems for empirical measures and Poissonization. In *Statistics and Probability: Essays in Honor of C.R. Rao* (G. Kallianpur, P.R. Krishnaiah and J.K. Ghosh, eds.), 455–463. North Holland, Dordrecht.

LeCam, L. (1986): *Asymptotic Methods in Statistical Decision Theory*. Springer Verlag, New York.

Lehmann, E.L. (1983): *Theory of Point Estimation*. Wiley, New York.

Lehmann, E.L. (1986): *Testing Statistical Hypotheses*, 2nd ed. Wiley, New York.

Levit, B.Ya. (1975): On the efficiency of a class of nonparametric estimates. *Theory Probab. Appl.* **20** 723–740.

Luenberger, D. (1969): *Optimization by Vector Space Methods*. Wiley, New York.

Maronna, R.A. (1976): Robust M estimators of multivariate location and scatter. *Ann. Statist.* **4** 51–67.

Maronna, R.A. and Yohai, V.J. (1981): Asymptotic behavior of general M estimates for regression and scale with random carriers. *Z. Wahrsch. verw. Gebiete* **58** 7–20.

Martin, R.D., Yohai, V.J. and Zamar, R.H. (1989): Minmax bias robust regression. *Ann. Statist.* **17** 1608–1630.

Millar, P.W. (1979): Robust tests of statistical hypotheses. Unpublished.

Millar, P.W. (1981): Robust estimation via minimum distance methods. *Z. Wahrsch. verw. Gebiete* **55** 73–89.

Millar, P.W. (1982): Optimal estimation of a general regression function. *Ann. Statist.* **10** 717–740.

Millar, P.W. (1983): The minimax principle in asymptotic statistical theory. In *Ecole d'Eté de Probabilités de Saint Flour XI 1981* (P.L. Hennequin, ed.), 75–266. Lecture Notes in Mathematics #976. Springer-Verlag, Berlin.

Millar, P.W. (1984): A general approach to the optimality of minimum distance estimators. *Trans. Amer. Math. Soc.* **286** 377–418.

Millar, P.W. (1985): Nonparametric Applications of an Infinite-Dimensional Convolution Theorem. *Z. Wahrsch. verw. Gebiete* **68** 545–555.

von Mises, R. (1947): On the asymptotic distribution of differentiable statistical functionals. *Ann. Math. Stat.* **18** 309–348.

Moore, D.S. (1968): An elementary proof of asymptotic normality of linear functions of order statistics. *Ann. Math. Stat.* **39** 263–265.

Müller, C. (1987): *Optimale Versuchspläne für Robuste Schätzfunktionen in Linearen Modellen.* Dissertation, Freie Universität Berlin.

Müller, C. (1993): One-step M-estimators in conditionally contaminated linear models. To appear in *Statistics and Decisions.*

Noelle, G. and Plachky, D. (1968): Zur schwachen Folgenkompaktheit von Testfunktionen. *Z. Wahrsch. verw. Gebiete* **8** 182–184.

O'Reilly, N.E. (1974): On the weak convergence of empirical processes in sup norm metrics. *Ann. Probability* **2** 642–651.

Parpola, S. (1970): *Letters from Assyrian Scholars to the Kings Esarhaddon and Assurbanipal.* Part I: *Texts.* Part II: *Commentary and Appendices.* Verlag Butzon & Bercker, Kevelaer.

Parthasarathy, K.R. (1967): *Probability Measures on Metric Spaces.* Academic Press, New York.

Pfanzagl, J. and Wefelmeyer, W. (1982): *Contributions to a General Asymptotic Statistical Theory*. Lecture Notes in Statistics #13. Springer-Verlag, Berlin.

Prokhorov, Yu.V. (1956): Convergence of random processes and limit theorems in probability theory. *Theory Probab. Appl.* **1** 157–214.

Pyke, R. and Shorack, G. (1968): Weak convergence of a two-sample empirical process and a new approach to Chernoff–Savage theorems. *Ann. Math. Statist.* **39** 755-771.

Reeds, J.A. (1976): *On the Definition of von Mises Functionals.* Ph.D. Thesis, Harvard University, Cambridge.

Rieder, H. (1977): Least favorable pairs for special capacities. *Ann. Statist.* **5** 909–921.

Rieder, H. (1978): A robust asymptotic testing model. *Ann. Statist.* **6** 1080–1094.

Rieder, H. (1980): Estimates derived from robust tests. *Ann. Statist.* **8** 106–115.

Rieder, H. (1981): On local asymptotic minimaxity and admissibility in robust estimation. *Ann. Statist.* **9** 266–277.

Rieder, H. (1983): Robust estimation of one real parameter when nuisance parameters are present. *Transactions of the Ninth Prague Conference on Information Theory, Statistical Decision Functions, and Random Processes* **A** 77–89. Reidel, Dordrecht.

Rieder, H. (1985): *Robust Estimation of Functionals*. Unpublished Technical Report, University of Bayreuth.

Rieder, H. (1987 a): Contamination games in a robust k sample model. *Statistics* **18** 527–562.

Rieder, H. (1987 b): Robust regression estimators and their least favorable contamination curves. *Statistics and Decisions* **5** 307–336.

Rieder, H. (1989): A finite-sample minimax regression estimator. *Statistics* **20** 211–221.

Rieder, H. (1991): Robust testing of functionals. In *Directions in Robust Statistics and Diagnostics* (W. Stahel and S. Weisberg, eds.), Part II, 159–183. The IMA Volumes in Mathematics and Its Applications #34. Springer-Verlag, New York.

Roussas, G.G. (1972): *Contiguity of Probability Measures*. Cambridge University Press.

Rudin, W. (1973): *Functional Analysis*. McGraw-Hill, New York.

Rudin, W. (1974): *Real and Complex Analysis*, 2nd ed. McGraw-Hill, New York.

Samarov, A.M. (1985): Bounded influence regression via local minimax mean squared error. *J. Amer. Statist. Assoc.* **80** 1032–1040.

Schlather, M. (1994): *Glattheit von Generalisierten Linearen Modellen und statistische Folgerungen*. Diplomarbeit, Universität Bayreuth.

Schönholzer, H. (1979): *Robuste Kovarianz*. Ph.D. Thesis, ETH Zürich.

Serfling, R.J. (1980): *Approximation Theorems of Mathematical Statistics*. Wiley, New York.

Sichelstiel, G. (1993): *Robuste Tests in Linearen Modellen*. Diplomarbeit, Universität Bayreuth.

Sova, M. (1966): Conditions of differentiability in linear topological spaces. *Czech. Math. J.* **16** 339–362.

Staab, M. (1984): *Robust parameter estimation for* ARMA *models*. Dissertation, Universität Bayreuth.

Stein, C. (1956): Efficient nonparametric testing and estimation. *Proc. Third Berkeley Symp. Math. Stat. Prob.* **1** 187–195. Univ. California Press, Berkeley.

Stigler, S.M. (1973): The asymptotic distribution of the trimmed mean. *Ann. Statist.* **1** 472–477.

Stigler, S.M. (1974): Linear functions of order statistics with smooth weight functions. *Ann. Statist.* **2** 676–693.

Stigler, S.M. (1990): A Galtonian perspective on shrinkage estimators. *Statist. Sci.* **5** 147–155.

Takeuchi, K. (1967): Robust estimation and robust parameter. Unpublished.

Vainberg, M.M. (1964): *Variational Methods for the Study of Nonlinear Operators*. Holden Day, San Francisco.

Wang, P.C.C. (1981): Robust asymptotic tests of statistical hypotheses involving nuisance parameters. *Ann. Statist.* **9** 1096–1106.

Witting, H. (1985): *Mathematische Statistik I*. B.G. Teubner, Stuttgart.

Index

Springer Series in Statistics

(continued from p. ii)

Reiss: A Course on Point Processes.
Reiss: Approximate Distributions of Order Statistics: With Applications to Non-parametric Statistics.
Ross: Nonlinear Estimation.
Sachs: Applied Statistics: A Handbook of Techniques, 2nd edition.
Salsburg: The Use of Restricted Significance Tests in Clinical Trials.
Särndal/Swensson/Wretman: Model Assisted Survey Sampling.
Seneta: Non-Negative Matrices and Markov Chains, 2nd edition.
Shedler: Regeneration and Networks of Queues.
Siegmund: Sequential Analysis: Tests and Confidence Intervals.
Tanner: Tools for Statistical Inference: Methods for the Exploration of Posterior Distributions and Likelihood Functions, 2nd edition.
Todorovic: An Introduction to Stochastic Processes and Their Applications.
Tong: The Multivariate Normal Distribution.
Vapnik: Estimation of Dependences Based on Empirical Data.
West/Harrison: Bayesian Forecasting and Dynamic Models.
Wolter: Introduction to Variance Estimation.
Yaglom: Correlation Theory of Stationary and Related Random Functions I: Basic Results.
Yaglom: Correlation Theory of Stationary and Related Random Functions II: Supplementary Notes and References.